Richard W. Compans • Walter A. Orenstein
Editors

# Vaccines for Pandemic Influenza

 Springer

*Editors*

Richard W. Compans
Department of Microbiology &
    Immunology and Emory Vaccine Center
School of Medicine, Emory University
Atlanta GA 30322,
Rollins Research Center 3001
USA
compans@microbio.emory.edu

Walter A. Orenstein
Emory University, School of Medicine
Department of Medicine
1510 Clifton Road
Atlanta GA 30322
USA
walter.orenstein@gatesfoundation.org

ISBN 978-3-642-24240-3          e-ISBN 978-3-540-92165-3
DOI 10.1007/978-3-540-92165-3
Springer Heidelberg Dordrecht London New York

*Cover illustration*: VLPs produced by Novavax, Inc., Rockville, MD; electron micrograph by Penny M. Heaton, MD and Gale Smith, PhD, Novavax, Inc.
Please see the chapter by John Treanor (this volume) for more information.

*Cover design*: WMX Design, Manfred Bender

Printed on acid-free paper

Springer is part of Springer Science+Business Media (www.springer.com)

# Preface

Recent years have seen unprecedented outbreaks of avian influenza A viruses. In particular, highly pathogenic H5N1 viruses have not only resulted in widespread outbreaks in domestic poultry, but have been transmitted to humans, resulting in numerous fatalities. The rapid expansion in their geographic distribution and the possibility that these viruses could acquire the ability to spread from person to person raises the risk that such a virus could cause a global pandemic with high morbidity and mortality. An effective influenza vaccine represents the best approach to prevent and control such an emerging pandemic. However, current influenza vaccines are directed at existing seasonal influenza viruses, which have little or no antigenic relationship to the highly pathogenic H5N1 strains. Concerns about pandemic preparedness have greatly stimulated research activities to develop effective vaccines for pandemic influenza viruses, and to overcome the limitations inherent in current approaches to vaccine production and distribution. These limitations include the use of embryonated chicken eggs as the substrate for vaccine production, which is time-consuming and could involve potential biohazards in growth of new virus strains. Other limitations include the requirement that the current inactivated influenza vaccines be administered using needles and syringes, requiring trained personnel, which could be a bottleneck when attempting to vaccinate large populations in mass campaigns. In addition, the current inactivated vaccines that are delivered by injection elicit limited protective immunity in the upper respiratory tract where the infection process is initiated. Most of these limitations of the current vaccines are being addressed by research on novel approaches to vaccine development that are described in many of the chapters in this volume.

As an introduction to the topic, H.L. Yen and R.G. Webster describe the reservoir of influenza viruses with pandemic potential present in aquatic birds, particularly focusing on the evolution of highly pathogenic H5N1 viruses in Asia. As these viruses have continued to spread geographically, they also continue to diversify genetically, raising a strain selection problem for vaccine development. However, A.C.M. Boon and R.J. Webby review recent studies that show that substantial levels of antigenic cross reactivity are exhibited among the surface antigens of H5N1 strains, and that they can elicit cross-protective immune responses. A better definition of antigenic epitopes involved in cross-protection will be an important advance in enabling the design of effective vaccines.

To put new approaches into perspective, several chapters are devoted to reviewing current methods of developing and evaluating seasonal and pandemic influenza vaccines. A.E. Fiore, C.B. Bridges, and N.J. Cox review current efforts to produce seasonal vaccines and the impact of these vaccines on preventing influenza and its complications. E. O'Neill and R.O. Donis describe how candidate vaccine strains are detected, processed, and evaluated, bringing together surveillance, genetic and antigenic characterization, production of reassortant vaccine strains, and analysis of their safety and growth. Live attenuated, cold-adapted, temperature-sensitive influenza vaccine strains (LAIV) have proved highly effective, particularly in young children, against seasonal influenza. G.L. Chen and K. Subbarao show how the lessons learned in developing LAIV can be used to develop effective pandemic vaccines.

In addition to human vaccines, there is high interest in developing vaccines to control infection in poultry. D.R. Kapczynski and D.E. Swayne review the production of inactivated vaccines for avian species, many of which are formulated with oil-based adjuvants. In addition to commercial poultry, such vaccines have also been used in exotic and endangered species. Live attenuated vaccines have not been utilized in birds because of their potential to reassort with other avian influenza viruses. The development and the application of avian H5N1 influenza vaccines in China are discussed by H. Chen and Z. Bu. These include inactivated vaccines as well as live-vectored vaccines based on recombinant Newcastle disease virus. These vaccines have been widely used in Southeast Asia as well as Egypt, and have played an important role in control disease outbreaks.

A number of novel approaches for pandemic influenza vaccine development are now being actively pursued. T. Horimoto and Y. Kawaoka review the use of reverse genetics to develop recombinant virus strains for use in vaccine development, and present an overview of alternative strategies that are available for the development of H5N1 influenza vaccines. Genetically modified viruses with alterations in the NS1 gene have been evaluated as attenuated vaccines. This approach is reviewed by J. Richt and A. Garcia-Sastre. These viruses exhibit reduced virulence because these NS1 mutants do not inhibit interferon responses, unlike the native NS1 protein, which enhances viral replication. These genetically altered viruses represent new live vaccine candidates that confer protection in several animal models. The development of DNA plasmids as vaccines is also being pursued for influenza viruses; strategies to improve the potency and efficacy of such vaccines are described in the chapter by J. Kim and J. Jacob. An attractive alternative to egg-based vaccine production is the use of cell culture systems, in which recombinant expression vectors can be used for antigen production. Vaccines consisting of the purified HA protein have been produced using recombinant baculovirus expression in insect cells; J. Treanor reviews clinical trial results which show that these recombinant vaccines are well tolerated and induce functional antibody responses. Although the HA protein is considered the major component of most vaccines, the neuraminidase (NA) protein is also able to elicit protective immunity, probably by inhibiting cell-to-cell spread of the virus. The role of the neuraminidase in influenza vaccines is the subject of the chapter by M. Sylte and D. Suarez. The use of recombinant virus vectors that express influenza antigens represents an attractive approach for rapid vaccine production; S.A. Kopecky-Bromberg and P. Palese discuss the advantages and limitations of several recombinant vectors that are currently under investigation as influenza vaccines. Another novel approach to vaccine development is the use of virus-like particles that are assembled through the expression of viral structural proteins, particularly the HA, NA and M1 proteins. These particles closely resemble the influenza virion but lack the viral genome, and thus have a high degree of safety. S.-M. Kang and co-workers describe recent studies that demonstrate the production and characterization of influenza VLPs and their evaluation in animal models. A major limitation of current influenza vaccines is their induction of neutralizing antibodies that are highly strain-specific and are thus not able to protect against newly arising variant strains; it is therefore highly desirable to develop vaccines that would induce immune responses with an enhanced breadth of immunity. L.J. DiMenna and H.C.J. Ertl describe some approaches that are under investigation to develop potential universal vaccines against influenza A viruses. Results of initial human trials of H5N1 vaccines have shown that these antigens elicit relatively low immune responses, and it was observed that two immunizations with high doses of antigen were needed to achieve satisfactory responses. Such studies have stimulated research on the use of adjuvants to enhance responses to such vaccines. R.L. Atmar and W.A. Keitel provide an overview of current research on a number of these candidate adjuvants being evaluated with influenza vaccines.

Inactivated influenza vaccines are now delivered by hypodermic needles and syringes. This is a time-consuming process that complicates the ability to rapidly deploy a new vaccine to immunize a large population. As an alternative approach to vaccine delivery, I. Skountzou and S.-M. Kang review vaccine delivery by transcutaneous immunization (i.e., the direct application of vaccines to the skin). Mild chemical or physical disruption of the stratum corneum allows macromolecules as well as large particulate antigens to penetrate the skin and elicit immune responses. Such topical delivery provides an alternative approach to vaccination that could potentially result in self-administered vaccines. Alternatively, vaccine delivery through the skin can be accomplished by using micron-scale needles, as reviewed by M.R. Prausnitz and colleagues. Microneedles of various designs have been successfully used to deliver a range of vaccine antigens, including proteins, DNA vaccines and recombinant viruses. This approach to vaccine delivery has a number of advantages, including little or no pain compared to hypodermic needles, possible dose sparing, and the potential for the development of a stable patch formulation that could be self-administered.

Animal models are a critical means of evaluating the effectiveness of pandemic influenza vaccines. R.A. Tripp and S.M. Tompkins review a variety of animal models used to study influenza, and their strengths and weaknesses. Current seasonal influenza vaccines have limited immunogenicity in the age group that is most at risk of influenza complications, the elderly. This age group suffered disproportionately during the influenza pandemics of 1957 and 1968. S. Sambhara and J.E. McElhaney describe what is known about the molecular mechanisms that lead to hyporesponse in the elderly as a potential guide to finding ways to strengthen the response. A variety of vaccines against potential avian influenza pandemic virus candidates have been developed and tested in human clinical trials. W.A. Keitel and R.L. Atmar discuss the results of these candidates in humans, including the effects of dose, number of doses, and both aluminum- and oil-in-water-containing adjuvants.

All potential influenza vaccines that could be used in humans in prepandemic preparedness efforts or in reaction to a pandemic must be approved by regulatory authorities. N.W. Baylor and F. Houn review some of the challenges that the Food and Drug Administration (FDA) faces in evaluating pandemic influenza vaccines for licensure, and describe some of the efforts being made by the FDA to speed up the development of such vaccines, such as accelerated approval and priority review. They comment on guidance documents that help manufacturers ensure that they collect the critical information needed for these reviews.

Pandemics have the potential for massive global impact. Thus, vaccines should ideally be available throughout the world. K.M. Edwards et al. discuss potential global needs and current global production capacity. It is likely that vaccine supply in the early phases of a pandemic will not be adequate to meet the needs of even an industrialized country such as the United States. B. Schwartz and W.A. Orenstein review efforts within the United States to set priorities for mass vaccination, including the criteria used and the public input process that went into establishing the current proposed priorities.

The editors hope that this volume will stimulate research on improved influenza vaccines, including those that will be able to effectively prevent the next pandemic. We thank all of the authors for their contributions. We are extremely indebted to Erin-Joi Collins for all she did to make this volume possible; this included helping to organize the chapters, communicating with the authors, tracking progress, identifying and resolving problems, and much more.

Atlanta, GA, USA                                                    Richard W. Compans
                                                                          Walter A. Orenstein

# Contents

# Contributors

Alexander K. Andrianov
Apogee Technology, 129 Morgan Drive, Norwood, MA 02062, USA
aandrianov@apogeemems.com

Robert L. Atmar
Departments of Medicine and Molecular Virology and Microbiology, Baylor
College of Medicine 280, One Baylor Plaza, Houston, TX 77030, USA
ratmar@bcm.edu

Norman W. Baylor
FDA/CBER/OVRR, HFM-405, 1401 Rockville Pike,
Rockville, MD 20852, USA
norman.baylor@fda.hhs.gov

Adrianus C.M. Boon
Division of Virology, Department of Infectious Diseases,
St. Jude Children's Research Hospital, 332 N Lauderdale St.,
Memphis, TN 38105, USA
Jacco.Boon@stjude.org

John W. Boslego
Director, Vaccine Development Global Program, PATH, 1800 K Street, NW,
Suite 800, Washington, DC 20006, USA
jboslego@path.org

C.B. Bridges
Centers for Disease Control and Prevention, 1600 Clifton Rd,
NE, Atlanta, GA 30333, USA

R.A. Bright
Novavax, Inc., Vaccine Technologies, 1 Taft Court, Rockville, MD, USA

Zhigao Bu
Harbin Veterinary Research Institute, 427 Maduan Street, Harbin 150001,
People's Republic of China

Grace L. Chen
NIAID, NIH, Building 33, 3E 13C.2, 33 North Drive, MSC 3203,
Bethesda, MD 20892-3203, USA
chengra@niaid.nih.gov

Hualan Chen
Harbin Veterinary Research Institute, 427 Maduan Street,
Harbin 150001, People's Republic of China
hlchen1@yahoo.com

Richard W. Compans
Department of Microbiology & Immunology and Emory Vaccine Center,
School of Medicine, Emory University, Atlanta GA 30322,
Rollins Research Center 3001, USA
compans@microbio.emory.edu

Michel Cormier
278 Andsbury Avenue, Mountain View, CA 94043, USA
michel.cormier@sbcglobal.net

N.J. Cox
Centers for Disease Control and Prevention, 1600 Clifton Rd, NE, Atlanta,
GA 30333, USA

Lauren J. DiMenna
The Wistar Institute, 3601 Spruce Street, Philadelphia, PA 19104, USA

Ruben Donis
Influenza Division, National Center for Immunization and Respiratory
Diseases, Coordinating Centers for Infectious Diseases, Centers for
Disease Control and Prevention, 1600 Clifton Road, Mail Stop G-16,
Atlanta, GA 30333, USA
rdonis@cdc.gov

Kathryn M. Edwards
Sarah H. Sell Professor of Pediatrics, Vanderbilt Vaccine Research Program,
Vanderbilt University School of Medicine, 1161 21st Avenue South,
CCC-5323 Medical Center North, Nashville, TN 37232, USA
kathryn.edwards@vanderbilt.edu

Hildegund C.J. Ertl
The Wistar Institute, 3601 Spruce Street, Philadelphia, PA 19104, USA

A.E. Fiore
Centers for Disease Control and Prevention, 1600 Clifton Rd, NE, Atlanta,
GA 30333, USA

Adolfo García-Sastre
Department of Microbiology, Department of Medicine, Division of Infectious
Diseases and Emerging Pathogens Institute, Mount Sinai School of Medicine,
New York, NY, 10029, USA
adolfo.garcia-sastre@mssm.edu

T. Horimoto
Division of Virology, Department of Microbiology and Immunology,
Institute of Medical Science, University of Tokyo, 4-6-1 Shirokanedai,
Minato-ku, Tokyo 108-8639, Japan

Florence Houn
FDA/CBER/OVRR, HFM-405, 1401 Rockville Pike, Rockville,
MD 20852, USA
florence.houn@fda.hhs.gov

Joshy Jacob
Department of Microbiology & Immunology and Emory Vaccine Center,
School of Medicine, Emory University, 954 Gatewood Road, Atlanta,
GA 30329, USA
joshy.jacob@emory.edu

S.M. Kang
Department of Microbiology & Immunology and Emory Vaccine Center,
School of Medicine, Emory University, 1510 Clifton Road, Atlanta,
GA 30322, USA
skang2@emory.edu

Darrell R. Kapczynski
Exotic and Emerging Avian Viral Diseases Research Unit, Southeast Poultry
Research Laboratory
USDA—Agricultural Research Service—South Atlantic Area, 934 College
Station Road, Athens, Georgia 30605, USA
darrell.kapczynski@ars.usda.gov

Y. Kawaoka
Department of Pathological Sciences, School of Veterinary Medicine,
University of Wisconsin–Madison, Madison, WI 53706, USA

Wendy A. Keitel
Molecular Virology & Microbiology and Medicine, Baylor College of
Medicine 280, One Baylor Plaza, Houston, TX 77030, USA
wkeitel@bcm.edu

Jin Hyang Kim
Department of Microbiology & Immunology and Emory Vaccine Center,
School of Medicine, Emory University, 954 Gatewood Road, Atlanta,
GA 30329, USA

Sarah A. Kopecky-Bromberg
Department of Microbiology, Mount Sinai School of Medicine, New York,
NY 10029-6574, USA

Janet E. McElhaney
Department of Medicine, University of British Columbia, 9B St. Paul's Hospital,
1081 Burrard Street, Vancouver, BC V6Y 1Y6, Canada
JMcElhaney@providencehealth.bc.ca

John A. Mikszta
BD Technologies, 21 Davis Drive, Research Triangle Park, NC 27709, USA
john_mikszta@bd.com

Eduardo O'Neill
Influenza Division, National Center for Immunization and Respiratory Diseases,
Coordinating Centers for Infectious Diseases, Centers for Disease
Control and Prevention 1600 Clifton Road, Mail Stop G-16, Atlanta,
GA 30333, USA
eoneill@cdc.gov

Walter Orenstein
Emory University, School of Medicine, Department of Medicine,
1510 Clifton Road, Atlanta GA 30322, USA
walter.orenstein@gatesfoundation.org

Peter Palese
Department of Medicine, Mount Sinai School of Medicine, New York,
NY 10029-6574, USA
Peter.Palese@mssm.edu

Andrew Pasternak
Oliver Wyman, 10 South Wacker Drive, 13th Floor, Chicago, IL 60606, USA
andy.pasternak@oliverwyman.com

Mark R. Prausnitz
School of Chemical and Biomolecular Engineering, Georgia Institute
of Technology, 311 Ferst Drive Atlanta, GA 30332-0100, USA
prausnitz@gatech.edu

P. Pushko
Novavax, Inc., Vaccine Technologies, 1 Taft Court, Rockville, MD, USA

Juergen A. Richt
Virus and Prion Diseases of Livestock Research Unit, National Animal Disease
Center, Ames, IA, 5001, USA

Adam Sabow
Oliver Wyman, 10 South Wacker Drive, 13th Floor, Chicago, IL 60606, USA
adam.sabow@oliverwyman.com

Suryaprakash Sambhara
Influenza Division, Centers for Disease Control and Prevention,
1600 Clifton Road, Atlanta, GA 30333, USA
ssambhara@cdc.gov

Benjamin Schwartz
Centers for Disease Control and Prevention, 1600 Clifton Rd NE,
Mailstop E-05, Atlanta, GA 30333, USA
bxs1@cdc.gov

Ioanna Skountzou
Department of Microbiology & Immunology and Emory Vaccine Center,
School of Medicine, Emory University, 1510 Clifton Road, Atlanta,
GA 30322, USA
iskount@emory.edu

G. Smith
Novavax, Inc., Vaccine Technologies, 1 Taft Court, Rockville, MD, USA

David L. Suarez
Southeast Poultry Research Laboratory, Agricultural Research Service,
US Dept. of Agriculture, 934 College Station Road, Athens, GA 30605, USA
David.Suarez@ars.usda.gov

Kanta Subbarao
Laboratory of Infectious Diseases, NIAID, NIH, Bldg 33, Room 3E13C.1,
33 North Drive, MSC 3203, Bethesda, MD 20892-3203, USA
ksubbarao@niaid.nih.gov

David E. Swayne
Exotic and Emerging Avian Viral Diseases Research Unit, Southeast Poultry
Research Laboratory, USDA-Agricultural Research Service—South Atlantic
Area, 934 College Station Road, Athens, GA 30605, USA

Matthew J. Sylte
Department of Infectious Diseases, College of Veterinary Medicine,
University of Georgia, Athens, GA 30602, USA

S. Mark Tompkins
University of Georgia, Center for Disease Intervention, Animal Health Research
Center, 111 Carlton Street, Athens, GA 30602, USA
Tompkins@vet.uga.edu

John Treanor
Infectious Diseases Division, Department of Medicine, University of Rochester
Medical Center, 601 Elmwood Avenue, Rochester, NY 14642, USA
John_Treanor@urmc.rochester.edu

Ralph A. Tripp
University of Georgia, Center for Disease Intervention, Animal Health Research
Center, 111 Carlton Street, Athens, GA 30602, USA
rtripp@vet.uga.edu

Richard J. Webby
Division of Virology, Department of Infectious Diseases, St. Jude Children's
Research Hospital, 332 N Lauderdale St., Memphis, TN 38105, USA
Richard.Webby@stjude.org

Robert G. Webster
Division of Virology, Department of Infectious Diseases, St. Jude Children's
Research Hospital, 332 N. Lauderdale, Memphis, TN 38105, USA
robert.webster@stjude.org

Hui-Ling Yen
Division of Virology, Department of Infectious Diseases, St. Jude Children's
Research Hospital, 332 N. Lauderdale, Memphis, TN 38105, USA

# Part I
# Pandemic Influenza Overview

# Pandemic Influenza as a Current Threat

Hui-Ling Yen and Robert G. Webster

## Contents

**Abstract** Pandemics of influenza emerge from the aquatic bird reservoir, adapt to humans, modify their severity, and cause seasonal influenza. The catastrophic Spanish H1N1 virus may have obtained all of its eight gene segments from the avian reservoir, whereas the Asian H2N2 and the Hong Kong H3N2 pandemics emerged by reassortment between the circulating human virus and an avian H2 or

H.-L. Yen
Division of Virology, Department of Infections Diseases, St. Jude children's Research Hospital, 262, Danny Thomas Place, Memphis, TN 38105, USA
Present Address: Department of Microbiology, The Unviersity of Hong Kong, Hong Kong

R.G. Webster (✉)
Division of Virology, Department of Infectious Diseases, St. Jude Children's Research Hospital, 262 Danny Thomas Place, Memphis, TN 38105, USA
e-mail: robert.webster@stjude.org

R.W. Compans and W.A. Orenstein (eds.), *Vaccines for Pandemic Influenza*,
Current Topics in Microbiology and Immunology 333,
DOI 10.1007/978-3-540-92165-3_1, © Springer-Verlag Berlin Heidelberg 2009

H3 donor. Of the 16 hemagglutinin subtypes, the H2, H5, H6, H7, and H9 viruses are considered to have pandemic potential. While this chapter focuses on the evolution of the Asian highly pathogenic (HP) H5N1 influenza virus, other subtypes are also considered. The unique features of the HP H5N1 viruses that have devastated the domestic poultry of Eurasia are discussed. Although they transmit poorly to humans, they continue to kill more than 60% of infected persons. It is unknown whether HP H5N1 will acquire human pandemic status; if it does not, another subtype eventually will do so, for a future influenza pandemic is inevitable.

# 1 Influenza Virus as a Noneradicable Zoonosis

## 1.1 Natural Reservoirs for Influenza A Virus

The established reservoirs of all 16 hemagglutinin (HA) and nine neuraminidase (NA) subtypes of influenza A viruses are the aquatic birds of the world (Fouchier et al. 2005; Webster et al. 1992). In this reservoir, the low pathogenic avian influenza viruses replicate in the respiratory tract and the intestine and live in apparent harmony with their hosts, causing no apparent disease signs (Webster et al. 1978; Kida et al. 1980).

In addition to aquatic birds, diverse animal species are susceptible to influenza A virus infection in nature or under laboratory conditions. Current information suggests host-specific lineages have been established in birds, pigs, and horses, as well as humans. Phylogenetic analyses suggest that mammalian influenza viruses all are derived from the avian influenza reservoir (Webster et al. 1992). However, the possibility exists that established host-specific influenza viruses may be introduced and further established in other species, as was observed with the equine H3N8 virus in dogs (Crawford et al. 2005).

The clinical outcome of influenza A virus infection depends on the host and the virus. Domestic poultry are susceptible to most subtypes of avian influenza virus infection. Intensive surveillance activities in the United States during 2002–2005 detected avian influenza virus or specific antibodies to H1–H13 subtypes and all nine NA subtypes (Senne 2007). Of the 16 HA subtypes, only two (H5 and H7) subtypes are known to have the capacity to become highly pathogenic (HP) in chickens and other gallinaceous birds. The HP H5 and H7 viruses usually produce asymptomatic to mild clinical infection in ducks or wild birds and are rarely lethal to wild birds, with the exception of the HP H5N1 virus that has emerged in Asia since 1997. The HP phenotype is related, but not restricted, to the presence of multiple basic amino acids at the HA cleavage site (Rott et al. 1995; Horimoto and Kawaoka 2001).

The error-prone viral RNA polymerase, the segmented RNA genome that allows dynamic genetic reassortment within the large gene pool perpetuated in aquatic birds, and the existence of multiple natural reservoirs all point to the influenza A virus as a noneradicable zoonosis.

## 1.2   Ecology of Influenza A Virus in Asia

Southern China is the hypothetical pandemic epicenter of influenza, as this environment may have provided the conditions for the emergence of 1957 Asian and 1968 Hong Kong pandemic influenza viruses (Shortridge and Stuart-Harris 1982). In tropical and subtropical areas, human influenza can be detected year-round. The warm winter in Southeast Asia attracts migratory birds from northern climes to spend the winter in this region. The high density of human population and prevalence of backyard poultry (ducks, geese, and chickens) and pigs provide the opportunity for close interaction between these influenza reservoir animals and the possibility of interspecies transmission and genetic reassortment. Pigs that possess both receptors for avian (sialic acids with $\alpha$-2,3-galactose linkage) and human (sialic acids with $\alpha$-2,6-galactose linkages) influenza viruses were considered "mixing vessels" for generating reassortant viruses (Scholtissek 1995). In addition, the live-poultry market ("wet market") system provides optimal conditions for influenza virus evolution, with transmission between avian species and possible infection of humans (Shortridge et al. 1998; Peiris et al. 2007). Transmission between different host species and serologic evidence of human infection with H4, H5, H6, H7, H10, and H11 subtypes of avian influenza virus were documented in this region prior to the 1997 Hong Kong H5N1 outbreak (Shortridge 1992; Peiris et al. 2007).

## 2   Human Influenza Epidemics and Pandemics

### 2.1   Epidemiology of Human Influenza

Humans can be infected with influenza A, B, or C viruses, all of which belong to the *Orthomyxoviridae* family and are distinguished by serologic reactions of conserved viral nucleoprotein or matrix protein (Beard 1970). Influenza in humans may occur in two epidemiologic forms: pandemics and epidemics (Nicholson 1998). An influenza pandemic is a large-scale global outbreak of the disease, while an epidemic is more sporadic and localized, as seen with seasonal influenza outbreaks. Influenza epidemics result from newly immune-selected variant strains that contain accumulated point mutations that result in amino acid changes in the antigenic sites in the HA glycoprotein (predominantly in HA1) as well as NA glycoprotein (antigenic drift) (Fig. 1a). Current epidemics are caused by antigenic variants of influenza A viruses of the H1N1 H3N2 or their reassortant H1N2 subtypes as well as influenza B viruses. Because most of the population possesses cross-reacting antibodies for recent antigenic variants, severe clinical signs and death are observed mostly among young children, the elderly, and people with other underlying diseases.

Pandemic influenza results from the emergence of a new subtype of influenza A virus (antigenic shift) (Fig. 1b). Because the population does not possess immunity to the new subtype of influenza A virus, the new subtype may spread globally with

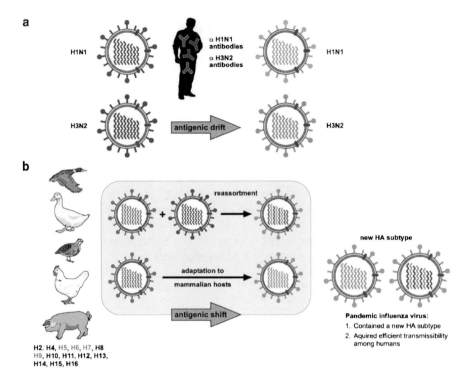

**Fig. 1** **a–b** Antigenic drift and antigenic shift of influenza virus. **a** Pre-existing antibody response against the HA and NA glycoproteins of influenza A virus of H1N1 or H3N2 subtypes or influenza B virus selects antigenic variants with amino acid changes modifying the antigenic structure that allow influenza virus to evade immunity. Antigenic drift is a result of both immune and natural selection. **b** Reassortment between avian and human influenza A virus or continued adaptation of an avian influenza virus may result in a new subtype of influenza A virus with sustained human-to-human transmissibility. Pre-existing antibody response provides little or no cross-protection for this major change in the HA (and NA); however, cytotoxic T lymphocyte responses that target the conserved peptides encoded in viral internal proteins may provide protection

high attack rates and may cause significant morbidity and mortality (Nicholson 1998). However, the severity of a pandemic may be dependent on the composition of the virus, as cytotoxic T lymphocyte responses that target the relatively conserved internal proteins may provide protection (Rimmelzwaan et al. 2007). A mild pandemic is possible when the pandemic virus emerges through genetic reassortment (see below) and by acquiring internal gene segments from previously circulated human influenza virus. During the twentieth century, there were three global pandemics. These pandemics occurred in 1918 (Spanish pandemic, H1N1 subtype), 1957 (Asian pandemic, H2N2 subtype), and 1968 (Hong Kong pandemic, H3N2 subtype). In addition, in 1977 there was a reemergence of the H1N1 subtype (Russian pandemic). With the emergence of a new subtype, the old subtype is usually replaced. The exception was the 1977 H1N1 pandemic virus, which continues to circulate along with the H3N2 subtype.

## 2.2 Molecular Requirements for a Pandemic Strain: Emergence of 1918, 1957, and 1968 Pandemic Strains

The minimum molecular requirement for a pandemic influenza strain is a new HA subtype derived from the avian reservoir with sustained human-to-human transmissibility. Influenza pandemics that occurred during the last century suggest that such a virus may emerge in two ways: (1) genetic reassortment between avian influenza virus and circulating human influenza A viruses (as seen with the 1957 and 1968 pandemic viruses) and (2) interspecies transmission from an avian reservoir into an intermediate host, followed by continued adaptations (as seen with the 1918 pandemic virus) (Horimoto and Kawaoka 2005; Webby et al. 2004; Belshe 2005) (Fig. 1b).

Genetic analyses showed that the H2N2 1957 Asian pandemic virus acquired three gene segments from an avian reservoir (PB1, HA, and NA) and kept five other gene segments from the H1N1 human strain circulating prior to 1957. Similarly, the 1968 H3N2 Hong Kong pandemic virus acquired two gene segments from an avian reservoir (PB1 and HA) and kept six other gene segments from the H2N2 human strain that circulated in 1957–1968 (Webster et al. 1992). Pig tracheae, which have sialyl receptors for avian and human influenza viruses, have been proposed as the site for genetic reassortment (Ito et al. 2000; Scholtissek 1995). Additionally, the intermediate host may not be restricted to pigs. A report also demonstrated the presence of sialyl receptors with α-2,3- and α-2,6-galactose linkages in chicken and quail intestines (Guo et al. 2007). Unlike the 1957 or 1968 pandemic viruses, genetic analysis of the 1918 pandemic virus suggests that all of its eight gene segments originated from the avian reservoir without genetic reassortment (Belshe 2005; Taubenberger et al. 2005). However, it is not clear how long it took for an avian-originated influenza virus to become adapted in mammals or in which mammalian reservoirs the adaptations occurred.

The HA glycoprotein of avian and human influenza viruses preferentially recognizes sialic acids with α-2,3- or α-2,6-galactose linkages, respectively. As the HA segments of the 1918, 1957, and 1968 pandemic strains were derived from the avian reservoir, one common feature between the pandemic strains is the acquisition of amino acid changes in the receptor binding site of the HA glycoprotein that alter the virus' receptor binding specificity from the α-2,3 to the α-2,6 linkage between sialic acid and galactose (Matrosovich et al. 2000; Stevens et al. 2006). The switch to a predominantly α-2,6-linked sialyl receptor specificity facilitated transmission of 1918 pandemic virus (Tumpey et al. 2007) and likely the 1957 and 1968 pandemic viruses. The effect of the switch in receptor specificity on viral pathogenicity is less understood. Theoretically, the changes in receptor specificity may result in a change in target cells from the lung epithelial cells (exhibit α-2,3-linked sialyl receptor) to the epithelial cells lining the upper respiratory tract (exhibit α-2,6-linked sialyl receptor), thereby reducing the occurrence of pneumonia.

Another molecular characteristic observed in the 1918 and 1957 pandemic influenza viruses is the loss of the secondary sialic acid binding site with hemadsorbing activity in NA (Matrosovich 2008), which is a molecular signature for NA derived from avian influenza viruses (Hausmann et al. 1995; Varghese et al. 1997). In addition to the surface glycoproteins, genetic analyses of influenza viruses isolated from different hosts have identified 32 residues from PB2, PA, NP, M1, and NS1 proteins as host-specific markers differentiating human and avian influenza viruses (Finkelstein et al. 2007). Among these 32 residues, 13 were conserved among the 1918, 1957, and 1968 pandemic influenza viruses (Finkelstein et al. 2007). The clear genetic difference between avian and human influenza viruses in these gene segments may be functionally related to the differences in cooperation with avian and human cellular machinery. It is likely that a pandemic strain should contain some of the human-specific markers that allow efficient replication and transmission.

## 3   H5N1 Virus as a Pandemic Threat

### 3.1   Emergence and Spread of H5N1 Virus

Before 1996, low-pathogenic H5 avian influenza viruses had been isolated from domestic ducks and geese in Southeastern China but not from chickens (Shortridge et al. 1998), and neutralizing antibodies to H5 virus were detected in pig sera from Southeastern China collected in 1977–1982 (2 of 127 samples) and 1998 (10 of 101 samples) (Peiris et al. 2007).

Genetic evidence showed that the precursor (A/Goose/Guangdong/1/96) of the currently circulating HP H5N1 virus was first detected in domestic geese in Guangdong, China, in 1996 (Peiris et al. 2007). To date, the precursor of this virus is unknown, although eight gene segments are closely related to those from low-pathogenic H5 viruses isolated from migratory birds and wild ducks in Hokkaido, Japan (Okazaki et al. 2000; Duan et al. 2007).

The index human case of H5N1 influenza occurred in May 1997, and the causative virus was identified in August 1997 (de Jong et al. 1997) as the first HP avian influenza virus known to cause lethal infection in humans. During the remainder of 1997, 17 additional human cases were detected, and six patients succumbed to H5N1 infection. Surveillance and epidemiologic studies established that poultry markets were the source of human H5N1 infection, as H5N1 virus was isolated from approximately 20% of fecal samples from chickens and from approximately 2% of fecal samples from ducks and geese in the market (Shortridge et al. 1998). Subsequent genetic analysis of the index human virus revealed that six internal genes were closely related to those in A/Quail/Hong Kong/G1/97 (H9N2) and that the NA gene was genetically similar to that of A/Teal/Hong Kong/W312/97 (H6N1), raising the possibility that reassortment between these viruses was involved in the genesis of the HP H5N1 virus (Peiris et al. 2007).

The culling of all poultry in Hong Kong effectively eradicated that particular genotype of HP H5N1 influenza virus. There were no more human cases in Hong Kong, but H5N1 viruses continued to circulate among apparently healthy domestic ducks in the coastal provinces of China between 1999 and 2002 (Chen et al. 2004). HP H5N1 viruses were also detected in geese in a live-poultry market in Vietnam in 2001 and from duck meat exported from China to Korea and Japan in 2001 and 2003 (Peiris et al. 2007). During 2001 and 2002, multiple H5N1 genotypes were detected in poultry in Southern China (Li et al. 2004). These viruses had HA typical of the A/Goose/Guangdong/1/96-like lineage but with a plethora of different internal genes. In addition, the NA genes of these variant H5N1 viruses were typical of that of A/Goose/Guangdong/1/96 but frequently had deletions of amino acids in the stalk region (Li et al. 2004). In 2002, H5N1 outbreaks of lethal disease in waterfowl occurred in Penfold Park and Kowloon Park in Hong Kong; many aquatic species as well as tree sparrows and pigeons were killed (Ellis et al. 2004).

The next key event in the development of H5N1 viruses was its re-emergence in humans in 2003. The daughter of a Hong Kong family died while visiting the Fujian province of China in February 2003. On their return to Hong Kong, H5N1 infection was diagnosed in her father and brother (Peiris et al. 2004); the father subsequently died, but the brother recovered.

In late 2003 to early 2004, outbreaks of HP H5N1 viruses in domestic poultry were reported in South Korea, Japan, Vietnam, Laos, Cambodia, and Indonesia. During this period, avian-to-human transmission resulted in lethal H5N1 human infection in Vietnam and Thailand. Serologic evidence suggests that limited human infections occurred in Japan and South Korea during the 2003–2004 H5N1 outbreaks. Genetic analysis showed that the viruses that spread to Japan and South Korea (genotype V) differed in the PA gene from the viruses that became dominant in Vietnam, Thailand, Cambodia, Indonesia, and Southern China (genotype Z) (Li et al. 2004).

Qinghai Lake in Western China is a leading breeding site of migratory waterfowl. In May 2005, a lethal outbreak of HP H5N1 influenza occurred at Qinghai Lake that affected bar-headed geese (*Anser indicas*), great black-headed gulls (*Larus ichthyaetus*), brown-headed gulls (*Larus brunnicephalus*), ruddy shelducks (*Tadorna ferruginea*), and great cormorants (*Phalacrocorax carbo*) and killed more than 6,000 migratory waterfowl (Chen et al. 2006; Peiris et al. 2007). Other wild birds that have been affected by H5N1 include whooper swans (*Cygnus cygnus*), black-necked cranes (*Grus nigricollis*), and pochards (diving ducks that belong to the subfamily *Aythyinae*) (Peiris et al. 2007). This event was the first major outbreak of H5N1 influenza virus in wild migratory birds. The precursors of the dominant Qinghai H5N1 virus were detected in mallard ducks at Poyang Lake, China, in March 2005 (Chen et al. 2006) and may have came from domestic poultry. During the outbreak at Qinghai Lake, at least four genotypes of H5N1 virus were detected in the waterfowl, but one genotype became dominant and rapidly spread to wild and domestic birds in Siberia (July 2005), Mongolia and Kazakhstan (August 2005), Croatia, Romania, and Turkey (October 2005), Middle Eastern and European countries (2006), and Nigeria and India (February 2006) (Chen et al. 2006; Peiris et al. 2007). Although the Qinghai-like H5N1 virus can transiently infect migratory waterfowl, available surveillance evidence does not indicate the perpetuation of this virus in this natural influenza reservoir.

In 2005, two major clades with no overlapping geographic distributions were identified on the basis of HA sequence analysis (World Health Organization 2005). Viruses isolated from Thailand, Cambodia, and Vietnam during the 2004–2005 outbreaks were clustered into clade 1, whereas viruses isolated from China, Indonesia,

Korea, and Japan during the 2003–2004 outbreaks were clustered into clade 2. In 2005, human infection with H5N1 viruses continued to be reported in Vietnam and Thailand, and new cases were reported in Cambodia, China, and Indonesia. Effective control measures taken by Vietnam (vaccination of poultry) and Thailand (stamping out) since 2006 have significantly reduced the number of outbreaks in these two countries as well as the circulation of clade 1 virus. On the other hand, clade 2 viruses continued to evolve into three major subclades that differ in geographic distribution. Indonesian H5N1 viruses isolated since 2003 continue to cluster into one sublineage (subclade 2.1, which can be further grouped into subclades 2.1.1, 2.1.2, and 2.1.3), suggesting the possibility of a single introduction of the virus into Indonesia and its continued evolution within the region since 2003 (World Health Organization 2005, 2006; Peiris et al. 2007; Smith et al. 2006b). Subclade 2.2 contains the H5N1 virus that caused the large-scale lethal outbreak in wild birds at Qinghai Lake during summer 2005 and the H5N1 viruses that subsequently spread to the Middle East, Europe, and Africa, suggesting a potential role for migratory birds in spreading the virus (World Health Organization 2005, 2006). Surveillance in Southern China from July 2005 to June 2006 identified a dominant sublineage that had replaced most of the previously established sublineages. These Fujian-like viruses formed a separate subclade 2.3 (which can be further grouped into subclades 2.3.1, 2.3.2, 2.3.3, and 2.3.4) and further spread to Hong Kong, Malaysia, Laos, Vietnam, and Thailand, causing outbreaks in wild birds and domestic poultry in 2006, 2007, and 2008 (World Health Organization 2005; Smith et al. 2006a).

## 3.2   Unique Features of H5N1 Viruses: Changing Patterns

As the H5N1 viruses continued to spread and evolve during the past decade, we have learned of and observed several unique features about the virus. The first feature noted was the ability of the H5N1 virus to cause lethal infection in wild birds, including waterfowl, after the outbreak in Hong Kong in winter 2002 (Ellis et al. 2004). These H5N1 isolates were highly lethal to mallard ducks and caused neurologic symptoms (Sturm-Ramirez et al. 2004). Although HP H5 viruses are highly lethal in chickens and other gallinaceous birds, they had rarely been reported to be pathogenic in wild birds. The only recorded incident prior to the Hong Kong H5N1 event was reported in 1961, when an H5N3 virus (A/Tern/South Africa/61) caused deaths in terns. We have further learned that, although some of the H5N1 viruses isolated since 2002 were initially highly lethal to mallard ducks, antigenic variants with decreased pathogenicity can be selected rapidly in this natural influenza reservoir (Hulse-Post et al. 2005). In addition, waterfowl (including domestic ducks) have exhibited higher resistance than chickens and other gallinaceous birds to H5N1 infection and thus can serve as hidden sources ("Trojan horses") for the maintenance and spread of the virus (Hulse-Post et al. 2005).

Unique features were also noted among the clade 2.2 H5N1 viruses, which spread widely to the Middle East, Europe, and Africa. The spread of this lineage of H5N1 virus is considered to have occurred partly due to the migration of the birds. Experimental infection with six wild duck species (*Anas* and *Aythya* species) revealed differences in susceptibility to H5N1 virus (Keawcharoen et al. 2008). In addition, it was noted among these wild duck species that virus shedding from the throat was higher and of longer duration than from the cloaca (Keawcharoen et al. 2008). This property of respiratory shedding must be considered when studying the ecology of this H5N1 virus in migratory birds. The collection of both oral and cloacal samples from birds is therefore critical for surveillance purposes. Another notable feature of the dominant Qinghai H5N1 virus is that it had a mutation of the PB2 gene (E→K at residue 627) that is one of the conserved host markers (E627 for avian and K627 for human influenza viruses) and is associated with increased viral virulence in mice (Chen et al. 2006; Hatta et al. 2001).

The re-emergence of human H5N1 infections in 2004 was accompanied by several unique characteristics of the virus, including an increased host range and increased pathogenicity in mammalian species. Although cats can be infected with influenza virus experimentally, the first report of natural influenza virus infection in felids was caused by the HP H5N1 virus in a zoo in Thailand: tigers and leopards that were fed H5N1-infected poultry carcasses showed severe pneumonia and succumbed to infection (Keawcharoen et al. 2004). Further laboratory study confirmed the susceptibility of domestic cats to HP H5N1 infection as well as experimental transmission among cats (Kuiken et al. 2005; Rimmelzwaan et al. 2006). In addition to cats, the fatal infection of a dog fed H5N1-infected duck carcasses in Thailand was reported (Songserm et al. 2006). Stone martens, a wild mammalian species that, like ferrets, belong to the *Mustelidae* family, were also infected during an H5N1 outbreak in wild birds in Germany, and H5N1 infection in Owston's palm civet (*Chrotagale owstoni*) was reported in Vietnam (Peiris et al. 2007). These cases highlight the potential threat of H5N1 in wild mammalian species.

Additionally, increased viral pathogenicity in mammalian species is associated with H5N1 viruses isolated from human infection (Govorkova et al. 2005). Characterization of an HP H5N1 virus isolated from a fatal human case in Vietnam showed that viral polymerase activity is a key factor for increased pathogenicity in mammals (Salomon et al. 2006). Other factors that may determine the host range and the pathogenicity of H5N1 viruses include viral surface glycoproteins, the presence of K at residue 627 in the PB2 protein, and the ability to evade the host innate immune response through viral NS1 protein (Neumann and Kawaoka 2006).

Overall, the widespread HP H5N1 virus has several unique characteristics that should be taken into account in any attempts to control the virus. First are the virus's abilities to replicate in both the respiratory and gastrointestinal tracts and cause lethal infection in waterfowl reservoirs. Second is that, in domestic duck and waterfowl reservoirs, selection of antigenic variants with decreased pathogenicity to these species may occur. Domestic ducks and waterfowl that harbor the selected variants without apparent symptoms may transmit the virus to chickens and other wild birds (such as geese and swans) that are highly susceptible to infection, thus causing outbreaks (Fig. 2). Third, HP H5N1 virus with an increased host range to

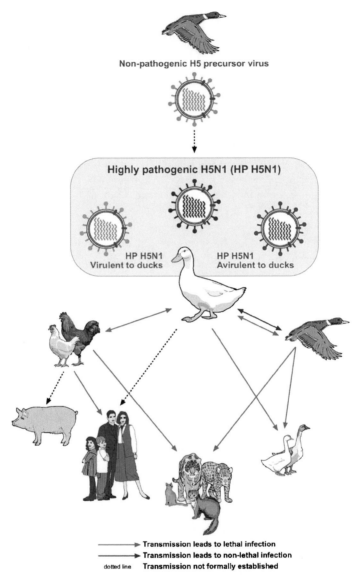

**Fig. 2** Drivers of diversity for H5N1 virus. Highly pathogenic (HP) H5N1 influenza viruses (*shown in purple*) evolved from nonpathogenic H5 precursors (*shown in green*) preserved in wild aquatic-bird reservoirs, with eight gene segments derived from the Eurasia influenza gene pool. While universally highly lethal to chickens, the HP H5N1 isolates demonstrate variable pathogenicity in mammals and ducks. In domestic ducks and mallards, inoculation of HP H5N1 viruses that are virulent to the duck species (*shown in red*) may lead to selection of antigenic variants with decreased pathogenicity in ducks (*shown in blue*). Domestic ducks or waterfowl that harbor the selected variants of HP H5N1 virus without apparent symptoms may transmit the virus to chickens or other wild birds (geese or swans) that are highly susceptible to infection, thus causing outbreaks. The proximity of multiple influenza reservoirs and the endemicity of the H5N1 avian influenza virus in Southeast Asia since 1996 have provided numerous opportunities for the viruses to interact with various avian and mammalian species. Because the selection pressure on H5N1 viruses varies with the host, interspecies transmission events may have driven both the antigenic and the host range diversity of the virus

felids or ferret species may provide the virus with opportunities to further adapt in mammals, including humans (Fig. 2).

## 3.3   Human Infection with H5N1

Although transmission of the H5N1 viruses among avian species is highly efficient, interspecies transmission from avian species to mammalians remains infrequent. After a decade of continued circulation, H5N1 viruses have resulted in more than 380 human infections with an approximately 60% case fatality rate. The highest fatality is observed among patients 10–19 years old, and the lowest rate is among patients 50 years or older (bdel-Ghafar et al. 2008). The typical clinical manifestation of human H5N1 infection is severe pneumonia that may progress to acute respiratory distress syndrome, although mild upper respiratory illness without pneumonia has been reported (Beigel et al. 2005; bdel-Ghafar et al. 2008). Depending on the clade of H5N1 virus, gastrointestinal symptoms have been reported among 5–52% of patients (bdel-Ghafar et al. 2008). Encephalopathy has been reported in one human case (de Jong et al. 2005a). High viral load, lymphopenia, increased levels of lactate dehydrogenase, and certain chemokine and cytokine levels correlate with fatal outcome after infection (de Jong et al. 2006; bdel-Ghafar et al. 2008). Seroepidemiology results among high-risk groups with close contact to infected poultry or patients suggest that asymptomatic infection is rare (Beigel et al. 2005; bdel-Ghafar et al. 2008).

Direct avian-to-human transmission as a result of close contact with H5N1-infected poultry, a contaminated environment, or consumption of undercooked poultry products is the predominant cause of human infection (Beigel et al. 2005; bdel-Ghafar et al. 2008). Vertical viral transmission from infected mother to fetus has been reported (Gu et al. 2007). Limited and nonsustained human-to-human infections have been reported from family members attending H5N1 patients (Ungchusak et al. 2005; Kandun et al. 2006). The observation that 90% of case clusters occur among blood-related family members also suggests the possibility of genetic susceptibility (bdel-Ghafar et al. 2008).

## 4   Other Subtypes with Pandemic Potential

### 4.1   H9N2 Viruses

Surveillance studies revealed that H9N2 avian influenza virus has become established in chickens and quails and has been detected in pigs in Southern China since the mid-1990s (Guan et al. 1999, 2000; Xu et al. 2007; Peiris et al. 2001). Genetic analysis of the circulating H9N2 avian influenza viruses in China suggested two major lineages in terrestrial poultry: A/Duck/Hong Kong/Y280/97-like and A/Quail/Hong Kong/G1/97-like (Guan et al. 2000; Xu et al. 2007).

The A/Quail/Hong Kong/G1/97 virus shared the six internal genes with H5N1 human isolates in Hong Kong in 1997. The continued circulation of H9N2 viruses as well as H5N1 viruses in Southern China has resulted in multiple reassortment genotypes in recent years (Xu et al. 2007). In Korea, the Middle East, and Europe, H9N2 outbreaks in poultry have also been reported since late 1990 (Alexander 2003, 2007; Cameron et al. 2000). The H9N2 viruses that circulated in the Middle East were genetically related to the A/Quail/Hong Kong/G1/97-like viruses (Aamir et al. 2007; Cameron et al. 2000).

In 1999, human infection with H9N2 avian influenza virus was first documented in two children with mild upper respiratory symptoms in Hong Kong (Peiris et al. 1999), followed by subsequent reports of human infections in mainland China (Guo et al. 2001). The H9N2 human isolates from Hong Kong were genetically related to the A/Quail/Hong Kong/G1/97-like lineage. In 2003, human infection with H9N2 virus was identified again in Hong Kong, and the human H9N2 isolate was genetically more related to the A/Duck/Hong Kong/Y280/97-like lineage (Butt et al. 2005). In December 2008 an H9N2 infection was reported from a two-month old in Hong Kong. To date, there has been little evidence of human-to-human transmission of H9N2 virus. However, H9N2 virus with dual or human-like receptor specificity (Matrosovich et al. 2001) is now prevalent in many Eurasian countries, and the probability of the H9N2 subtype continuing to evolve into a pandemic strain is high.

## 4.2   H7 Viruses

Self-limited human infections with H7 subtype of avian influenza viruses have been documented since 1970 (Campbell et al. 1970; Kurtz et al. 1996; Webster et al. 1981). Between February and May 2003, outbreaks of the HP H7N7 subtype were reported in the Netherlands, Germany, and Belgium (Alexander 2007). More than 25 million birds were slaughtered during the outbreaks, and H7 virus was detected in at least 86 human infections. Infection with the H7N7 viruses resulted in conjunctivitis in 83 of 89 confirmed cases and one fatal case with pneumonia in combination with acute respiratory distress syndrome (Fouchier et al. 2004). In 2004, HP H7N3 outbreaks were reported in British Columbia, Canada, resulting in the slaughter of more than 19 million domestic poultry and causing two human infections with conjunctivitis (Hirst et al. 2004). Genetic evidence showed that the HP H7N3 virus evolved from a low-pathogenic H7N3 virus by obtaining a 21-nucleotide insertion (derived from the M gene) at the HA cleavage site (Hirst et al. 2004). The continued incidence of human infection with the H7 subtype and the high frequency of human cases associated with conjunctivitis showed that the H7 virus could infect humans without prior adaptation.

## 4.3   H6 Viruses

Outbreaks of the H6 subtype in domestic poultry have been reported in many Eurasian countries in recent years (Alexander 2007). Surveillance studies in

Southern China during 2000–2005 revealed that the H6N1 and H6N2 viruses were prevalent in terrestrial minor poultry and domestic ducks (Cheung et al. 2007). The significance of the prevalent H6 viruses in Southern China is that these viruses were descendants of the A/teal/Hong Kong/W312/97 (H6N1) virus, with 97% homology in six internal genes and the NA gene to the H5N1 human index isolate A/Hong Kong/156/97 (Hoffmann et al. 2000). During the past decade, coevolution and cocirculation of H5N1, H9N2, and H6N1 viruses in Southern China have generated reassortant viruses between these subtypes, including the H6N2 viruses as well as multiple genotypes of H6N1 and H6N2 viruses (Chin et al. 2002; Cheung et al. 2007).

## *4.4 H2 Viruses*

The H2N2 subtype was prevalent in the human population after the 1957 Asian pandemic but has disappeared since the 1968 Hong Kong pandemic. Although the H2 subtype is currently detected mostly in avian species, the H2N3 virus, which can be transmitted among pigs and ferrets, has been isolated in pigs in the United States in 2006 (Ma et al. 2007). The H2N3 swine isolates contained gene segments derived from swine and avian origin, as well as molecular changes in the HA that may confer increased receptor binding affinity toward human-like sialyl receptors (Ma et al. 2007). Continued surveillance to monitor the evolution of the H2N3 viruses in the swine population should be a priority in America.

## 5 The Use of Antivirals for Pandemic Influenza

Vaccination and antiviral treatment are the two major options for controlling influenza. As the use of vaccines for pandemic influenza will be discussed extensively in subsequent chapters, here we will only briefly address the use of antivirals as a control measure for influenza pandemics.

M2 ion channel blockers (amantadine and rimantadine) and NA inhibitors (oseltamivir and zanamivir) are the two classes of drugs currently available for prophylaxis and treatment of seasonal influenza infections (Monto 2003). Due to the severity of H5N1 infection in humans, the current World Health Organization (WHO) guidelines for clinical management of human H5N1 infection recommend the use of oseltamivir as the primary treatment (Schunemann et al. 2007; World Health Organization 2007).

Amantadine and rimantadine are adamantane derivatives. They block virus replication by targeting the M2 protein of the influenza A virus to prevent the uncoating of ribonucleoproteins from the M1 protein (early antiviral effect) and causing early HA conformational change in the trans-Golgi (late antiviral effect). Prophylactic amantadine or rimantadine treatment was effective against pandemic influenza in 1968 Hong Kong and the 1977 Russian pandemics (Monto et al. 1979;

Smorodintsev et al. 1970). The primary limitation of applying ion-channel blockers for influenza virus treatment or prophylaxis is that fully pathogenic and transmissible resistant variants with single amino acid substitutions at residue 26, 27, 30, 31, or 34 of the M2 protein may emerge rapidly (Hayden et al. 1989, 1997). Surveillance results have shown that clade 1 and most of the clade 2.1 viruses are fully resistant to M2 ion channel blockers (bdel-Ghafar et al. 2008). When NA inhibitors are available, the use of adamantane monotherapy against H5N1 human infection is currently not recommended by WHO (World Health Organization 2007).

NA inhibitors were designed on a structural basis to target the conserved NA enzymatic site of influenza viruses (von Itzstein et al. 1993). They interrupt the virus replication cycle by preventing the release of virus from the infected cells and causing the viruses to clump (Gubareva et al. 2000). While both oseltamivir and zanamivir are approved for the treatment of seasonal influenza infection, oseltamivir is currently suggested by WHO as the primary treatment of choice for human H5N1 infection due to differences in drug delivery. Oseltamivir is available only in an oral formulation. The prodrug is adsorbed in the gastrointestinal tract and converted into the active form (oseltamivir carboxylate) by hepatic esterases (Gubareva et al. 2000). In contrast, zanamivir was designed to be a direct competitive inhibitor of sialic acid, the NA substrate, and is administered by inhalation. The adequacy of inhaled zanamivir delivery in patients with serious lower respiratory tract or extrapulmonary disease is a major concern in human H5N1 infections (World Health Organization 2007). However, both oseltamivir and zanamivir are recommended by WHO as chemoprophylaxis for high- or medium-risk H5N1 exposure groups (Schunemann et al. 2007). The data from uncontrolled clinical trials suggest that early treatment with oseltamivir (<5 days of disease onset) can improve the survival of H5N1 patients (bdel-Ghafar et al. 2008; de Jong 2008). H5N1 viral virulence has been shown to affect the necessary oseltamivir treatment dosage and duration in a mouse model (Yen et al. 2005). A standard dose of oseltamivir treatment is 75 mg twice daily for five days; however, higher treatment dosage (150 mg twice daily) for an increased duration (up to ten days) can be considered on a case-by-case basis. Oseltamivir-resistant variants with amino acid substitution at NA residues 274 (H→Y) or 294 (N→S) have been reported in H5N1 patients who were receiving oseltamivir treatment (Le et al. 2005; de Jong et al. 2005b) and, in one report, in a patient who had received prior treatment (Saad et al. 2007). Oseltamivir-resistant mutants with substitutions at NA residue 274 (H to Y) have also been detected in H1N1 influenza viruses in humans (Gubareva 2004; World Health Organization 2008). During the 2008–2009 influenza season a high prevalence of oseltamivir-resistance (98%) in human H1N1 influenza virus was detected in many regions of the world.

When the H5N1 virus is known or is likely to be susceptible to admantanes, combination therapy with an adamantine and oseltamivir may be considered (World Health Organization 2007). Immunotherapy using anti-H5N1-specific antibodies (monoclonal antibodies or polyclonal sera) is effective in small animal models (Hanson et al. 2006; Simmons et al. 2007) and has been evaluated in two H5N1 patients (who received both oseltamivir and convalescent plasma from H5N1 patients who survived) (World Health Organization 2007). Novel influenza anti-virals that are currently under development include CS-8958 along-acting inhaled NA inhibitor, DAS181 that removes sialic acids from respiratory epithelium, and T-705 that targets polymerase of influenza as well as some other RNA viruses.

# 6 Concluding Remarks

At the time of writing, it had been more than 40 years since the A/Hong Kong/1/68 (H3N2) human pandemic influenza virus emerged, and more than 30 years since the A/USSR/1/77 (H1N1) virus reappeared. It therefore has been an appreciably long interval since a pandemic strain of influenza has emerged, and there is increasing concern based on historical records (Potter 1998) that a pandemic is imminent. In the past decade, we have identified several subtypes of influenza A virus with increased pandemic potential, including: direct transmission of avian influenza virus of H5, H7, and H9 subtypes that resulted in mild to severe human infections; the emergence of H2 subtype virus in the swine population in the US and its ability to transmit among pigs and ferrets; the H6 influenza viruses that are now endemic in minor poultry in the live bird markets in Asia and continue to reassort with the H5N1 and H9N2 viruses.

The question frequently asked is whether the H5N1 influenza virus that has infected 409 humans from 15 countries of the world and killed 256 (March 2009) will achieve consistent human-to-human transmission and be declared a pandemic strain. From the perspective of the chicken farmers of Eurasia, this HP H5N1 virus has already achieved catastrophic status and has devastated the poultry industry in affected countries (Capua and Marangon 2007). From the perspective of its human pandemic potential, as long as the H5N1 virus continues to circulate and cause incidents of interspecies transmission, the possibility for the virus to become further adapted in the human population should be taken into consideration when assessing its pandemic threat. To date, the H5N1 virus has not yet acquired many changes at host-specific markers that differentiate human and avian influenza viruses and have been observed from the Spanish 1918, Asian 1957, and Hong Kong 1968 pandemic viruses (Finkelstein et al. 2007). However, the possibility that adaptation of H5N1 virus in human species may not follow the same pattern seen with previous pandemic viruses should not be excluded.

It would be premature to become complacent about the Asian H5N1 influenza viruses and their human pandemic potential, for they continue to evolve rapidly, both by accumulation of mutations and by reassortment. A recent reassortant in nature between an HP H5N1 and a nonpathogenic duck H3N8 in Laos PDR generated a virus with seven genes from H5N1 and the PB2 of the H3N8 virus (Boltz et al. unpublished data). This reassortant killed all mallard ducks inoculated in four days and transmitted efficiently to contact mallards. This is the first H5N1 genotype that consistently kills all infected mallard ducks, illustrating the continued evolution of these HP H5N1 viruses. While experimental generation of reassortants between human seasonal H3N2 and avian Asian H5N1 reassortants has so far not generated a transmissible virus in ferrets (Maines et al. 2006), this does not necessarily reflect the potential pattern of different reassortants that could be generated in nature. With increased host range to felids, dogs, and wild mammalian species (Owston's palm civet and stone martens) (Kuiken et al. 2004; Songserm et al. 2006; Peiris et al. 2007), the possibility exists for the H5N1 viruses to further adapt in mammals.

There is another school of thought supporting the contention that only H1, H2, and H3 influenza viruses have the capacity for human transmissibility (Palese 2004). Historical records and seroarchaeology of humans does support the presence of H1, H2, and H3 viruses in humans in earlier times (Potter 1998), and serologic evidence from the other 13 HA subtypes, including H4, H5, H6, H7, H10 and H11, is fragmentary at best (Shortridge 1992; Peiris et al. 2007). Thus, it can be argued that from serologic data and the outbreaks in the twentieth century that the strongest support is for a pandemic of H2 subtype rather than H5, H6, H7, or H9; however, we do not have sufficient knowledge to rule out the latter subtypes.

The WHO has established phases of pandemic alert for influenza viruses with six levels of preparedness, and the level has remained at phase 3 since 2005: a new influenza virus subtype is causing disease in humans but is not yet spreading efficiently and sustainably among humans (see http://www.who.int/csr/resources/publications/influenza/GIP_2005_5Eweb.pdf). The continued circulation of the HP H5N1 virus in Eurasia with peaks of activity in the cooler months and continued high activity in domestic poultry in Indonesia, India, and Bangladesh is of continuing concern. The continued occurrence of HP H5N1 in Thailand and Vietnam with peaks of activity in cool months over the past four years has been associated with grazing ducks, abundant human population and rice cropping intensity, and surprisingly not with chicken population numbers (Gilbert et al. 2008). Whether similar finds apply to China, Indonesia, and Nigeria merits intensified study. The different strategies used in Thailand and Vietnam to attempt to control and eradicate HP H5N1 have reduced the incidence of H5N1 in people and in poultry but have not yet been fully successful. In Vietnam, a massive poultry vaccination program successfully reduced the number of repeat cases in humans and chickens to undetectable levels in 2006, but the virus re-emerged in 2007 in both poultry and humans. In Thailand, massive surveillance and stamping out without the use of a vaccine have markedly reduced the incidence of H5N1 outbreaks, but isolated re-emerging outbreaks occurred in 2007 and 2008. While some of these re-emerging outbreaks may occur by the introduction of poultry smuggled across borders, phylogenetic analysis of the H5N1 viruses also supports the concept of local persistence. The identification of the reservoirs in the "evolutionary sink" of HP H5N1 is essential if this virus is to be successfully eradicated.

If migratory waterfowl are perpetuating HP H5N1, then eradication may not be feasible in the long term, and biosecurity will have to be the key strategy. The role of migratory waterfowl in perpetuating the HP H5N1 is unresolved; migratory waterfowl, particularly ducks, can spread the virus, but there is no convincing evidence of perpetuation of HP H5N1 in the breeding grounds and transmission to the next generation.

If we accept the hypothesis that migratory waterfowl are not the reservoir species for perpetuating HP H5N1, then it can be proposed that HP H5N1 is still an eradicable disease. There is precedent for eradicating HP H5N1 by intensive surveillance and stamping out without the use of a vaccine; however, if a vaccine is used in adjacent countries, disease signs can be masked, and continued intensive prospective surveillance will be necessary. The alternative strategy for controlling HP H5N1 is to use the differentiating infected from vaccinated animals (DIVA) strategy (Capua and Marangon 2007), quality poultry vaccines, as well as sentinel chickens to monitor viral shedding. The knowledge required to achieve the goal of eradication is available, but the global political will has not focused on this possibility. Prospective surveillance is absolutely essential, for it is becoming more and more apparent that HP H5N1 does not cause disease signs in all species and that the "Trojan horse" problem in duck species is still not fully appreciated.

**Acknowledgments** The authors thank David Galloway for editorial assistance and Betsy Williford for illustrations. This work was funded in whole or in part with funds from the National Institute Allergy and Infectious Diseases, National Institutes of Health, under contract no. HHSN266200700005C, and by the American Lebanese Syrian Associated Charities.

# References

Aamir UB, Wernery U, Ilyushina N, Webster RG (2007) Characterization of avian H9N2 influenza viruses from United Arab Emirates 2000–2003. Virology 361:45–55

Alexander DJ (2003) Report on avian influenza in the Eastern Hemisphere during 1997–2002. Avian Dis 47:792–797

Alexander DJ (2007) Summary of avian influenza activity in Europe, Asia, Africa, and Australasia, 2002–2006. Avian Dis 51:161–166

bdel-Ghafar AN, Chotpitayasunondh T, Gao Z, Hayden FG, Nguyen DH, de Jong MD, Naghdaliyev A, Peiris JSM, Shindo N, Soeroso S, Uyeki TM (2008) Update on avian influenza A (H5N1) virus infection in humans. N Engl J Med 358:261–273

Beard CW (1970) Demonstration of type-specific influenza antibody in mammalian and avian sera by immunodiffusion. Bull World Health Organ 42:779–785

Beigel JH, Farrar J, Han AM, Hayden FG, Hyer R, de Jong MD, Lochindarat S, Nguyen TK, Nguyen TH, Tran TH, Nicoll A, Touch S, Yuen KY (2005) Avian influenza A (H5N1) infection in humans. N Engl J Med 353:1374–1385

Belshe RB (2005) The origins of pandemic influenza—lessons from the 1918 virus. N Engl J Med 353:2209–2211

Butt KM, Smith GJ, Chen H, Zhang LJ, Leung YH, Xu KM, Lim W, Webster RG, Yuen KY, Peiris JSM, Guan Y (2005) Human infection with an avian H9N2 influenza A virus in Hong Kong in 2003. J Clin Microbiol 43:5760–5767

Cameron KR, Gregory V, Banks J, Brown IH, Alexander DJ, Hay AJ, Lin YP (2000) H9N2 subtype influenza A viruses in poultry in Pakistan are closely related to the H9N2 viruses responsible for human infection in Hong Kong. Virology 278:36–41

Campbell CH, Webster RG, Breese SS Jr. (1970) Fowl plague virus from man. J Infect Dis 122:513–516

Capua I, Marangon S (2007) Control and prevention of avian influenza in an evolving scenario. Vaccine 25:5645–5652

Chen H, Deng G, Li Z, Tian G, Li Y, Jiao P, Zhang L, Liu Z, Webster RG, Yu K (2004) The evolution of H5N1 influenza viruses in ducks in southern China. Proc Natl Acad Sci USA 101:10452–10457

Chen H, Li Y, Li Z, Shi J, Shinya K, Deng G, Qi Q, Tian G, Fan S, Zhao H, Sun Y, Kawaoka Y (2006) Properties and dissemination of H5N1 viruses isolated during an influenza outbreak in migratory waterfowl in western China. J Virol 80:5976–5983

Cheung CL, Vijaykrishna D, Smith GJ, Fan XH, Zhang JX, Bahl J, Duan L, Huang K, Tai H, Wang J, Poon LL, Peiris JSM, Chen H, Guan Y (2007) Establishment of influenza A virus (H6N1) in minor poultry species in southern China. J Virol 81:10402–10412

Chin PS, Hoffmann E, Webby R, Webster RG, Guan Y, Peiris JSM, Shortridge KF (2002) Molecular evolution of H6 influenza viruses from poultry in Southeastern China: prevalence of H6N1 influenza viruses possessing seven A/Hong Kong/156/97 (H5N1)-like genes in poultry. J Virol 76:507–516

Crawford PC, Dubovi EJ, Castleman WL, Stephenson I, Gibbs EP, Chen L, Smith C, Hill RC, Ferro P, Pompey J, Bright RA, Medina MJ, Johnson CM, Olsen CW, Cox NJ, Klimov AI, Katz JM, Donis RO (2005) Transmission of equine influenza virus to dogs. Science 310:482–485

de Jong MD (2008) Drug failure in H5N1 treatment: causes and implications. In: Bangkok International Conference on Avian Influenza 2008: Integration from Knowledge to Control, Bangkok, Thailand, 23–25 Jan 2008

de Jong JC, Claas EC, Osterhaus AD, Webster RG, Lim WL (1997) A pandemic warning? Nature 389:554

de Jong MD, Bach VC, Phan TQ, Vo MH, Tran TT, Nguyen BH, Beld M, Le TP, Truong HK, Nguyen VV, Tran TH, Do QH, Farrar J (2005a) Fatal avian influenza A (H5N1) in a child presenting with diarrhea followed by coma. N Engl J Med 352:686–691

de Jong MD, Tran TT, Truong HK, Vo MH, Smith GJ, Nguyen VC, Bach VC, Phan TQ, Ha DQ, Guan Y, Peiris JSM, Tran TH, Farrar J (2005b) Oseltamivir resistance during treatment of influenza A (H5N1) infection. N Engl J Med 353:2667–2672

de Jong MD, Simmons CP, Thanh TT, Hien VM, Smith GJ, Chau TN, Hoang DM, Chau NV, Khanh TH, Dong VC, Qui PT, Cam BV, Ha DQ, Guan Y, Peiris JSM, Chinh NT, Hien TT, Farrar J (2006) Fatal outcome of human influenza A (H5N1) is associated with high viral load and hypercytokinemia. Nat Med 12:1203–1207

Duan L, Campitelli L, Fan XH, Leung YH, Vijaykrishna D, Zhang JX, Donatelli I, Delogu M, Li KS, Foni E, Chiapponi C, Wu WL, Kai H, Webster RG, Shortridge KF, Peiris JSM, Smith GJ, Chen H, Guan Y (2007) Characterization of low pathogenic H5 subtype influenza viruses from Eurasia: Implications for the origin of highly pathogenic H5N1 viruses. J Virol 81:7529–7539

Ellis TM, Bousfield RB, Bissett LA, Dyrting KC, Luk GS, Tsim ST, Sturm-Ramirez K, Webster RG, Guan Y, Peiris JSM (2004) Investigation of outbreaks of highly pathogenic H5N1 avian influenza in waterfowl and wild birds in Hong Kong in late 2002. Avian Pathol 33:492–505

Finkelstein DB, Mukatira S, Mehta PK, Obenauer JC, Su X, Webster RG, Naeve CW (2007) Persistent host markers in pandemic and H5N1 influenza viruses. J Virol 81:10292–10299

Fouchier RA, Schneeberger PM, Rozendaal FW, Broekman JM, Kemink SA, Munster V, Kuiken T, Rimmelzwaan GF, Schutten M, Van Doornum GJ, Koch G, Bosman A, Koopmans M, Osterhaus AD (2004) Avian influenza A virus (H7N7) associated with human conjunctivitis and a fatal case of acute respiratory distress syndrome. Proc Natl Acad Sci USA 101:1356–1361

Fouchier RA, Munster V, Wallensten A, Bestebroer TM, Herfst S, Smith D, Rimmelzwaan GF, Olsen B, Osterhaus AD (2005) Characterization of a novel influenza A virus hemagglutinin subtype (H16) obtained from black-headed gulls. J Virol 79:2814–2822

Gilbert M, Xiao X, Pfeiffer DU, Epprecht M, Boles S, Czarnecki C, Chaitaweesub P, Kalpravidh W, Minh PQ, Otte MJ, Martin V, Slingenbergh J (2008) Mapping H5N1 highly pathogenic avian influenza risk in Southeast Asia. Proc Natl Acad Sci USA 105:4769–4774

Govorkova EA, Rehg JE, Krauss S, Yen HL, Guan Y, Peiris JSM, Nguyen DT, Hanh TH, Puthavathana P, Hoang TL, Buranathai C, Lim W, Webster RG, Hoffmann E (2005) Lethality to ferrets of H5N1 influenza viruses isolated from humans and poultry in 2004. J Virol 79:2191–2198

Gu J, Xie Z, Gao Z, Liu J, Korteweg C, Ye J, Lau LT, Lu J, Gao Z, Zhang B, McNutt MA, Lu M, Anderson VM, Gong E, Yu AC, Lipkin WI (2007) H5N1 infection of the respiratory tract and beyond: a molecular pathology study. Lancet 370:1137–1145

Guan Y, Shortridge KF, Krauss S, Webster RG (1999) Molecular characterization of H9N2 influenza viruses: were they the donors of the "internal" genes of H5N1 viruses in Hong Kong? Proc Natl Acad Sci USA 96:9363–9367

Guan Y, Shortridge KF, Krauss S, Chin PS, Dyrting KC, Ellis TM, Webster RG, Peiris JSM (2000) H9N2 influenza viruses possessing H5N1-like internal genomes continue to circulate in poultry in southeastern China. J Virol 74:9372–9380

Gubareva LV (2004) Molecular mechanisms of influenza virus resistance to neuraminidase inhibitors. Virus Res 103:199–203

Gubareva LV, Kaiser L, Hayden FG (2000) Influenza virus neuraminidase inhibitors. Lancet 355:827–835

Guo Y, Dong J, Wang M, Zhang Y, Guo J, Wu K (2001) Characterization of hemagglutinin gene of influenza A virus subtype H9N2. Chin Med J 114:76–79

Guo CT, Takahashi N, Yagi H, Kato K, Takahashi T, Yi SQ, Chen Y, Ito T, Otsuki K, Kida H, Kawaoka Y, Hidari KI, Miyamoto D, Suzuki T, Suzuki Y (2007) The quail and chicken intestine have sialyl-galactose sugar chains responsible for the binding of influenza A viruses to human type receptors. Glycobiology 17:713–724

Hanson BJ, Boon AC, Lim AP, Webb A, Ooi EE, Webby RJ (2006) Passive immunoprophylaxis and therapy with humanized monoclonal antibody specific for influenza A H5 hemagglutinin in mice. Respir Res 7:126

Hatta M, Gao P, Halfmann P, Kawaoka Y (2001) Molecular basis for high virulence of Hong Kong H5N1 influenza A viruses. Science 293:1840–1842

Hausmann J, Kretzschmar E, Garten W, Klenk HD (1995) N1 neuraminidase of influenza virus A/FPV/Rostock/34 has haemadsorbing activity. J Gen Virol 76:1719–1728

Hayden FG, Belshe RB, Clover RD, Hay AJ, Oakes MG, Soo W (1989) Emergence and apparent transmission of rimantadine-resistant influenza A virus in families. N Engl J Med 321:1696–1702

Hayden FG, Osterhaus AD, Treanor JJ, Fleming DM, Aoki FY, Nicholson KG, Bohnen AM, Hirst HM, Keene O, Wightman K (1997) Efficacy and safety of the neuraminidase inhibitor zanamivir in the treatment of influenza virus infections. GG167 Influenza Study Group. N Engl J Med 337:874–880

Hirst M, Astell CR, Griffith M, Coughlin SM, Moksa M, Zeng T, Smailus DE, Holt RA, Jones S, Marra MA, Petric M, Krajden M, Lawrence D, Mak A, Chow R, Skowronski DM, Tweed SA, Goh S, Brunham RC, Robinson J, Bowes V, Sojonky K, Byrne SK, Li Y, Kobasa D, Booth T, Paetzel M (2004) Novel avian influenza H7N3 strain outbreak, British Columbia. Emerg Infect Dis 10:2192–2195

Hoffmann E, Stech J, Leneva I, Krauss S, Scholtissek C, Chin PS, Peiris M, Shortridge KF, Webster RG (2000) Characterization of the influenza A virus gene pool in avian species in southern China: was H6N1 a derivative or a precursor of H5N1? J Virol 74:6309–6315

Horimoto T, Kawaoka Y (2001) Pandemic threat posed by avian influenza A viruses. Clin Microbiol Rev 14:129–149

Horimoto T, Kawaoka Y (2005) Influenza: lessons from past pandemics, warnings from current incidents. Nat Rev Microbiol 3:591–600

Hulse-Post DJ, Sturm-Ramirez KM, Humberd J, Seiler P, Govorkova EA, Krauss S, Scholtissek C, Puthavathana P, Buranathai C, Nguyen TD, Long HT, Naipospos TS, Chen H, Ellis TM, Guan Y, Peiris JSM, Webster RG (2005) Role of domestic ducks in the propagation and biological evolution of highly pathogenic H5N1 influenza viruses in Asia. Proc Natl Acad Sci USA 102:10682–10687

Ito T, Suzuki Y, Suzuki T, Takada A, Horimoto T, Wells K, Kida H, Otsuki K, Kiso M, Ishida H, Kawaoka Y (2000) Recognition of N-glycolylneuraminic acid linked to galactose by the alpha-2,3 linkage is associated with intestinal replication of influenza A virus in ducks. J Virol 74:9300–9305

Kandun IN, Wibisono H, Sedyaningsih ER, Yusharmen, Hadisoedarsuno W, Purba W, Santoso H, Septiawati C, Tresnaningsih E, Heriyanto B, Yuwono D, Harun S, Soeroso S, Giriputra S, Blair PJ, Jeremijenko A, Kosasih H, Putnam SD, Samaan G, Silitonga M, Chan KH, Poon LL, Lim W, Klimov A, Lindstrom S, Guan Y, Donis R, Katz J, Cox N, Peiris JSM, Uyeki TM (2006) Three Indonesian clusters of H5N1 virus infection in 2005. N Engl J Med 355:2186–2194

Keawcharoen J, Oraveerakul K, Kuiken T, Fouchier RA, Amonsin A, Payungporn S, Noppornpanth S, Wattanodorn S, Theambooniers A, Tantilertcharoen R, Pattanarangsan R, Arya N, Ratanakorn P, Osterhaus AD, Poovorawan Y (2004) Avian influenza H5N1 in tigers and leopards. Emerg Infect Dis 10:2189–2191

Keawcharoen J, van Riel D, van Amerongen G, Bestebroer T, Beyer WE, van Lavieren R, Osterhaus AD, Fouchier RA, Kuiken T (2008) Wild ducks as long-distance vectors of highly pathogenic avian influenza virus (H5N1). Emerg Infect Dis 14:600–607

Kida H, Yanagawa R, Matsuoka Y (1980) Duck influenza lacking evidence of disease signs and immune response. Infect Immun 30:547–553

Kuiken T, Rimmelzwaan G, van Riel D, van Amerongen G, Baars M, Fouchier RA, Osterhaus AD (2005) Avian H5N1 influenza in cats. Science 306:241

Kurtz J, Manvell RJ, Banks J (1996) Avian influenza virus isolated from a woman with conjunctivitis. Lancet 348:901–902

Le QM, Kiso M, Someya K, Sakai YT, Nguyen TH, Nguyen KH, Pham ND, Ngyen HH, Yamada S, Muramoto Y, Horimoto T, Takada A, Goto H, Suzuki T, Suzuki Y, Kawaoka Y (2005) Avian flu: isolation of drug-resistant H5N1 virus. Nature 437:1108

Li KS, Guan Y, Wang J, Smith GJ, Xu KM, Duan L, Rahardjo AP, Puthavathana P, Buranathai C, Nguyen TD, Estoepangestie AT, Chaisingh A, Auewarakul P, Long HT, Hanh NT, Webby RJ, Poon LL, Chen H, Shortridge KF, Yuen KY, Webster RG, Peiris JSM (2004) Genesis of a highly pathogenic and potentially pandemic H5N1 influenza virus in eastern Asia. Nature 430:209–213

Ma W, Vincent AL, Gramer MR, Brockwell CB, Lager KM, Janke BH, Gauger PC, Patnayak DP, Webby RJ, Richt JA (2007) Identification of H2N3 influenza A viruses from swine in the United States. Proc Natl Acad Sci USA 104:20949–20954

Maines TR, Chen LM, Matsuoka Y, Chen H, Rowe T, Ortin J, Falcon A, Nguyen TH, Mai LQ, Sedyaningsih ER, Harun S, Tumpey TM, Donis RO, Cox NJ, Subbarao K, Katz JM (2006) Lack of transmission of H5N1 avian-human reassortant influenza viruses in a ferret model. Proc Natl Acad Sci USA 103:12121–12126

Matrosovich MN (2008) What changes in the hemagglutinin and neuraminidase are required for the emergence of a pandemic influenza virus? In: Bangkok International Conference on Avian Influenza 2008: Integration from Knowledge to Control, Bangkok, Thailand, 23–25 January 2008

Matrosovich MN, Tuzikov A, Bovin N, Gambaryan A, Klimov A, Castrucci MR, Donatelli I, Kawaoka Y (2000) Early alterations of the receptor-binding properties of H1, H2, and H3 avian influenza virus hemagglutinins after their introduction into mammals. J Virol 74:8502–8512

Matrosovich MN, Krauss S, Webster RG (2001) H9N2 influenza A viruses from poultry in Asia have human virus-like receptor specificity. Virology 281:156–162

Monto AS (2003) The role of antivirals in the control of influenza. Vaccine 21:1796–1800

Monto AS, Gunn RA, Bandyk MG, King CL (1979) Prevention of Russian influenza by amantadine. JAMA 241:1003–1007

Neumann G, Kawaoka Y (2006) Host range restriction and pathogenicity in the context of influenza pandemic. Emerg Infect Dis 12:881–886

Nicholson KG (1998) Human influenza. In: Nicholson KG, Webster RG, Hay AJ (eds) Textbook of influenza. Blackwell, London, pp 219–264

Okazaki K, Takada A, Ito T, Imai M, Takakuwa H, Hatta M, Ozaki H, Tanizaki T, Nagano T, Ninomiya A, Demenev VA, Tyaptirganov MM, Karatayeva TD, Yamnikova SS, Lvov DK, Kida H (2000) Precursor genes of future pandemic influenza viruses are perpetuated in ducks nesting in Siberia. Arch Virol 145:885–893

Palese P (2004) Influenza: old and new threats. Nat Med 10:S82–S87

Peiris JSM, Yuen KY, Leung CW, Chan KH, Ip PL, Lai RW, Orr WK, Shortridge KF (1999) Human infection with influenza H9N2. Lancet 354:916–917

Peiris JSM, Guan Y, Markwell D, Ghose P, Webster RG, Shortridge KF (2001) Cocirculation of avian H9N2 and contemporary "human" H3N2 influenza A viruses in pigs in southeastern China: potential for genetic reassortment? J Virol 75:9679–9686

Peiris JSM, Yu WC, Leung CW, Cheung CY, Ng WF, Nicholls JM, Ng TK, Chan KH, Lai ST, Lim WL, Yuen KY, Guan Y (2004) Re-emergence of fatal human influenza A subtype H5N1 disease. Lancet 363:617–619

Peiris JSM, de Jong MD, Guan Y (2007) Avian influenza virus (H5N1): a threat to human health. Clin Microbiol Rev 20:243–267

Potter CW (1998) Chronicle of influenza pandemics. In: Nicholson KG, Webster RG, Hay AJ (eds) Textbook of influenza. Blackwell, London, pp 3–26

Rimmelzwaan GF, van Riel D, Baars M, Bestebroer TM, van Amerongen G, Fouchier RA, Osterhaus AD, Kuiken T (2006) Influenza A virus (H5N1) infection in cats causes systemic disease with potential novel routes of virus spread within and between hosts. Am J Pathol 168:176–183

Rimmelzwaan GF, Fouchier RA, Osterhaus AD (2007) Influenza virus-specific cytotoxic T lymphocytes: a correlate of protection and a basis for vaccine development. Curr Opin Biotechnol 18:529–536

Rott R, Klenk HD, Nagai Y, Tashiro M (1995) Influenza viruses, cell enzymes, and pathogenicity. Am J Respir Crit Care Med 152:S16-S19

Saad MD, Boynton BR, Earhart KC, Mansour MM, Niman HL, Elsayed NM, Nayel AL, AbdelghaniAS, EssmatHM, LabibEM, AyoubEA, MontevilleMR (2007) Detection of oseltamivir resistance mutation N294S in humans with influenza A H5N1. In: Options for the Control of Influenza VI, Toronto, Canada, 17 June 2007

Salomon R, Franks J, Govorkova EA, Ilyushina NA, Yen HL, Hulse-Post DJ, Humberd J, Trichet M, Rehg JE, Webby RJ, Webster RG, Hoffmann E (2006) The polymerase complex genes contribute to the high virulence of the human H5N1 influenza virus isolate A/Vietnam/1203/04. J Exp Med 203:689–697

Scholtissek C (1995) Molecular evolution of influenza viruses. Virus Genes 11:209–215

Schunemann HJ, Hill SR, Kakad M, Bellamy R, Uyeki TM, Hayden FG, Yazdanpanah Y, Beigel J, Chotpitayasunondh T, Del MC, Farrar J, Tran TH, Ozbay B, Sugaya N, Fukuda K, Shindo N, Stockman L, Vist GE, Croisier A, Nagjdaliyev A, Roth C, Thomson G, Zucker H, Oxman AD (2007) WHO Rapid Advice Guidelines for pharmacological management of sporadic human infection with avian influenza A (H5N1) virus. Lancet Infect Dis 7:21–31

Senne DA (2007) Avian influenza in North and South America, 2002–2005. Avian Dis 51:167–173

Shortridge KF (1992) Pandemic influenza: a zoonosis? Semin Respir Infect 7:11–25

Shortridge KF, Stuart-Harris CH (1982) An influenza epicentre? Lancet 2:812–813

Shortridge KF, Zhou NN, Guan Y, Gao P, Ito T, Kawaoka Y, Kodihalli S, Krauss S, Markwell D, Murti KG, Norwood M, Senne D, Sims L, Takada A, Webster RG (1998) Characterization of avian H5N1 influenza viruses from poultry in Hong Kong. Virology 252:331–342

Simmons CP, Bernasconi NL, Suguitan AL, Mills K, Ward JM, Chau NV, Hien TT, Sallusto F, Ha DQ, Farrar J, de Jong MD, Lanzavecchia A, Subbarao K (2007) Prophylactic and therapeutic efficacy of human monoclonal antibodies against H5N1 influenza. PLoS Med 4:e178

Smith GJ, Fan XH, Wang J, Li KS, Qin K, Zhang JX, Vijaykrishna D, Cheung CL, Huang K, Rayner JM, Peiris JSM, Chen H, Webster RG, Guan Y (2006a) Emergence and predominance of an H5N1 influenza variant in China. Proc Natl Acad Sci USA 103:16936–16941

Smith GJ, Naipospos TS, Nguyen TD, de Jong MD, Vijaykrishna D, Usman TB, Hassan SS, Nguyen TV, Dao TV, Bui NA, Leung YH, Cheung CL, Rayner JM, Zhang JX, Zhang LJ, Poon LL, Li KS, Nguyen VC, Hien TT, Farrar J, Webster RG, Chen H, Peiris JSM, Guan Y (2006b) Evolution and adaptation of H5N1 influenza virus in avian and human hosts in Indonesia and Vietnam. Virology 350:258–268

Smorodintsev AA, Karpuhin GI, Zlydnikov DM, Malyseva AM, Svecova EG, Burov SA, Hramcova LM, Romanov JA, Taros LJ, Ivannikov JG, Novoselov SD (1970) The prophylactic effectiveness of amantadine hydrochloride in an epidemic of Hong Kong influenza in Leningrad in 1969. Bull World Health Organ 42:865–872

Songserm T, Amonsin A, Jam-on R, Sae-Heng N, Pariyothorn N, Payungporn S, Theamboonlers A, Chutinimitkul S, Thanawongnuwech R, Poovorawan Y (2006) Fatal avian influenza A H5N1 in a dog. Emerg Infect Dis 12:1744–1747

Stevens J, Blixt O, Glaser L, Taubenberger JK, Palese P, Paulson JC, Wilson IA (2006) Glycan microarray analysis of the hemagglutinins from modern and pandemic influenza viruses reveals different receptor specificities. J Mol Biol 355:1143–1155

Sturm-Ramirez KM, Ellis T, Bousfield B, Bissett L, Dyrting K, Rehg JE, Poon LL, Guan Y, Peiris JSM, Webster RG (2004) Reemerging H5N1 influenza viruses in Hong Kong in 2002 are highly pathogenic to ducks. J Virol 78:4892–4901

Taubenberger JK, Reid AH, Lourens RM, Wang R, Jin G, Fanning TG (2005) Characterization of the 1918 influenza virus polymerase genes. Nature 437:889–893

Tumpey TM, Maines TR, van Hoeven N, Glaser L, Solorzano A, Pappas C, Cox NJ, Swayne DE, Palese P, Katz JM, Garcia-Sastre A (2007) A two-amino acid change in the hemagglutinin of the 1918 influenza virus abolishes transmission. Science 315:655–659

Ungchusak K, Auewarakul P, Dowell SF, Kitphati R, Auwanit W, Puthavathana P, Uiprasertkul M, Boonnak K, Pittayawonganon C, Cox NJ, Zaki SR, Thawatsupha P, Chittaganpitch M, Khontong R, Simmerman JM, Chunsutthiwat S (2005) Probable person-to-person transmission of avian influenza A (H5N1). N Engl J Med 352:333–340

Varghese JN, Colman PM, van Donkelaar A, Blick TJ, Sahasrabudhe A, Kimm-Breschkin JL (1997) Structural evidence for a second sialic acid binding site in avian influenza virus neuraminidases. Proc Natl Acad Sci USA 94:11808–11812

von Itzstein M, Wu WY, Kok GB, Pegg MS, Dyason JC, Jin B, Van PT, Smythe ML, White HF, Oliver SW (1993) Rational design of potent sialidase-based inhibitors of influenza virus replication. Nature 363:418–423

Webby R, Hoffmann E, Webster RG (2004) Molecular constraints to interspecies transmission of viral pathogens. Nat Med 10:S77–S81

Webster RG, Yakhno M, Hinshaw VS, Bean W, Murti G (1978) Intestinal influenza: replication and characterization of influenza viruses in ducks. Virology 84:268–278

Webster RG, Geraci J, Petursson G, Skirnisson K (1981) Conjunctivitis in human beings caused by influenza A virus of seals. N Engl J Med 304:911

Webster RG, Bean WJ, Gorman OT, Chambers TM, Kawaoka Y (1992) Evolution and ecology of influenza A viruses. Microbiol Rev 56:152–179

World Health Organization (2005) Evolution of H5N1 avian influenza viruses in Asia. Emerg Infect Dis 11:1515–1521

World Health Organization (2006) Antigenic and genetic characteristics of H5N1 viruses and candidate H5N1 vaccine viruses developed for potential use as pre-pandemic vaccines. Wkly Epidemiol Rec 81:328–330

World Health Organization (2007) Clinical management of human infection with avian influenza (H5N1) virus. http://www.who.int/csr/disease/avian_influenza/guidelines/ClinicalManagement07.pdf

World Health Organization (2008) Influenza A (H1N1) virus resistance to oseltamivir. http://www.who.int/csr/disease/influenza/H1N1ResistanceWeb20080425.pdf

Xu KM, Li KS, Smith GJ, Li JW, Tai H, Zhang JX, Webster RG, Peiris JSM, Chen H, Guan Y (2007) Evolution and molecular epidemiology of H9N2 influenza A viruses from quail in southern China, 2000–2005. J Virol 81:2635–2645

Yen HL, Monto AS, Webster RG, Govorkova E (2005) Virulence may determine the necessary duration and dosage of oseltamivir treatment for highly pathogenic A/Vietnam/1203/04 influenza virus in mice. J Infect Dis 192:665–672

# Antigenic Cross-Reactivity Among H5N1 Viruses

**Adrianus C.M. Boon and Richard J. Webby**

## Contents

**Abstract** The unprecedented spread of highly pathogenic H5N1 viruses since 1996 has had public health and scientific entities scrambling to prepare for a possible pandemic. Central to many of these efforts has been the development of vaccines. As the viruses have continued to spread, however, they have continued to diversify genetically, complicating vaccine strain selection. The key to successful vaccine design is understanding the cross-reactivity between these genetically distinct H5N1 strains. Studies conducted to date show encouraging amounts of cross-reaction and cross-protection between various H5N1 strains, although our ability to predict one based upon the other is poor. Understanding the targets and mechanisms behind this cross-protection should be a key focus of pandemic preparedness.

A.C.M. Boon and R.J. Webby (✉)

Division of Virology, Department of Infectious Diseases, St. Jude Children's Research Hospital, 332 N Lauderdale St, Memphis, TN, 38105, USA

Richard.Webby@stjude.org

R.W. Compans and W.A. Orenstein (eds.), *Vaccines for Pandemic Influenza*,      25
Current Topics in Microbiology and Immunology 333,
DOI 10.1007/978-3-540-92165-3_2, © Springer-Verlag Berlin Heidelberg 2009

# 1   Introduction

The biggest challenge faced in preparing vaccines for influenza viruses, be they pandemic (as defined in this review as a virus that has not yet gained the capacity for sustained human-to-human transmission) or seasonal strains, is that the best protection, neutralizing antibodies, target the most variable protein, hemagglutinin (HA). H5N1 viruses are certainly no exception, and in the past decade we have witnessed the divergence of the A/goose/Guangdong/1/96-like H5 HA into a number of different variants. Indeed, ten genetically defined clades of this virus now exist (http://www.who.int/csr/disease/avian_influenza/guidelines/nomenclature/en/index.html). This divergence poses many difficulties for those involved in pandemic preparedness, especially in the absence of our ability to assign pandemic risk to a given isolate; in other words, we have no way of knowing which, if any, of the ten clades is likely to provide the strain that transforms into a human-adapted pathogen. As such, pandemic preparedness must include provisions to cover the genetic and antigenic diversity represented by the current and future H5N1 strains. In this report we will review the antigenic cross-reactivity among H5N1 viruses as measured by a number of in vitro and in vivo parameters. The importance of this information, in part, lies in the justification, or lack thereof, for creating vaccine stockpiles in anticipation of an H5N1 pandemic. Preparing vaccine is not an inexpensive endeavor, and it is difficult to justify this expenditure if the stockpiled vaccine offers protection against only a fraction of circulating strains.

# 2   Polyclonal Antibodies to H5N1 Viruses

The benchmark for assessing the antigenic relationships between influenza viruses is hemagglutination inhibition (HI) assays using postinfection ferret serum. With seasonal influenza viruses, a vaccine strain change is considered when an emerging viral lineage has a reciprocal fourfold-reduced HI titer to the current vaccine strain. Considering that antigenic analysis of currently circulating H5N1 viruses typically demonstrates at least a fourfold drop in HI titer between clades (Fig. 1), using these criteria would mean that the impetus for stockpiling H5N1 vaccine is not great. The cons become somewhat less dominant, however, if the criteria for pandemic vaccination (i.e., vaccination while a matching vaccine is in development) are re-evaluated. The goal for seasonal influenza vaccines in most age groups is to reduce the consequence of viral infection to an almost asymptomatic level. In the context of an emerging pandemic situation, this is probably an unrealistic goal, and instead the vaccine should be used with a view to reducing mortality and severe disease. With these as the criteria, the desired levels of cross-reactivity become less clear.

The difficulty when generalizing about the antigenic relatedness of H5N1 strains is that for most isolates this information is simply not available. In addition, comparisons performed across HI assays done in different labs with different reagents are not always advisable. Nevertheless, using an example HI table as reported by the

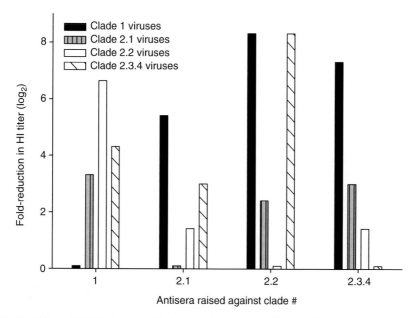

**Fig. 1** Ability of H5N1 virus-specific ferret antisera to neutralize H5N1 viruses from different clades. Figure adapted from WHO (H5N1 vaccine update February 2008; http://www.who.int), depicting the fold reduction in HI titer between the homologous strain(s) of highly pathogenic H5N1 influenza A virus and H5N1 strain(s) from different clades. Fold reduction was calculated from the median HI titer of ferret antisera against one or more H5N1 viruses from a particular clade. Clade 1: *A/Vietnam/1203/04*, A/Thailand/676/05, A/duck/VN/NCVD16/07, and A/muscovy duck/Vietnam/33/07; Clade 2.1: *A/Indonesia/5/2005*, A/duck/Hunan/795/2002, A/Indonesia/6/2005, A/Indonesia/CDC1031/2007, and A/Indonesia/CDC625L/2006; Clade 2.2: *A/whooper swan/ Mongolia/244/2005*, *A/chicken/India/NIV-33487/2006*, A/Turkey/65–596/2006, A/egret/Egypt/ 9402Namru3/2007, A/chicken/Egypt/9403Namru3/2007; Clade 2.3.4: *A/Anhui/1/2005, A/japanese white-eye/Hong Kong/1038/2006*, A/Anhui/2/2005, A/duck/Laos/3295/2006, A/house crow/Hong Kong/719/2007, A/Laos/JP058/2007, A/chicken/Vietnam/NCVD74/2007, A/duck/Vietnam/ NCVD81/2007, A/Vietnam/HN31203I/2007, and A/chicken/Malaysia/935/2006 (the viruses used to raise ferret antisera are shown in *italics*)

World Health Organization, some generalizations can be made. Again, it is important to state that these trends are very much oversimplifications and, due to the need to select representative viruses from each clade, they could be misleading. Analysis of the H5N1 clades that include human isolates shows that the observed genetic diversity is concomitant with an antigenic diversity. In this particular example, some clade representatives induce a relatively broad cross-reactive response after infection of ferrets. For example, postinfection sera from the clade 2.1 virus A/ Indonesia/5/05 generally reacts well with clade 2.2 and 2.3 strains and, although to a lesser degree, also to clade 1 viruses. In contrast, post infection sera specific for the clade 2.2 virus A/whooperswan/Mongolia/244/05 reacts well with clade 2.1 viruses, but relatively poorly with representatives from clades 1 and 2.3. Although, as stressed, overinterpreting such limited data is not advisable, this information can help guide the necessary preclinical and clinical paths. The information also suggests that a single vaccine antigen is unlikely to produce the desired coverage of circulating strains, and that multiple vaccines (or antigens) or different approaches to induce cross-reactive responses will be necessary.

## 3 Monoclonal Antibodies to H5N1 Viruses

In addition to the antigenic characterization of circulating H5N1 strains using postinfection or postimmunization polyclonal serum, monoclonal antibodies (mAbs) have also been utilized for this purpose. In the most exhaustive of these studies, a panel of seventeen different H5-specific mouse mAbs was used to antigenically group H5N1 viruses isolated from Asia from 2002 to 2007. The mAbs were prepared from mice immunized with five representative H5N1 viruses from clades 1, 2.1, 2.2, 2.3, and 8 and were tested for reactivity against 41 strains of H5N1 virus selected from ten different genetic clades by HI assay (Wu et al. 2008). The mAb panel was able to separate the 41 viruses into four reactivity groups (A–D). One set of six mAbs neutralized all but one (A/Chinese pond heron/Hong Kong/18/05) of the H5N1 viruses, demonstrating, perhaps not too surprisingly, that at least one antigenic epitope is very well conserved across the genetic clades. This result is consistent with many of the preclinical results shown below, where relatively good protection can be seen across clades. However, because this protection is often seen in the absence of a detectable cross-reactive polyclonal response, these conserved epitopes may not necessarily be immunodominant. A second set of mAbs neutralized viruses from clades 1, 2.1, and a number of other minor clades (again consistent with good protection against clade 2.1 viruses by clade 1 vaccines, see below). The third set of mAbs neutralizes viruses from clades 2.1, 2.2, 2.4, and a number, but not all, of group 2.3 viruses. Although there is insufficient data to make any firm conclusions, this mAb study suggests that some emphasis should be placed on examining the ability of clade 1-based vaccines to protect against clade 2.3 viruses.

In a second murine mAb study, investigators identified a mAb that neutralized viruses from clades 1, 2.1, 2.2, and 2.3 (Kaverin et al. 2002). Analysis of escape mutants identified positions 131 and 156 (H3 numbering) as important residues for binding, and sequence analysis confirmed the conservation of these residues among H5N1 strains. If the conservation of these residues is imparted by a functional or structural constraint, the intriguing possibility arises that a vaccine designed to target this epitope could induce very broad cross-reactivity. If, however, the conservation is simply due to a lack of antibody-driven selection pressure (i.e., the epitope is subdominant), then this approach would be hampered by virus escape. Some preliminary data has been generated from looking at the effects of single amino acid changes on H5N1 antigenicity and immunogenicity; however, this approach has not been fully explored in terms of generating a more broadly reactive antigen (Hoffmann et al. 2005; Wu et al. 2008; Yang et al. 2007).

Although the in vitro studies described above are based on the neutralizing ability of antibodies, it should also be noted that nonneutralizing antibodies may play an important role in protection from, or clearance of, the virus. In one report, a H1 HA-specific mouse mAb was shown to have neutralizing activity in vivo but not in vitro (Feng et al. 2002; Mozdzanowska et al. 2006). The addition of C1q and another yet to be identified serum factor enhanced the in vitro neutralizing activity. A similar phenomenon was recently reported for a human mAb specific for H5 HA. This particular antibody neutralized H5N1 viruses from clades 1 and 2.1 in vivo, but only neutralized viruses from clade 1 in vitro (Simmons et al. 2007). While these findings indicate that our current assays for detecting virus-neutralizing antibodies (HI and VN) are far from perfect and possibly require the addition of C1q and other serum factors, it may also indicate that nonneutralizing antibodies can be very effective for prophylaxis or treatment of influenza virus infection.

# 4 Evidence for Cross-Reactivity from Human Clinical Trials

Although we are in no position to accurately predict the real risk associated with the current H5N1 viruses, the spread of the virus has resulted in a substantial boost in funding for pandemic preparedness activities. Highlighting these activities is the fact that over 60 H5N1 human vaccine clinical trials have either been completed or are in progress (most using antigen derived from clade 1 viruses). Although the first reports of the immunogenicity of vaccines derived from the contemporary strains of H5N1 were disappointing, although not unexpected, the use of novel adjuvants has seen the required antigen content reduced markedly (Leroux-Roels et al. 2007, 2008; Treanor et al. 2006). With this in mind, a number of recent studies have reported on the ability of serum from vaccinated individuals to cross-react with viruses from other genetic clades. Although multiple studies have been, or will soon be, completed, it is worth mentioning that although it is tempting to do so, it can be misleading to compare across different trials when assessing cross-reactivity. Different studies use different parameters for measuring seroconversion, highlighting the need for international standardization of methodologies. Nevertheless, the general results of these studies are encouraging and suggest that despite the genetic diversification of H5N1 viruses, a degree of antigenic cross-reactivity is maintained.

In one of the first studies to look at the ability of H5N1 vaccines to induce cross-reactive antibodies, Stephenson and colleagues measured the ability of serum from individuals vaccinated with an A/duck/Singapore/97 (a clade 0 virus) based vaccine to react with clade 1 viruses (Stephenson et al. 2005). Despite the fact that a three-dose schedule and an adjuvant was required to induce good seroconversion (as measured by a microneutralization (MN) assay titer of ≥80 and at least a fourfold increase from prevaccination titers) to the homologous antigen, seroconversion rates against clade 1 viruses ranged from 100% (to A/Hong Kong/213/03) to 43% (A/Vietnam/1203/04). The key question with this and other such studies is whether the adjuvant (MF59 in this case) actually increases the diversity of the antibody response generated or whether the cross-reactivity seen with the adjuvanted formulation is simply a reflection of the greater titer of the response. Although it is not immediately obvious how an adjuvant could affect the immunodominance of an antibody response to an antigen, mechanistic studies of MF59 and other adjuvants do show that they are able to differentially stimulate antigen-presenting cells and potentially alter the resulting immune response (Kool et al. 2008; Seubert et al. 2008).

Following on from the study of Stephenson et al., others have looked at the cross-reactivity of the responses generated to clade 1 H5N1 strains. After a single dose of adjuvanted vaccine based on the clade 1 strain A/Vietnam/1194/04, Leroux-Roels and colleagues (Leroux-Roels et al. 2007) were able to show that 27–54% (depending on the vaccine dose) of recipients seroconverted to the clade 2.1 strain A/Indonesia/5/05, as measured by a fourfold rise in MN assay. After the second dose of vaccine, these percentages increased to 67–77%. In a follow-up study, these same investigators were able to show that two doses of this vaccine containing 3.8 µg of HA protein induced fourfold increases in MN titers against other clade 2 viruses; 85% against A/turkey/Turkey/1/05 (clade 2.2) and 75% against A/Anhui/1/05 (clade 2.3), and the geometric mean titer was similar for all three clades (Leroux-Roels et al. 2008). Even more encouraging was the finding that significant numbers of individuals retained these antibody titers for at least six months.

# 5  Evidence for H5N1 Cross-Reactivity from Preclinical Studies

Although the studies described above provide relatively convincing evidence that the different H5N1 clades share functionally relevant antigenic determinants, the real difficulty when extrapolating these data is a lack of good immune correlates of protection in humans. In the absence of these immune correlates and of a safe H5N1 challenge trial in humans, the field has turned to various animal models to understand H5N1 vaccine efficacy. In regards to seasonal influenza vaccines, protective immunity is generally associated with a neutralizing antibody titer, measured by HI, of ≥1:40. Preclinical studies in mice and ferrets have however shown that, despite the low immunogenicity of H5 candidate vaccines without adjuvant, animals are protected from lethal homologous challenges even in the absence of HI titers of 40 or above. More importantly, vaccinated animals are also partially or fully protected from challenge with H5N1 viruses from different clades. The following examples provide more details on the experiments leading to this statement.

## 5.1  Ferrets

Four studies have assessed the cross-protective potential of candidate H5 vaccines in ferrets (Table 1) (Baras et al. 2008; Govorkova et al. 2006; Lipatov et al. 2006; Suguitan et al. 2006). Fortuitously, in most cases the vaccines used for preclinical

**Table 1** H5N1 Antigenic cross-reactivity in preclinical studies in ferrets

| Type of vaccine | HA protein origin | Vaccination scheme | In vitro | Antigenic cross-reactivity Virus (dose) | Antigenic cross-reactivity Effect | References |
|---|---|---|---|---|---|---|
| Inactivated whole virus | A/duck/Singapore/3/97 | 7 μg HA (2 doses) + alum | HA ≤ 10<br>VN ≤ 40 | A/Vietnam/1203/04 ($10^6$ EID$_{50}$) | Protection from mortality and reduced virus titers in nasal washes | Lipatov et al. 2006 |
| | A/Hong Kong/213/03 | 7 or 15 μg HA + alum<br>7 μg HA (2 doses) | HA ≤ 10<br>VN ≤ 10<br>HA ≤ 10<br>VN ≤ 16 | A/Vietnam/1203/04 ($10^6$ EID$_{50}$) | Protection from mortality and reduced virus titers in nasal washes | Govorkova et al. 2006 |
| | A/Hong Kong/213/03 | 7 or 15 μg HA + alum<br>7 μg HA (2 doses) | HA ≤ 64<br>VN ≤ 10<br>HA = 64<br>VN = 256 | A/Hong Kong/156/97 ($10^6$ EID$_{50}$) | Reduced viral titers in nasal washes | Govorkova et al. 2006 |
| Split virus vaccine | A/Vietnam/1194/04 | 15 μg HA (2 doses) | VN ≤ 28 | A/Indonesia/5/05 ($10^5$ TCID$_{50}$) | No reduction in lung viral titer and no protection from mortality | Baras et al. 2008 |
| | | 1.5–15 μg HA (2 doses) + adjuvant | VN = 26–43 | | Reduced lung viral titers and shedding in URT[a] | Baras et al. 2008 |
| Live attenuated virus | A/Hong Kong/491/97 | $10^7$ TCID$_{50}$ (2 doses) | ND[1] | A/Vietnam/1203/04 ($10^7$ EID$_{50}$) | Reduced virus titer in lungs, brains, and nasal turbinates | Suguitan et al. 2006 |
| | | | ND | A/Indonesia/6/05 ($10^7$ EID$_{50}$) | Reduced virus titer in lungs, brains, and nasal turbinates | Suguitan et al. 2006 |
| | A/Vietnam/1203/04 | $10^7$ TCID$_{50}$ (2 doses) | ND | A/Hong Kong/156/97 ($10^7$ EID$_{50}$) | Reduced virus titer in lungs and nasal turbinates | Suguitan et al. 2006 |
| | | | ND | A/Indonesia/6/05 ($10^7$ EID$_{50}$) | Reduced virus titer in lungs, brains, and nasal turbinates | Suguitan et al. 2006 |

[a] Upper respiratory tract

studies closely resemble those that were used in human safety and immunogenicity trials, allowing for some comparisons. The one caveat to this statement is, however, that the formulations of the vaccines do not necessarily correspond. In the human studies described above, vaccines were formulated as surface antigen (in the A/duck/ Singapore/3/97 trials) or as split virion (in the A/Vietnam/1194/04 trials). In some of the preclinical trials described below, the vaccine antigen was used in the form of whole virus, which has been shown to be more immunogenic (Lin et al. 2006).

In the first of the ferret studies, two 7 μg doses of A/duck/Singapore/97 (H5N3) whole-virus vaccine were unable to induce detectable levels of neutralizing antibodies against the clade 1 virus A/Vietnam/1203/04. Nevertheless, upon challenge with a lethal dose of the same virus, all three vaccinated animals survived and had lower virus titers in the nasal washes as compared to mock-vaccinated controls (Lipatov et al. 2006). In a second study, the cross-protective potential of a formalin-inactivated clade 1 whole A/Hong Kong/213/03-based vaccine was assessed. Multiple vaccine regiments were taken; single 7 or 15 μg doses adjuvanted with aluminum hydroxide or two 7 μg doses without adjuvant. Vaccinated and control ferrets were challenged with clade 1 A/Vietnam/1203/04 or clade 0 A/Hong Kong/156/97 viruses (Govorkova et al. 2006). Neither single-dose regimens resulted in detectable HI or MN titers to A/Vietnam/1203/04, although low but measurable titers were induced after two doses. Neutralizing antibodies to A/Hong Kong/156/97 could be detected by HI assay after one vaccination and by MN after two doses of vaccine. Again, against predictions based upon the in vitro data and regardless of the vaccine schedule used, all vaccinated ferrets were protected from the lethal effects of A/Vietnam/1203/04 challenge. Although protected from mortality, ferrets vaccinated with the single adjuvanted doses (at either antigen amount) presented with clinical symptoms including fever. In all vaccinated animals, virus titers in nasal washes were significantly reduced on days 5 and 7 postchallenge as compared to the control ferrets. The third ferret study assessed the ability of the clade 1 A/Vietnam/1194/04-based vaccine to protect against the clade 2.1 A/Indonesia/5/05 virus. The strength of this study is that the vaccine used was the same formulation as that used in the human clinical trial described above. Groups of ferrets were vaccinated twice with four dose levels (1.7–15 μg HA) of inactivated split A/Vietnam/1194/2004 virus formulated with the proprietary oil-in-water emulsion-based adjuvant that was shown to be immunogenic and safe in humans (Leroux-Roels et al. 2007). After two doses of vaccine, more than 50% of the vaccinated ferrets had detectable MN antibody titers to the vaccine virus. More importantly, these same ferrets also had MN antibody titers specific for A/Indonesia/5/05 virus, albeit at reduced amounts. As seen in the human trials, vaccine without adjuvant was unable to induce seroconversion. Although the challenge virus in this study, A/Indonesia/5/05, did induce a lethal infection in this study, vaccinated animals had reduced virus titers in both the lung and upper respiratory tract as compared to control animals (Baras et al. 2008). The fourth ferret study is of note in that a different vaccine strategy was used; the vaccine was based on the live attenuated platform. Such vaccines are reputed to induce both antibody and cellular immunity, which could in theory improve the cross-reactivity of the vaccine-generated immunity. In this study, the cross-protective efficacy of two doses of live attenuated vaccines based on the clade 0 A/Hong Kong/491/97 or clade 1 A/Vietnam/1203/04 viruses was determined. Virus titers in the brain, lungs, and nasal turbinates of vaccinated and control ferrets were determined three days after heterologous virus challenge with A/Vietnam/1203/04 and clade 2.1 A/Indonesia/5/05. Compared to unvaccinated ferrets, virus titers in all three organs were significantly lower in the vaccinated ferrets (Suguitan et al. 2006), demonstrating a similar level of cross-protection afforded by this approach.

## 5.2 Mice

In addition to ferrets, mice have played an important role in the preclinical development of various vaccine approaches. The murine model is certainly the cheapest and easiest influenza animal model, but it is also relatively easy to induce protective immunity in this host, and the information gained should be considered somewhat preliminary. In mice, the immunogenicity of H5N1 virus vaccines has varied substantially between different studies. This variation is due to differences in vaccine formulation, presence of an adjuvant, amount of administered antigen, route of vaccination, and the number of booster doses used. The efficacy of the vaccine is also affected by the size of the challenge inoculum and the pathogenicity of the challenge virus strain. In studies where the H5 vaccine induces detectable neutralizing titers to the challenge strain, mice are more often than not well protected (Desheva et al. 2006; Kreijtz et al. 2007a,b; Lipatov et al. 2005; Lu et al. 1999, 2006; Schwartz et al. 2007; Takada et al. 1999). Following similar trends to those seen in ferrets, in some instances H5-vaccinated mice with low or undetectable levels of neutralizing antibodies to the challenge strain are also protected from mortality (Hoelscher et al. 2006; Ichinohe et al. 2007; Kreijtz et al. 2007b; Lu et al. 2006).

Perhaps the best evidence for HA-mediated cross-protection comes from mouse challenge studies with vectored vaccines. In one such study, mice were inoculated twice with an adenovirus expressing the HA gene of the clade 0 virus A/Hong Kong/156/97. Although vaccination did not induce significant amounts of neutralizing antibodies specific for the clade 1 virus A/Vietnam/1203/04, all of the mice survived a lethal challenge from this strain (Hoelscher et al. 2006). Recombinant HA protein (3 μg) from the clade 2.1 virus A/Indonesia/5/05 was also shown to induce cross-protective immunity to A/Vietnam/1203/04. Again, cross-protection was demonstrated in the absence of a detectable HI titer against the challenge virus although cross-reactive antibodies were detected by ELISA (Bright et al. 2008).

The challenge data presented to date tend to paint a very positive picture in terms of cross-protection amongst H5N1 viruses. Putting things a little more in perspective, however, are studies by Kreijtz et al. who vaccinated mice with modified vaccinia Ankara expressing the H5 protein of A/Hong Kong/156/97 or A/Vietnam/1194/04 virus. While the mice vaccinated with A/Vietnam/1194/04 HA protein were protected from challenge with A/Indonesia/6/05 virus, mice vaccinated with the clade 0 vaccine virus were not (Kreijtz et al. 2007b), showing that, as expected, there is a limit to cross-protection, even in the murine model. The power of the mouse is that reagents are available for dissecting the mechanisms behind this cross-protection, such as the role of T cell immunity, nonneutralizing antibodies, and immunity against other proteins (see below).

# 6 Antigenic Cross-Reactivity Mediated by Mechanisms Other than Virus-Neutralizing Antibodies

Current human vaccine strategies rely predominately on the induction of HA-directed virus-neutralizing antibodies for efficacy. As discussed earlier, the flaw in this approach is that the continuous drift of influenza viruses means that monitoring systems, such as the World Health Organization Global Influenza Surveillance Program, need to be in place to ensure optimal vaccine efficacy. As a consequence of this limitation, a number of alternative mechanisms for influenza vaccines have been explored. These approaches include targeting more conserved proteins such as matrix protein 2 (M2), NA, and nucleoprotein with humoral and cellular immunity. Most of these strategies do not prevent infection with H5 virus, but they significantly reduce clinical symptoms and reduce the burden of disease. While other chapters in this edition discuss these approaches in detail, a discussion of H5N1 antigenicity is not complete without briefly mentioning a few of these approaches.

## 6.1 Neuraminidase-Specific Antibodies

Protruding from the surfaces of influenza virions are two glycoproteins, HA and NA. Whereas HA is the viral protein that mediates the attachment and entry of the virus into a host cell, the NA acts at the other end of the replication cycle and via its sialidase activity releases the progeny virus from the cell surface. As such, whereas antibodies to HA are able to neutralize infection, antibodies to NA are limited to reducing cell-to-cell spread and enhancing viral agglutination. Antibodies directed against NA, although not able to provide sterilizing immunity, are able to protect from severe disease (Chen et al. 2000; Johansson et al. 1989, 1993; Qiu et al. 2006).

In terms of cross-reactive NA responses, the data is generally less conclusive. The 1918 Spanish influenza killed in the vicinity of 50 million lives, whereas the mortality associated with the subsequent 1957 H2N2 pandemic was at least an order of magnitude lower, and that associated with the 1968 H3N2 lower still (Kilbourne 2006; Viboud et al. 2005). Although we are beginning to learn that the virus responsible for the 1918 pandemic had an unusually high virulence (Kobasa et al. 2004; Tumpey et al. 2005), there is little to suggest that either of the H2N2 or H3N2 viruses did. It has been postulated that cross-reactive NA antibodies may have contributed to this phenomenon. With respect to HA, the H3N2 virus responsible for this outbreak was antigenically novel, but in terms of NA it could not be distinguished from preceding H2N2 strains that descended from the 1957 Asian influenza pandemic virus (Schulman and Kilbourne 1969). It has been proposed that NA-specific immunity against H2N2 virus moderated the virulence of H3N2 virus in humans. Evidence for this has been provided by epidemiological investigation (Monto and Kendal 1973), a human H2N2 vaccine study (Eickhoff and Meiklejohn 1969), and mouse prime/challenge experiments (Schulman and Kilbourne 1969). Although little has been done to explore the level of NA-directed antibodies between H5N1 strains, we have previously investigated the cross-reactivity between the N1 proteins of the H5N1 viruses and contemporary seasonal human H1N1 strains. The rationale for these studies was that previous exposure to seasonal strains may impart a level of antibody cross-reactivity to H5N1 strains. In these studies, mice vaccinated with N1 NA from A/New Caledonia/20/99 (H1N1) were partially protected from mortality after a lethal challenge with A/Vietnam/1203/04 virus (Sandbulte et al. 2007). This protection was attributed to NA-specific antibodies as adoptive transfer of immune serum to unvaccinated animals resulted in a similar level of protection to that seen in the vaccinated animals themselves. These data suggest that NA-specific antibodies can attribute to protection from infection and may be considered an alternative for inducing cross-protective H5N1 virus-specific antibodies. Future work will show if N1 NA from circulating H5N1 virus offers additional protection from infection with H5N1 viruses from different antigenic clades, and if the reactivity has any real clinical benefit.

## 6.2 M2e-Specific Antibodies

In addition to HA and NA, a portion of a third viral protein is also accessible to recognition by antibodies; the extra membrane 24 amino acids of the viral M2 protein, or M2e. The appeal of this protein in terms of vaccine targeting is that it is highly conserved between all viral subtypes because of a conserved function and a lack of pressure due to the fact that virtually no detectable antibodies are made against this protein during a natural infection. It has, however, been proposed that driving an immune response against the M2 extracellular domain (M2e) via vaccination could elicit a protective antibody response. Data from several studies partially support this hypothesis. An M2-specific monoclonal antibody administered to mice permitted more rapid elimination of influenza A virus from lungs during sublethal challenge (Treanor et al. 1990). Protection from the challenge was observed after vaccinations with several forms of M2, including baculovirus-expressed full-length protein (Slepushkin et al. 1995), M2 with deleted transmembrane region (Frace et al. 1999), and M2e conjugated with hepatitis B virus core (HBc), keyhole limpet hemocyanin, or *Neisseria meningitidis* outer membrane protein complex (OMPC) (Fan et al. 2004; Neirynck et al. 1999). Despite the apparent promise of these and other approaches, the viability of an M2-based influenza

vaccination strategy still remains to be determined, although its efficacy has also
been tested against H5N1 viruses in mice with some success (Tompkins et al. 2007).

One limitation of using M2-based vaccines may be the mechanism of action of
M2-based immunity, which has not been well resolved. The efficacy of M2-based
vaccines is antibody mediated, but the antibodies are not neutralizing. Studies by
Jegerlehner and colleagues consequently suggested that instead of binding to the
virus, these nonneutralizing antibodies bind to the infected cells, and protection is
mediated by antibody-dependent cell cytotoxicity (ADCC) of natural killer cells
(Jegerlehner et al. 2004). Results of the study suggest that resistance to viral challenge was comparatively weak: M2-specific antibodies reduced disease severity but
failed to prevent infection. The investigators also suggested that the mechanism of
M2 action is confined largely to the lung, as opposed to upper respiratory tract tissues
which, if accurate, may make this approach particularly applicable to H5N1 infection,
which can have a predilection for the lung, particularly in humans (van Riel et al.
2006; Yamada et al. 2006).

## 6.3   T Cell-Mediated Immunity

While antibodies are considered superior in protecting the host from infection with
influenza virus, including H5N1, perhaps the most cross-reactive immune responses
are mediated by T cells. In contrast to most antibodies, which recognize relatively
complex epitopes, T cells recognize short linear sections of proteins, which by
chance alone are more likely to be conserved across influenza strains. T cell-
mediated immunity has been shown to contribute to resolution of influenza virus
infection in various animal models, and as such can protect from mortality but not
infection. Although most extensively studied in other models, a protective role for
T cells in H5N1 infection in mice has also been demonstrated (Thomas et al. 2006).
T cells, and in particular the cytotoxic T lymphocytes, often recognize epitopes on
the more conserved internal genes of the virus. Analysis of all known T cell
epitopes on the influenza virus proteome, previously summarized by Bui et al.
(2007), demonstrated that approximately 30% of all the T cell epitopes presented
by H1 or H3 viruses circulating between 1999 and 2004 are identical in most of the
H5N1 viruses from all four major clades. These data, although by no means unexpected, clearly show the benefits of T cell responses in generating cross-reactive
immunity. The major challenge for vaccine approaches that hope to exploit T cell-
based cross-reactivity is providing convincing evidence for their role in protecting
humans from influenza virus infection and/or mortality.

## 7   Concluding Remarks

Human seasonal strains aside, as a scientific community we know more about the
natural history of the current highly pathogenic H5N1 viruses than perhaps any
other subtype. Starting with the first characterization of a virus from a goose in

Guangdong Province in China in 1996, the spread and genetic evolution of the virus has been relatively closely followed. Lagging somewhat behind the genetic characterization of these isolates have been the antigenic analyses. This lag is due to a number of issues, such as proper assay choice, assay standardization, access to challenge models, and availability of viable virus strains. Considerable attention must be directed towards addressing these issues, as although genetic data can guide, antigenic data and the cross-reactivity between viruses that are actually central to optimal vaccine strain selection. A number of studies have shown that our in vitro assays fail miserably to predict H5 vaccine efficacy. Without good immune correlates of protection, efforts to assess the coverage provided by available and pending vaccines are severely hamstrung. A review of the information currently available is encouragingly optimistic in terms of the ability of a vaccine derived from one H5N1 virus to protect against another. There are, however, some exceptions, and the longer the H5N1 viruses continue to circulate the more of these exceptions there will be.

**Acknowledgments** This project has been funded in whole or in part with federal funds from the National Institute of Allergy and Infectious Diseases, National Institutes of Health, Department of Health and Human Services, under Contract No. HHSN266200700005C, and by the American Lebanese Syrian Associated Charities.

# References

Baras B, Stittelaar KJ, Simon JH, Thoolen RJ, Mossman SP, Pistoor FH, van Amerongen G, Wettendorff MA, Hanon E, Osterhaus AD (2008) Cross-protection against lethal H5N1 challenge in ferrets with an adjuvanted pandemic influenza vaccine. PLoS ONE 3:e1401

Bright RA, Carter DM, Crevar CJ, Toapanta FR, Steckbeck JD, Cole KS, Kumar NM, Pushko P, Smith G, Tumpey TM, Ross TM (2008) Cross-clade protective immune responses to influenza viruses with H5N1 HA and NA elicited by an influenza virus-like particle. PLoS ONE 3:e1501

Bui HH, Peters B, Assarsson E, Mbawuike I, Sette A (2007) Ab and T cell epitopes of influenza A virus, knowledge and opportunities. Proc Natl Acad Sci USA 104:246–251

Chen Z, Kadowaki S, Hagiwara Y, Yoshikawa T, Matsuo K, Kurata T, Tamura S (2000) Cross-protection against a lethal influenza virus infection by DNA vaccine to neuraminidase 2. Vaccine 18:3214–3222

Desheva JA, Lu XH, Rekstin AR, Rudenko LG, Swayne DE, Cox NJ, Katz JM, Klimov AI (2006) Characterization of an influenza A H5N2 reassortant as a candidate for live-attenuated and inactivated vaccines against highly pathogenic H5N1 viruses with pandemic potential. Vaccine 24:6859–6866

Eickhoff TC, Meiklejohn G (1969) Protection against Hong Kong influenza by adjuvant vaccine containing A2-Ann Arbor-675. Bull World Health Organ 41:562–563

Fan J, Liang X, Horton MS, Perry HC, Citron MP, Heidecker GJ, Fu TM, Joyce J, Przysiecki CT, Keller PM, Garsky VM, Ionescu R, Rippeon Y, Shi L, Chastain MA, Condra JH, Davies ME, Liao J, Emini EA, Shiver JW (2004) Preclinical study of influenza virus A M2 peptide conjugate vaccines in mice, ferrets, and rhesus monkeys. Vaccine 22:2993–3003

Feng JQ, Mozdzanowska K, Gerhard W (2002) Complement component C1q enhances the biological activity of influenza virus hemagglutinin-specific antibodies depending on their fine antigen specificity and heavy-chain isotype. J Virol 76:1369–1378

Frace AM, Klimov AI, Rowe T, Black RA, Katz JM (1999) Modified M2 proteins produce heterotypic immunity against influenza A virus. Vaccine 17:2237–2244

Govorkova EA, Webby RJ, Humberd J, Seiler JP, Webster RG (2006) Immunization with reverse-genetics-produced H5N1 influenza vaccine protects ferrets against homologous and heterologous challenge. J Infect Dis 194:159–167

Hoelscher MA, Garg S, Bangari DS, Belser JA, Lu X, Stephenson I, Bright RA, Katz JM, Mittal SK, Sambhara S (2006) Development of adenoviral-vector-based pandemic influenza vaccine against antigenically distinct human H5N1 strains in mice. Lancet 367:475–481

Hoffmann E, Lipatov AS, Webby RJ, Govorkova EA, Webster RG (2005) Role of specific hemagglutinin amino acids in the immunogenicity and protection of H5N1 influenza virus vaccines. Proc Natl Acad Sci USA 102:12915–12920

Ichinohe T, Kawaguchi A, Tamura S, Takahashi H, Sawa H, Ninomiya A, Imai M, Itamura S, Odagiri T, Tashiro M, Chiba J, Sata T, Kurata T, Hasegawa H (2007) Intranasal immunization with H5N1 vaccine plus Poly I:Poly C12U, a Toll-like receptor agonist, protects mice against homologous and heterologous virus challenge. Microbes Infect 9:1333–1340

Jegerlehner A, Schmitz N, Storni T, Bachmann MF (2004) Influenza A vaccine based on the extracellular domain of M2: weak protection mediated via antibody-dependent NK cell activity. J Immunol 172:5598–5605

Johansson BE, Bucher DJ, Kilbourne ED (1989) Purified influenza virus hemagglutinin and neuraminidase are equivalent in stimulation of antibody response but induce contrasting types of immunity to infection 12. J Virol 63:1239–1246

Johansson BE, Grajower B, Kilbourne ED (1993) Infection-permissive immunization with influenza virus neuraminidase prevents weight loss in infected mice. Vaccine 11:1037–1039

Kaverin NV, Rudneva IA, Ilyushina NA, Varich NL, Lipatov AS, Smirnov YA, Govorkova EA, Gitelman AK, Lvov DK, Webster RG (2002) Structure of antigenic sites on the haemagglutinin molecule of H5 avian influenza virus and phenotypic variation of escape mutants. J Gen Virol 83:2497–2505

Kilbourne ED (2006) Influenza pandemics of the 20th century. Emerg Infect Dis 12:9–14

Kobasa D, Takada A, Shinya K, Hatta M, Halfmann P, Theriault S, Suzuki H, Nishimura H, Mitamura K, Sugaya N, Usui T, Murata T, Maeda Y, Watanabe S, Suresh M, Suzuki T, Suzuki Y, Feldmann H, Kawaoka Y (2004) Enhanced virulence of influenza A viruses with the haemagglutinin of the 1918 pandemic virus. Nature 431:703–707

Kool M, Soullie T, van Nimwegen NM, Willart MA, Muskens F, Jung S, Hoogsteden HC, Hammad H, Lambrecht BN (2008) Alum adjuvant boosts adaptive immunity by inducing uric acid and activating inflammatory dendritic cells. J Exp Med 205:869–882

Kreijtz JH, Bodewes R, van Amerongen G, Kuiken T, Fouchier RA, Osterhaus AD, Rimmelzwaan GF (2007a) Primary influenza A virus infection induces cross-protective immunity against a lethal infection with a heterosubtypic virus strain in mice. Vaccine 25:612–620

Kreijtz JH, Suezer Y, van Amerongen AG, de MG, Schnierle BS, Wood JM, Kuiken T, Fouchier RA, Lower J, Osterhaus AD, Sutter G, Rimmelzwaan GF (2007b) Recombinant modified vaccinia virus Ankara-based vaccine induces protective immunity in mice against infection with influenza virus H5N1. J Infect Dis 195:1598–1606

Leroux-Roels I, Borkowski A, Vanwolleghem T, Drame M, Clement F, Hons E, Devaster JM, Leroux-Roels G (2007) Antigen sparing and cross-reactive immunity with an adjuvanted rh5n1 prototype pandemic influenza vaccine: a randomised controlled trial. Lancet 370:580–589

Leroux-Roels I, Bernhard R, Gerard P, Drame M, Hanon E, Leroux-Roels G (2008) Broad Clade 2 cross-reactive immunity induced by an adjuvanted clade 1 rh5n1 pandemic influenza vaccine. PLoS ONE 3:e1665

Lin J, Zhang J, Dong X, Fang H, Chen J, Su N, Gao Q, Zhang Z, Liu Y, Wang Z, Yang M, Sun R, Li C, Lin S, Ji M, Liu Y, Wang X, Wood J, Feng Z, Wang Y, Yin W (2006) Safety and immunogenicity of an inactivated adjuvanted whole-virion influenza A (H5N1) vaccine: a phase I randomised controlled trial. Lancet 368:991–997

Lipatov AS, Webby RJ, Govorkova EA, Krauss S, Webster RG (2005) Efficacy of H5 influenza vaccines produced by reverse genetics in a lethal mouse model. J Infect Dis 191:1216–1220

Lipatov AS, Hoffmann E, Salomon R, Yen HL, Webster RG (2006) Cross-protectiveness and immunogenicity of influenza A/Duck/Singapore/3/97(H5) vaccines against infection with A/Vietnam/1203/04(H5N1) virus in ferrets. J Infect Dis 194:1040–1043

Lu X, Tumpey TM, Morken T, Zaki SR, Cox NJ, Katz JM (1999) A mouse model for the evaluation of pathogenesis and immunity to influenza A (H5N1) viruses isolated from humans. J Virol 73:5903–5911

Lu X, Edwards LE, Desheva JA, Nguyen DC, Rekstin A, Stephenson I, Szretter K, Cox NJ, Rudenko LG, Klimov A, Katz JM (2006) Cross-protective immunity in mice induced by live-attenuated or inactivated vaccines against highly pathogenic influenza A (H5N1) viruses. Vaccine 24:6588–6593

Monto AS, Kendal AP (1973) Effect of neuraminidase antibody on Hong Kong influenza 4. Lancet 1:623–625

Mozdzanowska K, Feng J, Eid M, Zharikova D, Gerhard W (2006) Enhancement of neutralizing activity of influenza virus-specific antibodies by serum components. Virology 352:418–426

Neirynck S, Deroo T, Saelens X, Vanlandschoot P, Jou WM, Fiers W (1999) A universal influenza A vaccine based on the extracellular domain of the M2 protein. Nat Med 5:1157–1163

Qiu M, Fang F, Chen Y, Wang H, Chen Q, Chang H, Wang F, Wang H, Zhang R, Chen Z (2006) Protection against avian influenza H9N2 virus challenge by immunization with hemagglutinin- or neuraminidase-expressing DNA in BALB/c mice 1. Biochem Biophys Res Commun 343:1124–1131

Sandbulte MR, Jimenez GS, Boon AC, Smith LR, Treanor JJ, Webby RJ (2007) Cross-reactive neuraminidase antibodies afford partial protection against H5N1 in mice and are present in unexposed humans. PLoS Med 4:e59

Schulman JL, Kilbourne ED (1969) Independent variation in nature of hemagglutinin and neuraminidase antigens of influenza virus: distinctiveness of hemagglutinin antigen of Hong Kong-68 virus 1. Proc Natl Acad Sci USA 63:326–333

Schwartz JA, Buonocore L, Roberts A, Suguitan A, Jr., Kobasa D, Kobinger G, Feldmann H, Subbarao K, Rose JK (2007) Vesicular stomatitis virus vectors expressing avian influenza H5 HA induce cross-neutralizing antibodies and long-term protection. Virology 366:166–173

Seubert A, Monaci E, Pizza M, O'Hagan DT, Wack A (2008) The adjuvants aluminum hydroxide and MF59 induce monocyte and granulocyte chemoattractants and enhance monocyte differentiation toward dendritic cells. J Immunol 180:5402–5412

Simmons CP, Bernasconi NL, Suguitan AL, Mills K, Ward JM, Chau NV, Hien TT, Sallusto F, Ha DQ, Farrar J, de J, Lanzavecchia A, Subbarao K (2007) Prophylactic and therapeutic efficacy of human monoclonal antibodies against H5N1 influenza. PLoS Med 4:e178

Slepushkin VA, Katz JM, Black RA, Gamble WC, Rota PA, Cox NJ (1995) Protection of mice against influenza A virus challenge by vaccination with baculovirus-expressed M2 protein. Vaccine 13:1399–1402

Stephenson I, Bugarini R, Nicholson KG, Podda A, Wood JM, Zambon MC, Katz JM (2005) Cross-reactivity to highly pathogenic avian influenza H5N1 viruses after vaccination with nonadjuvanted and MF59-adjuvanted influenza A/Duck/Singapore/97 (H5N3) vaccine: a potential priming strategy. J Infect Dis 191:1210–1215

Suguitan AL Jr, McAuliffe J, Mills KL, Jin H, Duke G, Lu B, Luke CJ, Murphy B, Swayne DE, Kemble G, Subbarao K (2006) Live, attenuated influenza A H5N1 candidate vaccines provide broad cross-protection in mice and ferrets. PLoS Med 3:e360

Takada A, Kuboki N, Okazaki K, Ninomiya A, Tanaka H, Ozaki H, Itamura S, Nishimura H, Enami M, Tashiro M, Shortridge KF, Kida H (1999) Avirulent avian influenza virus as a vaccine strain against a potential human pandemic. J Virol 73:8303–8307

Thomas PG, Keating R, Hulse-Post DJ, Doherty PC (2006) Cell-mediated protection in influenza infection. Emerg Infect Dis 12:48–54

Tompkins SM, Zhao ZS, Lo CY, Misplon JA, Liu T, Ye Z, Hogan RJ, Wu Z, Benton KA, Tumpey TM, Epstein SL (2007) Matrix protein 2 vaccination and protection against influenza viruses, including subtype H5N1. Emerg Infect Dis 13:426–435

Treanor JJ, Tierney EL, Zebedee SL, Lamb RA, Murphy BR (1990) Passively transferred monoclonal antibody to the M2 protein inhibits influenza A virus replication in mice. J Virol 64:1375–1377

Treanor JJ, Campbell JD, Zangwill KM, Rowe T, Wolff M (2006) Safety and immunogenicity of an inactivated subvirion influenza A (H5N1) vaccine. N Engl J Med 354:1343–1351

Tumpey TM, Basler CF, Aguilar PV, Zeng H, Solorzano A, Swayne DE, Cox NJ, Katz JM, Taubenberger JK, Palese P, Garcia-Sastre A (2005) Characterization of the reconstructed 1918 Spanish influenza pandemic virus. Science 310:77–80

van Riel D, Munster VJ, de Wif E, Rimmelzwaan GF, Fouchier RA, Osterhaus AD, Kuiken T (2006) H5N1 virus attachment to lower respiratory tract. Science 312:399

Viboud C, Grais RF, Lafont BA, Miller MA, Simonsen L (2005) Multinational impact of the 1968 Hong Kong influenza pandemic: evidence for a smoldering pandemic. J Infect Dis 192:233–248

Wu WL, Chen Y, Wang P, Song W, Lau SY, Rayner JM, Smith GJ, Webster RG, Peiris JS, Lin T, Xia N, Guan Y, Chen H (2008) Antigenic profile of avian H5N1 viruses in Asia from 2002–2007. J Virol 82(4):1798–1807

Yamada S, Suzuki Y, Suzuki T, Le MQ, Nidom CA, Sakai-Tagawa Y, Muramoto Y, Ito M, Kiso M, Horimoto T, Shinya K, Sawada T, Kiso M, Usui T, Murata T, Lin Y, Hay A, Haire LF, Stevens DJ, Russell RJ, Gamblin SJ, Skehel JJ, Kawaoka Y (2006) Haemagglutinin mutations responsible for the binding of H5N1 influenza A viruses to human-type receptors. Nature 444:378–382

Yang ZY, Wei CJ, Kong WP, Wu L, Xu L, Smith DF, Nabel GJ (2007) Immunization by avian H5 influenza hemagglutinin mutants with altered receptor binding specificity. Science 317:825–828

# Part II
# Current Approaches for Human and Avian Vaccine Production

# Seasonal Influenza Vaccines

**Anthony E. Fiore, Carolyn B. Bridges, and Nancy J. Cox**

**Contents**

A.E. Fiore, C.B. Bridges, and N.J. Cox (✉)
Centers for Disease Control and Prevention, 1600 Clifton Rd, NE-Atlanta, GA 30333, USA

R.W. Compans and W.A. Orenstein (eds.), *Vaccines for Pandemic Influenza*,
Current Topics in Microbiology and Immunology 333,
DOI 10.1007/978-3-540-92165-3_3, © Springer-Verlag Berlin Heidelberg 2009

**Abstract**  Influenza vaccines are the mainstay of efforts to reduce the substantial health burden from seasonal influenza. Inactivated influenza vaccines have been available since the 1940s and are administered via intramuscular injection. Inactivated vaccines can be given to anyone six months of age or older. Live attenuated, cold-adapted influenza vaccines (LAIV) were developed in the 1960s but were not licensed in the United States until 2003, and are administered via nasal spray. Both vaccines are trivalent preparations grown in eggs and do not contain adjuvants. LAIV is licensed for use in the United States for healthy nonpregnant persons 2–49 years of age.

Influenza vaccination induces antibodies primarily against the major surface glycoproteins hemagglutinin (HA) and neuraminidase (NA); antibodies directed against the HA are most important for protection against illness. The immune response peaks at 2–4 weeks after one dose in primed individuals. In previously unvaccinated children <9 years of age, two doses of influenza vaccine are recommended, as some children in this age group have limited or no prior infections from circulating types and subtypes of seasonal influenza. These children require both an initial priming dose and a subsequent booster dose of vaccine to mount a protective antibody response.

The most common adverse events associated with inactivated vaccines are sore arm and redness at the injection site; systemic symptoms such as fever or malaise are less commonly reported. Guillian–Barré Syndrome (GBS) was identified among approximately 1 per 100,000 recipients of the 1976 swine influenza vaccine. The risk of influenza vaccine-associated GBS from seasonal influenza vaccine is thought to be at most approximately 1–2 cases per 1 million vaccinees, based on a few studies that have found an association; other studies have found no association.

The most common adverse events associated with LAIV are nasal congestion, headache, myalgias or fever. Studies of the safety of LAIV among young children suggest an increased risk of wheezing in some young children, and the vaccine is not recommended for children younger than 2 years old, ages 2-4 old with a history of recurrent wheezing or reactive airways disease, or older persons who have any medical condition that confers an increased risk of influenza-related complications.

The effectiveness of influenza vaccines is related predominantly to the age and immune competence of the vaccinee and the antigenic relatedness of vaccine strains to circulating strains. Vaccine effectiveness in preventing laboratory-confirmed influenza illness when the vaccine strains are well matched to circulating strains is 70–90% in randomized, placebo-controlled trials conducted among children and young healthy adults, but is lower among elderly or immunocompromised persons. In years with a suboptimal match, vaccine benefit is likely to be lower, although the vaccine can still provide substantial benefit, especially against more severe outcomes. Live, attenuated influenza vaccines have been most extensively studied among children, and have been shown to be more effective than inactivated vaccines in several randomized controlled trials among young children.

Influenza vaccination is recommended in the United States for all children six months or older, all adults 50 years or older, all persons with chronic medical conditions, and pregnant women, and contacts of these persons, including healthcare workers. The global disease burden of influenza is substantial, and the World Health Organization has indicated that member states should evaluate the cost-effectiveness of introducing influenza vaccination into national immunization programs. More research is needed to develop more effective seasonal influenza vaccines that provide long-lasting immunity and broad protection against strains that differ antigenically from vaccine viruses.

# 1 Introduction and Background

Epidemics of illness consistent with influenza have been identified for centuries (Creighton 1891; Thompson 1852). However, the causative agent was not identified until the 1930s, when influenza virus was first isolated from pigs and then from humans (Shope 1931; Smith et al. 1933). Influenza vaccines were first developed and tested shortly after the identification of the influenza virus, with most initial studies conducted in military populations (Anom 1944; Davenport 1962; Dowdle 1981; Salk et al. 1945; Stokes et al. 1937).

Influenza viruses cause yearly epidemics with wide variation in timing and severity (Barker 1986; Choi and Thacker 1982; Lui and Kendal 1987; Simonsen et al. 1997, 2000; Thompson et al. 2003, 2004). In temperate climates, seasonal influenza activity occurs predominantly during the winter months, but may occur year-round or in twice-yearly peaks in tropical climates (Brooks et al. 2007; Shek and Lee 2003; Yang et al. 2008).

The incubation period for seasonal influenza averages two days, but ranges from one to four days (Fox et al. 1982a; Gregg 1980; Johnson et al. 1985; Morris et al. 1966). Adults infected with influenza generally shed influenza virus for up to five days after illness onset and one day before symptoms start Carrat et al. 2007. Children may shed virus for ten days to two weeks, and—rarely—severely immunocompromised persons can shed virus for months (Johnson et al. 1986; Rocha et al. 1991).

During seasonal epidemics, infection rates are generally highest among younger children, with illness rates that can reach 30%. Influenza illness rates in younger adults and the elderly are generally less than 10% (Bridges et al. 2000; Edwards et al. 1994; Fox et al. 1982b; Hall et al. 1973; Keitel et al. 1997; Monto and Kioumehr 1975; Wilde et al. 1999). In addition to those that become ill, 30–50% of people infected with influenza develop either no symptoms or very mild symptoms (Edwards et al. 1994; Fox et al. 1982a; Keitel et al. 1997; Morris et al. 1966; Weingarten et al. 1988; Wilde et al. 1999). Typical symptoms of seasonal influenza include elevated temperature, cough, body aches and extreme fatigue, coryza, and sore throat (Monto et al. 2000; Morris et al. 1966; Neuzil et al. 2003; Poehling et al. 2006). Gastrointestinal symptoms including vomiting, abdominal pain, and diarrhea occur more frequently in children than adults. Most symptoms resolve within a week, but coughing can persist for two weeks or more. Although high fever is often a good predictor of influenza, fever is often absent in the elderly and presenting signs may include anorexia, lassitude, or confusion (Babcock et al. 2006; Neuzil et al. 2003).

In most people, influenza infection results in a self-limited respiratory illness lasting 5–7 days, but serious secondary complications can develop in some of those infected. Complications of seasonal influenza include secondary bacterial pneumonia, primary viral pneumonia, otitis media and sinusitis, encephalopathy and encephalitis, myocarditis and myositis, croup and bronchitis, sepsis-like syndrome and febrile seizures in young children, worsening of underlying conditions such as chronic obstructive pulmonary disease, diabetes and congestive heart failure, and possibly Reye's syndrome (Bhat et al. 2005; Connolly et al. 1993; Corey et al. 1976; Dietzman et al. 1976; Heikkinen et al. 1999; Keren et al. 2005; Kim et al. 1979; Poehling et al. 2006; Ruuskanen et al. 1989; Simon et al. 1970; Surtees and DeSousa 2006; Togashi et al. 2000).

In the USA, during the 1990s, an annual average of 226,000 hospitalizations and 36,000 deaths were attributed to influenza (Thompson et al. 2003, 2004). The highest rates of influenza-related hospitalization occur among those less than two years old and those greater than 64 years of age, and the highest rates of death occur among those greater than 64 years of age. Other groups at increased risk of influenza hospitalization and/or death from seasonal influenza include people of any age with chronic pulmonary, cardiovascular, renal, hepatic, hematological, immunologic (including HIV and immunosuppression caused by medications), neurologic, neuromuscular or metabolic disorders, and pregnant women (Barker 1986; Choi and Thacker 1982; Irving et al. 2000; Lui and Kendal 1987; Neuzil et al. 1998; Perrotta et al. 1985; Poehling et al. 2006; Schoenbaum and Weinstein 1979; Simonsen et al. 1997, 2000; Thompson et al. 2003, 2004). Even among healthy adults and children, though, influenza illness can result in lost work and school days, medical visits, and occasional severe complications, including death (Bhat et al. 2005; Bridges et al. 2000; Neuzil et al. 2002; Nichol et al. 1995; Poehling et al. 2006). Seasonal epidemics can lead to overwhelmed hospitals and regional medical care systems (Glaser et al. 2002). During uncontrolled outbreaks of influenza in nursing homes, illness rates can be as high as 60% and influenza-related mortality of >10% has been reported (Arden et al. 1986; CDC 1992; Coles et al. 1992; Patriarca et al. 1985).

Two influenza virus types, A and B, cause substantial morbidity and mortality among humans during seasonal influenza outbreaks. Influenza A viruses are subtyped based on the surface glycoproteins hemagglutinin (Gerhard et al. 2001) and neuraminidase (NA); influenza B viruses are not subtyped. Both A and B viruses undergo continual antigenic change, referred to as "drift," through the accumulation of point mutations in viral RNA during virus replication, which leads to the emergence of new strains. This constant drift plus variable circulation of different types and subtypes of influenza complicates both the epidemiology of influenza as well as yearly vaccine development and vaccine effectiveness. The influenza virus surface protein of most importance for antigenic response is the hemagglutinin (Gerhard et al. 2001), followed by the surface protein NA (Doherty et al. 2006). The ongoing process of antigenic drift in these surface proteins ensures a constantly renewed pool of susceptible hosts and annual epidemics. Drift also necessitates annual review of and frequent changes in vaccine strains. The constant emergence of new strains and resulting changes in the vaccine composition as well as waning immunity during the year after vaccination necessitate the annual administration of inactivated influenza vaccine for seasonal influenza (Beyer et al. 1998; Hoskins et al. 1973, 1979).

Currently circulating influenza A subtypes in humans are influenza A (H3N2) and influenza A (H1N1) viruses. Antibody against one subtype of influenza A confers little to no protection against different subtypes. However, vaccination with one strain can result in antibody with some cross-protection against related strains within the same subtype. Influenza B viruses are not subtyped, but two distinct lineages of influenza B are currently in global circulation: B/Yamagata-like and B/Victoria-like viruses (CDC 2007). Similar to different A virus subtypes, immunity against viruses from one of these B lineages provides little to no protection against the other.

During a given year, the predominant circulating influenza types, subtypes and even strains can vary geographically and temporally, even within the same community. Also, while all strains and subtypes can result in epidemic disease, the highest levels of influenza-associated mortality and hospitalization in the United States have occurred most often during seasons predominated by A (H3N2) viruses (Simonsen et al. 1997, 2000; Thompson et al. 2003, 2004). Current influenza vaccine is trivalent and contains representative influenza A (H3N2), influenza A (H1N1), and influenza B viruses thought most likely to provide the broadest protection against influenza strains anticipated to circulate in the upcoming year.

The influenza vaccination program in the USA and the programs in most other countries have focused on vaccinating those persons at greatest risk of influenza-related complications. Vaccination of close contacts of high-risk persons, including healthcare workers and household contacts of high-risk persons, is also recommended as a way to decrease transmission of influenza to high-risk persons (Fiore et al. 2008) (Table 1). In the USA, with the recent expansion of influenza vaccination recommendations to include all children six months through 18 years old, adults 50–64 years of age, and household contacts or persons less than 5 years old or more than 49 years old, more than 80% of the United States population is currently recommended for annual influenza vaccination. The ACIP has recently recommended universal vaccination for all school-aged children starting no later than the 2009–2010 season (see http://www.cdc.gov/mmwr/preview/mmwrhtml/rr5707a1.htm). In Ontario, Canada, universal vaccination was implemented in 2000 and has resulted in higher rates of vaccination, particularly among those <65 years of age (Kwong et al. 2007; Kwong et al. 2008).

Although vaccination is the mainstay of efforts to reduce the substantial health burden from influenza, two classes of prescription medications, adamantanes (amantadine and rimantadine, which have activity against influenza A viruses) and neuramindase (NA) inhibitors (oseltamivir and zanamivir, which have activity against both influenza A and B viruses), are available for the treatment and prevention of influenza. Widespread adamantane resistance, mostly among seasonal influenza A (H3N2) viruses, led to new recommendations that these medications should not be used as single agent for the treatment of influenza (Bright et al. 2006; Fiore et al. 2008). In 2008, a large increase in the prevalence of resistance to oseltamivir was observed among influenza A (H1N1) viruses in many countries, complicating treatment and chemoprophylaxis recommendations.

**Table 1** Summary of 2008 Advisory Committee on Immunization Practices (ACIP) influenza vaccination recommendations for the United States

Annual vaccination against influenza is recommended for:

- All persons who want to reduce the risk of becoming ill with influenza or of transmitting influenza to others
- All children aged 6 months through age 18 years
- All persons aged ≥50 years
- Children and adolescents (aged 6 months to 18 years) receiving long-term aspirin therapy who therefore might be at risk for experiencing Reye's syndrome after influenza virus infection
- Women who will be pregnant during the influenza season
- Adults and children who have chronic pulmonary (including asthma), cardiovascular (except hypertension), renal, hepatic, neurological, hematological or metabolic disorders (including diabetes mellitus)
- Adults and children who have immunosuppression (including immunosuppression caused by medications or by human immunodeficiency virus
- Adults and children who have any condition (e.g., cognitive dysfunction, spinal cord injuries, seizure disorders, or other neuromuscular disorders) that can compromise respiratory function or the handling of respiratory secretions or that can increase the risk for aspiration
- Residents of nursing homes and other chronic-care facilities
- Health-care personnel
- Healthy household contacts (including children) and caregivers of children aged <5 years and adults aged ≥50 years, with particular emphasis on vaccinating contacts of children aged <6 months
- Healthy household contacts (including children) and caregivers of persons with medical conditions that put them at higher risk for severe complications from influenza

The epidemiology and clinical presentation of oseltamivir-resistant influenza A (H1N1) viruses is similar to oseltamivir-sensitive viruses, and does not appear to be related to previous oseltamivir exposure or prevalence of use (Dharan et al. 2009; Hauge et al. 2009; Meijer et al. 2009; Kramarz et al. Euro Surveill). As of April 2009, nearly all H1N1 viruses are resistant to oseltamivir, but all remain sensitive to zanamivir. Interestingly, oseltamivir-resistant H1N1 viruses have typically been sensitive to adamantanes. However, all recently circulating H3N2 viruses remain resistant to adamantanes, and B viruses and adamantanes have no activity against influenza B viruses. Updated recommendations for use of antivirals in the control and treatment of influenza have recently been published (Harper et al. Clin Infect Dis 2009).

Two different influenza vaccines are available in the USA for seasonal influenza: inactivated influenza vaccine, administered via intramuscular injection; and live attenuated, cold-adapted influenza vaccine, administered via nasal spray. Both vaccines are trivalent preparations grown in eggs and do not contain adjuvants. In the rest of this chapter, we will describe these seasonal vaccines in more detail. Other influenza vaccine formulations are available in other countries or are in development, including MF-59 adjuvanted inactivated vaccine, purified HA-based vaccines, cell-culture-based vaccines, adenovirus-vectored vaccine, and virosome-based vaccine (Frey et al. 2003; Halperin et al. 2002; Hoelscher et al. 2006; Kistner et al. 1999; Palache et al. 1997, 1999; Percheson et al. 1999). These formulations are discussed in other chapters.

# 2 Inactivated Influenza Vaccine

## 2.1 History of Vaccine Development and Current Vaccine

The first commercial vaccines were approved for use in the United States in 1945, based on efficacy studies performed in military recruits and college students using whole-virus inactivated influenza vaccines (Francis 1945; Francis et al. 1946; Salk et al. 1945; Stokes et al. 1937). An influenza vaccine was of particular interest to the US military during World War II, in part because of the devastation caused in both military and civilian populations by the 1918–1919 influenza pandemic during the late stages of World War I. Interest in the use of influenza vaccine in the US military continues with large-scale vaccination programs (Russell et al. 2005).

All of the influenza vaccine viruses manufactured for the US market are replicated individually in eggs. To prepare the inactivated vaccines, the harvested influenza viruses are inactivated using either formalin or β-propiolactone. Several purification steps reduce nonviral proteins and other materials introduced during the manufacturing process. The monovalent vaccines are combined to formulate the final trivalent bulk vaccines.

Although whole-virus inactivated influenza vaccines are still in use in some countries and are highly effective, most vaccines manufactured since the 1970s have been subvirion (sometimes referred to as "split") preparations. These vaccines retain the immunogenic properties of the viral proteins but are associated with reduced reactogenicity compared to whole virus vaccines (Cate et al. 1977, 1983; Quinnan et al. 1983; Wright et al. 1977, 1983). Subvirion vaccines are prepared by using a solvent to disrupt the viral lipid envelope.

Influenza A vaccine viruses are "high-growth" influenza A reassortant viruses that contain internal genes from A/Puerto Rico/8/34 (PR8), a strain adapted to replication in eggs, and the HA and NA from wild-type viruses (Baez et al. 1980; Kilbourne and Murphy 1960). These reassortants help maximize virus growth and yield of virus for vaccine production. There is no high-growth donor strain for influenza B viruses for vaccine production, however.

## 2.2 Vaccine Constituents, Including Antibiotics and Preservatives

Hemagglutinin (Gerhard et al. 2001) is the main immunogen in inactivated influenza vaccines, and the amount of HA administered is correlated with the immunogencity of the vaccine (Cate et al. 1983; Keitel et al. 2006; LaMontagne et al. 1983; Quinnan et al. 1983). Aminoglycoside antibiotics are used by some manufacturers to reduce bacterial growth in eggs during processing, but are reduced to trace or undetectable amounts during the purification process. Minimal amounts of the detergents or solvents used for virus disruption may remain in the final vaccine preparation, but purification and dilution steps often reduce the amount to the limits of detection.

Thimerosal, a mercury-containing compound, is used in multidose vials and some single-dose preparations to reduce the total amount of bacteria and fungi during production of influenza vaccines in eggs and/or as a preservative to prevent growth of bacteria and fungi in the final vaccine formulation. Efforts to limit the amount of mercury present in vaccines of all kinds have encouraged greater availability of inactivated influenza vaccines that are thimerosal-free or contain only trace amounts.

Either formalin or β-propiolactone is used to inactivate influenza vaccine viruses. β-propiolactone is chemically degraded during vaccine processing so that levels in the final vaccine product are below the limits of detection (LoGrippo 1960). Although detectable quantities of formaldehyde persist in inactivated vaccines when formalin is used, the steps used for purification also reduce the amount of free formaldehyde.

## 2.3   Dosage, Administration, and Storage

Inactivated influenza virus vaccines are recommended for intramuscular administration. Current doses recommended for inactivated influenza vaccines are 15 μg of HA for each vaccine strain for persons three years of age and older and 7.5 μg of HA for each vaccine strain per vaccine dose for children 6–35 months. Children less than nine years of age who have not been previously vaccinated against influenza require two doses of vaccine in their first year with the doses separated by four or more weeks. Both immunogenicity studies and more recent vaccine effectiveness studies have shown that immune naïve children require two doses to maximize the immune response and clinical benefit of inactivated influenza vaccine. In studies of children less than five years of age, very low or no clinical effectiveness has been demonstrated after only one dose (Allison et al. 2006; Englund et al. 2005; Neuzil et al. 2006; Ritzwoller et al. 2005; Shuler et al. 2007). The ACIP also recommends that children aged six months to eight years who received only one dose in their first year of vaccination receive two doses the following year, as the immune response is superior if the child first received two doses with the same antigen (Englund et al. 2005; Fiore et al. 2008).

It is recommended that inactivated influenza vaccines should be stored at 4–8°C.

## 2.4   Immune Response to Inactivated Influenza Vaccination

Influenza vaccination primarily induces antibodies against the major surface glycoproteins HA and NA; antibodies directed against the HA are most important for protection against illness, while antibodies directed against the NA may reduce the severity of disease (Gerhard 2001; Kilbourne et al. 1968). The antibody response to inactivated vaccines is correlated with age, with lower responses among the elderly and those who are multiply vaccinated, as well as the immune competence of the vaccinee. In addition, persons without existing antibody from prior infections with influenza or prior vaccination need two doses of vaccine to obtain a maximal response. The immune response peaks at 2–4 weeks after vaccination with one dose in primed individuals (Brokstad et al. 1995; el-Madhun et al. 1998; Gross et al. 1997). In children <9 years of age, two doses of inactivated vaccine are recommended, as some children in this age group have limited or no prior infections from circulating types and subtypes of seasonal influenza. Several recent studies have demonstrated that receipt of only one dose of vaccine in previously unvaccinated children less than five years of age resulted in substantially reduced or no clinical effectiveness (Allison et al. 2006; Englund et al. 2005; LaMontagne et al. 1983; Neuzil et al. 2006; Ritzwoller et al. 2005; Shuler et al. 2007).

Serum antibody titers based on HI testing generally correlate with protection against influenza, and HI titers of 1:32 to 1:40 are often used as benchmarks for an adequate immune response to inactivated influenza vaccine (Demicheli et al. 2000; Dowdle et al. 1973; Gross et al. 1995; Monto et al. 1970). However, this is not an absolute level and some persons who achieve such titers or higher can remain susceptible to influenza infection (Davies and Grilli 1989; Dowdle et al. 1973; Hobson et al. 1972).

## 2.5 Efficacy and Effectiveness of Inactivated Influenza Vaccine

The effectiveness of the inactivated influenza vaccine is predominantly affected by the age and immune competence of the vaccinee and the antigenic relatedness of vaccine strains to circulating strains. In years with a suboptimal match, vaccine benefit is likely to be lower, although the vaccine still can provide substantial benefit in most years, especially against more severe outcomes (Edwards et al. 1994; Herrera et al. 2007; Nichol et al. 2007; Ritzwoller et al. 2005; Russell et al. 2005). An important factor affecting the measurement of vaccine efficacy and effectiveness is the specificity of the outcome measure used in the study (Orenstein et al. 1988). When less specific outcomes are used, such as non-laboratory-confirmed respiratory illness, the calculated effectiveness will be reduced compared to using laboratory-confirmed influenza as the outcome. For example, one study found vaccine efficacy of 86% against serologically confirmed influenza illness, 34% against febrile respiratory illness, and 10% against upper respiratory infection in the same year and population (Bridges et al. 2000). This is because the nonspecific outcomes include illnesses caused by agents other than the influenza virus. Influenza vaccine would not be expected to protect against these agents. Influenza illness rates can also vary substantially from year to year, and, in years with low rates, the power of smaller studies to detect a difference in vaccine effectiveness may be compromised, even when using a lab-confirmed outcome (Hoberman et al. 2003; Shuler et al. 2007; Szilagyi et al. 2008).

Many different outcome measures, study designs and populations have been used to assess inactivated influenza vaccine effectiveness. Some meta-analyses have been done, although the differences in study designs (most of which are not randomized trials) and study outcomes provide great challenges when conducting and also interpreting meta-analyses of vaccine effectiveness (Gross et al. 1995; Rivetti et al. 2006; Smith et al. 2006b; Vu et al. 2002; Jefferson et al. 2007).

Another major difficulty in assessing influenza vaccine benefit in nonrandomized studies is addressing confounding by indication and biases in who is selected to receive influenza vaccine. These issues are even more difficult in nonrandomized studies that use non-laboratory-confirmed outcomes, as is the case with most studies in the elderly (Hak et al. 2002b; Jackson et al. 2006; Nichol et al. 2007; Simonsen et al. 2007).

## 2.6   Influenza Vaccine Effectiveness Among Adults ≥65 Years of Age and Those with Chronic Conditions

Ironically, the greatest number and quality of influenza vaccine efficacy studies have been conducted among younger, healthy adults, a group for whom vaccination efforts are not targeted. In contrast, only one randomized trial has been conducted among older adults, the group at highest risk of complications and death from influenza. This study, which included persons aged 60 years or older without underlying health conditions, found that influenza vaccination reduced symptomatic laboratory-confirmed influenza by 58% (Govaert et al. 1994). No randomized studies or studies using a laboratory-confirmed influenza outcome have assessed the effectiveness of inactivated influenza against influenza-related complications, such as hospitalization or death, in the elderly. Nonrandomized studies, particularly in the elderly, present a challenge in estimating vaccine effectiveness and also accounting for differences in the age and health status of vaccinated and unvaccinated populations and different tendencies to seek medical care (Hak et al. 2002b; Jackson et al. 2006). Although the immune response to the vaccine and vaccine effectiveness is lower in the elderly than younger persons, the body of evidence finding reductions in influenza complications after vaccination clearly supports the use of influenza vaccine in this age group (Treanor 2007).

The greatest benefit of influenza vaccination among the elderly is the reduction in influenza-related hospitalizations and deaths with studies generally showing lower levels of effectiveness against respiratory illness. Several meta-analyses have been done, again, almost exclusively including nonrandomized cohort and case–control studies among those aged 65 and older. These analyses have found reductions in pneumonia and influenza hospitalization by 27–52% and death from all causes by 47–68% (Gross et al. 1995; Rivetti et al. 2006; Vu et al. 2002).

Most influenza vaccination studies in the elderly have involved the retrospective analysis of large administrative databases, investigations of outbreaks, or case–control studies. One administrative database study covering ten influenza seasons (1990–2000) estimated that vaccine reduced influenza-related pneumonia and influenza hospitalizations by 27%, and influenza-related deaths from all causes by 48% (Nichol et al. 2007) while adjusting for several biases inherent in such studies. Although effectiveness was substantially reduced in years with a poor antigenic match, the vaccine continued to provide some protection against influenza-related deaths. Another study using a large database found that influenza vaccine reduced influenza hospitalization by 18% over an eight-year period, but that the benefit appeared to be explained by biases in the study population, in that persons who were vaccinated tended to be at a lower risk of the outcome measured, even when influenza was not circulating (Jackson et al. 2006).

Several case–control studies have demonstrated the ability of influenza vaccine to prevent hospitalization. During the 1989–1990 influenza season in the United Kingdom, one study estimated that vaccine reduced influenza, pneumonia, bronchitis, and emphysema combined by 63% (Ahmed et al. 1997), while a Spanish study estimated that vaccine reduced radiologically confirmed pneumonia by 79% (Puig-Barbera et al. 1997). In the United States, a two-year study conducted in 1990–1992 (Ohmit and Monto 1995) estimated that vaccine reduced influenza-related hospitalization by 31% during an influenza A outbreak and by 32% during an influenza B outbreak. One study using nine years of data from a health insurance plan estimated that vaccine reduced pneumonia and influenza hospitalization by 30–33% overall and by 51–83% in years when A(H3N2) viruses predominated (Mullooly et al. 1994). Another US study during 1989–1990 estimated that the vaccine reduced pneumonia and influenza hospitalization by 45% (Foster et al. 1992).

The effectiveness of vaccine against death has also been examined in case–control studies. A 1989–1990 UK study estimated that vaccine reduced influenza-related deaths by 75% (Fleming et al. 1995), while a Canadian study estimated that deaths during hospitalizations from any cause were reduced by 27–30% (Fedson et al. 1993).

Vaccine effectiveness estimates in nursing home residents have also varied widely, with generally low effectiveness in preventing influenza-like illness but continued benefits in preventing hospitalization and death (Arden et al. 1986; Deguchi et al. 2000; Meiklejohn and Hall 1987; Monto et al. 2001; Rivetti et al. 2006; Staynor et al. 1994). For example, in years where the vaccine and circulating strains were well matched, studies have estimated that vaccine reduces influenza-like illness during nursing home outbreaks by zero to 80%, with most estimates approximating 40% (Arden et al. 1986; Deguchi et al. 2000; Meiklejohn and Hall 1987; Monto et al. 2001; Morens and Rash 1995; Ohmit et al. 1999; Staynor et al. 1994). One meta-analysis of influenza vaccine among elderly in nursing homes found an effectiveness of 23% against influenza-like illness, 46% against pneumonia, 45% against hospitalization, and 42% against influenza and pneumonia deaths (Rivetti et al. 2006). However, when the authors restricted studies included in the meta-analysis to those that used laboratory-confirmed influenza as an outcome, vaccination effectiveness was not demonstrated (Rivetti et al. 2006). These findings again illustrate the difficulty in demonstrating vaccine effectiveness among the elderly in nursing homes as well as among community-dwelling elderly, and suggests that biases in the study populations, such as differences in underlying health and use of healthcare services among vaccinated and unvaccinated elderly in the community and among nursing home residents, might account for some of the effect of vaccination in nonrandomized studies (Jackson et al. 2006). Vaccination with a second dose of vaccine has not been demonstrated to be of benefit in improving the immune response in this population and is not recommended (Fiore et al. 2008).

Some studies suggest that vaccination rates of 80% or more of nursing home residents can induce herd immunity and decrease the risk of influenza outbreaks in nursing homes (Arden et al. 1986; Oshitani et al. 2000). Vaccination of health care workers also may reduce morbidity and mortality in nursing home residents by decreasing the risk of influenza exposure among nursing home residents (Carman et al. 2000; Potter et al. 1997; Saito et al. 2002; Hayward et al. BMJ 2006).

A limited number of influenza vaccine studies have assessed the effectiveness of the vaccine among groups with underlying medical conditions (Fine et al. 2001; Hak et al. 2005; Herrera et al. 2007). In general, these studies show either moderate or no reductions in effectiveness compared to their use in age-matched groups. Two large administrative database studies (Hak et al. 2002a; Nichol et al. 1998) of persons 65 years and older with different medical conditions found no statistically significant differences in vaccine effectiveness among subgroups of elderly persons (healthy elderly; those with heart disease, lung disease, diabetes; immune-compromised persons; or those with either dementia, stroke, vasculitis, or rheumatologic conditions). Very few studies have been conducted among HIV-infected persons, but available data suggest that the vaccine offers clinical benefit in this group (Tasker et al. 1999).

## 2.7   Efficacy and Effectiveness in Adults <65 Years of Age

Initial studies of inactivated influenza vaccine were conducted among military populations and found to reduce influenza illness by 70% to over 90% (Williams and Wood 1993). More recent studies among military populations continue to find similarly high vaccine effectiveness against laboratory-confirmed influenza (92–94%) (Russell et al. 2005; Strickler et al. 2007). Estimates of vaccine effectiveness against non-laboratory-confirmed outcomes are, as expected, substantially lower (0–52%) in these same studies with high effectiveness against laboratory-confirmed outcomes.

Additional studies have also been conducted among the general population, including a five-year placebo-controlled trial that found reductions in culture-confirmed influenza by 70–79% (Edwards et al. 1994). Another five-year placebo-controlled trial among 30–60 year olds from 1983 to 1988 showed reductions of 47–73% against serologically or culture-confirmed influenza, including one year when the antigenic match between the vaccine and circulating strains was suboptimal (Keitel et al. 1997).

Two randomized, placebo-controlled studies have been conducted among healthy adult workers in the United States. One study did not include a laboratory-confirmed outcome, but did find reductions of upper respiratory illness by 25%, illness-related work absenteeism by 43%, and physician visits by 44% (Nichol et al. 1995). In the first year of the second study, the vaccine and circulating strains were not matched well. The estimated vaccine efficacy against serologically confirmed febrile respiratory illness was 50% but was not statistically significant with a low attack rate; no reductions in overall febrile respiratory illness, physician visits, or work absenteeism were found. However, in the second year, vaccine effectiveness was 86% against laboratory-confirmed influenza, 34% against febrile respiratory illness, 42% against physician visits, and 32% against work absenteeism (Bridges et al. 2000).

Three randomized trials have been conducted among health care workers (Saxen and Virtanen 1999; Weingarten et al. 1988; Wilde et al. 1999). A one-year study conducted during a year with a poor vaccine match found no reductions of influenza-like illness or work absenteeism (Weingarten et al. 1988). A three-year study found reductions in serologically confirmed influenza by 88–89% (Wilde et al. 1999), febrile respiratory illness by 29%, and illness-related work absenteeism by 53%, but the findings were not statistically significant in this small study. Another one-year study without an influenza laboratory-confirmed outcome found no reduction in respiratory illnesses, but work absenteeism related to respiratory illness was reduced by 28% (Saxen and Virtanen 1999).

A meta-analysis of studies in healthy adults estimated that, overall, inactivated influenza vaccine reduced laboratory-confirmed influenza by 80% in years when the vaccine and circulating strains were well matched, and 50% when the match was suboptimal (Jefferson et al. 2007).

## 2.8   Efficacy and Effectiveness in Children

Inactivated influenza vaccine effectiveness studies among school-aged children generally show results similar to those in young adults, although overall effectiveness appears to be lower in young children. In a randomized trial conducted over five influenza seasons in the United States among children 1–15 years of age, vaccine reduced influenza illness by 77–91% (Neuzil et al. 2001). A one-year study reported vaccine efficacies of 56% among healthy 3–6 year olds and 100% among healthy 10–18 year olds (Clover et al. 1991). In Japan, efficacy studies based on serologic confirmation of influenza infection were carried out among children during 1982–1984 (Oya and Nerome 1986). For the 1983 influenza season, one or two subcutaneous doses of vaccine reduced illness by 76% among elementary school children and by 83% among junior and senior high-school children (Oya and Nerome 1986). In Russia, inactivated influenza vaccine was estimated to reduce respiratory illness by 24% in school children aged 7–10 years and by 30% in children aged 11–14 years in the 1989–1990 influenza season, and by 27% in both groups in the 1990–1991 season (Rudenko et al. 1993). In Italy, a one-year randomized trial of 344 children aged 1–6 years estimated that influenza-like illness was reduced by 67% (Colombo et al. 2001).

For young children, results are more varied and few randomized trials have been done which have the power to assess vaccine effectiveness in the youngest age groups. In the USA, a randomized study of 127 children aged 24–60 months attending day care estimated that vaccine reduced serologically identified influenza infection by 45% (95% CI: 5–66%) for influenza A and B combined, influenza B by 45% (95% CI: −20% to +69%), influenza A (H3N2) by 31% (95% CI: −95% to +73%), and febrile influenza-like illness by 7% (95% CI: −30% to +23%) (Hurwitz et al. 2000b). In a two-year randomized study among children 6–24 months of age, the vaccine was estimated to reduce respiratory illness by 66% in year 1 ($n = 411$) when influenza viruses circulated widely. However, the vaccine was not efficacious in year 2 ($n = 375$), when influenza circulation was limited (attack rate of 3% in the unvaccinated) (Hoberman et al. 2003)

Several nonrandomized cohort studies have been done in recent years after the USA first encouraged in 2002 and then recommended influenza vaccination of children 6–23 months of age. A cohort study of children 6–23 months old evaluated the effectiveness of one vs. two doses of the inactivated vaccine against influenza-like illness (ILI) and pneumonia and influenza (Inouye and Kramer 2001). In a year with a suboptimal match between the vaccine and circulating strains, the vaccine was 25% against ILI and 49% against P&I for two doses, but for those who received only one dose among those that needed two doses, the vaccine was not effective against ILI and was 22% effective for P&I (Ritzwoller et al. 2005). Another study conducted the same year, but using laboratory-confirmed outcomes, also found no vaccine effectiveness among children <2 years who received only one dose, but it did find effectiveness among children who received two doses (Shuler et al. 2007).

Limited studies have been conducted among children with chronic medical conditions. In a nonrandomized study, vaccine reduced culture- or serologically confirmed influenza (against a drifted strain) by 22–54% in asthmatic children 2–6 years of age and by 60–78% in 7–14 year olds (Sugaya et al. 1994). A retrospective analysis using a computerized primary care database estimated that vaccine reduced medically attended visits for respiratory illness or otitis media among asthmatic children by 27% (95% CI: −7% to 51%) for those 0–12 years old, by 55% (20–75%) for those less than 6 years old, and by −5% (−81% to 39%) for those 6–12 years old (Smits et al. 2002). Overall, vaccine efficacy may be lower among high-risk children, particularly immune-suppressed children, compared with healthy adults or older healthy children. One randomized controlled trial conducted in Bangladesh that provided vaccination to pregnant women during the third trimester demonstrated a 29% reduction in respiratory illness with fever, and 36% reduction in respiratory illness with fever among their infants during the first 6 months after birth. In addition, infants born to vaccinated women had a 63% reduction in laboratory confirmed influenza illness during the first 6 months of life (Zaman et al. 2008).

# 3 Safety

## 3.1 Common Adverse Events

In more recent randomized trials of influenza vaccine among adults with split virus vaccine, the only difference in side effects among vaccinated vs. placebo recipients was sore arm and redness at the injection site (Bridges et al. 2000; Nichol et al. 1995), or red eyes (Ohmit et al. 2008). Data from older studies are less informative given the improvements in vaccine purification steps and use of split virus vaccines over whole-virus inactivated influenza vaccines (Cate et al. 1983; Quinnan et al.

1983; Wright et al. 1983). These local reactions are generally mild and resolve generally within two days. Systemic reactions, which include fever, myalgia, arthralgia and headache, have been reported, but occur much less frequently than local reactions; they are more likely to occur in very young children and in others exposed to influenza virus vaccines or to one of the antigens in the vaccine for the first time (Barry et al. 1976; Cate et al. 1977; Ennis et al. 1977; Wright et al. 1977).

Guillian–Barré Syndrome (GBS), a rare neurological syndrome characterized by ascending paralysis, paresthesia and dysesthesia, has been associated with receipt of the 1976 swine influenza vaccine and also with many respiratory and gastrointestinal illnesses and, in particular, infection with *Campylobacter species* (Green 2002; Prevots and Sutter 1997; Ropper 1992; Schonberger et al. 1979; Willison and Yuki 2002). The increase in GBS cases above the background rate was approximately one case for every 100,000 persons vaccinated with the swine influenza vaccine (Schonberger et al. 1979).

Subsequent observational studies in the United States during 1977–1991 did not find similar increases in GBS cases among recipients of inactivated vaccine (Hurwitz et al. 1981; Kaplan et al. 1982). However, a study of the 1992–1994 influenza seasons estimated an increased incidence of approximately one GBS case per million recipients of inactivated influenza vaccine among adults (Lasky et al. 1998), which is substantially less than the risk of developing severe complications from influenza infection. A more recent study from Canada also found an increased risk of GBS among adult vaccinees, with a similar incidence of approximately one case per million persons vaccinated (Juurlink et al. 2006). No experience similar to the 1976 vaccine campaign has since occurred, and the Advisory Committee on Immunization Practices (ACIP) has stated that "the potential benefits of influenza vaccination in preventing serious illness, hospitalization, and death greatly outweigh the possible risks for developing vaccine-associated GBS" (Fiore et al. 2008).

Based on data from the swine influenza vaccine campaign of 1976, the risk of anaphylaxis from influenza vaccine is estimated to be one in four million vaccinees. Persons with a history of anaphylactic hypersensitivity to eggs or to previous influenza vaccination should not be vaccinated with inactivated influenza vaccine until they have been evaluated by a physician. In addition to egg antigens, it is possible (as with any vaccine) that other components may be sensitizing. Allergic reactions to the preservative thimerosal may also occur and are characterized by a delayed-type hypersensitivity that most often results in a local inflamed or indurated lesion (Audicana et al. 2002). Influenza vaccines containing no thimerosal and formulations containing only trace amounts of thimerosal are available for those with thimerosal allergies.

Influenza vaccine is recommended for women who will be pregnant during the influenza season based on their increased risk of hospitalization during seasonal influenza outbreaks and their increased risk of death and fetal loss during pandemics. Although few studies of inactivated influenza vaccine safety have been conducted among pregnant women, the vaccine has been used in the USA among pregnant women for many years (Fiore et al. 2008; Freeman and Barno 1959; Harris 1919; Heinonen et al. 1973; Mak et al. 2008; Neuzil et al. 1998; Schanzer et al. 2007; Widelock et al. 1963).

# 4    Live Attenuated Influenza Vaccines

## 4.1    History of Vaccine Development and Current Vaccines

The development of stable, immunogenic, and safe live attenuated influenza virus vaccines (LAIV, also sometimes referred to as cold-adapted, live-attenuated influenza vaccine-trivalent [CAIV-T]) was an important scientific achievement. Wild-type influenza viruses replicate most efficiently at 37–39°C, but poorly at 25°C (Maassab 1967). Cold-adapted influenza viruses were developed in the 1960s by serially pas-saging wild-type viruses in primary chick kidney cells and gradually lowering the temperature to 25°C (Maassab 1969). With continued passaging, these cold-adapted viruses became attenuated because of mutations in multiple genes, and unlike wild-type viruses do not cause influenza-like illness in ferrets (Maassab et al. 1982). These strains are also temperature sensitive (replicating poorly at 37°C), rendering them unable to replicate efficiently in the lower airways, and therefore unlikely to cause systemic symptoms of influenza. Cold-adapted, attenuated mutants of influenza A/ Ann Arbor/6/60 (H2N2) and influenza B/Ann Arbor/1/66 serve as the master donor strains for vaccine reassortants used in LAIV available in the United States (Belshe and Mendelman 2003). The genetic changes in master donor strains have been char-acterized, and include mutations in the polymerase genes (*PB1*, *PB2*) and the nucleo-capsid structural component (NP) gene for A/Ann Arbor/6/60 (H2N2) (Snyder et al. 1988), and in a viral polymerase subunit gene (Palache et al. 1997), the nucleoprotein (NP) and the matrix protein (M1) for B/Ann Arbor/1/66 (Chen et al. 2006; Donabedian et al. 1988). To make reassortants for vaccines, hemagglutinin HA and NA genes from a currently circulating wild-type virus are introduced by culturing the wild-type virus together with a master donor strain, resulting in attenuated, cold-adapted, tem-perature-sensitive reassortants containing wild-type HA and NA genes. LAIV reas-sortants have been shown to be stable during the manufacture and clinical use of LAIV (Belshe and Mendelman 2003). Cold-adapted master donor strains have also been developed and characterized in Russia (Cox et al. 1985; Ghendon et al. 1984), and LAIV has been used since the 1980s (Alexandrova et al. 1986).

## 4.2    Vaccine Constituents, Including
##           Antibiotics and Preservatives

Reassortants containing the six master donor strain genes necessary for replication and assembly as well as contemporary wild-type HA and NA genes were  made using classical viral genetics reassortment techniques; in recent years, reverse genetics techniques have been substituted. The resulting 6:2 reassortant strains are grown in pathogen-free embryonated hen's eggs in the same way that the compo-nents of inactivated influenza vaccines are manufactured. Allantoic fluid from inocu-lated eggs is harvested, pooled and clarified by filtration, and virus is concentrated by ultracentrifugation. Each dose of LAIV (manufactured in the United States as FluMist, MedImmune Vaccines Inc., Gaithersburg, MD) contains the same three antigens used in inactivated influenza vaccine for the influenza season. The antigen component of the vaccine consists of $10^{6.5-7.5}$ fluorescent focus units (FFU) of each of the three strains. Additional components of LAIV have changed in recent formulations to improve stability at 2–8°C and allow for reduced volumes of vaccine that increase acceptability of intranasal delivery. Components in the 2008–2009 influenza season formulation include stabilizing buffers containing monosodium glutamate (0.188 mg/dose), hydrolyzed porcine gelatin (2.00 mg/dose), arginine (2.42 mg/dose), sucrose (13.68 mg/dose), dibasic potassium phosphate (2.26 mg/dose), monosodium phosphate (0.96 mg/dose), and gentamicin sulfate (<0.015 mcg/mL). LAIV does not contain thimerosal or any other preservatives (MedImmune 2008).

## 4.3   Dosage, Administration, and Storage

Recommendations for the use of LAIV and inactivated influenza vaccines in the United States are compared in Table 2. LAIV is recommended for use in the United States for healthy nonpregnant persons 2–49 years of age (Fiore et al. 2008). LAIV is only intended for intranasal administration and should not be administered by the intramuscular, intradermal, or intravenous route. LAIV (FluMist) is supplied in a prefilled, single-use sprayer containing 0.2 mL of vaccine. Approximately 0.1 mL (i.e., half of the total sprayer contents) is sprayed into one nostril while the recipient is in an upright position. An attached dose-divider clip is removed from the sprayer to administer the second half of the dose into the other nostril. LAIV is shipped to end users at 35–46°F (2–8°C). LAIV should be stored at 35–46°F (2–8°C) upon receipt, and can remain at that temperature until the expiration date is reached (MedImmune 2007). LAIV should not be administered until at least 48 h after the use of an influenza antiviral medication has ceased, because of the potential for reduction in immunogenicity with antiviral activity against the live virus vaccine. If a vaccine recipient sneezes after administration, the dose should not be repeated. However, if nasal congestion is present that might impede delivery of the vaccine to the nasopharyngeal mucosa, deferral of administration should be considered until resolution of the illness, or inactivated influenza vaccine should be administered instead. Use of LAIV concurrently with MMR and varicella vaccines among 12–15 month olds has been studied, and no interference with the immunogenicity of either vaccine was observed. Concurrent use of other active or inactivated vaccines has not been studied (MedImmune 2008). No data are available regarding concomitant use of nasal corticosteroids or other intranasal medications.

LAIV should be administered annually according to the following schedule:

- Children aged 2–8 years previously unvaccinated at any time with either LAIV or inactivated influenza vaccine should receive two doses of LAIV separated by at least four weeks.

**Table 2** Live attenuated influenza vaccine (LAIV) compared with inactivated influenza vaccine (TIV); United States formulations

| Factor | LAIV | TIV |
|---|---|---|
| Route of administration | Intranasal spray | Intramuscular injection |
| Type of vaccine | Live virus | Killed virus |
| No. of included virus strains | 3 (2 influenza A, 1 influenza B) | 3 (2 influenza A, 1 influenza B) |
| Vaccine virus strains updated | Annually | Annually |
| Frequency of administration | Annually[a] | Annually[a] |
| Approved age and risk groups[b] | Healthy non-pregnant persons aged 2 through 49 years | Persons aged ≥6 months |
| Interval between two doses recommended for children aged ≥6 months through 8 years who are receiving influenza vaccine for the first time | 4 weeks | 4 weeks |
| Can be administered to family members or close contacts of immunosuppressed persons not requiring a protected environment? | Yes | Yes |
| Can be administered to family members or close contacts of immunosuppressed persons requiring a protected environment (e.g., hematopoietic stem cell transplant recipient)? | No | Yes |
| Can be administered to family members or close contacts of persons at high risk but not severely immunosuppressed? | Yes | Yes |
| Can be simultaneously administered with other vaccines? | Yes[c] | Yes[d] |
| If not simultaneously administered, can be administered within four weeks of another live vaccine? | Prudent to space 4 weeks apart | Yes |
| If not simultaneously administered, can be administered within four weeks of an inactivated vaccine? | Yes | Yes |

[a] Children aged six months through eight years who have never received influenza vaccine before should receive two doses. Those who only receive one dose in their first year of vaccination should receive two doses in the following year

[b] Populations at higher risk from complications of influenza infection include persons aged ≥50 years, residents of nursing homes and other chronic-care facilities that house persons with chronic medical conditions, adults and children with chronic disorders of the pulmonary or cardiovascular systems, adults and children with chronic metabolic diseases (including diabetes mellitus), renal dysfunction, hemoglobinopathies, or immunnosuppression, children and adolescents receiving long-term aspirin therapy (at risk for developing Reye's syndrome after wild-type influenza infection), pregnant women, and children aged <5 years. However, no influenza vaccine is approved for children <6 months of age in the United States

[c] Live attenuated influenza vaccine coadministration has been evaluated systematically only among children aged 12–15 months with measles, mumps and rubella (MMR) and varicella vaccines

[d] Inactivated influenza vaccine coadministration has been evaluated systematically only among adults with pneumococcal polysaccharide and zoster vaccine

- Children aged 2–8 years previously vaccinated at any time with either LAIV or inactivated influenza vaccine should receive one dose of LAIV. However, a child of this age who received influenza vaccine for the first time last season and did not receive two doses last season should receive two doses (as above) during the current season.
- Persons aged 9–49 years should receive one dose of LAIV. Children aged 2–8 years who received one or more doses of either inactivated influenza vaccine or LAIV two or more seasons previously should be given one dose of LAIV.

## 4.4  Shedding, Transmission, and Stability of Vaccine Viruses

After vaccination, LAIV viruses establish infection and replicate in the mucosa of the upper respiratory tract. Systemic symptoms of influenza are not observed among vaccine recipients, although some experience effects of intranasal vaccine administration or local viral replication (e.g., nasal congestion) (Belshe et al. 1998; Nichol et al. 1999). Vaccinated persons have been shown to shed vaccine viruses, although in lower amounts than occur typically with shedding of wild-type influenza viruses; shedding appears to be more common among children than adults (Moritz et al. 1980; Wright et al. 1982). Shed LAIV viruses are phenotypically stable and do not revert to virulent strains (Cha et al. 2000; Vesikari et al. 2006b), although minor sequence variations among shed viruses compared to vaccine strains have been observed (Buonagurio et al. 2006).

Among 345 subjects aged 5 through 49 years, 30% had detectable virus in nasal secretions obtained by nasal swabbing after receiving LAIV. The duration of virus shedding and the amount of virus shed was inversely correlated with age, and maximal shedding occurred within 2 days of vaccination. Symptoms reported after vaccination, including runny nose, headache, and sore throat, did not correlate with virus shedding (Block et al. 2008). Other smaller studies have reported similar findings. (Ali et al. 2004; Talbot et al. 2005). Limited data indicates that shedding among immunocompromised persons does not appear to be more common. Among HIV-infected adults, vaccine strain virus was detected from nasal secretions in one (2%) of 57 HIV-infected adults who received LAIV (King et al. 2000). Vaccine strain virus was detected from nasal secretions in three (13%) of 23 HIV-infected children compared with seven (28%) of 25 children who were not HIV-infected (King et al. 2001). No participants in these studies shed virus beyond ten days after receipt of LAIV. Whether LAIV is excreted in human milk is unknown. Positive rapid influenza tests have been reported up to seven days after receipt of LAIV (Ali et al. 2004).

In rare instances, vaccine viruses can be transmitted from vaccine recipients to unvaccinated persons. One study of children aged 8–36 months in a child-care center assessed the transmissibility of vaccine viruses from 98 vaccinated to 99 unvaccinated children who had received placebo; 80% of vaccine recipients shed one or more virus strains for an average of 7.6 days. One influenza type B vaccine strain isolate was recovered from a placebo recipient who had minor respiratory symptoms. This isolate retained the cold-adapted, temperature-sensitive, attenuated phenotype, and it had the same genetic sequence as a virus shed from a vaccine recipient who was in the same playgroup. The probability of acquiring vaccine virus after close contact with a single LAIV recipient in this child-care population was estimated to be 0.6–2.4% (Vesikari et al. 2006b). Serious illnesses have not been reported among unvaccinated persons who have been inadvertently infected with LAIV viruses.

## 4.5   Immune Response to LAIV

LAIV is administered intranasally, and the nasopharyngeal infection engendered by LAIV is assumed to mimic wild-type virus infection. Correlates of immune protection identified in challenge studies with vaccine strains include serum and nasal wash antibodies and serum hemagglutination inhibition (HAI) antibodies. LAIV-specific nasal IgA antibodies and cytotoxic T cells likely provide protection but are difficult to measure (Belshe et al. 2000b). Serum HAI antibody does not correlate with protective immunity as well for LAIV as it does for inactivated influenza vaccine (Clover et al. 1991). Some interference between vaccine strains has been noted, with previously unvaccinated seronegative children given monovalent H1N1 LAIV showing significantly higher rates of HAI antibody seroconversion compared to similar children given bivalent H1N1/H2N2; however, the clinical importance of this observation is unknown, and interference can be overcome by giving two doses to young vaccine-naïve children (Gruber et al. 1996).

## 4.6   Efficacy and Effectiveness of LAIV

### 4.6.1   Healthy Children

Several randomized, double-blind, placebo-controlled trials using laboratory-confirmed influenza infection endpoints have demonstrated LAIV efficacy among children. The first large randomized, double-blind, placebo-controlled clinical trial included 1,602 healthy children aged 15–71 months who received LAIV during each of two consecutive seasons (Belshe et al. 1998, 2000a). In season one (1996–1997), when vaccine and circulating virus strains were well matched, efficacy against culture-confirmed influenza was 94% for participants who received two doses of LAIV and 89% for those who received one dose. In season 2, when the A (H3N2) component in the vaccine was not well matched with circulating virus strains, efficacy was 86%, for an overall efficacy over two influenza seasons of 92%. In addition, 21% fewer febrile illnesses and a significant decrease in acute otitis media requiring antibiotics was observed (Belshe and Gruber 2000; Belshe et al. 1998). Randomized, placebo-controlled trials have demonstrated 85–89% efficacy against culture-confirmed influenza among children aged 6–35 months attending child care centers during consecutive influenza seasons (Vesikari et al. 2006a), and 64–70% efficacy against culture-confirmed influenza among children 12–36 months old living in Asia during consecutive influenza seasons (Tam et al. 2007). In one community-based, nonrandomized open-label study, reductions in medically attended acute respiratory infection (MAARI) were observed among children who received one dose of LAIV during the 1990–2000 and 2000–2001 influenza seasons, even though heterotypic variant influenza A/H1N1 and B were circulating during that season (Gaglani et al. 2004). Protection against heterotypic H3N2 strain was demonstrated in an open-label, nonrandomized, community-based influenza vaccine trial conducted among children 5–18 years old. In this study, significant protection against laboratory-confirmed influenza (37%) and pneumonia and influenza events (50%) was shown among children who received LAIV but not inactivated influenza vaccine. Interestingly, LAIV recipients had similar protection against influenza-positive illness within 14 days of vaccine receipt compared with >14 days after vaccination, a finding attributed by the investigators to protection provided by innate immune system mechanisms stimulated by LAIV but not inactivated influenza vaccine (Piedra et al. 2007b). Additional studies comparing the effectiveness of LAIV to that of inactivated influenza vaccine are discussed below (see "Comparisons of LAIV and Inactivated Influenza Vaccine Efficacy or Effectiveness").

### 4.6.2 Healthy Adults

A randomized, double-blind, placebo-controlled trial of LAIV effectiveness among 4,561 healthy working adults aged 18–64 years, which assessed multiple clinical endpoints (without laboratory confirmation of influenza virus infection), was conducted during the 1997–1998 influenza season, when the vaccine and circulating influenza A (H3N2) strains were not well matched. Multiple clinical endpoints were assessed, including self-reported respiratory tract illness without laboratory confirmation, work loss, healthcare visits, and medication use during influenza outbreak periods (Nichol et al. 1999). The frequency of febrile illnesses was not significantly decreased among LAIV recipients compared with those who received placebo. However, significant reductions in severe febrile illnesses (19% effectiveness) and in febrile upper respiratory tract illnesses (24% effectiveness) were observed, as well as significant reductions in days of illness, days of work lost, days with health-care-provider visits, and use of prescription antibiotics and over-the-counter medications (Nichol et al. 1999). The relatively low effectiveness estimates are likely a reflection of the use of nonspecific clinical endpoints that also include illnesses caused by other pathogens against which the vaccine would not be expected to protect. Efficacy against culture-confirmed influenza in a randomized, placebo-controlled study was 57% during the 2004-05 influenza season, and 43% during the 2005-06 influenza season, although efficacy in these studies was not demonstrated to be significantly greater than placebo (Ohmit et al. 2006; Ohmit J Infect Dis 2008).

## 4.7   Adverse Events After Receipt of LAIV

### 4.7.1   Children

LAIV is typically well tolerated by both children and adults; severe adverse events are rare, and more common adverse events are self-limited and of short duration. Runny nose or nasal congestion is fairly common (51–58%) and occurs significantly more often among LAIV recipients than among those who receive placebo in most studies. Runny nose or nasal congestion might be due to intranasal administration or viral replication in the nasopharynx. Other side effects reported include: headache (3–9%), fever >101°F (4%), decreased appetite (13–21%), irritability (12–21%), sore throat (5–11%) and myalgias (2–6%) (Belshe et al. 1998, 2007; MedImmune 2008; Neuzil et al. 2001; Vesikari et al. 2006a; Zangwill et al. 2001). These symptoms were associated more often with the first dose. In some studies, the rates of these adverse events were not significantly higher than the rates among placebo recipients.

In some studies, episodes of wheezing have been reported after LAIV administration among young children, including among children with no previous wheezing history. In a trial comparing LAIV with inactivated influenza vaccine, wheezing that required bronchodilator therapy or that was associated with significant respiratory symptoms occurred in 5.9% of FluMist recipients aged 6–23 months, compared with 3.8% of those who received inactivated influenza vaccine (risk ratio [RR] = 1.5, CI = 1.2–2.1). Wheezing was not greater among children aged 24–59 months who received FluMist (Belshe et al. 2007). Children with medically diagnosed or treated wheezing within 42 days before enrollment or a history of severe asthma were excluded from this study. In a randomized, placebo-controlled safety trial among children aged from 12 months to 17 years, an elevated risk for asthma events (RR = 4.06, CI = 1.29–17.86) was noted among 728 children aged 18–35 months who received FluMist; of the 16 children with asthma-related events, none required hospitalization, and elevated risks for asthma were not observed in other age groups (Bergen et al. 2004). In a multiyear open-label study, no increase in asthma visits within 15 days after vaccination compared with the prevaccination period was reported for children aged 18 months to four years; however, a significant increase in asthma events was reported 15–42 days after vaccination, but only in vaccine year 1 (Piedra et al. 2005). In all of these studies, wheezing requiring medical care was rare, and typically responded to administration of a metered dose inhaler. However, the ACIP advises healthcare practitioners administering LAIV to screen for possible reactive airways diseases when considering the use of FluMist for children aged 2–4 years, and to avoid the use of this vaccine in children with asthma or a recent wheezing episode (Fiore et al. 2008).

### 4.7.2   Adults

Among adults, runny nose or nasal congestion, headache, sore throat, tiredness/weakness, muscle aches, cough or chills have been reported more often among vaccinated adult recipients compared to adults who received placebo. However, rates among placebo recipients were nearly as high as those among LAIV recipients (MedImmune 2008). For example, among a subset of healthy adults aged 18–49 years, signs and symptoms reported more frequently among LAIV recipients than placebo recipients within 7 days after each dose included cough (14% and 11%, respectively), runny nose (45% and 27%, respectively), sore throat (28% and 17%, respectively), chills (9% and 6%, respectively), and tiredness/weakness (26% and 22%, respectively) (Belshe et al. 2004).

## 4.8 LAIV Use in Persons at Higher Risk from Influenza-Related Complications

LAIV is currently licensed for use only among nonpregnant persons aged 2–49 years, and the vaccine is not licensed for use in persons with chronic medical conditions that confer a higher risk of influenza-related complications. However, studies of LAIV use among persons at higher risk for influenza-related complications are available. Children aged 6–71 months with a history of recurrent respiratory infections and children aged 6–17 years with asthma have received LAIV in controlled studies and have not demonstrated differences in postvaccination wheezing or asthma exacerbations, respectively, compared to children who have received inactivated influenza vaccine (Ashkenazi et al. 2006; Fleming et al. 2006). No serious adverse events were reported among 54 HIV-infected persons aged 18–58 years with CD4 counts >200 cells/mm$^3$ (King et al. 2000), or among HIV-infected children aged 1–8 years on effective antiretroviral therapy who were administered LAIV, compared with HIV-uninfected persons receiving LAIV (King et al. 2001). LAIV was also well tolerated among adults aged greater than or equal to ≥65 years with chronic medical conditions (Jackson et al. 1999). These findings suggest that persons at risk for influenza complications who have inadvertent exposure to LAIV would not have significant adverse events, and those persons who have contact with persons at higher risk for influenza-related complications may receive LAIV.

## 4.9 Serious Adverse Events

Serious adverse events requiring medical attention among healthy children aged 5–17 years or healthy adults aged 18–49 years occur at a rate of <1% (MedImmune 2008). Reviews of reports to the Vaccine Adverse Events Reporting System (VAERS) after vaccination of approximately 2.5 million persons during the 2003–2004 and 2004–2005 influenza seasons did not indicate any new safety concerns (Izurieta et al. 2005). LAIV is not licensed for children younger than two years old, but serious adverse events might be more common in this age group. In one study, hospitalization for any cause within 180 days of vaccination was significantly more common among LAIV (6.1%) recipients aged 6–11 months compared with inactivated influenza vaccine recipients (2.6%) (Belshe et al. 2007). Healthcare professionals should report all clinically significant adverse events promptly to VAERS after LAIV administration.

## 4.10   Vaccination of Close Contacts of Immunocompromised Persons

LAIV can be used for vaccinating persons, including healthcare personnel, who have close contact with persons with lesser degrees of immunosuppression (e.g., persons with diabetes, persons with asthma who take corticosteroids, those who might have been cared for previously in a protective environment but who are no longer in that protective environment, or persons infected with HIV). LAIV transmission from a recently vaccinated person causing clinically important illness in an immunocompromised contact has not been reported. However, the ACIP has indicated that inactivated influenza vaccine is preferred for vaccinating household members, healthcare personnel (HCP), and others who have close contact with severely immunosuppressed persons (e.g., patients with hematopoietic stem cell transplants) during those periods in which the immunosuppressed person requires care in a protective environment (typically defined as a specialized patient-care area with a positive airflow relative to the corridor, high-efficiency particulate air filtration, and frequent air changes), because of the potential risk that a live, attenuated vaccine virus could be transmitted to the severely immunosuppressed person. Healthcare personnel or hospital visitors who receive LAIV should avoid contact with severely immunosuppressed patients for seven days after vaccination, but should not be restricted from contact or providing care for less severely immunosuppressed patients (Fiore et al. 2008).

Low-level introduction of vaccine viruses into the environment can occur when administering LAIV, but the risk for acquiring vaccine viruses from the environment is likely to be very low. The ACIP has recommended that severely immunosuppressed persons should not administer LAIV. Other persons at high risk for influenza complications, including persons with underlying medical conditions pregnant women, persons with asthma, and persons aged ≥50 years, may administer LAIV (Fiore et al. 2008).

## 4.11   Persons Who Should Not Be Vaccinated with LAIV

LAIV is not currently licensed for use in the following groups, and these persons should not be vaccinated with LAIV (Fiore et al. 2008):

- Persons with a history of hypersensitivity, including anaphylaxis, to any of the components of LAIV or to eggs
- Persons aged <2 years or those aged ≥50 years
- Persons with any of the underlying medical conditions that serve as an indication for routine influenza vaccination, including asthma, reactive airways disease, or other chronic disorders of the pulmonary or cardiovascular systems; other underlying medical conditions, including such metabolic diseases as diabetes, renal dysfunction, and hemoglobinopathies; neurologic or neuromuscular disorders; or known or suspected immunodeficiency diseases or immunosuppressed states

- Children or adolescents receiving aspirin or other salicylates (because of the association of Reye's syndrome with wild-type influenza virus infection)
- Persons with a history of GBS within six weeks of an influenza vaccination
- Pregnant women

# 5 Comparisons of LAIV and Inactivated Influenza Vaccine Efficacy or Effectiveness

Data directly comparing the efficacy of LAIV and inactivated influenza vaccine are limited, and have been obtained in a variety of settings and populations using several different clinical endpoints. One randomized, double-blind, challenge study among 92 healthy adults aged 18–41 years assessed the efficacy of LAIV and inactivated influenza vaccine compared to placebo in preventing influenza infection when challenged with wild-type strains that were antigenically similar to vaccine strains (Treanor et al. 1999). Upon challenge with wild-type influenza virus, laboratory-confirmed influenza was identified in 14 of 31 (45%) placebo recipients, four of 32 (13%) inactivated influenza vaccine recipients, and two of 29 (7%) LAIV recipients. The overall efficacy in preventing laboratory-documented influenza from all three influenza strains combined was 85% and 71%, respectively, when challenged 28 days after vaccination by viruses to which study participants were susceptible before vaccination; efficacy at timepoints later than 28 days after vaccination was not determined. The difference in efficacy between the two vaccines was not statistically significant.

Comparative studies have also been conducted using community-acquired influenza virus infection as an endpoint. During 2004–2005, an influenza season when the majority of circulating H3N2 viruses were antigenically drifted from that season's vaccine viruses, a randomized, double-blind, placebo-controlled trial conducted among 1,247 young adults demonstrated the efficacy of LAIV and inactivated influenza vaccine against culture-confirmed influenza to be 57% and 77%, respectively, compared to placebo. The difference in efficacy was not statistically significant and was based largely upon a difference in efficacy against influenza B (Ohmit et al. 2006). A similar study in 2005-06 again showed higher efficacy among persons who received TIV compared to LAIV recipients (Ohmit et al. 2008). However, among children, limited data suggest that LAIV is more effective than inactivated influenza vaccine. A study conducted among children aged 6–71 months during the 2004–2005 influenza season, when vaccine strains were well matched against circulating strains, demonstrated a 55% reduction in cases of culture-confirmed influenza among children who received LAIV compared with those who received inactivated influenza vaccine (Belshe et al. 2007). An open-label, nonrandomized, community-based influenza vaccine trial conducted among children 5–18 years old demonstrated significant protection against laboratory-confirmed influenza (37%) and pneumonia and influenza events (50%) among children who received LAIV but not inactivated influenza vaccine during the 2003–2004 influenza season, when the H3N2 vaccine strain was not well matched to circulating strains (Piedra et al. 2007b). A recent observational study conducted among military personnel aged 17-49 years over three influenza seasons found that persons who received TIV had a significantly lower incidence of healthcare encounters resulting in diagnostic coding for pneumonia and influenza, compared to those who received LAIV. However, among new recruits being vaccinated for the first time, the incidence of healthcare encounters among those received LAIV was similar to those receiving TIV (Wang et al. JAMA 2009.

Although LAIV is not currently licensed for use in persons with risk factors for influenza complications, several studies have compared the efficacy of LAIV to inactivated influenza vaccine in these groups. LAIV provided 32% increased protection in preventing culture-confirmed influenza compared with inactivated influenza vaccine in one study conducted among older children with asthma (Fleming et al. 2006), and 52% increased protection among children aged 6–71 months with recurrent respiratory tract infections (Ashkenazi et al. 2006).

To date, no preferential recommendation for LAIV over inactivated influenza vaccine has been made by the ACIP or other advisory bodies, and the two vaccine types are considered to be equally acceptable options for vaccinating healthy non-pregnant persons aged 2–49 years. Effectiveness and safety data from additional influenza seasons and in different study settings is needed in order to determine whether using LAIV might be advantageous for some persons, particularly young children (Cox and Bridges 2007).

## 6  Effectiveness of Vaccination for Decreasing Transmission to Contacts

Decreasing transmission of influenza from caregivers and household contacts to persons at higher risk of influenza complications would reduce influenza-related illness and deaths. The ACIP currently recommends that all household and close contacts of persons at higher risk for influenza complications, including contacts of persons younger than five years old or older than 49 years old and all healthcare personnel (HCP), receive influenza vaccine annually (Fiore et al. 2008).

Reducing the risk of introducing influenza to patients by increasing vaccine coverage of HCP is an important preventive measure (Smith et al. 2006a). Influenza virus infection and ILI are common among HCP (Elder et al. 1996; Lester et al. 2003; Wilde et al. 1999), and influenza outbreaks have been attributed to low vaccination rates among HCP in hospitals and long-term care facilities (Cunney et al. 2000; Salgado et al. 2004; Sato et al. 2005). Vaccination of HCP has been associated with decreased deaths among nursing home patients in several observational studies (Carman et al. 2000; Potter et al. 1997). In a large randomized controlled trial, significant decreases in mortality, ILI, and medical visits for ILI care were demonstrated among residents in nursing homes in which staff were offered influenza vaccination, compared with nursing homes in which staff were not provided with vaccination (Hayward et al. 2006). A recent review of these and other data concluded that vaccination of HCP in settings in which patients were also vaccinated provided significant reductions in deaths among elderly patients from all causes and deaths from pneumonia (Thomas et al. 2006).

Epidemiologic studies of community outbreaks of influenza have shown that children typically have the highest attack rates of influenza (Glezen and Couch 1978; Monto and Kioumehr 1975), suggesting that reducing transmission through routine universal vaccination of children might reduce transmission to their household contacts and possibly others in the community. This might indirectly reduce the risk of acquiring influenza among persons at higher risk of influenza complications, who typically do not respond as well to vaccination as do healthy children and young adults.

Recent data provide support for the idea that the benefits of vaccinating children might extend to protection of their adult contacts. A randomized, placebo-controlled trial among children with recurrent respiratory tract infections demonstrated that members of families with vaccinated children were significantly less likely to have respiratory tract infections and reported significantly fewer workdays lost compared with families with children who received placebo (Esposito et al. 2003). A single-blinded, randomized controlled study demonstrated that vaccinating healthy preschool-aged children reduced influenza-related morbidity among a subset of their household contacts (Hurwitz et al. 2000a). In Russia, vaccination of school children with LAIV appeared to reduce illness rates in teachers and staff (Rudenko et al. 1993). In a large nonrandomized study conducted among 5,840 children attending 28 schools in four states, households with children attending schools in which school-based LAIV immunization programs had been established reported less ILI and fewer physician visits during the predicted peak week of the influenza season compared with households with children in schools in which no LAIV immunization had been offered. In addition, ILI-related economic and medical consequences (e.g., workdays lost and number of healthcare provider visits) were reduced in households of vaccine recipients. However, a decrease in the overall rate of school absenteeism was not reported in intervention schools (King et al. 2006). These results demonstrate the difficulty of measuring changes in nonspecific outcomes such as respiratory illnesses or absenteeism as an assessment of indirect effects of vaccinating children against influenza (Fukuda and Kieny 2006).

Some studies have also indicated that the benefits of vaccinating children might extend to community members beyond household contacts, including to elderly persons. A modeling analysis predicted that a program of vaccinating as few as 20% of United States school children could reduce the number of influenza illnesses by 49% among those ≤18 years old, and by 43% among adults (Weycker et al. 2005). In a community-based observational study conducted during the 1968 influenza pandemic, a vaccination program targeting school-aged children that achieved vaccine coverage of 86% in one community reduced influenza rates within the community among all age groups compared with another community in which aggressive vaccination was not conducted among school-aged children (Monto et al. 1970). An observational study conducted in Russia demonstrated reductions in ILI among the community-dwelling elderly after implementation of a vaccination program for children aged 3–6 years (57% coverage) and children and adolescents aged 7–17 years (72% coverage) (Ghendon et al. 2006). In a nonrandomized community-based study conducted over three influenza seasons, 8–18% reductions in the incidence of MAARI during the influenza season among adults ≥35 years old were observed in communities in which LAIV was routinely offered to all children 18 months of age or older compared to communities without vaccination programs, even though coverage rates among children were estimated to be only 20–25% (Piedra et al. 2005a). In a subsequent influenza season, the same investigators showed a 9% reduction in MAARI rates during the influenza season among persons 35–44 years in intervention communities, where coverage was estimated at 31% among school children compared to control communities. However, rates among persons ≥45 years were lower in the intervention communities regardless of the presence of influenza in the community, suggesting that factors other than vaccination of school children against influenza may have contributed to the decreased rate of MAARI.

An ecologic analysis of the effects of a universal vaccination program in Japan for school-age children (estimated 80% coverage at peak vaccine use) that began in 1962 and was discontinued in 1994 concluded that excess mortality and mortality attributed to pneumonia and influenza from November to April dropped dramatically during the period when vaccination was recommended, and increased again after the vaccination program ceased. Mortality data from the United States did not show changes over these periods (Reichert et al. 2001b). These results have been criticized due to a failure to adjust for changes in demographics and the overly long definition of the influenza season (Fukuda et al. 2001; Inouye and Kramer 2001; Jordan et al. 2006; Yamazaki et al. 2001). Controlling for changes in the age distribution in the Japanese population over time did not substantially alter results (Reichert et al. 2001a), but the results of this study remain controversial because of an inability to attribute causality to specific agents in this ecologic analysis of all-cause mortality data (Glezen 2006; Jordan et al. 2006). The largest study to examine the community effects of increasing overall vaccine coverage was an eco-logic study that described the experience in Canada, where Ontario was the only province to implement a universal influenza immunization program beginning in 2000. After program introduction, influenza-associated mortality, hospitalizations, emergency department use, and physicians' office visits decreased significantly more in Ontario, based on models developed from administrative and viral surveil-lance data, than in other provinces, with the largest reductions observed in younger age groups (Kwong et al. 2008).

# 7   Expanding Use of Influenza Vaccines

As the substantial annual burden of influenza-related morbidity and mortality has become increasingly well known, interest in expanding recommendations for annual vaccination to broader population groups such as healthy children and increasing the number of countries that have vaccination programs has also grown, particularly in developed countries (Abramson et al. 2006). In addition, the World Health Assembly has stated that "…influenza may be a larger public health problem in poor societies than realized…" and it "strongly encourages the implementation of epidemiological surveillance, disease burden assessments and, where appropriate infrastructure is available, demonstration projects to estimate the impact of vaccination on disease in poor countries." It also has urged Member States with influenza vaccination policies to increase vaccination coverage of all people at high risk, setting a goal for vaccination coverage of elderly people of at least 50% by 2006 and 75% by 2010. Furthermore, WHO has indicated that "exploration of the safety and cost-effectiveness of introducing influenza vaccination into national immunization programmes is clearly warranted" (WHO 2005). Most recently, the Strategic Advisory Group of Experts on immunization, a committee periodically convened by WHO to examine immunization priorities, included control of seasonal influenza as a "high priority," putting influenza into the same priority category as cholera, typhoid fever and yellow fever (WHO 2008).

The ACIP has indicated that recommendations to continue expanding the population groups in the United States who are recommended for vaccination, with an ultimate goal of universal vaccination, will be considered over the next five years. The ACIP recommended annual vaccination for all children aged 6 months through 18 years beginning in 2008, with the expectation that this measure will reduce infections among children and their contacts, and indirectly reduce the risk of influenza virus exposure for persons at a higher risk for influenza complications (see http://www.cdc.gov/mmwr/preview/mmwrhtml/rr5707a1.htm). However, enthusiasm for additional recommendation expansion is tempered by the considerable implementation challenges that universal vaccination would entail (Edwards and Griffin 2006; Fiore et al. 2008; Schwartz et al. 2006). Plans to improve surveillance systems capable of monitoring vaccine effectiveness and safety must be developed and maintained; these will be particularly challenging if, as expected, many healthy older children and adults are vaccinated outside of traditional medical settings. Evaluating the impact of expanding recommendations will require large multiyear, multisite studies to account for the expected variations in influenza epidemiology, strain circulation and vaccine matching (Halloran and Longini 2006).

Additional challenges remain unmet. More effective vaccines are needed. Ideally, better vaccines would stimulate longer-lasting cross-reactive immunity against multiple strains. Most importantly, vaccines must be effective in protecting the very young, the chronically ill, and the elderly, who bear the largest burden of influenza illness. Efforts to develop more effective influenza vaccines and document their safety must be accompanied by a better understanding of how to motivate persons at risk to seek annual influenza vaccination. Relatively low coverage has thus far been achieved among many groups who are already recommended for vaccination, such as persons with chronic medical conditions, HCP, and young children (Fiore et al. 2008). In addition, improvements in vaccine financing, particularly for adults, will be needed (Helms et al. 2005; Orenstein et al. 2007). Although these actions pose considerable challenges, the experience gained could prove to be of even greater societal value than just reducing the considerable morbidity caused by seasonal influenza. For example, immunization programs capable of delivering annual influenza vaccination to a broad range of the population could potentially serve as a resilient and sustainable platform for delivering vaccines and monitoring outcomes for other urgently required public health interventions (Fiore et al. 2008; Mair et al. 2006).

# References

Abramson JS, Neuzil KM, Tamblyn SE (2006) Annual universal influenza vaccination: ready or not? Clin Infect Dis 42:132–135

Ahmed AH et al. (1997) Effectiveness of influenza vaccine in reducing hospital admissions during the 1989–90 epidemic. Epidemiol Infect 118:27–33

Alexandrova GI et al. (1986) Study of live recombinant cold-adapted influenza bivalent vaccine of type A for use in children: an epidemiological control trial. Vaccine 4:114–118

Ali T et al. (2004) Detection of influenza antigen with rapid antibody-based tests after intranasal influenza vaccination (Flumist). Clin Infect Dis 38:760–762

Allison MA et al. (2006) Influenza vaccine effectiveness in healthy 6- to 21-month-old children during the 2003–2004 season. J Pediatr 149:755–762

Anon (1944) A clinical evaluation of vaccination against influenza. JAMA 124:982–985

Arden NH, Patriarca PA, Kendal AP (1986) Experiences in the use and efficacy of inactivated influenza vaccine in nursing homes. In: Kendal AP, Patriarca PA (eds) Options for the control of influenza. Alan R. Liss, New York, pp 155–168

Ashkenazi S et al. (2006) Superior relative efficacy of live attenuated influenza vaccine compared with inactivated influenza vaccine in young children with recurrent respiratory tract infections. Pediatr Infect Dis J 25:870–879

Audicana MT et al. (2002) Allergic contact dermatitis from mercury antiseptics and derivatives: study protocol of tolerance to intramuscular injections of thimerosal. Am J Contact Dermat 13:3–9

Babcock HM, Merz LR, Fraser VJ (2006) Is influenza an influenza-like illness? Clinical presentation of influenza in hospitalized patients. Infect Control Hosp Epidemiol 27:266–270

Baez M, Palese P, Kilbourne ED (1980) Gene composition of high-yielding influenza vaccine strains obtained by recombination. J Infect Dis 141:362–365

Barker WH (1986) Excess pneumonia and influenza associated hospitalization during influenza epidemics in the United States, 1970–78. Am J Public Health 76:761–765

Barry DW et al. (1976) Comparative trial of influenza vaccines. II. Adverse reactions in children and adults. Am J Epidemiol 104:47–59

Belshe RB, Gruber WC (2000) Prevention of otitis media in children with live attenuated influenza vaccine given intranasally. Pediatr Infect Dis J 19:S66–S71

Belshe RB, Mendelman PM (2003) Safety and efficacy of live attenuated, cold-adapted, influenza vaccine-trivalent. Immunol Allergy Clin North Am 23:745–767

Belshe RB et al. (1998) The efficacy of live attenuated, cold-adapted, trivalent, intranasal influenzavirus vaccine in children. N Engl J Med 338:1405–1412

Belshe RB et al. (2000a) Efficacy of vaccination with live attenuated, cold-adapted, trivalent, intranasal influenza virus vaccine against a variant (A/Sydney) not contained in the vaccine. J Pediatr 136:168–175

Belshe RB et al. (2000b) Correlates of immune protection induced by live, attenuated, cold-adapted, trivalent, intranasal influenza virus vaccine. J Infect Dis 181:1133–1137

Belshe RB et al. (2004) Safety, efficacy, and effectiveness of live, attenuated, cold-adapted influenza vaccine in an indicated population aged 5–49 years. Clin Infect Dis 39:920–927

Belshe RB et al. (2007) Live attenuated versus inactivated influenza vaccine in infants and young children. N Engl J Med 356:685–696

Bergen R et al. (2004) Safety of cold-adapted live attenuated influenza vaccine in a large cohort of children and adolescents. Pediatr Infect Dis J 23:138–144

Beyer WE et al. (1998) The plea against annual influenza vaccination? "The Hoskins' Paradox" revisited. Vaccine 16:1929–1932

Bhat N et al. (2005) Influenza-associated deaths among children in the United States, 2003–2004. N Engl J Med 353:2559–2567

Block SL et al. (2008) Shedding and immunogenicity of live attenuated influenza vaccine virus in subjects 5-49 years of age. Vaccine 26:4940–6

Bridges CB et al. (2000) Effectiveness and cost-benefit of influenza vaccination of healthy working adults: a randomized controlled trial. JAMA 284:1655–1663

Bright RA et al. (2006) Adamantane resistance among influenza A viruses isolated early during the 2005–2006 influenza season in the United States. JAMA 295:891–894

Brokstad KA et al. (1995) Parenteral influenza vaccination induces a rapid systemic and local immune response. J Infect Dis 171:198–203

Brooks WA et al. (2007) Human metapneumovirus infection among children, Bangladesh. Emerg Infect Dis 13:1611–1613

Buonagurio DA et al. (2006) Genetic and phenotypic stability of cold-adapted influenza viruses in a trivalent vaccine administered to children in a day care setting. Virology 347:296–306

Carrat F, et al. (2008) Time lines of infection and disease in human influenza: a review of volunteer challenge studies. Am J Epidemiol 167:775–85

Carman WF et al. (2000) Effects of influenza vaccination of health-care workers on mortality of elderly people in long-term care: a randomised controlled trial. Lancet 355:93–97

Cate TR et al. (1977) Clinical trials of monovalent influenza A/New Jersey/76 virus vaccines in adults: reactogenicity, antibody response, and antibody persistence. J Infect Dis 136(Suppl):S450–S455

Cate TR et al. (1983) Reactogenicity, immunogenicity, and antibody persistence in adults given inactivated influenza virus vaccines—1978. Rev Infect Dis 5:737–747

CDC (1992) Outbreak of influenza A in a nursing home—New York, December 1991–January 1992. MMWR 41:129–131

CDC (2007) Update: influenza activity—United States and Worldwide, May 20–September 17, 2007. MMWR 56:1001–1004

Cha TA et al. (2000) Genotypic stability of cold-adapted influenza virus vaccine in an efficacy clinical trial. J Clin Microbiol 38:839–845

Chen Z et al. (2006) Genetic mapping of the cold-adapted phenotype of B/Ann Arbor/1/66, the master donor virus for live attenuated influenza vaccines (flumist). Virology 345:416–423

Choi K, Thacker SB (1982) Mortality during influenza epidemics in the United States, 1967–1978. Am J Public Health 72:1280–1283

Clover RD et al. (1991) Comparison of heterotypic protection against influenza A/Taiwan/86 (H1N1) by attenuated and inactivated vaccines to A/Chile/83-like viruses. J Infect Dis 163:300–304

Coles FB, Balzano GJ, Morse DL (1992) An outbreak of influenza A (H3N2) in a well immunized nursing home population. J Am Geriatr Soc 40:589–592

Colombo C et al. (2001) Influenza vaccine in healthy preschool children. Rev Epidemiol Sante Publique 49:157–162

Connolly AM et al. (1993) What are the complications of influenza and can they be prevented? Experience from the 1989 epidemic of H3N2 influenza A in general practice. BMJ 306:1452–1454

Corey L et al. (1976) A nationwide outbreak of Reye's syndrome. Its epidemiologic relationship to influenza B. Am J Med 61:615–625

Cox NJ, Bridges CB (2007) Inactivated and live attenuated influenza vaccines in young children—how do they compare? N Engl J Med 356:729–731

Cox NJ et al. (1985) Comparative studies of A/Leningrad/134/57 wild-type and 47-times passaged cold-adapted mutant influenza viruses: oligonucleotide mapping and RNA-RNA hybridization studies. J Gen Virol 66(Pt 8):1697–1704

Creighton C (1891) A History of epidemics in Britain, AD 664–1666. Cambridge University Press, New York

Cunney RJ et al. (2000) An outbreak of influenza A in a neonatal intensive care unit. Infect Control Hosp Epidemiol 21:449–454

Davenport FM (1962) Current knowledge of influenza vaccine. JAMA 182:121–123

Davies JR, Grilli EA (1989) Natural or vaccine-induced antibody as a predictor of immunity in the face of natural challenge with influenza viruses. Epidemiol Infect 102:325–333

Deguchi Y, Takasugi Y, Tatara K (2000) Efficacy of influenza vaccine in the elderly in welfare nursing homes: reduction in risks of mortality and morbidity during an influenza A (H3N2) epidemic. J Med Microbiol 49:553–556

Demicheli V et al. (2000) Prevention and early treatment of influenza in healthy adults. Vaccine 18:957–1030

Demicheli V et al. (2004) Vaccines for preventing influenza in healthy adults. Cochrane Database Syst Rev CD001269

Dharan NJ et al. (2009) Infections with oseltamivir-resistant influenza A(H1N1) viruses in the United States. JAMA 301:1034–41

Dietzman DE et al. (1976) Acute myositis associated with influenza B infection. Pediatrics 57:255–258

Doherty PC et al. (2006) Influenza and the challenge for immunology. Nat Immunol 7:449–455

Donabedian AM et al. (1988) A mutation in the PA protein gene of cold-adapted B/Ann Arbor/1/66 influenza virus associated with reversion of temperature sensitivity and attenuated virulence. Virology 163:444–451

Dowdle WR (1981) Influenza immunoprophylaxis after 30 years' experience. In: Genetic variation among influenza viruses. Academic, New York, pp 525–534

Dowdle WR et al. (1973) Inactivated influenza vaccines. 2. Laboratory indices of protection. Postgrad Med J 49:159–163

Edwards KM, Griffin MR (2006) Influenza vaccination of children: can it be accomplished? J Infect Dis 194:1027–1029

Edwards KM et al. (1994) A randomized controlled trial of cold-adapted and inactivated vaccines for the prevention of influenza A disease. J Infect Dis 169:68–76

el-Madhun AS et al. (1998) Systemic and mucosal immune responses in young children and adults after parenteral influenza vaccination. J Infect Dis 178:933–939

Elder AG et al. (1996) Incidence and recall of influenza in a cohort of Glasgow healthcare workers during the 1993–94 epidemic: results of serum testing and questionnaire. BMJ 313:1241–1242

Englund JA et al. (2005) A comparison of 2 influenza vaccine schedules in 6- to 23-month-old children. Pediatrics 115:1039–1047

Ennis FA et al. (1977) Correlation of laboratory studies with clinical responses to A/New Jersey influenza vaccines. J Infect Dis 136(Suppl):S397–S406

Esposito S et al. (2003) Effectiveness of influenza vaccination of children with recurrent respiratory tract infections in reducing respiratory-related morbidity within the households. Vaccine 21:3162–3168

Fedson DS et al. (1993) Clinical effectiveness of influenza vaccination in Manitoba. JAMA 270:1956–1961

Fine AD et al. (2001) Influenza A among patients with human immunodeficiency virus: an outbreak of infection at a residential facility in New York City. Clin Infect Dis 32:1784–1791

Fiore AE et al. (2008) Prevention and control of influenza. Recommendations of the Advisory Committee on Immunization Practices (ACIP), 2008. MMWR Recomm Rep 57:1–60

Fleming DM et al. (1995) Study of the effectiveness of influenza vaccination in the elderly in the epidemic of 1989–90 using a general practice database. Epidemiol Infect 115:581–589

Fleming DM et al. (2006) Comparison of the efficacy and safety of live attenuated cold-adapted influenza vaccine, trivalent, with trivalent inactivated influenza virus vaccine in children and adolescents with asthma. Pediatr Infect Dis J 25:860–869

Foster DA et al. (1992) Influenza vaccine effectiveness in preventing hospitalization for pneumonia in the elderly. Am J Epidemiol 136:296–307

Fox JP et al. (1982a) Influenzavirus infections in Seattle families, 1975–1979. I. Study design, methods and the occurrence of infections by time and age. Am J Epidemiol 116:212–227

Fox JP et al. (1982b) Influenzavirus infections in Seattle families, 1975–1979. II. Pattern of infection in invaded households and relation of age and prior antibody to occurrence of infection and related illness. Am J Epidemiol 116:228–242

Francis T (1945) The development of the 1943 vaccination study of the Commission on Influenza. Am J Hyg 42:1–11

Francis T, Salk JE, Brace WM (1946) The protective effect of vaccination against epidemic influenza B. JAMA 131:275–278

Freeman DW, Barno A (1959) Deaths from Asian influenza associated with pregnancy. Am J Obstet Gynecol 78:1172–1175

Frey S et al. (2003) Comparison of the safety, tolerability, and immunogenicity of a MF59-adjuvanted influenza vaccine and a non-adjuvanted influenza vaccine in non-elderly adults. Vaccine 21:4234–4237

Fukuda K, Kieny MP (2006) Different approaches to influenza vaccination. N Engl J Med 355:2586–2587

Fukuda K, Thompson WW, Cox NJ (2001) To the editor: (vaccinating Japanese schoolchildren against influenza). N Engl J Med 344:1946–1947

Gaglani MJ et al. (2004) Direct and total effectiveness of the intranasal, live-attenuated, trivalent cold-adapted influenza virus vaccine against the 2000–2001 influenza A(H1N1) and B epidemic in healthy children. Arch Pediatr Adolesc Med 158:65–73

Gerhard W (2001) The role of the antibody response in influenza virus infection. Curr Top Microbiol Immunol 260:171–190

Ghendon YZ et al. (1984) Recombinant cold-adapted attenuated influenza A vaccines for use in children: molecular genetic analysis of the cold-adapted donor and recombinants. Infect Immun 44:730–733

Ghendon YZ, Kaira AN, Elshina GA (2006) The effect of mass influenza immunization in children on the morbidity of the unvaccinated elderly. Epidemiol Infect 134:71–78

Glaser CA et al. (2002) Medical care capacity for influenza outbreaks, Los Angeles. Emerg Infect Dis 8:569–574

Glezen WP (2006) Herd protection against influenza. J Clin Virol 37:237–243

Glezen WP, Couch RB (1978) Interpandemic influenza in the Houston area, 1974–76. N Engl J Med 298:587–592

Govaert TM et al. (1994) The efficacy of influenza vaccination in elderly individuals. A randomized double-blind placebo-controlled trial. JAMA 272:1661–1665

Green DM (2002) Advances in the management of guillain-barre syndrome. Curr Neurol Neurosci Rep 2:541–548

Gregg MB (1980) The epidemiology of influenza in humans. Ann N Y Acad Sci 353:45–53

Gross PA et al. (1995) The efficacy of influenza vaccine in elderly persons. A meta-analysis and review of the literature. Ann Intern Med 123:518–527

Gross PA et al. (1997) Time to earliest peak serum antibody response to influenza vaccine in the elderly. Clin Diagn Lab Immunol 4:491–492

Gruber WC et al. (1996) Evaluation of live attenuated influenza vaccines in children 6–18 months of age: safety, immunogenicity, and efficacy. National Institute of Allergy and Infectious Diseases, Vaccine and Treatment Evaluation Program and the Wyeth-Ayerst ca Influenza Vaccine Investigators Group. J Infect Dis 173:1313–1319

Gubareva LV et al. (1998) Evidence for zanamivir resistance in an immunocompromised child infected with influenza B virus. J Infect Dis 178:1257–1262

Gubareva LV et al. (2001) Selection of influenza virus mutants in experimentally infected volunteers treated with oseltamivir. J Infect Dis 183:523–531

Hak E et al. (2002a) Influence of high-risk medical conditions on the effectiveness of influenza vaccination among elderly members of 3 large managed-care organizations. Clin Infect Dis 35:370–377

Hak E et al. (2002b) Confounding by indication in non-experimental evaluation of vaccine effectiveness: the example of prevention of influenza complications. J Epidemiol Commun Health 56:951–955

Hak E et al. (2005) Clinical effectiveness of influenza vaccination in persons younger than 65 years with high-risk medical conditions: the PRISMA study. Arch Intern Med 165:274–280

Hall CE, Cooney MK, Fox JP (1973) The Seattle virus watch. IV. Comparative epidemiologic observations of infections with influenza A and B viruses, 1965–1969, in families with young children. Am J Epidemiol 98:365–380

Halloran ME, Longini IM, Jr. (2006) Public health. Community studies for vaccinating schoolchildren against influenza. Science 311:615–616

Halperin SA et al. (2002) Safety and immunogenicity of a trivalent, inactivated, mammalian cell culture-derived influenza vaccine in healthy adults, seniors, and children. Vaccine 20:1240–1247

Harper SA et al. (2009) Seasonal influenza in adults and children--diagnosis, treatment, chemoprophylaxis, and institutional outbreak management: clinical guidelines of the Infectious Diseases Society of America 48:1003–32

Harris JW (1919) Influenza occurring in pregnant women: a statistical study of thirteen hundred and fifty cases. JAMA 72:978–980

Hauge SH et al. (2009) Oseltamivir-resistant influenza viruses A (H1N1), Norway, 2007-08. Emerg Infect Dis 15:155–62

Hayward AC et al. (2006) Effectiveness of an influenza vaccine programme for care home staff to prevent death, morbidity, and health service use among residents: cluster randomised controlled trial. BMJ 333:1241

Heikkinen T, Thint M, Chonmaitree T (1999) Prevalence of various respiratory viruses in the middle ear during acute otitis media. N Engl J Med 340:260–264

Heinonen OP et al. (1973) Immunization during pregnancy against poliomyelitis and influenza in relation to childhood malignancy. Int J Epidemiol 2:229–235

Helms CM et al. (2005) Strengthening the nation's influenza vaccination system: a National Vaccine Advisory Committee assessment. Am J Prev Med 29:221–226

Herrera GA et al. (2007) Influenza vaccine effectiveness among 50–64-year-old persons during a season of poor antigenic match between vaccine and circulating influenza virus strains: Colorado, United States, 2003–2004. Vaccine 25:154–160

Hoberman A et al. (2003) Effectiveness of inactivated influenza vaccine in preventing acute otitis media in young children: a randomized controlled trial. JAMA 290:1608–1616

Hobson D et al. (1972) The role of serum haemagglutination-inhibiting antibody in protection against challenge infection with influenza A2 and B viruses. J Hyg (Lond.) 70:767–777

Hoelscher MA et al. (2006) Development of adenoviral-vector-based pandemic influenza vaccine against antigenically distinct human H5N1 strains in mice. Lancet 367:475–481

Hoskins TW et al. (1973) Controlled trial of inactivated influenza vaccine containing the a-Hong Kong strain during an outbreak of influenza due to the a-England-42–72 strain. Lancet 2:116–120

Hoskins TW et al. (1979) Assessment of inactivated influenza-A vaccine after three outbreaks of influenza A at Christ's Hospital. Lancet 1:33–35

Hurwitz ES et al. (1981) Guillain–Barre syndrome and the 1978–1979 influenza vaccine. N Engl J Med 304:1557–1561

Hurwitz ES et al. (2000a) Effectiveness of influenza vaccination of day care children in reducing influenza-related morbidity among household contacts. JAMA 284:1677–1682

Hurwitz ES et al. (2000b) Studies of the 1996–1997 inactivated influenza vaccine among children attending day care: immunologic response, protection against infection, and clinical effectiveness. J Infect Dis 182:1218–1221

Inouye S, Kramer MH (2001) To the editor: (vaccinating Japanese schoolchildren against influenza). N Engl J Med 344:1946

Irving WL et al. (2000) Influenza virus infection in the second and third trimesters of pregnancy: a clinical and seroepidemiological study. BJOG 107:1282–1289

Izurieta HS et al. (2005) Adverse events reported following live, cold-adapted, intranasal influenza vaccine. JAMA 294:2720–2725

Jackson LA et al. (1999) Safety of a trivalent live attenuated intranasal influenza vaccine, flumist, administered in addition to parenteral trivalent inactivated influenza vaccine to seniors with chronic medical conditions. Vaccine 17:1905–1909

Jackson LA et al. (2006) Evidence of bias in estimates of influenza vaccine effectiveness in seniors. Int J Epidemiol 35:337–344

Jefferson TO et al. (2007) Vaccines for preventing influenza in healthy adults. Cochrane Database Syst Rev CD001269

Johnson PR, Jr. et al. (1985) Comparison of long-term systemic and secretory antibody responses in children given live, attenuated, or inactivated influenza A vaccine. J Med Virol 17:325–335

Johnson PR et al. (1986) Immunity to influenza A virus infection in young children: a comparison of natural infection, live cold-adapted vaccine, and inactivated vaccine. J Infect Dis 154:121–127

Jordan R et al. (2006) Universal vaccination of children against influenza: are there indirect benefits to the community? A systematic review of the evidence. Vaccine 24:1047–1062

Juurlink DN et al. (2006) Guillain–Barre syndrome after influenza vaccination in adults: a population-based study. Arch Intern Med 166:2217–2221

Kaplan JE et al. (1982) Guillain–Barre syndrome in the United States, 1979–1980 and 1980–1981. Lack of an association with influenza vaccination. JAMA 248:698–700

Keitel WA et al. (1997) Efficacy of repeated annual immunization with inactivated influenza virus vaccines over a five year period. Vaccine 15:1114–1122

Keitel WA et al. (2006) Safety of high doses of influenza vaccine and effect on antibody responses in elderly persons. Arch Intern Med 166:1121–1127

Keren R et al. (2005) Neurological and neuromuscular disease as a risk factor for respiratory failure in children hospitalized with influenza infection. JAMA 294:2188–2194

Kilbourne ED, Murphy JS (1960) Genetic studies of influenza viruses. I. Viral morphology and growth capacity as exchangeable genetic traits. Rapid in-ovo adaptation of early passage Asian strain isolates by combination with PRS. J Exp Med 111:387–406

Kilbourne ED et al. (1968) Antiviral activity of antiserum specific for an influenza virus neuraminidase. J Virol 2:281–288

Kim HW et al. (1979) Influenza A and B virus infection in infants and young children during the years 1957–1976. Am J Epidemiol 109:464–479

King JC et al. (2000) Comparison of the safety, vaccine virus shedding, and immunogenicity of influenza virus vaccine, trivalent, types A and B, live cold-adapted, administered to human immunodeficiency virus (HIV)-infected and non-HIV-infected adults. J Infect Dis 181:725–728

King JC, Jr. et al. (2001) Safety, vaccine virus shedding and immunogenicity of trivalent, cold-adapted, live attenuated influenza vaccine administered to human immunodeficiency virus-infected and noninfected children. Pediatr Infect Dis J 20:1124–1131

King JC, Jr. et al. (2006) Effectiveness of school-based influenza vaccination. N Engl J Med 355:2523–2532

Kiso M et al. (2004) Resistant influenza A viruses in children treated with oseltamivir: descriptive study. Lancet 364:759–765

Kistner O et al. (1999) A novel mammalian cell (Vero) derived influenza virus vaccine: development, characterization and industrial scale production. Wien Klin Wochenschr 111:207–214

Kramarz P et al. (2009) Use of oseltamivir in 12 European countires between 2002 and 2007–lack of association with the appearance of oseltamivir-resistant influenza A(H1N1) viruses

Kwong JC et al. (2006) The effect of universal influenza immunization on vaccination rates in Ontario. Health Rep 17:31–40

Kwong JC, Rosella LC, Johansen H (2007) Trends in influenza vaccination in Canada, 1996/1997 to 2005. Health Rep 18:9–19

Kwong JC et al. (2008) The effect of universal influenza immunization on mortality and health care use. PLoS Med 5:e211

LaMontagne Jr et al. (1983) Summary of clinical trials of inactivated influenza vaccine – 1978. Rev Infect Dis 5:723–736

Lasky T et al. (1998) The Guillain Barre syndrome and the 1992–1993 and 1993–1994 influenza vaccines. N Engl J Med 339:1797–1802

Lester RT et al. (2003) Use of, effectiveness of, and attitudes regarding influenza vaccine among house staff. Infect Control Hosp Epidemiol 24:839–844

LoGrippo GA (1960) Investigations of the use of beta propiolactone in virus inactivation. Ann NY Acad Sci 83:578–594

Lui KJ, Kendal AP (1987) Impact of influenza epidemics on mortality in the United States from October 1972 to May 1985. Am J Public Health 77:712–716

Maassab HF (1967) Adaptation and growth characteristics of influenza virus at 25°C. Nature 213:612–614

Maassab HF (1969) Biologic and immunologic characteristics of cold-adapted influenza virus. J Immunol 102:728–732

Maassab HF et al. (1982) Evaluation of a cold-recombinant influenza virus vaccine in ferrets. J Infect Dis 146:780–790

Mair M et al. (2006) Universal influenza vaccination: the time to act is now. Biosecur Bioterror 4:20–40

Mak TK et al. (2008) Influenza vaccination in pregnancy: current evidence and selected national policies. Lancet Infect Dis 8:44–52

MedImmune (2008) Flumist (package insert). MedImmune, Gaithersburg (see http://www.med immune.com/pdf/products/flumist_pi.pdf)

Meijer A. et al. (2009) Oseltamivir-resistant influenza virus, Europe, 2007-08 season. Emerg Infect Dis Available from http://www.cdc.gov/EID/content/15/4/552.htm

Meiklejohn G, Hall H (1987) Unusual outbreak of influenza A in a Wyoming nursing home. J Am Geriatr Soc 35:742–746

Monto AS et al. (1970) Modification of an outbreak of influenza in Tecumseh, Michigan by vaccination of schoolchildren. J Infect Dis 122:16–25

Monto AS et al. (2000) Clinical signs and symptoms predicting influenza infection. Arch Intern Med 160:3243–3247

Monto AS, Kioumehr F (1975) The Tecumseh study of respiratory illness. IX. Occurrence of influenza in the community, 1966–1971. Am J Epidemiol 102:553–563

Monto AS, Hornbuckle K, Ohmit SE (2001) Influenza vaccine effectiveness among elderly nursing home residents: a cohort study. Am J Epidemiol 154:155–160

Morens DM, Rash VM (1995) Lessons from a nursing home outbreak of influenza A. Infect Control Hosp Epidemiol 16:275–280

Moritz A et al. (1980) Studies with a cold-recombinant A/Victoria/3/75 (H3N2) virus. II. Evaluation in adult volunteers. J Infect Dis 142:857–860

Morris JA et al. (1966) Immunity to influenza to antibody levels. N Engl J Med 274:527–535

Mullooly JP et al. (1994) Influenza vaccination programs for elderly persons: cost-effectiveness in a health maintenance organization. Ann Intern Med 121:947–952

Neuzil KM et al. (1998) Impact of influenza on acute cardiopulmonary hospitalizations in pregnant women. Am J Epidemiol 148:1094–1102

Neuzil KM et al. (2001) Efficacy of inactivated and cold-adapted vaccines against influenza A infection, 1985 to 1990: the pediatric experience. Pediatr Infect Dis J 20:733–740

Neuzil KM, Hohlbein C, Zhu Y (2002) Illness among schoolchildren during influenza season: effect on school absenteeism, parental absenteeism from work, and secondary illness in families. Arch Pediatr Adolesc Med 156:986–991

Neuzil KM et al. (2003) Recognizing influenza in older patients with chronic obstructive pulmonary disease who have received influenza vaccine. Clin Infect Dis 36:169–174

Neuzil KM, Jackson LA, Nelson J (2006) Immunogenicity and reactogenicity of one versus two doses of trivalent inactivated influenza vaccine in vaccine-naive 5–8-year-old children. J Infect Dis

Nichol KL et al. (1995) The effectiveness of vaccination against influenza in healthy, working adults. N Engl J Med 333:889–893

Nichol KL, Wuorenma J, von Sternberg T (1998) Benefits of influenza vaccination for low-, intermediate-, and high-risk senior citizens. Arch Intern Med 158:1769–1776

Nichol KL et al. (1999) Effectiveness of live, attenuated intranasal influenza virus vaccine in healthy, working adults: a randomized controlled trial. JAMA 282:137–144

Nichol KL et al. (2007) Effectiveness of influenza vaccine in the community-dwelling elderly. N Engl J Med 357:1373–1381

Ohmit SE, Monto AS (1995) Influenza vaccine effectiveness in preventing hospitalization among the elderly during influenza type A and type B seasons. Int J Epidemiol 24:1240–1248

Ohmit SE, Arden NH, Monto AS (1999) Effectiveness of inactivated influenza vaccine among nursing home residents during an influenza type A (H3N2) epidemic. J Am Geriatr Soc 47:165–171

Ohmit SE et al. (2006) Prevention of antigenically drifted influenza by inactivated and live attenuated vaccines. N Engl J Med 355:2513–2522

Ohmit SE et al. (2008) Prevention of symptomatic seasonal influenza in 2005-06 by inactivated and live attenuated vaccines. J Infect Dis 198:312–7

Orenstein WA, Bernier RH, Hinman AR (1988) Assessing vaccine efficacy in the field. Further observations. Epidemiol Rev 10:212–241

Orenstein WA et al. (2007) Financing immunization of adults in the United States. Clin Pharmacol Ther 82:764–768

Oshitani H et al. (2000) Influenza vaccination levels and influenza-like illness in long-term-care facilities for elderly people in Niigata, Japan, during an influenza A (H3N2) epidemic. Infect Control Hosp Epidemiol 21:728–730

Oya A, Nerome K (1986) Experiences with mass vaccination of young age groups with inactivated vaccines. In: Kendal AP, Patriarca PA (eds) Options for the control of influenza. Alan R. Liss, New York, pp. 183–192

Palache AM, Brands R, van Scharrenburg GJ (1997) Immunogenicity and reactogenicity of influenza subunit vaccines produced in MDCK cells or fertilized chicken eggs. J Infect Dis 176(Suppl 1):S20–S23

Palache AM et al. (1999) Safety, reactogenicity and immunogenicity of Madin Darby canine kidney cell-derived inactivated influenza subunit vaccine. A meta-analysis of clinical studies. Dev Biol Stand 98:115–125

Patriarca PA et al. (1985) Efficacy of influenza vaccine in nursing homes reduction in illness and complications during an Influenza A (H3N2) epidemic. JAMA 253:1136–1139

Percheson PB et al. (1999) A Phase I, randomized controlled clinical trial to study the reactogenicity and immunogenicity of a new split influenza vaccine derived from a non-tumorigenic cell line. Dev Biol Stand 98:127–132

Perrotta DM, Decker M, Glezen WP (1985) Acute respiratory disease hospitalizations as a measure of impact of epidemic influenza. Am J Epidemiol 122:468–476

Piedra PA et al. (2005a) Herd immunity in adults against influenza-related illnesses with use of the trivalent-live attenuated influenza vaccine (CAIV-T) in children. Vaccine 23:1540–1548

Piedra PA et al. (2005b) Live attenuated influenza vaccine, trivalent, is safe in healthy children 18 months to 4 years, 5 to 9 years, and 10 to 18 years of age in a community-based, nonrandomized, open-label trial. Pediatrics 116:e397–e407

Piedra PA et al. (2007) Trivalent live attenuated intranasal influenza vaccine administered during the 2003–2004 influenza type A (H3N2) outbreak provided immediate, direct, and indirect protection in children. Pediatrics 120:e553–e564

Poehling KA et al. (2006) The underrecognized burden of influenza in young children. N Engl J Med 355:31–40

Potter J et al. (1997) Influenza vaccination of health care workers in long-term-care hospitals reduces the mortality of elderly patients. J Infect Dis 175:1–6

Prevots DR, Sutter RW (1997) Assessment of Guillain–Barre syndrome mortality and morbidity in the United States: implications for acute flaccid paralysis surveillance. J Infect Dis 175:S151–S155

Puig-Barbera J et al. (1997) Reduction in hospital admissions for pneumonia in non-institutionalised elderly people as a result of influenza vaccination: a case–control study in Spain. J Epidemiol Community Health 51:526–530

Quinnan GV et al. (1983) Serologic responses and systemic reactions in adults after vaccination with monovalent A/USSR/77 and trivalent A/USSR/77, A/Texas/77, B/Hong Kong/72 influenza vaccines. Rev Infect Dis 5:748–757

Reichert TA et al. (2001a) To the editor: (vaccinating Japanese schoolchildren against influenza). N Engl J Med 344:1947–1948

Reichert TA et al. (2001b) The Japanese experience with vaccinating schoolchildren against influenza. N Engl J Med 344:889–896

Ritzwoller DP et al. (2005) Effectiveness of the 2003–2004 influenza vaccine among children 6 months to 8 years of age, with 1 vs 2 doses. Pediatrics 116:153–159

Rivetti D et al. (2006) Vaccines for preventing influenza in the elderly. Cochrane Database Syst Rev 3:CD004876

Rocha E et al. (1991) Antigenic and genetic variation in influenza A (H1N1) virus isolates recovered from a persistently infected immunodeficient child. J Virol 65:2340–2350

Ropper AH (1992) The Guillain–Barre syndrome. N Engl J Med 326:1130–1136

Rudenko LG et al. (1993) Efficacy of live attenuated and inactivated influenza vaccines in schoolchildren and their unvaccinated contacts in Novgorod, Russia. J Infect Dis 168:881–887

Russell KL et al. (2005) Effectiveness of the 2003–2004 influenza vaccine among US military basic trainees: a year of suboptimal match between vaccine and circulating strain. Vaccine 23:1981–1985

Ruuskanen O et al. (1989) Acute otitis media and respiratory virus infections. Pediatr Infect Dis J 8:94–99

Saito R et al. (2002) The effectiveness of influenza vaccine against influenza a (H3N2) virus infections in nursing homes in Niigata, Japan, during the 1998–1999 and 1999–2000 seasons. Infect Control Hosp Epidemiol 23:82–86

Salgado CD et al. (2004) Preventing nosocomial influenza by improving the vaccine acceptance rate of clinicians. Infect Control Hosp Epidemiol 25:923–928

Salk JE, Menke WJ, Francis T (1945) A clinical, epidemiological and immunological evaluation of vaccination against epidemic influenza. Am J Hyg 42:57–93

Sato M et al. (2005) Antibody response to influenza vaccination in nursing home residents and healthcare workers during four successive seasons in Niigata, Japan. Infect Control Hosp Epidemiol 26:859–866

Saxen H, Virtanen M (1999) Randomized, placebo-controlled double blind study on the efficacy of influenza immunization on absenteeism of health care workers. Pediatr Infect Dis J 18:779–783

Schanzer DL, Langley JM, Tam TW (2007) Influenza-attributed hospitalization rates among pregnant women in Canada 1994–2000. J Obstet Gynaecol Can 29:622–629

Schoenbaum SC, Weinstein L (1979) Respiratory infection in pregnancy. Clin Obstet Gynecol 22:293–300

Schonberger LB et al. (1979) Guillain–Barre syndrome following vaccination in the National Influenza Immunization Program, United States, 1976–1977. Am J Epidemiol 110:105–123

Schwartz B et al. (2006) Universal influenza vaccination in the United States: are we ready? Report of a meeting. J Infect Dis 194(Suppl 2):S147–S154

Shek LP, Lee BW (2003) Epidemiology and seasonality of respiratory tract virus infections in the tropics. Paediatr Respir Rev 4:105–111

Shope RE (1931) Swine influenza. I. Experimental transmission and pathology. J Exp Med 54:349–359

Shuler CM et al. (2007) Vaccine effectiveness against medically attended, laboratory-confirmed influenza among children aged 6 to 59 months, 2003–2004. Pediatrics 119:e587–e595

Simon NM, Rovner RN, Berlin BS (1970) Acute myoglobinuria associated with type A2 (Hong Kong) influenza. JAMA 212:1704–1705

Simonsen L et al. (1997) The impact of influenza epidemics on mortality: introducing a severity index. Am J Public Health 87:1944–1950

Simonsen L et al. (2000) The impact of influenza epidemics on hospitalizations. J Infect Dis 181:831–837

Simonsen L, Viboud C, Taylor RJ (2007) Effectiveness of influenza vaccination. N Engl J Med 357:2729–2730

Smith W, Andrewes CH, Laidlaw PP (1933) A virus obtained from influenza patients. Lancet 2:66–68

Smith NM et al. (2006a) Prevention and control of influenza: recommendations of the Advisory Committee on Immunization Practices (ACIP). MMWR Recomm Rep 55:1–42

Smith S et al. (2006b) Vaccines for preventing influenza in healthy children. Cochrane Database Syst Rev CD004879

Smits AJ et al. (2002) Clinical effectiveness of conventional influenza vaccination in asthmatic children. Epidemiol Infect 128:205–211

Snyder MH et al. (1988) Four viral genes independently contribute to attenuation of live influenza A/Ann Arbor/6/60 (H2N2) cold-adapted reassortant virus vaccines. J Virol 62:488–495

Staynor K et al. (1994) Influenza A outbreak in a nursing home: the value of early diagnosis and the use of amantadine hydrochloride. Can J Infect Control 9:109–111

Stokes J et al. (1937) Results of immunization by means of active virus of human influenza. J Clin Invest 16:237–243

Strickler JK et al. (2007) Influenza vaccine effectiveness among US military basic trainees, 2005–06 season. Emerg Infect Dis 13:617–619

Sugaya N et al. (1994) Efficacy of inactivated vaccine in preventing antigenically drifted influenza type A and well-matched type B. JAMA 272:1122–1126

Surtees R, DeSousa C (2006) Influenza virus associated encephalopathy. Arch Dis Child 91:455–456

Szilagyi P et al. (2008) Influenza vaccine effectiveness among children 6 to 59 months of age during two influenza seasons: a case-cohort study. Arch Med (in press)

Talbot TR et al. (2005) Duration of virus shedding after trivalent intranasal live attenuated influenza vaccination in adults. Infect Control Hosp Epidemiol 26:494–500

Tam JS et al. (2007) Efficacy and safety of a live attenuated, cold-adapted influenza vaccine, trivalent against culture-confirmed influenza in young children in Asia. Pediatr Infect Dis J 26:619–628

Tasker SA et al. (1999) Efficacy of influenza vaccination in HIV-infected persons. A randomized, double-blind, placebo-controlled trial. Ann. Intern Med 131:430–433

Thomas RE et al. (2006) Influenza vaccination for health-care workers who work with elderly people in institutions: a systematic review. Lancet Infect Dis 6:273–279

Thompson T (1852) Annals of influenza or epidemic catarrhal fever in Great Britain from 1510 to 1837. Sydenham Society, London

Thompson WW et al. (2003) Mortality associated with influenza and RSV. JAMA 289:179–186

Thompson WW et al. (2004) Influenza-associated hospitalizations in the United States. JAMA 292:1333–1340

Togashi T, Matsuzono Y, Narita M (2000) Epidemiology of influenza-associated encephalitis-encephalopathy in Hokkaido, the northernmost island of Japan. Pediatr Int 42:192–196

Treanor JD (2007) Influenza—the goal of control. N Engl J Med 357:1439–1441

Treanor JJ et al. (1999) Evaluation of trivalent, live, cold-adapted (CAIV-T) and inactivated (TIV) influenza vaccines in prevention of virus infection and illness following challenge of adults with wild-type influenza A (H1N1), A (H3N2), and B viruses. Vaccine 18:899–906

Vesikari T et al. (2006a) Safety, efficacy, and effectiveness of cold-adapted influenza vaccine-trivalent against community-acquired, culture-confirmed influenza in young children attending day care. Pediatrics 118:2298–2312

Vesikari T et al. (2006b) A randomized, double-blind study of the safety, transmissibility and phenotypic and genotypic stability of cold-adapted influenza virus vaccine. Pediatr Infect Dis J 25:590–595

Vu T et al. (2002) A meta-analysis of effectiveness of influenza vaccine in persons aged 65 years and over living in the community. Vaccine 20:1831–1836

Wang Z, et al. (2009) Live attenuated or inactivated influenza vaccines and medical encounters for respiratory illnesses among US military personnel. JAMA 301:1066–9

Weingarten S et al. (1988) Do hospital employees benefit from the influenza vaccine? A placebo-controlled clinical trial. J Gen Intern Med 3:32–37

Weycker D et al. (2005) Population-wide benefits of routine vaccination of children against influenza. Vaccine 23:1284–1293

WHO (2005) H5N1 avian influenza: first steps towards development of a human vaccine. Wkly Epidemiol Rec 80:277–278

WHO (2008) Meeting of the immunization strategic advisory group of experts, November 2007—conclusions and recommendations. Wkly Epidemiol Rec 83:1–16

Widelock D, Csizmas L, Klein S (1963) Influenza, pregnancy, and fetal outcome. Public Health Rep 78:1–11

Wilde JA et al. (1999) Effectiveness of influenza vaccine in health care professionals: a randomized trial. JAMA 281:908–913

Williams MS, Wood JM (1993) A brief history of inactivated influenza virus vaccines. In: Hannoun C et al. (eds) Elsevier, Amsterdam, pp 169–171

Willison HJ, Yuki N (2002) Peripheral neuropathies and anti-glycolipid antibodies. Brain 125:2591–2625

Wright PF et al. (1977) Trials of influenza A/New Jersey/76 virus vaccine in normal children: an overview of age-related antigenicity and reactogenicity. J Infect Dis 136(Suppl):S731–S741

Wright PF et al. (1982) Cold-adapted recombinant influenza A virus vaccines in seronegative young children. J Infect Dis 146:71–79

Wright PF et al. (1983) Antigenicity and reactogenicity of influenza A/USSR/77 virus vaccine in children—a multicentered evaluation of dosage and safety. Rev Infect Dis 5:758–764

Yamazaki T, Suzuki T, Yamamoto K (2001) To the editor: (vaccinating Japanese schoolchildren against influenza). N Engl J Med 344:1947

Yang L et al. (2008) Synchrony of clinical and laboratory surveillance for influenza in Hong Kong. PLoS ONE 3:e1399

Zaman K, et al. (2008) Effectiveness of maternal influenza immunization in mothers and infants. N Engl J Med 359:1555–64

Zangwill KM et al. (2001) Prospective, randomized, placebo-controlled evaluation of the safety and immunogenicity of three lots of intranasal trivalent influenza vaccine among young children. Pediatr Infect Dis J 20:740–746

# Generation and Characterization of Candidate Vaccine Viruses for Prepandemic Influenza Vaccines

**Eduardo O'Neill and Ruben O. Donis**

## Contents

**Abstract** Vaccination will be a critical public health intervention to mitigate the next influenza pandemic. Its effectiveness will depend on preparedness at multiple levels, from the laboratory bench to the population. Here we describe a global approach to ensure that appropriate candidate vaccine viruses are produced, evaluated, and made available to vaccine manufacturers in a timely fashion.

E. O'Neill and R.O. Donis (✉)

Influenza Division, National Center for Immunization and Respiratory Diseases, Coordinating Centers for Infectious Diseases, Centers for Disease Control and Prevention, 1600 Clifton Road—Mail Stop G-16, Atlanta, GA 30333, USA
e-mail: rdonis@cdc.gov

R.W. Compans and W.A. Orenstein (eds.), *Vaccines for Pandemic Influenza*, Current Topics in Microbiology and Immunology 333, DOI 10.1007/978-3-540-92165-3_4, © Springer-Verlag Berlin Heidelberg 2009

This is an integrated activity involving global virologic and epidemiologic surveillance, genetic and antigenic characterization of influenza viruses, pandemic risk assessments, selection of appropriate virus strains for vaccines, production of reassortant viruses by reverse genetics, and finally, analysis of their safety and growth characteristics prior to distribution. These procedures must comply with national and international regulations governing vaccine and environmental safety.

# 1 The Pandemic Threat Posed by Highly Pathogenic Avian Influenza

Annual global mortality due to seasonal human influenza is estimated at 250,000, mostly among infant or geriatric age groups (WHO 2003). However, these numbers pale in comparison to the much larger threat to public health that pandemic influenza represents. The 1918–1919 influenza pandemic caused an estimated 50 million deaths, including young and adult population age groups (Johnson and Mueller 2002). Influenza pandemics occurred three times in the previous century (Kilbourne 2006). The critical nexus of previous pandemics was the introduction of antigenically novel viruses with genes derived from an animal influenza reservoir (Webster and Laver 1972). Novel viruses transmitted from animals to humans can acquire the capacity to transmit horizontally from person to person and spread in the population.

A large proportion of the human population has immunologic memory of influenza as a result of annual epidemics of type A and B viruses. Immunologic memory and protection from reinfection with the same strains of influenza is mediated primarily by antibodies to the viral hemagglutinin (HA) (Couch et al. 1979; Smith et al. 1935). However, there are 16 different alleles of the HA gene (i.e., subtypes H1 through H16) in the pool of influenza viruses circulating in birds (Blok and Air 1982; Fouchier et al. 2005). The antigenic structures of the 16 HAs have diverged extensively, and so solid antibody-mediated protection is specific for the immunizing HA subtype (Couch et al. 1986; Hobson et al. 1972). Furthermore, those in the population born after 1968 have humoral immunity to only the H1 and H3 subtypes of influenza A. This indicates that novel viruses with HA genes from any of the remaining 14 HA subtypes circulating in animal reservoirs could initiate a pandemic unhindered by immunity to HA, whereas morbidity and mortality by re-emerging subtypes could be dampened by partial immunity in the population.

# 2 Role of Vaccines in Pandemic Preparedness

Although advances in public health have converted other scourges of mankind into historical notes, pandemic influenza remains a major threat. Modeling studies revealed that uncontrolled spread of an even moderately severe pandemic could overwhelm the health care delivery systems globally (Medema et al. 2004). Rapid mobilization of effective antiviral drugs from national and international stockpiles may help contain the spread of an emerging pandemic and reduce the peak demand for medical care (Ferguson et al. 2005; Longini et al. 2004). However, long-term antiviral prophylaxis will be unsustainable in the course of a pandemic because antiviral drug manufacturers will be unable to meet the surge in product demand. In contrast, vaccination provides durable protection and is considered one of the most effective interventions to mitigate serious public health consequences and prevent the collapse of health care infrastructures. National pandemic plans generally heed World Health Organization (WHO) recommendations and contemplate the use of stockpiled prepandemic vaccines as well as promote the development of manufacturing and distribution capacity for rapid vaccine deployment when a pandemic is imminent or declared (WHO 2005b, 2007c).

The WHO has recommended the development and stockpiling of prepandemic influenza vaccines that are prepared in a similar type of formulation as the current seasonal influenza trivalent inactivated vaccines (TIV) which have been approved for human use by the respective national drug regulatory agencies (EMEA 2004; Nichol and Treanor 2006; WHO 2005a; Wood and Robertson 2004). Seasonal TIVs contain strains of the two influenza A subtypes that are circulating along with a representative type B strain. The strains used for the type A components are high-growth reassortant viruses that usually contain six genes (PB2, PB1, PA, NP, M, and NS) derived from the laboratory-adapted A/Puerto Rico/8/34 (H1N1) (PR8) strain together with the HA and NA surface antigens from the recommended influenza strain (Nichol and Treanor 2006). A candidate vaccine virus is characterized antigenically, genetically and phenotypically, and is provided by a WHO Collaborating Center (CC) or by an approved reference laboratory. In the case of prepandemic vaccines, the candidate vaccine virus will be derived by one of three approaches: (1) reverse genetics derivation of a reassortant containing the HA (lacking the multibasic cleavage site; see Sect. 3 for definition) and NA from the appropriate novel subtype (e.g., H5N1) virus and the remaining six genes from PR8; (2) egg-based classical reassortant with the HA and NA from a low pathogenic avian or mammalian influenza virus and the remaining six genes from PR8; or (3) a fully avian or animal influenza virus of low virulence without any genetic manipulation. The focus of this chapter is on reverse genetics (rg) derivation of prepandemic reassortant candidate vaccine viruses; herein termed "rg candidate vaccine viruses."

The rg candidate vaccine viruses provided by the WHO Collaborating Centers are expanded and optimized by vaccine manufacturers to generate seed viruses for industrial-scale production in chicken embryos (WHO 2007b). Vaccines are produced by propagating the reassortant seed virus in the allantoic cavity of embryonated hens' eggs and then purified and concentrated by zonal centrifugation and/or column chromatography followed by chemical inactivation with formalin (Matthews 2006). The immunogenic component of current TIV licensed in many countries globally can consist of inactivated whole-virus, subvirion, or surface antigen. Whole-virus vaccines contain intact inactivated virions; subvirion vaccines contains detergent-disrupted inactivated virus; and surface antigen vaccines contain purified or highly enriched HA and NA proteins (Nichol and Treanor 2006; Subbarao and Joseph 2007). The antigen content of seasonal TIV is routinely standardized by single radial immunodiffusion to contain 15 μg per dose of HA from each virus strain (Wood et al. 1981).

The potential collapse of the fertile egg supply during a pandemic has provided a strong rationale for using alternative substrates (e.g., cell culture) in vaccine production. Clinical evaluation of TIV vaccines produced using cell culture substrate systems showed satisfactory immunogenicity, and some have been submitted for regulatory approval (Percheson et al. 1999). Nevertheless, both egg or cell culture-based vaccine production systems depend on candidate vaccine virus seeds that multiply with high efficiency and do not pose a hazard to vaccine manufacturing workers and their communities. Two general types of vaccines should be considered with regards to preparedness for the next influenza pandemic: (1) Prepandemic vaccine. This term refers to a product that can be used safely during the interpandemic phase (WHO phase 1–3) as well as during higher levels of alert (phases 4–5) preceding declaration of a uncontrolled pandemic (Phase 6) (WHO 2008b); and (2) pandemic vaccine which could be used only during a pandemic, a period in which certain risks (e.g., progressive reversion of the virus to the wild type) are of lesser importance than the public health impact of the pandemic virus.

# 3   Risk Assessment to Inform Pandemic Vaccine Preparedness

Effective planning is necessary to prioritize the investment of limited resources in stockpiling of prepandemic vaccines against influenza virus subtypes and strains with the greatest probability of causing a pandemic (Dowdle 1997; Fauci 2006b; Fedson 2004; Gerberding 2006; WHO 2006a; Peiris et al. 2007; Tam et al. 2005). Although we cannot be certain of the HA subtype of the next pandemic influenza virus, a multifactorial risk assessment can identify which viruses deserve the greatest consideration (Table 1). The main factors are those directly or indirectly affecting transmission to humans and the subsequent acquisition of sustained transmissibility in humans (Ferguson et al. 2003; Fraser et al. 2004; Garcia-Sastre and Whitley 2006). Other important factors include severity of disease, virus load in the environment and geographic dispersion.

Influenza A viruses that circulate strictly in waterfowl seem to lack critical functions required to infect humans. In contrast, influenza viruses that have become adapted to poultry have caused human infections with some frequency, such as subtypes H5, H7, and H9 (CDC 1997; Butt et al. 2005; Fouchier et al. 2004; Lin et al. 2000; Subbarao et al. 1998; Tweed et al. 2004). Waterfowl viruses may be transmitted to swine, where they become adapted to this species and subsequently become a source of infection for humans (Rimmelzwaan et al. 2001). However, sustained transmission with any of these viruses in humans has not been observed.

Human infections with highly pathogenic avian influenza (HPAI) viruses have caused severe and even fatal disease in humans (Fouchier et al. 2004; Subbarao et al. 1998). HPAI viruses are derived from low-pathogenic avian influenza viruses (LPAI) by mutations in the HA that generally result in the accumulation of four or more basic

**Table 1** Pandemic risk assessment of contemporary animal influenza viruses

| Criterion | Subtypes to consider [a] | References |
|---|---|---|
| Suspected or confirmed transmission from human to human (clusters) | H5, H7 | Du Ry van Beest Holle et al. (2005); Olsen et al. (2005); Ungchusak et al. (2005) |
| Suspected or confirmed transmission from animals to humans | H5, H6, H7, H9 | Buchy et al. (2007); CDC (1997); Fauci (2006a); Lin et al. (2000); Mumford et al. (2007); Myers et al. (2007); Peiris et al. (2004); Perdue and Swayne (2005); Saito et al. (2001); Webster et al. (2006) |
| Circulation in swine and other food mammals | H2, H4, H9 | Karasin et al. (2000a,b); Ma et al. (2007); Ninomiya et al. (2002); Peiris et al. (2001) |
| Circulation in companion pet animals | H3 | Crawford et al. (2005) |
| Receptor binding specificity changes | H9, H7, H5 | Glaser et al. (2005); Glaser et al. (2006); Rogers et al. (1985); Stevens et al. (2006a); Suzuki (2001) |
| Transmission to terrestrial poultry | H5, H9, H6, H7, H4 | Campitelli et al. (2004); Guan et al. (2002); Ito and Kawaoka (2000); Li et al. (2003); Mannelli et al. (2006); Wallensten et al. (2007); Xu et al. (2007) |
| Transmission to swine or other farm/domestic or companion mammals | H9, H5 | Ninomiya et al. (2002); Songserm et al. (2006a,b) |
| Experimental replication, virulence, and transmission in ferret and other carnivores | H9, H7, H5 | Belser et al. (2007); Zitzow et al. (2002) |
| Experimental replication and virulence in the mouse animal model | H5, H7, H9 | Belser et al. (2007); Gabriel et al. (2005); Joseph et al. (2007); Li et al. (2005); Lu et al. (2003); Shinya et al. (2005, 2007) |
| High pathogenicity in chickens | H5, H7 | Perdue et al. (1997); Suarez et al. (1998) |
| Geographic spread | H9, H7, H7 | Chen et al. (2006); Ducatez et al. (2007); Wallace et al. (2007) |

[a]H1 strains related to contemporary human viruses were excluded because pre-existing immunity would hinder the emergence of a severe pandemic

amino acids at the cleavage site of HA0 (the precursor polypeptide encoded by the HA gene that yields HA1 and HA2 fragments upon cleavage by cellular proteases) (Garten et al. 1982; Horimoto and Kawaoka 1995; Kawaoka et al. 1984). N-linked glycosylation in the vicinity of the HA cleavage site may modulate protease access and cleavage efficiency (Kawaoka et al. 1984). Cleavage of HA0 is required for virions to become infectious (Klenk et al. 1975; Lazarowitz and Choppin 1975). In vivo replication of LPAI viruses is restricted to the airway and intestinal epithelia because only these tissues produce the necessary protease (Steinhauer 1999). In contrast, the HAs of HPAI have a multibasic cleavage site that can be cleaved by ubiquitous intracellular furin-like proteases, which enables systemic spread of the virus in chickens and, in some cases, in mammalian species as well. Conversion from LPAI to HPAI has been demonstrated after sustained transmission in terrestrial poultry (Horimoto and Kawaoka 1995). So far, only subtypes H5 or H7 LPAI have acquired high pathogenicity. Interestingly, Eurasian HPAI H5 or H7 viruses have resulted in fatal infections in humans, whereas the American lineage counterparts have not (Fouchier et al. 2004; Subbarao et al. 1998; Tweed et al. 2004). In contrast, human infections in Asia with LPAI, subtype H9, have not been associated with fatalities (Butt et al. 2005; Lin et al. 2000).

# 4 Impact of Virus Evolution on Pandemic Preparedness

The rapid evolution of the influenza A HA translates into cumulative amino acid changes responsible for a progressive drift in antigenicity (Webster et al. 1982). Consequently, host-protective antibodies elicited in response to infection or vaccination with a given influenza strain lose the ability to neutralize the antigenically drifted variant viruses. The rapid mutation and antigenic drift of the HA have been recognized since the initiation of influenza vaccination programs (Pereira et al. 1969). Mutations in HA that mediate escape from neutralizing antibodies in the human population are thus positively selected and become fixed (Fitch et al. 1997). In the case of seasonal human influenza viruses, a dominant antigenic variant strain survives immune surveillance and seeds subsequent seasons, whereas many subdominant alternative variants become extinct (Ferguson and Bush 2004). Therefore, only closely related influenza viruses circulate in the population at any given time (Russell et al. 2008). This simplifies influenza vaccine formulation; virus seeds for vaccine manufacturing include just one virus strain matching the prevailing virus of that subtype. Global virologic surveillance and antigenic characterization of circulating viruses by the WHO Global Influenza Surveillance Network (GISN) are crucial to define the representative strain in order to update vaccine composition and compensate for the effects of antigenic drift.

The HA of animal influenza viruses evolves into multiple cocirculating lineages with unique antigenic characteristics (Wang et al. 2008; WHO 2005c; Wu et al. 2008). For example, many antigenically distinct viruses have evolved from the A/goose/Guangdong/1/96 H5N1 virus that emerged in South China more than a decade ago (Wu et al. 2008). The evolution of the H5N1 viruses currently in circulation in birds in Asia, Africa, and Europe reveals that, although some lineages have apparently become extinct, several different HA clades continue to evolve and diversify as a result of species differences, geographic isolation and human interventions, among other factors (Fig. 1) (Wu et al. 2008). The impact of viral evolution on the antigenic structure of HA becomes evident upon comparing the amino acid sequences of the HA from the

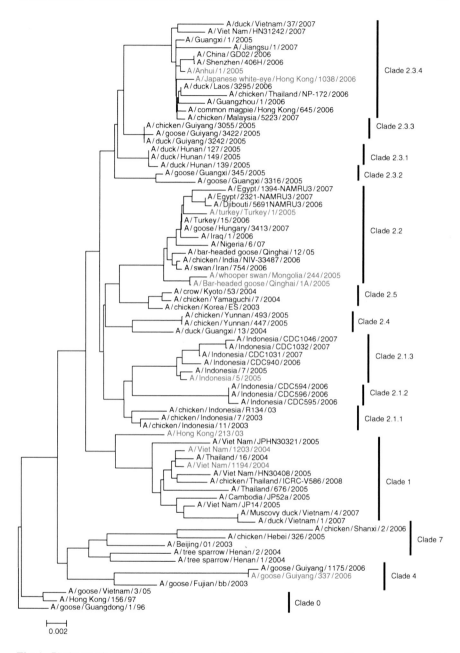

**Fig. 1** Phylogenetic tree of the H5 hemagglutinin lineage derived from A/goose/Guangdong/1/96 (H5N1) reveals the cocirculation of multiple virus clades. Neighbor-joining tree of selected HA genes from representative H5N1 viruses isolated from humans since 1997 and closely related viruses from birds, constructed using PAUP* v4.0b10. The tree was rooted to A/goose/Guangdong/1/96. *Scale bar* represents nucleotide changes per site. Candidate vaccine viruses are denoted in *red font*

major clades of H5N1; 14 of the 68 amino acids that constitute putative antigenic sites in the HA have drifted among the four major clades (Stevens et al. 2006b; Wilson and Cox 1990) (Table 2). Furthermore, clade-specific changes can be observed amongst these 14 positions. The consequence of such structural evolution is reflected in the HI reaction of these viruses with panels of ferret antisera (Table 3). The antigenic changes measured by HI identify major drift variants, but comparisons among multiple clades to identify minor differences are difficult to judge reliably. Analysis of HI data by mathematical methods implemented in new computer software

**Table 2** Variation in the antigenic sites of H5 hemagglutinin clades comprising viruses isolated from humans[a]

| Antigenic site | H3 num | H5 num | Clade O | Clade 1[g] | Clade 2.1 | Clade 2.2 | Clade 2.3.4 |
|---|---|---|---|---|---|---|---|
| E | 80 | 71 | I | – | – | L | – |
| E | 91 | 83 | A | – | – | I | – |
| E | 92 | 84 | S | N | N | N | N |
| E[b] | 94 | 86 | A | V | T | – | – |
| A | 131 | 126 | D | E | E | E | E |
| A | 134 | 129 | S | L | – | – | – |
| A | 142 | 138 | H | Q | L | Q | Q |
| A | 144 | 140 | K | – | S | R | T |
| A[c] | 145 | 141 | S | – | P | – | P |
| B | 158 | 154 | N | – | – | D | – |
| B | 159 | 155 | S | – | – | N | N |
| B | 160 | 156 | A | T | T | – | T |
| B | 185 | 181 | P | – | – | – | S |
| B[d,e,f] | 193 | 189 | K | – | R | R | – |

[a]Stevens et al. (2006b), and reviewed in Wilson and Cox (1990)
[b]Changes at H5 a.a. 48, 50, 53, 54, 58, 66, 69, 75, 78, 80, 102, 257, 259, and 262 occur in <50% of viruses
[c]Changes at H5 a.a. 128, 133, 136, 137, and 142 occur in <50% of viruses
[d]Changes at H5 a.a. 123, 151–153, 182–188, and 190–193 occur in <50% of viruses
[e]Changes at site C (a.a. 40–45 and 273–277) occur in <50% of viruses
[f]Changes at site D (a.a. 167–168, 210, 214, 215, and 222–225) occur in <50% of viruses
[g]Letter indicates a.a. change from the modal in Clade 0 or conservation (–)

**Table 3** Hemagglutination inhibition titers of H5N1 viruses from selected HA clades[a]

| Reference antigens | Clade | VN/1203 | Reference antisera | | | |
| | | | IND/5 | MG/244 | NIV | ANH/1 |
|---|---|---|---|---|---|---|
| A/Vietnam/1203/2004 (VN/1203) | 1 | **320** | 20 | <10 | 40 | 40 |
| A/Indonesia/5/2005 (IND/5) | 2.1 | 10 | **640** | 80 | 40 | 320 |
| A/whooper swan/Mongolia/244/05 (MG/244) | 2.2 | 20 | 160 | **320** | 320 | 40 |
| A/chicken/India/NIV-33487/2006 (NIV) | 2.2 | 10 | 320 | 320 | **320** | 20 |
| A/Anhui/1/05 (ANH/1) | 2.3.4 | 40 | 320 | <10 | 20 | **640** |

[a]From WHO (2008a). (–) not tested. Homologous titers are given in bold

programs have been applied to increase the efficiency and resolution of the analysis as well as improve visualization of the antigenic relationships among many strains (Smith et al. 2004). The HI antigenic maps generated by these programs may help standardize antigenic surveillance data analysis. Thus, continued surveillance, including genetic and antigenic characterization of the H5N1 viruses by the WHO, the Food and Agriculture Organization and national government agencies, is essential to inform risk assessment and vaccine composition decisions.

# 5 Selection and Preparation of Reference Viruses for Prepandemic Vaccine Manufacturing

The selection and preparation of prepandemic vaccine seed viruses for vaccine manufacturing mimics the process used to update the composition of seasonal influenza vaccines (WHO 2005e, 2007b). The guidelines developed under the coordination of the WHO, with the participation of international experts, provide scientific and operational guidance to national public health authorities (Wood 2002). The major steps are as follows (see also Fig. 2).

## 5.1 Surveillance to Detect and Isolate Novel Influenza Viruses that May Infect Humans

Influenza surveillance programs have been established to identify the viruses that cause influenza-like illnesses. The process begins with the collection of clinical specimens for virus detection by rapid antigen tests or molecular assays, virus isolation and/or serologic detection by health care providers. Clinical and epidemiological data are also collected by healthcare workers. Influenza virus specimens are subsequently sent to state and provincial public health laboratories for further characterization. Influenza viruses that are identified as type A but not readily identified as one of the circulating subtypes H1 or H3 should be sent to a National Influenza Center (NIC) for further characterization. The protocol for handling specimens suspected to contain for H5N1 or other influenza A subtype viruses, such as H9N2, which are potential pandemic viruses, differ from the routine virus isolation for seasonal influenza. When an avian influenza infection such as H5N1 is suspected in a human, the samples are usually sent to NICs (or their designated referral laboratories). The NIC or other national influenza reference laboratories conduct preliminary diagnostic testing, most often using polymerase chain reaction (PCR). Specimens that reveal possibly novel influenza by PCR are sent to laboratories with appropriate high-level biosafety facilities (BSL3 and above) to isolate the novel (e.g., H5N1) viruses in MDCK cells or eggs and conduct partial or whole genome sequencing. Suspected novel influenza viruses are cultured in MDCK or chicken embryos and the resulting isolates are characterized by nucleotide sequencing

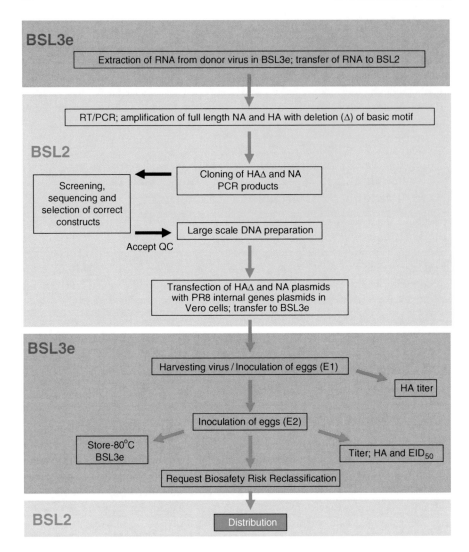

**Fig. 2** Flow chart depicting the major steps in the generation of a prepandemic candidate vaccine rg virus from a HPAI. Initial handling of the donor HPAI virus (e.g., H5N1) is performed in a high containment laboratory. After RNA extraction and verification of sterility, the nucleic acid is transferred to a BSL2 laboratory for amplification and cloning. The HA cDNA clone with the desired deletion of the multibasic cleavage site is cotransfected in Vero cells certified for human vaccine use with the NA plasmid and six plasmids encoding the so-called "internal genes" of PR8. Recovery of infectious virus is performed in a BSL3-enhanced (BSL3-e) laboratory. Characterization of the rg candidate vaccine virus is also performed in BSL3e until the safety data presented to regulatory authorities is deemed acceptable for transfer to BSL2. The biosafety level requirement for each step is shown on the *left column*

and analyzed serologically with panels of reference antisera in hemagglutination inhibition (HI) to determine the HA subtype and by neuraminidase inhibition (NI) to determine NA subtype (WHO 2002).

## 5.2  Genetic and Antigenic Analyses

The HA and NA (and often the complete genome) of the novel influenza isolates are sequenced by National Influenza Centers and WHO Collaborating Centers to establish the ancestry of the viruses and the degree of relatedness to influenza viruses isolated previously.

To perform a complete antigenic analysis, ferrets are inoculated with the new isolates to produce antiserum. Early convalescent antiserum from ferrets is the most sensitive reagent for detecting antigenic variation among influenza viruses (Palmer et al. 1975). Ferret antiserum panels are used in HI tests with the virus panels used to generate the antisera (homologous titers) as well as new viruses for which antisera are not yet available (test panel). The relative HI titers of each virus in the HI test panel can be used as a quantitative measure to reveal the extent of antigenic difference (or relatedness) between new influenza isolates and other cocirculating viruses or vaccine candidate viruses. HI titers are reliable indicators of viral neutralization by antibodies and thought to reflect the level of cross-protection between influenza virus strains (Wood et al. 1994). Ultimately, antigenic characterization of all representative animal viruses that could potentially infect humans is needed to insure that the antigen used for prepandemic vaccine manufacturing closely represents them.

## 5.3  WHO Consultation on Vaccine Strain Selection

The information generated through antigenic characterization and that obtained through gene sequence analysis are mutually complementary and, combined with epidemiologic and serologic data, constitute the major criteria for selection of H5N1 reference viruses to be used as prepandemic candidate vaccines (WHO 2007b). To this end, WHO brings together representatives from the WHO CCs and Reference Laboratories twice each year for a comprehensive consultation on the cumulative data generated by individual laboratories. Once the viruses for vaccine use have been selected, a public meeting is held and the WHO recommendations are communicated to all participants. The information is also published on the WHO website and/or in the Weekly Epidemiologic Record (WHO 2006b, 2007b,d, 2008a).

## 5.4  Derivation of Reassortant Seed Viruses and Deletion of Virulence Determinants

HPAI viruses, including the H5N1 viruses currently affecting many countries, are usually rapidly lethal to chicken embryos. Early embryo lethality by wild-type HPAI is known to greatly reduce virus yield and is a major obstacle to vaccine production. In addition, HPAI viruses are a human health hazard and strict biocontainment

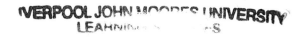

protocols and facilities are necessary for their propagation. Most vaccine manufacturers lack the facilities and equipment to propagate HPAI at industrial scale in chicken embryos in compliance with biosafety and biosecurity regulations. Consequently, industrial-scale manufacturing of vaccine by propagation of wild-type HPAI in eggs is generally not feasible; a LPAI must be identified for this purpose. In most cases, there is no LPAI virus with the appropriate HA to manufacture an effective prepandemic vaccine safely in eggs. Therefore, the development of high-growth reassortants from HPAI viruses, such as H5N1, for human vaccine development and production is more complicated than for seasonal influenza viruses and requires a different process. First, the H5N1 virus must be made less pathogenic. This is accomplished using a plasmid-based reverse genetics approach to engineer a virus with an HA gene lacking the stretch of basic amino acids at the cleavage site (Subbarao et al. 2003). The multibasic amino acid-deleted HA and the NA genes from the H5N1 (or other novel subtype) are used to generate a reassortant virus with the remaining six genes from the PR8 egg-adapted virus; these reassortants are termed 6:2 viruses, denoting their genotype. All of the manipulations required for HA and NA gene cloning, virus recovery and subsequent propagation in eggs are performed in dedicated facilities using vaccine-qualified cells and reagents under Good Laboratory Practices (GLP) and a quality management system (WHO 2005a). The resulting reassortant virus is expected to grow to high titers in chicken eggs, but depends on exogenous trypsin for propagation in cell cultures due to deletion of the multibasic cleavage site of HA (Tobita et al. 1975).

## 5.5   Characterization of the Reassortant Candidate Virus

Several analyses are performed to establish the safety and authenticity of the rg candidate vaccine virus, as follows.

### 5.5.1   Molecular Analysis

The sequence of the HA and NA genes of the reassortants is determined to ensure that any changes resulting from laboratory manipulation have not affected the amino acids critical to antigenicity or attenuation of pathogenicity. The absence of a polybasic cleavage site and the presence of stabilizing mutations in HA are recorded. The internal genes are sequenced to establish their derivation from the PR8 ancestor.

### 5.5.2   In Vitro Studies to Determine the Trypsin Dependency
of Multicycle Replication

This is established by plaque assays in chicken embryo fibroblasts that compare plaque formation efficiency in the presence or absence of trypsin in the agar overlay.

### 5.5.3  Embryo Lethality Test

The lethality of the reassortant reference virus and the parental viruses is evaluated by recording the survival of embryos after allantoic sac injection with a minimal embryo lethal dose. The HPAI virus but not the rg candidate vaccine virus is expected to spread systemically and kill the embryo within a 72-h period (Horimoto et al. 2006; Rott et al. 1980).

### 5.5.4  In Vivo Pathogenicity Studies in Chickens and Ferrets

The virulence properties of the rg vaccine viruses are established by inoculation into chickens and ferrets to determine if indeed virulence was eliminated; a LPAI-like phenotype is anticipated. To date, several H5N1 and H7N7 rg vaccine viruses derived from HPAI viruses and PR8 have demonstrated reduced virulence in chicken and ferret animal models as compared to HPAI (Nicolson et al. 2005; Webby et al. 2004).

### 5.5.5  Antigenic Characterization

The H5N1 rg candidate vaccine viruses are fully characterized antigenically to establish the degree of antigenic similarity to the original HPAI donor virus. To this end, the reference virus is compared to the parental donor strain by two-way HI cross-reactivity using ferret antisera to the rg candidate vaccine virus strain and the wild-type virus.

### 5.5.6  Progeny Virus Yield in Chicken Embryos

The replication properties of rg candidate vaccine virus in chicken eggs is evaluated to determine the yield of infectious virus and HA activity. Typically, the H5N1 reverse genetics modified candidate vaccine viruses replicate less efficiently than seasonal influenza A candidate vaccine viruses. This poses a challenge for influenza vaccine manufacturers if large quantities of vaccine must be prepared in a short period of time. However, an increased yield of the reassortant is observed relative to the wild-type virus, since the former does not affect embryo viability during virus growth.

### 5.5.7  Residual Plasmid DNA

Derivation of vaccine viruses by reverse genetics entails the use of plasmid DNA for transfection. The plasmids pose a hypothetical health risk associated with antibiotic resistance markers and strong viral promoters. Therefore, it is recommended that the reference virus should not contain detectable amounts of DNA. This determination is made by real-time PCR using primers that target unique sequences of the plasmid vector used.

## 5.6 Regulatory Compliance

In the USA, the rg candidate vaccine viruses are subject to scrutiny by two types of national regulatory agencies. Government entities concerned with protecting the safety of the environment, human, and animal health will require evidence demonstrating that the biohazard risk posed by the newly derived virus is minimal. Animal and human health protection agencies regulate the facilities and protocols for the safe use and handling of HPAI and its rg candidate vaccine virus derivatives. Therefore, each rg vaccine virus reference stock derived from HPAI must undergo a risk assessment by regulatory authorities before it is approved for use in BSL2 containment facilities that are compatible with processes used in seasonal influenza vaccine manufacturing plants. The WHO CC laboratory that produces the rg candidate vaccine virus prepares a dossier that provides all the information necessary for the authorities to make this determination.

In addition, consumer protection agencies have regulatory requirements to ensure the safety and efficacy of prepandemic vaccines. A dossier describing the biological properties of these rg candidate vaccine viruses is provided so that the agency can evaluate the level of risk associated with the cell lines used for derivation of the virus by reverse genetics, animal-derived products used for cell culture, the use of plasmid DNA with antibiotic resistance markers, and strong viral promoters among others.

## 5.7 Communication on Availability of Reference Viruses for Vaccines

The WHO Global Influenza Program (GIP) has coordinated the development of candidate H5N1 prepandemic vaccines (Table 4) and announced their availability through postings at the official website stohr (2003). A total of ten H5N1 rg candidate vaccine strains have been developed for the various clades of the H5 HA, and additional

**Table 4** H5N1 candidate vaccine viruses generated by reverse genetics[a]

| HPAI strain source of HA and NA | HA clade | References |
|---|---|---|
| A/Hong Kong/213/2003 | 1 | Webby et al. (2004) |
| A/Vietnam/1203/2004 | 1 | Hoffmann et al. (2005); WHO (2005c) |
| A/Vietnam/1194/2004 | 1 | Bresson et al. (2006); Wood and Robertson (2004) |
| A/Indonesia/5/2005 | 2.1.3 | Hoschler et al. (2007); WHO (2007d) |
| A/bar-headed goose/Qinghai/1A/2005 | 2.2 | WHO (2007d) |
| A/whooper swan/Mongolia/244/2005 | 2.2 | WHO (2007d) |
| A/turkey/Turkey/1/2005 | 2.2 | Hoschler et al. (2007); WHO (2007d) |
| A/Anhui/1/2005 | 2.3.4 | Hoschler et al. (2007); WHO (2007d) |
| A/Jap white-eye/Hong Kong/1038/2006 | 2.3.4 | WHO (2008a) |
| A/goose/Guyiang/337/2006 | 4 | WHO (2008a) |

[a] Current as of April 30, 2008

strains are under consideration (WHO 2008a). The HA and NA genes of viruses selected for vaccine development are derived from either human or avian isolates from each of the major H5 clades (including subclades within clade 2) that have infected humans since late 2003 (WHO 2007a).

# 6 Laboratory Procedures for the Derivation of rg Candidate Vaccine Viruses

Conventional reassortment in eggs has been the only approved method for the production of TIV for human use (Matthews 2006). Candidate prepandemic vaccines can be derived using this method whenever a suitable HA is available from a low pathogenic avian virus donor (Desheva et al. 2006; Jadhao et al. 2008; Pappas et al. 2007). However, when a suitable LP HA is not available to match a potentially pandemic virus, conventional reassortment is not an option since it does not provide a mechanism to modify the HA cleavage motif responsible for virulence. Therefore, seed viruses are engineered to lack the HPAI cleavage motif by reverse genetics (Nicolson et al. 2005; WHO 2005c). The above mentioned objectives are performed following GLP quality standards in approved and dedicated laboratory facilities (WHO 2005a,e). Important checkpoints are introduced to reduce the probability of errors, which could cause costly delays. Standard operating procedures are developed and statistically validated for all laboratory methods. Safety and environmental considerations are reviewed and documented. Data are subject to quality control and reported to an independent auditor for certification.

To generate rg candidate vaccine viruses, the HA and NA genes from the HPAI virus of interest are cloned into specialized plasmids for reverse genetics. RNA extracted from the H5N1 strain of interest is subjected to reverse transcription and polymerase chain reaction (RT/PCR) to amplify the surface glycoprotein genes, HA, and NA (WHO 2005c). Amplification of the HA gene is performed with two primer sets to secure the elimination of several (usually 2–4) basic amino acid residues at the cleavage site of the HA upon ligation (Subbarao et al. 2003). Primers are designed to include certain silent codon changes at the cleavage site to hinder the possible reversion of the virus by the reacquisition of sequences encoding multiple basic amino acids (Garcia et al. 1996; Perdue et al. 1997). The amplified PCR products for both the HA and NA genes are cloned into a dual-promoter reverse genetics plasmid vector that contains regulatory elements for the production of a complete set of influenza viral RNAs (genomic polarity) and proteins upon transfection into human cells (Fodor et al. 1999; Hoffmann et al. 2000; Hoffmann et al. 2001; Neumann et al. 1999). The plasmids containing the HA (modified) and NA genes from the strain of interest (e.g., H5N1) are combined with six plasmids carrying the remaining six genes of PR8 virus; namely PB2, PB1, PA, NP, M, and NS, and the mixture transfected into mammalian cells to produce a rg candidate vaccine virus with a 6:2 genotype. The internal genes of PR8 are used since they confer a high-yield phenotype required for vaccine production (Baez et al. 1980; Kilbourne 1969).

The cultured cells used for the derivation of reference viruses from plasmid DNA by reverse genetics merit special attention. Only primary cells or cell lines that have been previously used to manufacture licensed vaccines for human use are acceptable (WHO 2005e). In addition, human RNA polymerase I promoter elements in the reverse genetics plasmids are functionally restricted to primate cells. Consequently, only the Vero cell line is recommended for the derivation of rg candidate vaccine stocks by reverse genetics because of its longstanding safety record (Montagnon and Vincent-Falquet 1998). The particular Vero cells used in the process must originate from a fully characterized working cell bank established for vaccine production per WHO and national regulatory guidelines (WHO 2005e).

The newly produced recombinant virus is harvested and propagated in ten-day-old embryonated hen's eggs. The genetic and biologic properties of the newly generated rg candidate vaccine virus are analyzed in compliance with WHO and national regulations (Fig. 2).

# 7  Biosafety Considerations

All laboratory procedures involving live highly pathogenic influenza viruses should take place at a high level of biological containment (e.g., BSL3 and above, as recommended by WHO and national regulatory bodies) (WHO 2005d). In the United States, federal agencies provide specific facility and protocol recommendations for work with live HPAI (Centers for Disease Control and Prevention (US) and National Institutes of Health (US) 2007). In addition, the US Department of Agriculture Select Agent Program has direct oversight of all work with HPAI and requires specific enhancements; therefore, it is referred to as BSL3-enhanced (BSL3-e). Newly generated rg candidate vaccine viruses with the modified HA from the HPAI donor are classified in the same risk group as the parental HPAI viruses. Although the deletion of the multibasic cleavage site by reverse genetics and reassortment with PR8 is predicted to eliminate the pathogenicity of the virus, experimental demonstration by tests performed to that effect (including sequencing, trypsin-dependent plaque formation, lack of chicken embryo lethality, lack of chicken pathogenicity, and lack of ferret pathogenicity) are required to request its reclassification as a BSL-2 risk group agent. Regulatory authorities review these applications on a case-by-case basis.

# 8  Virulence Properties of rg Candidate Vaccine Viruses

HPAI viruses are a threat to human and animal health. The health risks are thought to be greatly reduced by deleting the multibasic cleavage site of HA and replacing six genes from the HPAI with those derived from the laboratory-adapted strain PR8.

However, consumer and environmental protection agencies require evidence that the newly created virus does not cause severe disease in a mammalian model; either mouse or ferret (WHO 2005a). In addition, animal health agencies charged with protecting poultry stocks require evidence that the rg candidate vaccine virus is not likely to replicate in birds in cases of accidental release (USDA 2005). Critical in vitro properties of the newly generated viruses that require experimental verification include testing for trypsin dependency of plaque formation in primary chicken embryo fibroblast cultures (Table 5). The in vivo properties of newly generated rg candidate vaccine viruses are evaluated in animal models (ferret and chicken) and compared to their parental wild-type HPAI counterparts (WHO 2005a). For this purpose, ferrets are inoculated with $10^6$ median egg infectious doses ($EID_{50}$) of the rg candidate vaccine virus and the clinical signs and virus replication in the upper and lower respiratory tract are determined and contrasted with those of the parental HPAI and PR8 viruses. A similar approach is used to assess pathogenicity in birds. One type of study is the standardized World Organization for Animal Health (OIE) pathogenicity index. Ten chickens 4–6 weeks of age are inoculated with the rg candidate vaccine virus (0.2 mL of 1/10 dilution of allantoic fluid stock), and mortality is recorded for ten days (OIE 2008; Pearson 2003). Virus shedding is evaluated by inoculating eight four-week-old white rock chickens intranasally with $10^6$ $EID_{50}$ of each virus and monitoring virus titers in cloacal and tracheal secretions as well as clinical signs for 14 days (Matsuoka et al. 2003). In all cases, the eight chickens inoculated with the candidate seed viruses should remain healthy throughout the 14-day observation period (Table 5). In contrast, all chickens inoculated with the wild-type parental HPAI viruses die within a few days. Results from these safety studies are compiled into two dossiers. One is submitted to animal and human health regulatory authorities to request approval to use the rg candidate vaccine virus as a BSL-2 risk group agent. The other dossier is submitted to consumer protection agencies to document the safety of the vaccine derived from this virus seed in the unlikely event of accidents resulting in incomplete virus inactivation.

**Table 5** Biological properties of prepandemic candidate vaccine viruses

| Virus | Clade | HA cleavage site[b] | Mortality[c] | Plaque formation[d] |
|---|---|---|---|---|
| A/VN/1203/2004 | 1 | GLRNSPQRERRRKKR↓GLF | 8/8 | 9.0(8.8) |
| VNH5N1-PR8[a] | 1 | GLRNSPQRETR↓GLF | 0/8 | 5.5[e](8.3) |
| A/Indonesia/05/2005 | 2.1 | GLRNSPQRESRRKKR↓GLF | 8/8 | 8.8(9.2) |
| Ind05-PR8 | 2.1 | GLRNSPQRESR↓GLF | 0/8 | ≤1.0(8.0) |
| A/Anhui/01/2005 | 2.3.4 | GLRNSPLRERRRKKR↓GLF | 8/8 | 9.3(9.3) |
| Anhui01-PR8 | 2.3.4 | GLRNSPLRER↓GLF | 0/8 | ≤1.0(9.0) |

[a] VNH5N1-PR8, Ind05-PR8 and Anhui01-PR8 are the reference viruses derived from A/VN/1203/2004, A/Indonesia/05/2005, and A/Anhui/01/2005, respectively

[b] ↓ Denotes the cleavage site

[c] Pathotyping was performed in 4-week-old white rock chickens

[d] Plaque formation was assessed in the absence/presence of 0.5 µg/ml of trypsin; expressed as log PFU/ml

[e] Fuzzy plaque morphology and reduced diameter relative to wild-type virus

# 9  Time Considerations for Pandemic Intervention Responses

Modeling studies have shown that vaccine interventions to mitigate the impact of an influenza pandemic must be deployed as early as possible to have a significant public health impact (Germann et al. 2006). After declaration of a pandemic (phase 6) by WHO, or immediately before (phases 4/5), it will be critical to immunize the entire population as soon as possible (Germann et al. 2006). Two doses of inactivated vaccine must be administered to elicit immunity to the new HA subtype (Bresson et al. 2006; Treanor et al. 2006). The use of stockpiled prepandemic vaccine allows for rapid immunologic priming of the population if the prepandemic vaccine HA is antigenically similar to that of the pandemic virus. Although some countries have stockpiled sufficient prepandemic H5N1 vaccine to immunize the whole population, the majority cannot afford such investment. Therefore, it will be necessary to rapidly manufacture pandemic vaccine. This process could be expedited if a 6:2 candidate vaccine virus with suitable antigenic characteristics was prepared beforehand for immediate use. Current vaccine preparedness efforts by the WHO Global Influenza Surveillance Network (GISN) have generated a collection of candidate vaccine strains that may increase the probability of finding an appropriate vaccine virus to allow the manufacture of a reasonably effective pandemic vaccine while the ideally matched rg candidate vaccine virus is being produced (WHO 2008a). With currently available technology, derivation of the rg candidate vaccine virus requires a minimum of 15 days from receipt of the HPAI virus (Table 6) (Nicolson et al. 2005). Safety studies in ferrets require an additional two weeks, assuming animals and BSL3-e facilities are immediately available. Virus characterization studies, including antigenic analyses, chick embryo lethality, and trypsin dependence for plaque formation on chicken embryo fibroblast, take approximately one week and can be performed in parallel with the animal pathogenicity studies.

**Table 6**  Chronology for derivation of candidate vaccine viruses by reverse genetics

| Activity | Required time (days) |
| --- | --- |
| Preparation of GLP laboratory | 2 |
| HA and NA sequencing | |
| Oligonucleotide synthesis | |
| Cloning of HA and NA genes | 4 |
| Revival of certified Vero cells | |
| Rescue transfectant virus in Vero cells | 3 |
| Preparation E2 stock and titration | 6 |
| Sequence analysis | |
| Total time | 15 |

# 10 Summary

Significant achievements in public health preparedness to contain and mitigate the next pandemic have been realized since the rapid spread of H5N1 HPAI after 2004. A global response coordinated by WHO has resulted in improved surveillance programs, the development of prepandemic rg candidate vaccine viruses, expanded vaccine manufacturing capacity, prepandemic vaccine clinical trials and international vaccine stockpiles. However, clinical trial data to support H5N1 vaccine use policy development is still insufficient. Multiple other challenges will need to be addressed, including the long-term stability of inactivated prepandemic vaccine stockpiles, continued emergence of H5N1 antigenic variants, the need for two-dose vaccine regimens to elicit protection, and the inelastic nature of the egg-based manufacturing capacity to meet a surge in demand.

**Acknowledgments** Derivation of prepandemic candidate vaccine viruses was supported in part by grants from the National Vaccine Program Office, Department of Health and Human Services. We thank all the members of the WHO Global Influenza Surveillance Network for continued support. We thank Elizabeth Mumford and Wenqing Zhang in the WHO Global Influenza Program for helpful discussions. We thank Todd Davis for phylogenetic analyses and James Stevens for very valuable suggestions. We thank our colleagues in the Influenza Division; in particular, the many contributions of the Virus Surveillance and Diagnosis Branch and the Molecular Virology and Vaccines Branch are greatly appreciated.

# References

Baez M, Palese P, Kilbourne ED (1980) Gene composition of high-yielding influenza vaccine strains obtained by recombination. J Infect Dis 141:362–365

Belser JA, Lu X, Maines TR, Smith C, Li Y, Donis RO, Katz JM, Tumpey TM (2007) Pathogenesis of avian influenza. (H7) virus infection in mice and ferrets: enhanced virulence of Eurasian H7N7 viruses isolated from humans. J Virol 81:11139–11147

Blok J, Air GM (1982) Sequence variation at the 3' end of the neuraminidase gene from 39 influenza type A viruses. Virology 121:211–229

Bresson JL, Perronne C, Launay O, Gerdil C, Saville M, Wood J, Hoschler K, Zambon MC (2006) Safety and immunogenicity of an inactivated split-virion influenza A/Vietnam/1194/2004. (H5N1) vaccine: phase I randomised trial. Lancet 367:1657–1664

Buchy P, Mardy S, Vong S, Toyoda T, Aubin JT, Miller M, Touch S, Sovann L, Dufourcq JB, Richner B, Tu PV, Tien NT, Lim W, Peiris JS, Van der Werf S (2007) Influenza A/H5N1 virus infection in humans in Cambodia. J Clin Virol 39:164–168

Butt KM, Smith GJ, Chen H, Zhang LJ, Leung YH, Xu KM, Lim W, Webster RG, Yuen KY, Peiris JS, Guan Y (2005) Human infection with an avian H9N2 influenza A virus in Hong Kong in 2003. J Clin Microbiol 43:5760–5767

Campitelli L, Mogavero E, De Marco MA, Delogu M, Puzelli S, Frezza F, Facchini M, Chiapponi C, Foni E, Cordioli P, Webby R, Barigazzi G, Webster RG, Donatelli I (2004) Interspecies transmission of an H7N3 influenza virus from wild birds to intensively reared domestic poultry in Italy. Virology 323:24–36

CDC. (1997) Isolation of avian influenza A. (H5N1) viruses from humans—Hong Kong, May–December 1997. MMWR Morb Mortal Wkly Rep 46:1204–1207

Centers for Disease Control and Prevention. (US), and National Institutes of Health. (US). (2007) Biosafety in microbiological and biomedical laboratories, 5th edn. (HHS publication; no. CDC 93-8395). US Department of Health and Human Services Public Health Service Centers for Disease Control and Prevention, National Institutes of Health, US GPO, Washington, DC

Chen H, Smith GJ, Li KS, Wang J, Fan XH, Rayner JM, Vijaykrishna D, Zhang JX, Zhang LJ, Guo CT, Cheung CL, Xu KM, Duan L, Huang K, Qin K, Leung YH, Wu WL, Lu HR, Chen Y, Xia NS, Naipospos TS, Yuen KY, Hassan SS, Bahri S, Nguyen TD, Webster RG, Peiris JS, Guan Y (2006) Establishment of multiple sublineages of H5N1 influenza virus in Asia: Implications for pandemic control. Proc Natl Acad Sci USA 103:2845–2850

Couch RB, Webster RG, Kasel JA, Cate TR (1979) Efficacy of purified influenza subunit vaccines and relation to the major antigenic determinants on the hemagglutinin molecule. J Infect Dis 140:553–559

Couch RRB, Kasel JJA, Glezen WWP, Cate TTR, Six HHR, Taber LLH, Frank AAL, Greenberg SSB, Zahradnik JJM, Keitel WWA (1986) Influenza: its control in persons and populations. J Infect Dis 153:431–440

Crawford PC, Dubovi EJ, Castleman WL, Stephenson I, Gibbs EP, Chen L, Smith C, Hill RC, Ferro P, Pompey J, Bright RA, Medina MJ, Johnson CM, Olsen CW, Cox NJ, Klimov AI, Katz JM, Donis RO (2005) Transmission of equine influenza virus to dogs. Science 310:482–485

Desheva JA, Lu XH, Rekstin AR, Rudenko LG, Swayne DE, Cox NJ, Katz JM, Klimov AI (2006) Characterization of an influenza A H5N2 reassortant as a candidate for live-attenuated and inactivated vaccines against highly pathogenic H5N1 viruses with pandemic potential. Vaccine 24:6859–6866

Dowdle WR (1997) Pandemic influenza: confronting a re-emergent threat. The 1976 experience. J Infect Dis 176(Suppl 1):S69–S72

Du Ry van Beest Holle M, Meijer A, Koopmans M, de Jager C (2005) Human-to-human transmission of avian influenza A/H7N7, The Netherlands, 2003. Euro Surveill 10:264–268

Ducatez MF, Olinger CM, Owoade AA, Tarnagda Z, Tahita MC, Sow A, De Landtsheer S, Ammerlaan W, Ouedraogo JB, Osterhaus AD, Fouchier RA, Muller CP (2007) Molecular and antigenic evolution and geographical spread of H5N1 highly pathogenic avian influenza viruses in western Africa. J Gen Virol 88:2297–2306

EMEA. (2004) Guideline on dossier structure and content for pandemic influenza vaccine marketing authorisation application. European Agency for the Evaluation of Medical Products Committee for Proprietary Medicinal Products, London

Fauci AS (2006a) Emerging and re-emerging infectious diseases: influenza as a prototype of the host–pathogen balancing act. Cell 124:665–670

Fauci AS (2006b) Seasonal and pandemic influenza preparedness: science and countermeasures. J Infect Dis 194Suppl 2:S73–S76

Fedson DS (2004) Vaccination for pandemic influenza: a six point agenda for interpandemic years. Pediatr Infect Dis J 23:S74–S77

Ferguson NM, Bush RM (2004) Influenza evolution and immune selection. International Congress Series 1263:12–16

Ferguson NM, Galvani AP, Bush RM (2003) Ecological and immunological determinants of influenza evolution. Nature 422:428–433

Ferguson NM, Cummings DA, Cauchemez S, Fraser C, Riley S, Meeyai A, Iamsirithaworn S, Burke DS (2005) Strategies for containing an emerging influenza pandemic in Southeast Asia. Nature 437:209–214

Fitch WM, Bush RM, Bender CA, Cox NJ (1997) Long term trends in the evolution of H(3) HA1 human influenza type A. Proc Natl Acad Sci USA 94:7712–7718

Fodor E, Devenish L, Engelhardt OG, Palese P, Brownlee GG, Garcia-Sastre A (1999) Rescue of influenza A virus from recombinant DNA. J Virol 73:9679–9682

Fouchier RA, Schneeberger PM, Rozendaal FW, Broekman JM, Kemink SA, Munster V, Kuiken T, Rimmelzwaan GF, Schutten M, Van Doornum GJ, Koch G, Bosman A, Koopmans M,

Osterhaus AD (2004) Avian influenza A virus. (H7N7) associated with human conjunctivitis and a fatal case of acute respiratory distress syndrome. Proc Natl Acad Sci USA 101:1356–1361

Fouchier RA, Munster V, Wallensten A, Bestebroer TM, Herfst S, Smith D, Rimmelzwaan GF, Olsen B, Osterhaus AD (2005) Characterization of a novel influenza A virus hemagglutinin subtype. (H16) obtained from black-headed gulls. J Virol 79:2814–2822

Fraser C, Riley S, Anderson RM, Ferguson NM (2004) Factors that make an infectious disease outbreak controllable. Proc Natl Acad Sci USA 101:6146–6151

Gabriel G, Dauber B, Wolff T, Planz O, Klenk HD, Stech J (2005) The viral polymerase mediates adaptation of an avian influenza virus to a mammalian host. Proc Natl Acad Sci USA 102:18590–18595

Garcia M, Crawford JM, Latimer JW, Rivera-Cruz E, Perdue ML (1996) Heterogeneity in the haemagglutinin gene and emergence of the highly pathogenic phenotype among recent H5N2 avian influenza viruses from Mexico. J Gen Virol 77(Pt 7):1493–1504

Garcia-Sastre A, Whitley RJ (2006) Lessons learned from reconstructing the 1918 influenza pandemic. J Infect Dis 194(Suppl 2):S127–S132

Garten W, Linder D, Rott R, Klenk HD (1982) The cleavage site of the hemagglutinin of fowl plague virus. Virology 122:186–190

Gerberding JL (2006) Pandemic preparedness: pigs, poultry, and people versus plans, products, and practice. J Infect Dis 194(Suppl 2):S77–S81

Germann TC, Kadau K, Longini IM Jr, Macken CA (2006) Mitigation strategies for pandemic influenza in the United States. Proc Natl Acad Sci USA 103:5935–5940

Glaser L, Stevens J, Zamarin D, Wilson IA, Garcia-Sastre A, Tumpey TM, Basler CF, Taubenberger JK, Palese P (2005) A single amino acid substitution in 1918 influenza virus hemagglutinin changes receptor binding specificity. J Virol 79:11533–11536

Glaser L, Zamarin D, Acland HM, Spackman E, Palese P, Garcia-Sastre A, Tewari D (2006) Sequence analysis and receptor specificity of the hemagglutinin of a recent influenza H2N2 virus isolated from chicken in North America. Glycoconj J 23:93–99

Guan Y, Peiris M, Kong KF, Dyrting KC, Ellis TM, Sit T, Zhang LJ, Shortridge KF (2002) H5N1 influenza viruses isolated from geese in Southeastern China: evidence for genetic reassortment and interspecies transmission to ducks. Virology 292:16–23

Hobson D, Curry RL, Beare AS, Ward-Gardner A (1972) The role of serum haemagglutination-inhibiting antibody in protection against challenge infection with influenza A2 and B viruses. J Hyg. (Lond) 70:767–777

Hoffmann E, Neumann G, Kawaoka Y, Hobom G, Webster RG (2000) A DNA transfection system for generation of influenza A virus from eight plasmids. Proc Natl Acad Sci USA 97:6108–6113

Hoffmann E, Stech J, Guan Y, Webster RG, Perez DR (2001) Universal primer set for the full-length amplification of all influenza A viruses. Arch Virol 146:2275–2289

Hoffmann E, Lipatov AS, Webby RJ, Govorkova EA, Webster RG (2005) Role of specific hemagglutinin amino acids in the immunogenicity and protection of H5N1 influenza virus vaccines. Proc Natl Acad Sci USA 102:12915–12920

Horimoto T, Kawaoka Y (1995) Molecular changes in virulent mutants arising from avirulent avian influenza viruses during replication in 14-day-old embryonated eggs. Virology 206:755–759

Horimoto T, Takada A, Fujii K, Goto H, Hatta M, Watanabe S, Iwatsuki-Horimoto K, Ito M, Tagawa-Sakai Y, Yamada S, Ito H, Ito T, Imai M, Itamura S, Odagiri T, Tashiro M, Lim W, Guan Y, Peiris M, Kawaoka Y (2006) The development and characterization of H5 influenza virus vaccines derived from a 2003 human isolate. Vaccine 24:3669–3676

Hoschler K, Gopal R, Andrews N, Saville M, Pepin S, Wood J, Zambon MC (2007) Cross-neutralisation of antibodies elicited by an inactivated split-virion influenza A/Vietnam/1194/2004. (H5N1) vaccine in healthy adults against H5N1 clade 2 strains. Influenza Other Respir Viruses 1:199–206

Ito T, Kawaoka Y (2000) Host-range barrier of influenza A viruses. Vet Microbiol 74:71–75

Jadhao SJ, Achenbach J, Swayne DE, Donis R, Cox N, Matsuoka Y (2008) Development of Eurasian H7N7/PR8 high growth reassortant virus for clinical evaluation as an inactivated pandemic influenza vaccine. Vaccine 26:1742–1750

Johnson NP, Mueller J (2002) Updating the accounts: global mortality of the 1918–1920 "Spanish" influenza pandemic. Bull Hist Med 76:105–115

Joseph T, McAuliffe J, Lu B, Jin H, Kemble G, Subbarao K (2007) Evaluation of replication and pathogenicity of avian influenza a H7 subtype viruses in a mouse model. J Virol 81:10558–10566

Karasin AI, Brown IH, Carman S, Olsen CW (2000a) Isolation and characterization of H4N6 avian influenza viruses from pigs with pneumonia in Canada. J Virol 74:9322–9327

Karasin AI, Olsen CW, Brown IH, Carman S, Stalker M, Josephson G (2000b) H4N6 influenza virus isolated from pigs in Ontario. Can Vet J 41:938–939

Kawaoka Y, Naeve CW, Webster RG (1984) Is virulence of H5N2 influenza viruses in chickens associated with loss of carbohydrate from the hemagglutinin. Virology 139:303–316

Kilbourne ED (1969) Future influenza vaccines and the use of genetic recombinants? Bull World Health Organ 41:643–645

Kilbourne ED (2006) Influenza pandemics of the 20th century. Emerg Infect Dis 12:9–14

Klenk HD, Rott R, Orlich M, Blodorn J (1975) Activation of influenza A viruses by trypsin treatment. Virology 68:426–439

Lazarowitz SG, Choppin PW (1975) Enhancement of the infectivity of influenza A and B viruses by proteolytic cleavage of the hemagglutinin polypeptide. Virology 68:440–454

Li C, Yu K, Tian G, Yu D, Liu L, Jing B, Ping J, Chen H (2005) Evolution of H9N2 influenza viruses from domestic poultry in Mainland China. Virology 340:70–83

Li KS, Xu KM, Peiris JS, Poon LL, Yu KZ, Yuen KY, Shortridge KF, Webster RG, Guan Y (2003) Characterization of H9 subtype influenza viruses from the ducks of southern China: a candidate for the next influenza pandemic in humans? J Virol 77:6988–6994

Lin YP, Shaw M, Gregory V, Cameron K, Lim W, Klimov A, Subbarao K, Guan Y, Krauss S, Shortridge K, Webster R, Cox N, Hay A (2000) Avian-to-human transmission of H9N2 subtype influenza A viruses: relationship between H9N2 and H5N1 human isolates. Proc Natl Acad Sci USA 97:9654–9658

Longini IM Jr, Halloran ME, Nizam A, Yang Y (2004) Containing pandemic influenza with antiviral agents. Am J Epidemiol 159:623–633

Lu XH, Cho D, Hall H, Rowe T, Mo IP, Sung HW, Kim WJ, Kang C, Cox N, Klimov A, Katz JM (2003) Pathogenesis of and immunity to a new influenza A. (H5N1) virus isolated from duck meat. Avian Dis 47:1135–1140

Ma W, Vincent AL, Gramer MR, Brockwell CB, Lager KM, Janke BH, Gauger PC, Patnayak DP, Webby RJ, Richt JA (2007) Identification of H2N3 influenza A viruses from swine in the United States. Proc Natl Acad Sci USA 104:20949–20954

Mannelli A, Ferre N, Marangon S (2006) Analysis of the 1999–2000 highly pathogenic avian influenza. (H7N1) epidemic in the main poultry-production area in northern Italy. Prev Vet Med 73(4):273–285

Matsuoka Y, Chen H, Cox N, Subbarao K, Beck J, Swayne D (2003) Safety evaluation in chickens of candidate human vaccines against potential pandemic strains of influenza. Avian Dis 47:926–930

Matthews JT (2006) Egg-based production of influenza vaccine: 30 years of commercial experience. In: Bugliarello G. (ed) The bridge. National Academy of Engineering, Washington, DC

Medema JK, Zoellner YF, Ryan J, Palache AM (2004) Modeling pandemic preparedness scenarios: health economic implications of enhanced pandemic vaccine supply. Virus Res 103:9–15

Montagnon BJ, Vincent-Falquet JC (1998) Experience with the Vero cell line. Dev Biol Stand 93:119–123

Mumford E, Bishop J, Hendrickx S, Embarek PB, Perdue M (2007) Avian influenza H5N1: risks at the human-animal interface. Food Nutr Bull 28:S357–S363

Myers KP, Setterquist SF, Capuano AW, Gray GC (2007) Infection due to three avian influenza subtypes in United States veterinarians. Clin Infect Dis 45:4–9

Neumann G, Watanabe T, Ito H, Watanabe S, Goto H, Gao P, Hughes M, Perez DR, Donis R, Hoffmann E, Hobom G, Kawaoka Y (1999) Generation of influenza A viruses entirely from cloned cDNAs. Proc Natl Acad Sci USA 96:9345–9350

Nichol KL, Treanor JJ (2006) Vaccines for seasonal and pandemic influenza. J Infect Dis 194(Suppl 2):S111–S118

Nicolson C, Major D, Wood JM, Robertson JS (2005) Generation of influenza vaccine viruses on Vero cells by reverse genetics: an H5N1 candidate vaccine strain produced under a quality system. Vaccine 23:2943–2952

Ninomiya A, Takada A, Okazaki K, Shortridge KF, Kida H (2002) Seroepidemiological evidence of avian H4, H5, and H9 influenza A virus transmission to pigs in southeastern China. Vet Microbiol 88:107–114

OIE. (2008) OIE manual of diagnostic tests and vaccines for terrestrial animals. Chapter 2.1.14

Olsen SJ, Ungchusak K, Sovann L, Uyeki TM, Dowell SF, Cox NJ, Aldis W, Chunsuttiwat S (2005) Family clustering of avian influenza A. (H5N1). Emerg Infect Dis 11:1799–1801

Palmer DF, Dowdle WR, Coleman MT, Schild GC (1975) Heamagglutination inhibition test. In: Advanced laboratory techniques for influenza diagnosis: procedural guide. US Department of Health, Education, and Welfare, Atlanta, GA, pp 25–62

Pappas C, Matsuoka Y, Swayne DE, Donis RO (2007) Development and evaluation of an Influenza virus subtype H7N2 vaccine candidate for pandemic preparedness. Clin Vaccine Immunol 14:1425–1432

Pearson JE (2003) International standards for the control of avian influenza. Avian Dis 47:972–975

Peiris JS, Guan Y, Markwell D, Ghose P, Webster RG, Shortridge KF (2001) Cocirculation of avian H9N2 and contemporary "human" H3N2 influenza A viruses in pigs in southeastern China: potential for genetic reassortment? J Virol 75:9679–9686

Peiris JS, Yu WC, Leung CW, Cheung CY, Ng WF, Nicholls JM, Ng TK, Chan KH, Lai ST, Lim WL, Yuen KY, Guan Y (2004) Re-emergence of fatal human influenza A subtype H5N1 disease. Lancet 363:617–619

Peiris JS, de Jong MD, Guan Y (2007) Avian influenza virus. (H5N1): a threat to human health. Clin Microbiol Rev 20:243–267

Percheson PB, Trepanier P, Dugre R, Mabrouk T (1999) A Phase I, randomized controlled clinical trial to study the reactogenicity and immunogenicity of a new split influenza vaccine derived from a non-tumorigenic cell line. Dev Biol Stand 98:127–132; discussion 133–134

Perdue ML, Garcia M, Senne D, Fraire M (1997) Virulence-associated sequence duplication at the hemagglutinin cleavage site of avian influenza viruses. Virus Res 49:173–186

Perdue ML, Swayne DE (2005) Public health risk from avian influenza viruses. Avian Dis 49:317–127

Pereira MS, Chakraverty P, Pane AR, Fletcher WB (1969) The influence of antigenic variation on influenza A2 epidemics. J Hyg. (Lond) 67:551–557

Rimmelzwaan GF, de Jong JC, Bestebroer TM, van Loon AM, Claas EC, Fouchier RA, Osterhaus AD (2001) Antigenic and genetic characterization of swine influenza A. (H1N1) viruses isolated from pneumonia patients in The Netherlands. Virology 282:301–306

Rogers GN, Daniels RS, Skehel JJ, Wiley DC, Wang XF, Higa HH, Paulson JC (1985) Host-mediated selection of influenza virus receptor variants. Sialic acid-alpha 2,6Gal-specific clones of A/duck/Ukraine/1/63 revert to sialic acid-alpha 2,3-Gal-specific wild type in ovo. J Biol Chem 260:7362–7367

Rott R, Reinacher M, Orlich M, Klenk HD (1980) Cleavability of hemagglutinin determines spread of avian influenza viruses in the chorioallantoic membrane of chicken embryo. Arch Virol 65:123–133

Russell CA, Jones TC, Barr IG, Cox NJ, Garten RJ, Gregory V, Gust ID, Hampson AW, Hay AJ, Hurt AC, de Jong JC, Kelso A, Klimov AI, Kageyama T, Komadina N, Lapedes AS, Lin YP, Mosterin A, Obuchi M, Odagiri T, Osterhaus AD, Rimmelzwaan GF, Shaw MW, Skepner E, Stohr K, Tashiro M, Fouchier RA, Smith DJ (2008) The global circulation of seasonal influenza A. (H3N2) viruses. Science 320:340–346

Saito T, Lim W, Suzuki T, Suzuki Y, Kida H, Nishimura SI, Tashiro M (2001) Characterization of a human H9N2 influenza virus isolated in Hong Kong. Vaccine 20:125–133

Shinya K, Suto A, Kawakami M, Sakamoto H, Umemura T, Kawaoka Y, Kasai N, Ito T (2005) Neurovirulence of H7N7 influenza A virus: Brain stem encephalitis accompanied with aspiration pneumonia in mice. Arch Virol 150:1653–1660

Shinya K, Watanabe S, Ito T, Kasai N, Kawaoka Y (2007) Adaptation of an H7N7 equine influenza A virus in mice. J Gen Virol 88:547–553

Smith W, Andrewes CH, Laidlaw PP (1935) Experiments on the immunization of Ferrets and Mice. Brit J Exp Path 16:291

Smith DJ, Lapedes AS, de Jong JC, Bestebroer TM, Rimmelzwaan GF, Osterhaus AD, Fouchier RA (2004) Mapping the antigenic and genetic evolution of influenza virus. Science 305:371–376

Songserm T, Amonsin A, Jam-on R, Sae-Heng N, Pariyothorn N, Payungporn S, Theamboonlers A, Chutinimitkul S, Thanawongnuwech R, Poovorawan Y (2006a) Fatal avian influenza A H5N1 in a dog. Emerg Infect Dis 12:1744–1747

Songserm T, Amonsin A, Jam-on R, Sae-Heng N, Meemak N, Pariyothorn N, Payungporn S, Theamboonlers A, Poovorawan Y (2006b) Avian influenza H5N1 in naturally infected domestic cat. Emerg Infect Dis 12:681–683

Steinhauer DA (1999) Role of hemagglutinin cleavage for the pathogenicity of influenza virus. Virology 258:1–20

Stevens J, Blixt O, Glaser L, Taubenberger JK, Palese P, Paulson JC, Wilson IA (2006a) Glycan microarray analysis of the hemagglutinins from modern and pandemic influenza viruses reveals different receptor specificities. J Mol Biol 355:1143–1155

Stevens J, Blixt O, Tumpey TM, Taubenberger JK, Paulson JC, Wilson IA (2006b) Structure and receptor specificity of the hemagglutinin from an H5N1 influenza virus. Science 312:404–410

Stohr K (2003) The WHO global influenza program and its animal influenza network. Avian Dis 47:934–938

Suarez DL, Perdue ML, Cox N, Rowe T, Bender C, Huang J, Swayne DE (1998) Comparisons of highly virulent H5N1 influenza A viruses isolated from humans and chickens from Hong Kong. J Virol 72:6678–6688

Subbarao K, Joseph T (2007) Scientific barriers to developing vaccines against avian influenza viruses. Nat Rev Immunol 7:267–278

Subbarao K, Klimov A, Katz J, Regnery H, Lim W, Hall H, Perdue M, Swayne D, Bender C, Huang J, Hemphill M, Rowe T, Shaw M, Xu X, Fukuda K, Cox N (1998) Characterization of an avian influenza A. (H5N1) virus isolated from a child with a fatal respiratory illness. Science 279:393–396

Subbarao K, Chen H, Swayne D, Mingay L, Fodor E, Brownlee G, Xu X, Lu X, Katz J, Cox N, Matsuoka Y (2003) Evaluation of a genetically modified reassortant H5N1 influenza A virus vaccine candidate generated by plasmid-based reverse genetics. Virology 305:192–200

Suzuki Y (2001) Host mediated variation and receptor binding specificity of influenza viruses. Adv Exp Med Biol 491:445–451

Tam T, Sciberras J, Mullington B, King A (2005) Fortune favours the prepared mind: a national perspective on pandemic preparedness. Can J Public Health 96:406–408

Tobita K, Sugiura A, Enomote C, Furuyama M (1975) Plaque assay and primary isolation of influenza A viruses in an established line of canine kidney cells. (MDCK) in the presence of trypsin. Med Microbiol Immunol. (Berl) 162:9–14

Treanor JJ, Campbell JD, Zangwill KM, Rowe T, Wolff M (2006) Safety and immunogenicity of an inactivated subvirion influenza A. (H5N1) vaccine. N Engl J Med 354:1343–1351

Tweed SASA, Skowronski DMDM, David STST, Larder AA, Petric MM, Lees WW, Li YY, Katz JJ, Krajden MM, Tellier RR, Halpert CC, Hirst MM, Astell CC, Lawrence DD, Mak AA (2004) Human illness from avian influenza H7N3, British Columbia. Emerg Infect Dis 10:2196–2199

Ungchusak K, Auewarakul P, Dowell SF, Kitphati R, Auwanit W, Puthavathana P, Uiprasertkul M, Boonnak K, Pittayawonganon C, Cox NJ, Zaki SR, Thawatsupha P, Chittaganpitch M, Khontong R, Simmerman JM, Chunsutthiwat S (2005) Probable person-to-person transmission of avian influenza A (H5N1). N Engl J Med 352:333–340

USDA (2005) Rules and regulations. Fed Reg 70:13252

Wallace RG, Hodac H, Lathrop RH, Fitch WM (2007) A statistical phylogeography of influenza A H5N1. Proc Natl Acad Sci USA 104:4473–4478

Wallensten A, Munster VJ, Latorre-Margalef N, Brytting M, Elmberg J, Fouchier RA, Fransson T, Haemig PD, Karlsson M, Lundkvist A, Osterhaus AD, Stervander M, Waldenstrom J, Bjorn O

(2007) Surveillance of influenza A virus in migratory waterfowl in northern Europe. Emerg Infect Dis 13:404–411

Wang JJ, Vijaykrishna DD, Duan LL, Bahl JJ, Zhang JJX, Webster RRG, Peiris JJSM, Chen HH, Smith GJGJD, Guan YY (2008) Identification of the progenitors of Indonesian and Vietnamese avian influenza A (H5N1) viruses from southern China. J Virol 82:3405–3414

Webby RJ, Perez DR, Coleman JS, Guan Y, Knight JH, Govorkova EA, McClain-Moss LR, Peiris JS, Rehg JE, Tuomanen EI, Webster RG (2004) Responsiveness to a pandemic alert: use of reverse genetics for rapid development of influenza vaccines. Lancet 363:1099–1103

Webster RG, Laver WG (1972) The origin of pandemic influenza. Bull World Health Organ 47:449–452

Webster RG, Laver WG, Air GM, Schild GC (1982) Molecular mechanisms of variation in influenza viruses. Nature 296:115–121

Webster RG, Peiris M, Chen H, Guan Y (2006) H5N1 outbreaks and enzootic influenza. Emerg Infect Dis 12:3–8

WHO. (2002) WHO manual on animal influenza diagnosis and surveillance. http://www.wpro.who.int/NR/rdonlyres/EFD2B9A7-2265-4AD0-BC98-97937B4FA83C/0/manualonanimalaidiagnosisandsurveillance.pdf

WHO. (2003) World Health Organization factsheet 211: influenza. http://www.who.int/mediacentre/factsheets/fs211/en/

WHO. (2005a) WHO biosafety risk assessment and guidelines for the production and quality control of human influenza pandemic vaccines. http://www.who.int/biologicals/publications/ECBS%202005%205%20Influenza.pdf. In: 56th Meeting of the WHO Expert Committee on Biological Standardization, Geneva, Switzerland, 24–28 Oct 2005, pp 1–33

WHO. (2005b) WHO global influenza preparedness plan. http://www.who.int/csr/resources/publications/influenza/WHO_CDS_CSR_GIP_2005_5.pdf

WHO (2005c) Evolution of H5N1 Avian influenza viruses in Asia. Emerg Infect Dis 11:1515–1521

WHO. (2005d) WHO laboratory biosafety guidelines for handling specimens suspected of containing avian influenza A virus. http://www.who.int/csr/disease/avian_influenza/guidelines/handlingspecimens/en/

WHO. (2005e) WHO guidance on development of influenza candidate vaccine viruses by reverse genetics. http://www.who.int/csr/resources/publications/influenza/WHO_CDS_CSR_GIP_2005_6.pdf

WHO. (2006a) Influenza research at the human and animal interface. http://www.who.int/csr/resources/publications/influenza/WHO_CDS_EPR_GIP_2006_3C.pdf

WHO (2006b) Antigenic and genetic characteristics of H5N1 viruses and candidate H5N1 vaccine viruses developed for potential use as pre-pandemic vaccines. Wkly Epidemiol Rec 81:328–330

WHO. (2007a) Towards a unified nomenclature system for the highly pathogenic H5N1 avian influenza viruses. http://www.who.int/csr/disease/avian_influenza/guidelines/nomenclature/en/

WHO. (2007b) A description of the process of seasonal and H5N1 influenza vaccine virus selection and development. http://www.who.int/csr/disease/avian_influenza/influenza_vaccine-Virus_slection/en/index.html

WHO. (2007c) Informal consultation on technical specifications for a WHO international H5N1 vaccine stockpile. http://www.who.int/vaccine_research/diseases/ari/final_report_stockpile_meeting.pdf

WHO (2007d) Antigenic and genetic characteristics of H5N1 viruses and candidate H5N1 vaccine viruses developed for potential use as pre-pandemic vaccines, March 2007. Relevé épidémiol Hebdom 82:164–167

WHO. (2008a) Antigenic and genetic characteristics of H5N1 viruses and candidate H5N1 vaccine viruses developed for potential use as human vaccines, February 2008. http://www.who.int/csr/disease/avian_influenza/guidelines/H5VaccineVirusUpdate20080214.pdf

WHO. (2008b) Current phase of alert in the WHO global influenza preparedness plan. http://www.who.int/csr/disease/avian_influenza/phase/en/index.html

Wilson IA, Cox NJ (1990) Structural basis of immune recognition of influenza virus hemagglutinin. Annu Rev Immunol 8:737–771

Wood JM (2002) Selection of influenza vaccine strains and developing pandemic vaccines. Vaccine 20:B40–B44

Wood JM, Robertson JS (2004) From lethal virus to life-saving vaccine: developing inactivated vaccines for pandemic influenza. Nat Rev Micro 2:842–847

Wood JM, Seagroatt V, Schild GC, Mayner RE, Ennis FA (1981) International collaborative study of single-radial-diffusion and immunoelectrophoresis techniques for the assay of hae-magglutinin antigen of influenza virus. J Biol Stand 9:317–330

Wood JJM, Gaines-Das RRE, Taylor JJ, Chakraverty PP (1994) Comparison of influenza serological techniques by international collaborative study. Vaccine 12:167–174

Wu WL, Chen Y, Wang P, Song W, Lau SY, Rayner JM, Smith GJ, Webster RG, Peiris JSM, Lin T, Xia N, Guan Y, Chen H (2008) Antigenic profile of avian H5N1 viruses in Asia from 2002 to 2007. J Virol 82:1798–1807

Xu KM, Smith GJ, Bahl J, Duan L, Tai H, Vijaykrishna D, Wang J, Zhang JX, Li KS, Fan XH, Webster RG, Chen H, Peiris JS, Guan Y (2007) The genesis and evolution of H9N2 influenza viruses in poultry from southern China, 2000 to 2005. J Virol 81:10389–10401

Zitzow LA, Rowe T, Morken T, Shieh WJ, Zaki S, Katz JM (2002) Pathogenesis of avian influenza A (H5N1) viruses in ferrets. J Virol 76:4420–4429

# Live Attenuated Vaccines for Pandemic Influenza

**Grace L. Chen and Kanta Subbarao**

**Contents**

**Abstract** In this chapter, we will review the development of and clinical experience with the currently licensed seasonal live attenuated influenza vaccines (LAIV) and preclinical studies of H5, H7, and H9 live attenuated pandemic influenza vaccine candidates. Vectored vaccine approaches will not be reviewed in this chapter. Experience with seasonal influenza vaccination has demonstrated the safety and efficacy of LAIV in both children and adults; moreover, cross-protection among antigenically distinct viruses within the same subtype may be induced by LAIV. While clinical studies and further characterization of the immunologic response to avian influenza viruses are still needed, the experience with seasonal LAIV underscores the potential of live attenuated vaccines to play an important role in the event of a pandemic.

G.L. Chen
NIAID, NIH, Building 33, 3E 13C.2, 33 North Drive, MSC 3203, Bethesda,
MD, 20892-3203, USA

K. Subbarao (✉)
Senior Investigator, Laboratory of Infectious Diseases, NIAID, NIH, Building 33,
Room 3E13C.1, 33 North Drive, MSC 3203, Bethesda, MD, 20892-3203, USA
e-mail: ksubbarao@niaid.nih.gov

R.W. Compans and W.A. Orenstein (eds.), *Vaccines for Pandemic Influenza*,        109
Current Topics in Microbiology and Immunology 333,
DOI 10.1007/978-3-540-92165-3_5, © Springer-Verlag Berlin Heidelberg 2009

**Keywords** live attenuated influenza vaccines pandemic; Plasmid-based reverse genetics techniques

## Abbreviations

| | |
|---|---|
| AI | Avian influenza |
| AIV | Avian influenza virus |
| *ca* | Cold-adapted |
| $EID_{50}$ | 50% Egg infectious dose |
| HA | Hemagglutinin |
| HAI | Hemagglutination inhibition |
| $HID_{50}$ | 50% Human infectious dose |
| i.n. | Intranasal |
| i.m. | Intramuscular |
| LAIV | Live attenuated influenza vaccine |
| $LD_{50}$ | 50% Lethal dose |
| MDV | Master donor virus |
| $MID_{50}$ | 50% Mouse infectious dose |
| NA | Neuraminidase |
| PFU | Plaque-forming units |
| $TCID_{50}$ | 50% Tissue culture infectious dose |
| TIV | Trivalent inactivated vaccine |
| *ts* | Temperature-sensitive |
| *wt* | Wild-type |

## 1 Live Attenuated Vaccines for Seasonal Influenza

### 1.1 Generation of Seasonal LAIV

Currently, there are two LAIVs licensed for use around the world. Both vaccines are derived from highly stable, attenuated influenza A and B master donor viruses (MDV).

Our review of the experience with seasonal LAIV is drawn from the English literature, in which studies of the LAIV licensed in the US predominate.

#### 1.1.1 US Vaccines

The LAIV formulation used in the United States is a trivalent preparation composed of two influenza A and one influenza B reassortant viruses in which the six internal protein genes of the MDVs, the influenza A/Ann Arbor (AA)/6/60 (H2N2) virus and the influenza B/Ann Arbor/1/66 virus are reassorted annually with the HA and

NA gene segments from the predicted circulating wild-type (*wt*) influenza A (H3N2 and H1N1) and influenza B viruses.

Plasmid-based reverse genetics techniques are now used to generate the seasonal LAIV. These methods were first developed in a 12- or 17-plasmid-based system, with eight plasmids encoding virion sense RNA under the control of a human Pol I promoter and 4–9 plasmids encoding messenger RNAs for different influenza proteins under the control of a Pol II promoter (Fodor et al. 1999; Neumann et al. 1999). Subsequently, these techniques were modified and further simplified by engineering the influenza virus gene segments into a bidirectional vector with Pol I and Pol II promoters flanking each gene segment, thereby reducing the number of plasmids required from 12 or 17 to eight (Hoffmann et al. 2000). Each year, when a new vaccine is required, the vaccine composition is updated with the HA and NA genes of the newly-selected circulating *wt* virus, which are cloned into the appropriate vector and cotransfected with the six internal protein genes of the MDV to generate 6:2 reassortant vaccine viruses (Fig. 1) (Hoffmann et al. 2002 a,b).

The MDVs were initially isolated from throat washings on primary chick kidney tissue cultures at 36°C and adapted to growth at 25°C through serial passage at

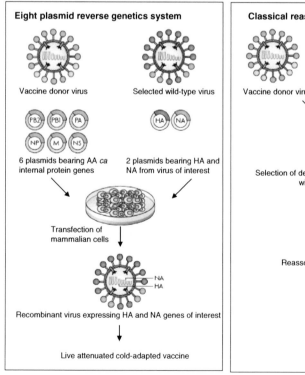
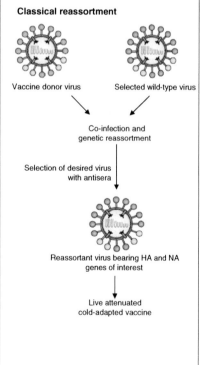

**Fig. 1**  Generation of live attenuated influenza virus vaccines (adapted from Subbarao and Joseph 2007)

progressively lower temperatures (Massab 1967, 1969). The resulting viruses display important phenotypes that have been crucial to their development for clinical use. First, they replicate efficiently at 25°C and 33°C but do not replicate efficiently at temperatures above 39°C; these phenotypic traits have respectively been designated cold-adapted (*ca*) and temperature-sensitive (*ts*) phenotypes. In addition, these viruses are attenuated in vivo (*att* phenotype); in humans and experimental animals, they are restricted in replication in the respiratory tract (Murphy and Coelingh 2002; Ali et al. 1982; Massab et al. 1982). These initial studies also demonstrated that the *ca*, *ts*, and *att* phenotypes conferred by the internal protein genes of the A/AA/6/60 *ca* MDV were reproducibly transferred to all 6:2 reassortant viruses (Keitel and Piedra 1998; Massab and Bryant, 1999; Murphy and Coelingh 2002). Thus, the 6:2 reassortant vaccines generated annually possess stable phenotypic traits of the donor virus and the antigenic properties conferred by the HA and NA of the recommended circulating viruses.

### 1.1.2  Russian Vaccines

Live attenuated influenza vaccines derived from *ca* MDVs have also been developed and licensed for use in Russia. Influenza A vaccines were based on the influenza A/Leningrad/134/57 (H2N2) virus which was passaged in embyronated hens' eggs at least 20 times, leading to attenuation; this strain subsequently underwent 17 or 47 further passages in eggs at reduced temperatures (primarily at 25–26°C), resulting in two *ca* donor viruses: the influenza A/Leningrad/137/17/57 *ca* (H2N2) (A/Len/17 *ca*) (Alexandrova and Smorodintsev 1965) virus and the influenza A/Leningrad/137/47/57 *ca* (H2N2) (A/Len/47 *ca*) (Garmashova et al. 1984) virus, respectively. The live attenuated vaccines based on these viruses and the influenza B donor virus, B/USSR/60/69, have been widely used in Russia for annual influenza vaccination in both children and adults (Kendal 1997).

## 1.2  *Genetic Characterization and Stability of Phenotypic Traits*

Sequence comparisons of the entire genomes of both US and Russian MDVs with their *wt* counterparts have identified coding changes in several gene products (see Table 1). Through the generation of single gene reassortants between the *wt* and MDV-A viruses, the phenotypic determinants of the donor viruses have been mapped to the internal protein gene segments. Using traditional single gene reassortment techniques, the PA and PB2 genes were identified as specifying the *ca* phenotype of the influenza A/Leningrad/137/17/57 *ca* (H2N2) donor virus, while the PB1 and PB2 genes specified the *ts* phenotype. The specific residues that confer these phenotypes have not been identified.

The application of reverse genetics techniques has led to the identification of five major loci on the PB2, PB1, and NP genes of the influenza A/Ann Arbor/6/60 *ca* (H2N2) virus that specify the *ts* phenotype (Jin et al. 2003) (see Table 1). Mutational analysis revealed that two amino acid loci in the NP gene and one amino acid loci in the PA gene controlled the expression of the *ts* phenotype in the influenza B/Ann Arbor/1/66 *ca* (H2N2) virus; these residues in combination with two additional residues on M1 controlled the *att* phenotype (Hoffmann et al. 2005). The polygenic basis for these phenotypes likely contributes to their stability.

The stability of these attenuating phenotypes has been a critical feature in the development of the LAIV for clinical use. Clinical studies have repeatedly demonstrated that the vaccine viruses shed in respiratory secretions retain the *att*, *ca*, and *ts* phenotypes (Keitel and Piedra 1998; Youngner et al. 1994; Buonagurio et al. 2006). The genetic stability of the A/AA/6/60 *ca*, A/Len/17 *ca* and A/Len/47 *ca* donor viruses isolated following vaccination and replication has also been demonstrated by sequence analysis (Klimov et al. 1995, 1996; Cha et al. 2000; Buonagurio et al. 2006).

**Table 1** Amino acid sequence comparison of influenza A *wt* and *ca* master donor viruses (from Subbarao and Katz 2004)

| | A/Ann Arbor/6/60[a] | | | A/Leningrad/137/57[b] | | | |
|---|---|---|---|---|---|---|---|
| Gene product | Amino acid residue | *wt* | *ca* | Amino acid residue | *wt* | *ca 17* | *ca 47* |
| PB2 | 265[c] | N | S | 478 | V | L | L |
| PB1 | 391[c] | K | E | 265 | L | N | N |
| | 457 | E | D | 317 | M | – | I |
| | 581[c] | E | G | 591 | V | I | I |
| | 661[c] | A | T | | | | |
| PA | 613 | K | E | 28 | L | P | P |
| | 715 | L | P | 341 | V | L | L |
| NP | 23 | T | N | 341 | I | – | I |
| | 34[c] | N | G | | | | |
| M1 | – | – | – | 15 | I | V | V |
| M2 | 86 | A | S | 86 | A | T | T |
| NS1 | 153 | A | T | – | – | – | – |
| NS2 | – | – | – | 100 | M | I | I |
| | B/Ann Arbor/1/66[d] | | | | | | |
| Gene product | Amino acid residue | *wt* | *ca* | | | | |
| PB2 | 630 | S | R | | | | |
| PA | 431 | V | M | | | | |
| | 497 | Y | H | | | | |
| NP | 55 | T | A | | | | |
| | 114 | V | A | | | | |
| | 410 | P | H | | | | |
| | 509 | A | T | | | | |
| M1 | 159 | H | Q | | | | |
| | 183 | M | V | | | | |

[a]Cox et al. (1988)
[b]Klimov et al. (1992)
[c]Amino acids associated with ts phenotype, as defined in Jin et al. (2003)
[d]Chen et al. (2006)

## 1.3   Clinical Trials of Safety, Immunogenicity, and Efficacy

### 1.3.1   Safety and Transmissibility

In the United States, during the development of the vaccine leading to licensure of the current trivalent LAIV, cold-adapted vaccines were administered in clinical studies to over 20,000 people aged six months to 93 years (Izurieta et al. 2005). In these studies, the vaccines were well tolerated and were associated with minimal side effects. Among the placebo-controlled trials leading up to licensure, in an effectiveness trial involving healthy adults 18–49 years of age, a higher rate of upper respiratory symptoms (i.e., sore throat and rhinorrhea) was reported among LAIV recipients compared to placebo recipients, but fever and other reactogenicity events occurred at a similar rate in both groups (Nichol et al. 1999). In a trial among children 15–70 months of age, higher rates of upper respiratory symptoms (e.g., rhinorrhea) and low-grade fever ($\geq$37.8°C) were noted among vaccinees after one dose, although this increase was not noted in a subgroup analysis of children $\geq$60 months of age (Belshe et al. 1998, 2004). While a high level of safety has been reported in both pediatric and adult populations receiving the Russian LAIV (Rudenko et al. 1996) and during postlicensure surveillance of the US LAIV thus far (Izurieta et al. 2005), a recent multinational study comparing the safety of the US LAIV formulation with the trivalent inactivated influenza vaccine (TIV) in children 6–59 months of age raised some questions regarding the incidence of medically significant wheezing among younger children receiving LAIV (Belshe et al. 2007). Overall, there was no significant difference in adverse events between the two groups; however, a trend towards an increased rate of medically significant wheezing was noted in LAIV recipients younger than 12 months of age after dose 1 (3.8 vs. 2.1%; $p = 0.08$); in a post-hoc analysis, a trend towards higher rates of hospitalization for any cause was also observed in this age group. These observations will likely need further study prior to expanding the use of LAIV in children less than one year of age.

Infectivity of the cold-adapted vaccines has been studied in trials using monovalent, bivalent, and trivalent vaccine preparations (containing A/H1N1, A/H3N2, and influenza B). The 50% human infectious dose ($HID_{50}$) for monovalent preparations ranges from $10^{4.9}$ to $10^{6.4}$ 50% tissue culture infectious doses ($TCID_{50}$) in healthy susceptible adults; in seronegative children, this range is approximately ten-fold lower (Keitel and Piedra 1998). Transmission of the vaccine viruses to susceptible contacts has not been documented in adults. Virus shedding in susceptible adult vaccine recipients is generally limited to 1–2 days within the first six days after vaccination. In addition, titers of the vaccine viruses in the respiratory secretions of adult vaccinees typically range from 1 to 100 $TCID_{50}$, well below the $HID_{50}$ of the vaccine for adults (Keitel and Piedra 1998). Given the higher levels of viral replication and shedding in the respiratory tracts of infants and children, establishing the nontransmissibility of LAIV viruses among this population has been critical. Lack of transmission of vaccine viruses between vaccinees and placebo recipients has been observed in all but one study that involved children aged 9–36 months

(Vesikari et al. 2006). Among this group, although 80% of the 98 vaccine recipients shed at least one vaccine strain, there was only one instance of confirmed transmission of an (influenza B) vaccine virus (an estimated risk of 0.58% based on the Reed–Frost model).

### 1.3.2 Immunogenicity and Efficacy

Immunity to influenza A viruses develops as a result of antibodies directed at the HA and NA glycoproteins (Wright et al. 2007). Both serum IgG and mucosal IgA antibodies contribute to resistance to infection in humans (Clements et al. 1986; Belshe et al. 2000a). A serum hemagglutination inhibition (HAI) titer of ≥40 or a fourfold rise in HAI titer is correlated with protection against seasonal influenza; correlates of protection are less well defined for mucosal antibodies, although these antibodies are likely important mediators of immune protection induced by LAIV.

LAIV elicits both serum and mucosal antibodies in children and adults (Clements et al. 1986; Johnson et al. 1986), although the immunogenicity of LAIV is dependent on several factors, including age of the recipient, presence of pre-existing antibody from previous influenza infection and the strain of virus used. Inactivated influenza virus vaccines induce higher serum HAI, IgG, and IgA antibodies compared with *ca* reassortant vaccines (Clements and Murphy 1986; Treanor et al. 1999), while LAIV induces significantly higher nasal wash IgA antibodies (Boyce et al. 1999; Beyer et al. 2002). In seronegative adults and in children older than six months of age, LAIV induces a protective serum and mucosal antibody response (Johnson et al. 1986); in infants younger than six months of age, antibody response is lower (Karron et al. 1995). In seropositive adults, LAIV augments serum and nasal antibody IgA titers and protects against viral shedding following homologous challenge (Clements et al. 1985; Treanor et al. 1990). LAIV is poorly immunogenic in the elderly, likely due to high levels of pre-existing serum and local antibody (Powers et al. 1992).

LAIV has been shown to be highly effective in children. Prior to licensure, several clinical trials demonstrated high rates of efficacy of the trivalent *ca* LAIV in preventing culture-confirmed influenza in healthy children >15 months of age (Belshe et al. 1998, 2000a,b). More recently, superior efficacy of LAIV compared with inactivated vaccine was observed in protecting against culture-proven influenza among children 6–59 months of age (Belshe et al. 2007). Among children, this protection has been observed for both antigenically well-matched and drifted viruses (Ashkenazi et al. 2006; Belshe et al. 2000a,b, 2007; Mendelman et al. 2004).

In adults <65 years of age, the effectiveness and efficacy of LAIV is comparable to the inactivated influenza virus vaccine. In two large-scale efficacy studies, the LAIV and inactivated influenza virus vaccine demonstrated comparable protection against culture-confirmed influenza (Edwards et al. 1994; Treanor et al. 1999). Notably, however, in a recent study conducted in a year in which the circulating virus was antigenically drifted from the vaccine virus, the protective efficacy of

LAIV as determined by culture, PCR, or serology was lower than the inactivated influenza virus vaccine, although the lower relative efficacy did not achieve statistical significance and appeared to be due to reduced protection against influenza B viruses (Ohmit et al. 2006).

## 2 Live Attenuated Pandemic Influenza Vaccine Candidates

### 2.1 Immunity Against Influenza

During an influenza infection, antibodies directed against the HA and NA glycoproteins are the major mediators of resistance to infection. In seasonal influenza, the levels of these serum antibodies are the most commonly used correlates of protection. The role of the innate and cellular immune response to influenza infection is not as well elucidated, but these mechanisms play a role in viral clearance and may play a role in protection from severe morbidity. Although LAIVs induce lower levels of serum antibody compared with inactivated influenza virus vaccines (Clements et al. 1986; Treanor et al. 1999), protective efficacy is comparable, underscoring a potentially significant role for mucosal and cellular immunity in protection. Furthermore, evidence of breadth of protection against drift variants in animal models (Armerding et al. 1982; Delem 1977) and clinical studies (Treanor et al. 1999; Ashkenazi et al. 2006; Belshe et al. 2007) of LAIV suggest that mucosal antibody and T cell responses (Powell et al. 2007) may play a role in cross-protection from antigenically distinct viruses within the same subtype. While much remains to be learned about the immune response to avian influenza (AI), the observations derived from experience with seasonal influenza suggest that live attenuated vaccines may be of great value in the event of a pandemic.

### 2.2 Generation of Pandemic LAIV Candidates

The live attenuated pandemic vaccine candidates that have been studied to date are generated by plasmid-based reverse genetics or reassortment (as shown in Fig. 1). Reassortant pandemic vaccine candidates based on the design of the seasonal LAIV consist of the six attenuating internal protein genes from a *ca* donor virus (either the influenza A/Ann Arbor/6/60 (H2N2) virus or the influenza A/Leningrad/57 (H2N2) virus) and the HA and NA genes of the AI subtype of interest. Using reverse genetics, live attenuated vaccine candidates have also been generated with M2 cytoplasmic tail deletions (Watanabe et al. 2007), elastase-dependent hemagglutinin cleavage sites (Gabriel et al. 2007) and truncated NS1 proteins (Vincent et al. 2007; Richt et al. 2006). Several preclinical studies have been conducted to evaluate these vaccine candidates (Table 2).

## 2.3 Preclinical Studies of Pandemic LAIV Candidates

### 2.3.1 H5 Vaccine Candidates

While several subtypes of AI have caused human illness, H5N1 viruses remain the most significant with regard to total number of cases as well as case fatality rate, resulting in a concern that this subtype represents the one most likely to cause a devastating pandemic. This subtype thus has been the main target for the generation of pandemic vaccine candidates.

H5N1Reassortant Vaccines

Several preclinical studies have been conducted to characterize the safety and immunogenicity of live attenuated vaccine H5N1 candidates. In 1999, Li and colleagues reported the generation of two live attenuated vaccine virus candidates based on two 1997 human H5N1 virus isolates (Li et al. 1999). The HA cleavage sites of these isolates were modified to resemble an avian virus of low pathogenicity by deleting five of the six basic amino acids at the cleavage site because the highly cleavable HA is a virulence determinant for pathogenicity in chickens (Horimoto and Kawaoka 1994; Klenk and Garten 1994; Perdue et al. 1997) and in mice (Hatta et al. 2001); additionally, a threonine was added upstream of the remaining arginine to stabilize a potential glycosylation site (Li et al. 1999). Li and colleagues noted that these recombinant viruses MVS156 (based on the influenza A/Hong Kong/156/97 (H5N1) virus) and MVS483 (based on the influenza A/Hong Kong/483/97 (H5N1) virus) replicated to high titer in embryonated eggs ($10^{9.4}$ and $10^{8.5}$ $EID_{50}$/mL, respectively) while retaining an attenuated phenotype in ferrets. Infectious virus was recovered from the nasal turbinates but not from the lungs of ferrets that were inoculated intranasally with $10^{7.0}$ $EID_{50}$ of the recombinant vaccine viruses. The antigenicity of both recombinant vaccine viruses resembled their *wt* parental viruses despite the modification of the HA cleavage site.

More recently, vaccine candidates have been developed from additional H5N1 viruses. Suguitan et al. generated 6:2 reassortant H5N1 vaccine candidates derived from three different *wt* H5N1 viruses (isolated in 1997, 2003, and 2004) and evaluated their pathogenicity in chickens and immunogenicity and efficacy in mice and ferrets (Suguitan et al. 2006). For each reassortant vaccine, two plasmids encoding a modified H5 HA and N1 NA gene from each of the *wt* viruses (the influenza A/VN/1203/2004 (H5N1) virus, the influenza A/HK/213/2003 (H5N1) virus and the influenza A/HK/491/1997 (HA) and A/HK/486/1997 (NA) viruses) were combined with the six internal protein genes of the influenza A/AA/6/60 *ca* (H2N2) virus. The HA of each reassortant virus was modified by a deletion of the sequence encoding the four basic amino acids at the HA cleavage site.

These three H5N1 *ca* vaccine viruses displayed trypsin dependency and a *ts* phenotype in vitro as well as attenuation in vivo in chickens following both intravenous

**Table 2** Summary of live attenuated pandemic influenza vaccine preclinical studies

| Vaccine candidate | Methods | Findings | Reference |
|---|---|---|---|
| **H5** | | | |
| H5N1 reassortant vaccines A/Hong Kong/156/97x A/Ann Arbor/6/60 *ca* A/Hong Kong/483/97x A/Ann Arbor/6/60 *ca* | Two reassortant 6:2 H5N1 vaccines were generated containing a modified HA and the NA gene of the Hong Kong H5N1 *wt* virus with the six internal protein gene segments from the influenza A/AA/6/60 *ca* virus | Reassortant viruses were of low pathogenicity for poultry and replicated to high titer in embryonated eggs Reassortant viruses protected against heterologous challenge in chickens Reassortant viruses were attenuated and immunogenic in ferrets | Li et al. (1999) |
| A/Hong Kong/213/03x A/Ann Arbor/6/60 *ca* A/Vietnam/1203/04x A/Ann Arbor/6/60 *ca* A/Hong Kong/491/97 (HA) x A/Hong Kong/486/97 (NA)x A/Ann Arbor/6/60 *ca* | Three reassortant 6:2 H5N1 vaccines were generating containing a modified HA and the NA gene of the H5N1 *wt* virus with the six internal protein gene segments from the influenza A/AA/6/60 *ca* virus | Reassortant viruses were of low pathogenicity in chickens and were attenuated in mice and ferrets In mice, inoculation with reassortant viruses protected against lethality from homologous and heterologous challenge after one dose and conferred protection against viral replication after two doses In ferrets, two doses of a reassortant virus provided complete protection from pulmonary viral replication of homologous and heterologous challenge viruses. | Suguitan et al. (2006) |
| H5N2 reassortant vaccines A/duck/Potsdam/1402-6/86 x A/Leningrad/134/17/57 *ca* | A 7:1 reassortant H5N2 vaccine containing the HA gene from a nonpathogenic H5N2 virus (A/duck/Potsdam/1402-6/86) and the NA and internal protein genes from A/Leningrad/134/17/57 *ca* | The reassortant virus was not pathogenic in chickens and was attenuated in mice Inoculation with the reassortant virus generated neutralizing antibody only against homologous virus Inoculation with the reassortant virus generated a cross-reactive IgG and nasal wash IgA response despite a lack of neutralizing antibody and conferred protection against lethality and viral replication following heterologous H5N1 virus challenge with HK/483/97 and HK/213/2003 | Desheva et al. (2006) |

| | | | |
|---|---|---|---|
| M2 cytoplasmic tail mutants A/Vietnam/1203/04 | A cDNA synthesized A/Vietnam/1203/04 M2 mutant (with an 11 amino acid deletion at the C-terminus) with a modified HA cleavage site | The VN1203-M2delII-HAavir virus replicated efficiently in vitro and was attenuated in mice<br>Inoculation with the VN1203-M2delII-HAavir virus provided protection from lethal challenge with homologous and heterologous H5N1 viruses (A/VN/1203/04 and A/Indonesia/7/05)<br>Serum IgG antibody response and tracheal wash IgG and IgA levels were higher in immunized mice compared with control groups | Watanabe et al. (2007) |
| **H7**<br>Elastase-dependent HA cleavage site mutants | Replacement of the polybasic HA cleavage site of a highly pathogenic H7N7 virus with an elastase motif (designated $SC35M_H$-E) | The $SC35M_H$-E mutants replicated to titers comparable to parental virus in the presence of elastase in vitro<br>The $SC35M_H$-E mutant virus was attenuated in mice<br>Four weeks after immunization with $10^4$, $10^5$, or $10^6$ PFU of the $SC35M_H$-E virus, all mice survived lethal challenge with 100 $LD_{50}$ of highly pathogenic parental $SC35M_H$ | Gabriel et al. (2007) |
| **H9**<br>H9N2 reassortant vaccines A/Chicken/Hong Kong/G9/97 x A/AA/6/60 *ca* | A 6:2 reassortant H9N2 vaccine containing the HA and NA from A/chicken/Hong Kong/G9/97 and the internal protein genes segments from the A/AA/6/60 *ca* | The G9/AA*ca* reassortant virus was not pathogenic in poultry<br>Reassortant virus displayed *ca*, *ts*, and *att* phenotypes of A/AA/6/60 *ca* parent virus<br>In mice, the reassortant virus provided protection from viral replication in the respiratory tract following homologous and heterologous H9N2 virus challenge | Chen et al. (2003) |

and intranasal administration and in mice and ferrets following intranasal administration. When administered intranasally, the H5N1 1997 and H5N1 2004 *wt* viruses were lethal for mice at 50% Lethal dose ($LD_{50}$) values of $10^2$ and $10^{0.4}$ $TCID_{50}$ respectively; the H5N1 2003 *wt* was lethal for a few mice only at a dose $\geq 10^6$ $TCID_{50}$. In contrast, lethality in mice was not observed for any of the three H5N1 *ca* reassortant viruses, even at the highest dose administered ($LD_{50} \geq 10^7$ $TCID_{50}$). Three days following inoculation in mice, the level of replication of the H5N1 1997 *ca* and the H5N1 2004 *ca* viruses was lower than the corresponding *wt* virus (with a mean reduction of approximately 800-fold) in the upper and lower respiratory tract; replication in the respiratory tract did not differ significantly between the H5N1 2003 *ca* and *wt* viruses. Following a single dose of $10^6$ $TCID_{50}$ of the H5N1 2004 *wt* virus administered intranasally, the virus was detected in the brains of mice on days 2 and 4 postinoculation, while the H5N1 2004 *ca* virus was not detected in the brain. A similar lack of neurotropism was also noted for the H5N1 1997 and 2003 *ca* viruses (unpublished data). In ferrets, the H5N1 2004 *wt* and the H5N1 1997 *wt* viruses replicated efficiently in the upper and lower respiratory tract, while the H5N1 2004 *ca* and H5N1 1997 *ca* viruses were not detected in the lower respiratory tract indicating that they were attenuated for ferrets. The H5N1 1997 *ca* and the H5N1 2004 *ca* were not detected in the brains of ferrets infected intranasally, but the H5N1 *wt* parent viruses disseminated to the brain following intranasal infection (Suguitan et al. 2006).

To assess the immunogenicity of the H5N1 *ca* candidate vaccines, the vaccines were administered to mice intranasally at a dose of $10^6$ $TCID_{50}$, and serum antibody responses were measured 28 days later by HAI and microneutralization assay. A single dose of vaccine did not induce a detectable neutralizing or HAI antibody response. However, four weeks following a second dose of vaccine, the titer and cross-reactivity of the neutralizing and HAI antibody response were significantly increased. Notably, a subsequent study demonstrated that the HAI antibody titer continued to rise following a single dose of the H5N1 2004 *ca* virus eight weeks after a single dose of vaccine, and geometric mean HAI titers were higher than titers observed at four weeks after vaccination, though a second dose of vaccine elicited a boost in HAI and neutralizing antibody titers (Suguitan et al. 2006).

Efficacy of each H5N1 *ca* vaccine candidate was evaluated with homologous and heterologous challenge with *wt* H5N1 viruses. Mice immunized with a H5N1 *ca* vaccine survived challenge with 50, 500, or 5,000 $LD_{50}$ of a homologous or heterologous H5N1 *wt* virus; this protection from lethality was observed in the absence of detectable serum neutralizing antibodies. Following a single dose of vaccine, challenge virus replicated in the respiratory tract, though at a lower titer than in mock-immunized control animals; after two doses of vaccine, complete protection from *wt* virus replication was observed in the respiratory tract. Upon challenge with newly emerged and antigenically distinct 2005 H5N1 viruses, a single dose of the H5N1 2004 *ca* candidate vaccine provided protection from lethality from both viruses; moreover, mice that received two doses of the 1997, 2003, or 2004 H5N1 *ca* vaccines were protected from replication of the influenza A/Indonesia/2005 (H5N1) virus in the lungs and brain. Similar efficacy was also noted in ferrets: two doses of the 1997 or 2004 H5N1 *ca* viruses conferred complete protection from viral replication in the lower respiratory tract against homologous and heterologous challenge with the influenza A/Indonesia/5/2005 (H5N1) virus. Taken together, the preclinical studies established that reassortant H5N1 vaccines with a modified multibasic cleavage site were safe and efficacious in protecting mice and ferrets from homologous and heterologous H5N1 virus challenge.

H5N2 Reassortant Vaccines

Another approach to the development of a live attenuated H5 pandemic vaccine has been the generation of a 7:1 reassortant H5N2 virus comprised of the HA gene from a nonpathogenic influenza A/duck/Potsdam/1402-6/86 (H5N2) (Pot/86) virus on the influenza A/Leningrad/134/17/57 (H2N2) (Len17 *ca*) virus backbone (Desheva et al. 2006).

This 7:1 reassortant H5N2 vaccine candidate (designated Len17/H5) was evaluated for pathogenicity in chickens and mice. Immunogenicity and protective efficacy of the vaccine were assessed in mice against challenge with two highly pathogenic *wt* H5N1 viruses (the influenza A/Hong Kong/483/97 and A/Hong Kong/213/03 viruses). Following intravenous or intranasal inoculation of chickens with the Len17/H5 virus, clinical signs of disease and death were not noted; three days postinfection, the reassortant virus was not isolated from the respiratory or gastrointestinal tract. In mice, the reassortant virus was restricted in replication in the lower respiratory tract and was not lethal. Virus was recovered from the lungs of only one of three mice (mean titer $10^{2.1}$ $EID_{50}$/ml) three days after they were infected intranasally with Len17/H5. The level of replication of the Len17/H5 virus was comparable to the Len17 *ca* parent (mean titer $10^{2.3}$ $EID_{50}$/ml), and this titer was lower than that of the Pot/86 parent virus (mean titer $10^{6.3}$ $EID_{50}$/ml). However, the Len17/H5 virus replicated efficiently in the upper respiratory tract and was recovered from the nasal turbinates at three days postinfection at higher titers than the Len17 *ca* and Pot/86 viruses (mean titer of $10^{3.5}$ $EID_{50}$/ml compared with $10^{2.7}$ $EID_{50}$/ml and $10^{1.6}$ $EID_{50}$/ml of the parental Len17 *ca* and Pot/86 viruses, respectively). These data suggest that the Len17/H5 virus was attenuated in the lower respiratory tract of mice.

Immunogenicity of the Len17/H5 reassortant was evaluated in mice by measuring neutralizing antibody and nasal and serum IgG and IgA ELISA antibody. Six weeks following intranasal administration of 300 50% mouse infectious doses ($MID_{50}$) of Len17/H5, neutralizing antibody to the homologous Pot/86 (H5N2) virus was detected. However, cross-reactive neutralizing antibodies to heterologous highly pathogenic H5N1 HK/156/97 and HK/483/97 viruses were not detected. In contrast, intramuscular vaccination with the Len17/H5 reassortant virus elicited cross-reactive neutralizing antibody to the HK/156/97 virus. Following intranasal inoculation with the Len17/H5 virus, substantial levels of H5N1 virus-specific (HK/156/97) serum IgG and respiratory tract IgA antibodies were detected. Notably, intranasal inoculation with the Len17 *ca* vaccine donor also induced a cross-reactive IgA in the nasal wash, suggesting that the local IgA response was more cross-reactive than the serum IgG response.

To assess the protective efficacy of Len17/H5 vaccine, mice were challenged with 50 $LD_{50}$ of the influenza A/HK/483/97 (H5N1) virus six weeks after receiving either PBS, 300 $MID_{50}$ Len17/H5 (i.n.), 10 mg of Len17/H5 (i.m.) or the parental Pot/86 (H5N2) and Len17 *ca* (H2N2) viruses intranasally. The mice that were mock-immunized with PBS died 5–9 days following H5N1 virus challenge. None of the mice inoculated with the parental *wt* Pot/86 (H5N2) virus displayed signs of disease or died over the 14-day postchallenge observation period; virus was not detected in the lungs, nose, brain, or thymus of mice that were sacrificed at day 6 postchallenge. Day 6 was selected as the time point to evaluate cross-protection because mice had previously been shown to have high titers of the HK/483/97 virus in the respiratory tract on day 6 postinfection. Among the mice that received the parental Len17 *ca* virus intranasally, six of eight mice died during the surveillance period, with a maximum weight loss of 19%; virus titers in the lungs were modestly but not significantly lower than in control mock-immunized mice but were significantly lower in the brain, thymus, and upper respiratory tract. The mice that received the Len17/H5 vaccine survived challenge with the HK/483/97 (H5N1) virus, although modest weight loss was noted. Low titers of virus were detected in the lungs of two of five of the Len17/H5-vaccinated mice; virus was not detected in the other organs. Subsequently, a second group of mice were challenged with 100 $LD_{50}$ of the influenza A/HK/213/03 (H5N1) virus six weeks after receiving either PBS, 300 $MID_{50}$ Len17/H5 (i.n.), 10 mg of Len17/H5 (i.m.), or the parental Pot/86 (H5N2) or Len17 *ca* viruses intranasally. Mice receiving the Len17/H5 vaccine intranasally were also protected from heterologous challenge with the influenza A/HK/213/2003 (H5N1) virus: nine of ten mice survived and lacked detectable challenge virus in the lungs. In summary, like the H5N1 AA *ca* reassortant viruses, the H5N2 Len17 *ca* reassortant viruses are attenuated and efficacious in protecting mice against homologous and heterologous *wt* virus challenge. While the generation of a live attenuated vaccine utilizing a nonpathogenic H5 virus may offer potential advantages with regard to containment facilities and manufacturing capacity, protection from challenge despite the lack of neutralizing antibody induced by immunization suggests that correlates of protection are yet to be defined. Delineating the role of NA in the immune response to avian influenza viruses may also be important in the further development of these candidate vaccines.

Vaccines Using M2 Cytoplasmic Tail Mutants

The influenza A virus M2 protein contains three structural domains: an extracellular domain of 24 amino acids, a transmembrane domain of 19 amino acids, and a cytoplasmic tail domain of 54 amino acids. The cytoplasmic tail domain plays an important role in viral assembly and morphogenesis (Watanabe et al. 2007). Live attenuated H5N1 vaccines with deletions in the M2 cytoplasmic tail have also recently been reported to show efficacy against homologous and heterologous *wt* virus H5N1 challenge. Using plasmid-based reverse genetics, a series of M2 cytoplasmic tail deletion mutants were generated in the highly pathogenic A/Vietnam/1203/04 (H5N1) virus. Of these mutants, one M2 mutant virus with an 11 amino acid deletion from the C-terminus (designated VN1203-M2del11) grew as well as the *wt* virus in vitro. Moreover, in vivo, the recombinant VN1203-M2del11 virus was reduced by more than ten-fold in the lungs and hundred-fold in nasal turbinates than the *wt* virus, and was not detected in the brains of mice. The HA gene of the VN1203-M2del11 mutant was subsequently modified by replacing sequences at the cleavage site with those from the HA of an avirulent virus, and the resulting virus (designated VN1203-M2del11-HAavir) was tested for attenuation and efficacy in mice.

In vivo, the VN1203-M2delll-HAavir virus was highly attenuated in mice. In mice infected with 100 plaque-forming units (PFU), the recombinant virus was recovered only from the lungs, and viral titers were ten-fold lower compared to the *wt* VN/1203/2004 virus. When 1,000 PFU of the virus was administered, virus was recovered from the lungs and nasal turbinates but the titers were lower than in mice infected with the *wt* virus. Even a dose of 100,000 PFU of the M2-del11-HAavir virus was not lethal for mice; the $LD_{50}$ of the VN1203-M2delll-HAavir virus was >100,000 PFU compared to 2.1 PFU for the *wt* VN/1203/2004 virus.

Three weeks after intranasal immunization with 100 or 1,000 PFU of the M2del11-HAavir virus, ELISA IgG levels in serum and IgG and IgA levels in tracheal washes were significantly higher than in mice that received PBS. However, no difference was detected in the antibody response in nasal washes of immunized mice compared with control mice, and neutralizing antibodies were not detected in the sera of immunized mice. To assess protective efficacy, mice were challenged with the homologous *wt* VN/1203/2004 (clade 1) or heterologous A/Indonesia/7/2005 (clade 2) virus 1 month following immunization. All of the M2del11-HAavir-immunized mice survived lethal challenge with either of the two highly pathogenic H5N1 viruses and did not show any signs of disease after challenge. In contrast, all control mice died or were euthanized due to symptoms by day 8 postchallenge. While high titers of *wt* virus were recovered from all of the organs of the mice in the control group, virus was not detected from any of the organs of the mice in the M2del11-HA virus vaccine group three days after challenge with VN/1203/2004 virus; a limited amount of virus ($10^{1.96}$ PFU/g) was detected in the nasal turbinates of one immunized mouse challenged with the Indonesia/7/2005 virus. Notably, the efficacy of protection from lethal challenge following immunization with 100 PFU or 1,000 PFU of the M2del11-HAavir virus was not associated with detectable neutralizing antibody response in the serum.

As seen in the studies with the H5N1 and H5N2 cold-adapted reassortant viruses, the M2 cytoplasmic tail mutant H5N1 viruses were attenuated and efficacious in protecting mice from lethal challenge with *wt* homologous and heterologous H5N1 viruses without eliciting a significant neutralizing antibody response. Watanabe et al. suggest that one dose provides complete protection because all eight gene segments of the vaccine virus match those of the *wt* H5N1 challenge virus. While this indeed may be a potential advantage in a live attenuated vaccine, safety concerns may exist regarding the potential for reversion and reassortment with circulating influenza viruses yielding pathogenic viruses.

### 2.3.2  H7 Vaccines

Direct transmission of other AI subtypes from infected poultry to humans has also occurred, highlighting the pandemic potential of this subtype. In addition to isolated cases of symptomatic human infection with highly pathogenic AI or low-pathogenic AI H7 following direct exposure (Webster et al. 1981; Kurtz et al. 1996; Banks et al. 1998; Kermode-Scott 2004), 89 cases of human H7N7 infection were reported in 2003 in the Netherlands during a poultry outbreak, and there was evidence suggestive of human to human transmission (Koopmans et al. 2004; Fouchier et al. 2004). A serologic analysis of 185 serum samples obtained from poultry workers in Italy revealed anti-H7 antibodies in 3.8% during a period in 2003 when H7N3 viruses were circulating in domestic poultry (Puzelli et al. 2005). More recently, an HP H7N3 outbreak emerged in domestic poultry in British Columbia, Canada, during which two cases of conjunctivitis were confirmed in humans (Hirst et al. 2004; Tweed et al. 2004).

Vaccines Using Elastase-Dependent HA Cleavage Site Mutants

Another strategy in the development of live attenuated pandemic vaccines has been the generation of HA cleavage site mutants. Using reverse genetics, the polybasic HA cleavage site of an H7N7 virus was replaced with an elastase motif. The $SC35M_H$ virus was derived from serial passage of the highly pathogenic A/Seal/Massachusetts/1/1980 (H7N7) virus in chick embryo fibroblasts and mouse lungs. An elastase motif was engineered into this virus and the elastase mutant virus recovered by reverse genetics was designated $SC35M_H$-E (Gabriel et al. 2007). In the presence of elastase, the $SC35M_H$-E mutant virus grew as well as the parent $SC35M_H$ virus in cell culture. However, in the absence of elastase in vivo, the $SC35M_H$-E virus was attenuated. Mice inoculated intranasally with $10^3$–$10^6$ PFU of the $SC35M_H$-E virus survived, while the parental $SC35M_H$ virus was lethal for mice at an $LD_{50}$ of $10^{1.4}$ PFU. In addition, the $SC35M_H$-E mutant virus was restricted in replication in vivo. Following intranasal inoculation in mice, the elastase-dependent $SC35M_H$-E virus was detected on day 1 in the lungs at a titer close to the inoculum dose, but the virus was undetectable by day 7. In comparison, the parental $SC35M_H$ replicated to high titer from days 1–7 postinfection in the lungs, heart, and brain of mice.

The investigators generated a 6:2 reassortant virus designated WSN-H7N7-E with the six internal protein genes of A/WSN/33 and the surface glycoprotein genes of $SC35M_H$-E, and an elastase-dependent virus designated SC35M-H1N1-E, a 6:2 reassortant virus with the six internal protein genes of SC35M and the HA and NA of WSN-E, an elastase-dependent mutant of A/WSN/33. One month following infection, mice vaccinated with $10^6$ PFU of any elastase-dependent virus including the SC35M$_H$-E, SC-35M-E, WSN-H7N7-E, and SC35M-H1N1-E developed high titers of serum IgG and mucosal IgA antibodies against the influenza A/Seal/Massachussetts/1/80 (H7N7) virus.

Four weeks following intranasal immunization, mice that were immunized with $10^5$ or $10^6$ PFU SC35M$_H$-E, SC35-E, or the WSN-H7N7-E virus survived challenge with 100 LD$_{50}$ of the highly pathogenic parental SC35M$_H$ virus. Mice immunized with lower doses of $10^3$ or $10^4$ PFU of SC35M$_H$-E, temporarily lost up to 20% of body weight, and 25% of the mice that received $10^3$ PFU SC35M$_H$-E died following challenge. None of the mice in the control group that received PBS survived the challenge. Notably, challenge virus replicated in the lungs of mice immunized with SC35M$_H$-E three days after challenge. Neutralizing antibodies against SC35M$_H$ were present in sera from mice immunized with $10^6$ and $10^5$ PFU SC35-E. Given the safety concerns over reassortment between vaccine and circulating influenza viruses, the concept of an exogenously controlled HA cleavage site is an intriguing one. The data from studies evaluating the SC35M$_H$-E mutant viruses suggest that these viruses are appropriately attenuated at a dose of $10^6$ PFU and were immunogenic and efficacious.

### 2.3.3   H9 Vaccines

In 1999, two cases of H9N2 infection associated with mild influenza symptoms were reported in Hong Kong (Peiris et al. 1999), and as many as five additional cases of H9N2 infection were reported from mainland China (Guo et al. 1999). Genetic analysis of these viruses revealed that the internal protein gene segments were closely related to those of the H5N1 viruses isolated in Hong Kong in 1997 (Lin et al. 2000). In 2003, a third case of an H9N2 infection was identified in a child in Hong Kong with influenza symptoms (Butt et al. 2005). Evidence of the ability of these viruses to cross the species barrier in conjunction with surveillance data revealing the cocirculation of H9N2 and H5N1 viruses in Asian poultry markets suggests that H9 viruses may also have pandemic potential (Horimoto and Kawaoka 2001).

H9N2 Reassortant Vaccine Candidates

Chen and colleagues (Chen et al. 2003) developed a 6:2 reassortant H9N2 vaccine candidate (designated G9/AA *ca*) that derived the HA and NA genes from the influenza A/chicken/Hong Kong/G9/97 (G9) (H9N2) virus and internal protein gene segments from the influenza A/AA/6/60 *ca* (H2N2) virus. The reassortant G9/AA *ca* virus retained the *ca*, *ts*, and *att* phenotypes of the parental influenza A/AA/6/60 *ca* (H2N2) virus. The antigenicity of the G9/AA *ca* resembled that of the parental G9 virus. The G9/AA *ca* virus was administered to chickens intravenously to assess its pathogenicity for poultry. When administered intravenously, although no clinical signs of disease or death were observed, the *wt* G9 virus replicated to high titer in the oropharynx and cloaca of chickens. The G9/AA *ca* virus was not recovered from the oropharynx or cloaca of chickens and did not cause clinical signs of disease or death.

The G9/AA *ca* virus was restricted in replication in the respiratory tract of mice (Chen et al. 2003) and ferrets (unpublished data). In mice, the G9/AA *ca* virus elicited HAI titers of 320 and 40 against homologous and heterologous A/Hong Kong/1073/99 (H9N2) virus, while mice that received the G9 parental virus developed HAI titers of 1,280 and 80 to these viruses. The level of HAI antibodies generated by the G9/AA *ca* virus were four-fold lower than the level generated by the parental G9 virus, and this may reflect the lower replication of the reassortant virus compared to the *wt* parental virus.

The reassortant vaccine virus demonstrated protective efficacy in mice against homologous and heterologous H9N2 virus challenge. Three days after challenge with two heterologous *wt* H9N2 viruses (A/Hong Kong 1073/99 or A/chicken/Korea/96323/96), viral replication was not detected in the upper and lower respiratory tracts of mice immunized with the G9/AA *ca* or the parental AA *ca* or G9 viruses. In summary, the G9/AA *ca* virus demonstrated low pathogenicity in poultry, an attenuated phenotype in mice and ferrets, and immunogenicity and efficacy in mice.

### 2.3.4    Truncated NS1 Modified Live Virus Vaccines

Attenuated vaccines with modified NS1 proteins are discussed in depth in another chapter. These vaccines have also been proposed as live attenuated vaccines because the NS1 proteins of *wt* influenza A viruses are potent interferon antagonists, and truncations of the NS1 protein that express the first 99 or 126 amino acids of the NS1 protein can attenuate these viruses (Talon et al. 2000). Intranasal administration of a recombinant H3N2 swine influenza virus expressing a truncated NS1 protein (TX90-NS1D126 virus) has been shown to be highly attenuated in swine and protected pigs from homologous and heterologous (A/SW/CO/23619/99) virus challenge (Vincent et al. 2007). When protective efficacy against heterosubtypic challenge with H1N1 viruses was assessed, decreased fever and viral shedding were observed, but no significant difference was noted in lung lesions and pathogenicity.

## 2.4   Clinical Studies of LAIV Vaccines

Phase I trials of LAIV vaccines are currently underway based on preclinical data on several vaccines (see http://www.who.int/vaccine_research/diseases/influenza/flu_trials_tables).

# 3 Considerations Surrounding the Use of LAIV Pandemic Vaccines

Several questions surrounding the development of pandemic LAIV remain. Most significantly, we have yet to fully understand the immunologic response to AI viruses and the correlates of protection. Defining these correlates of protection will be crucial in interpreting outcomes from the nascent clinical trials. In addition, safety concerns exist regarding the use of LAIVs as pandemic vaccine candidates, and include the theoretical potential for reassortment with a circulating *wt* influenza virus and transmission of a novel reassortant virus to a wider population. However, the experience with seasonal LAIV points to both phenotypic and genotypic stability of the vaccine virus and an extremely low level of transmissibility of the vaccine virus.

The safety, efficacy, and effectiveness of LAIV for seasonal influenza also underscore several other potential advantages of live attenuated vaccines in the event of a pandemic. The robust protective efficacy of LAIV in children may be indicative of the multifaceted immunologic response generated by LAIV in an immunologically naïve population. In addition, given the low likelihood that the vaccine virus will exactly match the pandemic virus, evidence pointing to greater cross-protection from LAIV also argues in favor of the development of pandemic live attenuated vaccines. Furthermore, logistical considerations such as a single-dose administration and lack of injections may make the LAIV an attractive vaccine as well. As many of the pandemic LAIV candidates are currently based on licensed *ca* vaccines, the infrastructure for manufacture and distribution already exists (Luke and Subbarao 2006). While further characterization of the immunologic response to AI viruses is still needed, the experience with seasonal LAIV underscores the potential of live attenuated vaccines to play an important role in the event of a pandemic.

**Acknowledgment** This research was supported in part by the Intramural Research Program of the NIAID, NIH.

# References

Alexandrova G, Smorodintsev A. (1965) Obtaining of an additionally attenuated vaccinating cryophilic influenza strain. Rev Rown Inframicrobiol 2:179–189

Ali M, Maassab HF, Jennings R, Potter CW. (1982) Infant rat model of attenuation for recombinant influenza viruses prepared from cold-adapted attenuated A/Ann Arbor/6/60. Infect Immun 38(2):610–619

Armerding D, Rossiter H, Ghazzouli I, Liehl E. (1982) Evaluation of live and inactivated influenza A virus vaccines in a mouse model. J Infect Dis 145(3):320–323

Ashkenazi S, Vertruyen A, Arístegui J, Vertruyen A, Ashkenazi S, Rappaport R, Skinner J, Saville MK, Gruber WC, Forrest BD. (2006) Superior relative efficacy of live attenuated influenza vaccine compared with inactivated influenza vaccine in young children with recurrent respiratory tract infections. Pediatr Infect Dis J 25:870

Banks J, Speidel E, Alexander DJ. (1998) Characterisation of an avian influenza A virus isolated from a human is an intermediate host necessary for the emergence of pandemic influenza viruses? Arch. Virol 143:781–787

Belshe RB, Mendelman PM, Treanor J, King J, Gruber WC, Piedra P, Bernstein DI, Hayden FG, Kotloff K, Zangwill K, Iacuzio D, Wolff M. (1998) The efficacy of live attenuated, cold adapted, trivalent, intranasal influenzavirus vaccine in children. N Engl J Med 338:1405–1412

Belshe RB, Gruber WC, Mendelman PM, Mehta HB, Mahmood K, Reisinger K, Treanor J, Zangwill K, Hayden FG, Bernstein DI, Kotloff K, King J, Piedra PA, Block SL, Yan L, Wolff M. (2000a) Correlates of immune protection induced by live, attenuated, cold-adapted, trivalent, intranasal influenza virus vaccine. J Infect Dis 181(3):1133–1137

Belshe RB, Gruber WC, Mendelman PM. (2000b) Efficacy of vaccination with live attenuated, cold-adapted, trivalent, intranasal influenza virus vaccine against a variant (A/Sydney) not contained in the vaccine. J Pediatr 136:168–175

Belshe RB, Nichol KL, Black SB, Shinefield H, Cordova J, Walker R, Hessel C, Cho I, Mendelman PM. (2004) Safety, efficacy, and effectiveness of live attenuated, cold-adapted influenza vaccine in an indicated population aged 5–49 years. Clin Infect Dis 39(7):920–927

Belshe RB, Edwards KM, Vesikari T, Black SV, Walker RE, Hultquist M, Kemble G, Connor E. (2007) Live attenuated versus inactivated influenza vaccine in infants and young children. N Engl J Med 356(7):685–696

Beyer WE, Palache AM, de Jong JC, Osterhaus AD. (2002) Cold-adapted live influenza vaccine versus inactivated vaccine: systemic vaccine reactions, local and systemic antibody response, and vaccine efficacy. A meta-analysis. Vaccine 20(9–10):1340–1353

Boyce TG, Gruber WC, Coleman-Dockery SD, Sannella EC, Reed GW, Wolff M, Wright PF. (1999) Mucosal immune response to trivalent live attenuated intranasal influenza vaccine in children. Vaccine 18(1–2):82–88

Buonagurio DA, O'Neill RE, Shutyak L, D'Arco GA, Bechert TM, Kazachkov Y, Wang HP, DeStefano J, Coelingh KL, August M, Parks CL, Zamb TJ, Sidhu MS, Udem SA. (2006) Genetic and phenotypic stability of cold-adapted influenza viruses in a trivalent vaccine administered to children in a day care setting. Virology 347(2):296–306

Butt KM, Smith GJ, Chen H, Zhang LJ, Leung YH, Xu KM, Lim W, Webster RG, Yuen KY, Peiris JS, Guan Y. (2005) Human infection with an avian H9N2 influenza A virus in Hong Kong in 2003. J Clin Microbiol 43(11):5760–5767

Cha TA, Kao K, Zhao J, Fast PE, Mendelman PM, Arvin A. (2000) Genotypic stability of cold-adapted influenza virus vaccine in an efficacy clinical trial. J Clin Microbiol 38(2):839–845

Chen H, Matsuoka Y, Swayne D, Chen Q, Cox NJ, Murphy BR, Subbarao K. (2003) Generation and characterization of a cold-adapted influenza A H9N2 reassortant as a live pandemic influenza virus vaccine candidate. Vaccine 21(27–30):4430–4436

Chen Z, Aspelund A, Kemble G, Jin H. (2006) Genetic mapping of the cold-adapted phenotype of B/Ann Arbor/1/66, the master donor virus for live attenuated influenza vaccines (FluMist). Virology 345(2):416–4123

Clements ML, Murphy BR. (1986) Development and persistence of local and systemic antibody responses in adults given live attenuated or inactivated influenza A virus vaccine. J Clin Microbiol 23(1):66–72

Clements ML, Tierney EL, Murphy BR. (1985) Response of seronegative and seropositive adult volunteers to live attenuated cold-adapted reassortant influenza A virus vaccine. J Clin Microbiol 21(6):997–999

Clements ML, Betts RF, Tierney EL, Murphy BR. (1986) Serum and nasal wash antibodies associated with resistance to experimental challenge with influenza A wild-type virus. J Clin Microbiol 24(1):157–160

Cox NJ, Kitame F, Kendal AP, Massab HF, Naeve C. (1988) Identification of sequence changes in the cold-adapted, live attenuated influenza vaccine strain, A/Ann Arbor/6/60 (H2N2). Virology 167:553–567

Delem A. (1977) Protective efficacy of RIT 4025, a live attenuated influenza vaccine strain, and evaluation of heterotypic immunity to influenza A viruses in ferrets. J Hyg 79(2):203–208

Desheva JA, Lu XH, Rekstin AR, Rudenko LG, Swayne DE, Cox NJ, Katz JM, Klimov AI. (2006) Characterization of an influenza A H5N2 reassortant as a candidate for live-attenuated and inactivated vaccines against highly pathogenic H5N1 viruses with pandemic potential. Vaccine 24(47–48):6859–6866

Edwards KM, Dupont WD, Westrich MK, Plummer WD Jr, Palmer PS, Wright PF. (1994) A randomized controlled trial of cold-adapted and inactivated vaccines for the prevention of influenza A disease. J Infect Dis 169(1):68–76

Fodor E, Devenish L, Engelhardt OG, Palese P, Brownlee GG, Garcia-Sastre A. (1999) Rescue of influenza A virus from recombinant DNA. J Virol 73:9679–9682

Fouchier RA, Schneeberger PM, Rozendaal FW, Broekman JM, Kemink SA, Munster V, Kuiken T, Rimmelzwaan GF, Schutten M, Van Doornum GJ, Koch G, Bosman A, Koopmans M, Osterhaus AD. (2004) Avian influenza A virus (H7N7) associated with human conjunctivitis and a fatal case of acute respiratory distress syndrome. Proc Natl Acad Sci USA 101(5):1356–1361

Gabriel G, Garn H, Wegmann M, Renz H, Herwig A, Klenk HD, Stech J. (2007) The potential of a protease activation mutant of a highly pathogenic avian influenza virus for a pandemic live vaccine. Vaccine 26(7):956–965

Garmashova L, Plezhaev F, Alexandrova G. (1984) Cold-adapted A/Leningrad/134/47/57 (H2N2) strain, a specific donor of attenuation of live influenza vaccine for children, and recombinants based on its base. Vopr Virusol 29:28–31

Guo Y, Li J, Cheng X. (1999) Discovery of men infected by avian influenza H9N2 virus. China J Exp Virology 13(2):105–108

Hatta M, Gao P, Halfmann P, Kawaoka Y. (2001) Molecular basis for high viruleunce of Hong Kong H5N1 influenza A viruses. Science 293(5536):1840–1842

Hirst M, Astell CR, Griffith M, Coughlin SM, Moksa M, Zeng T, Smailus DE, Holt RA, Jones S, Marra MA, Petric M, Krajden M, Lawrence D, Mak A, Chow R, Skowronski DM, Tweed SA, Goh S, Brunham RC, Robinson J, Bowes V, Sojonky K, Byrne SK, Li Y, Kobasa D, Booth T, Paetzel M. (2004) Novel avian influenza H7N3 strain outbreak, British Columbia. Emerg Infect Dis 10(12):2192–2195

Hoffmann E, Neumann G, Hobom G, Webster RG, Kawaoka Y. (2000) "Ambisense" approach for the generation of influenza A virus: vRNA and mRNA synthesis from one template. Virology 267(2):310–317

Hoffmann E, Krauss S, Perez D, Webby R, Webster RG. (2002a) Eight-plasmid system for rapid generation of influenza virus vaccines. Vaccine 20(25–26):3165–3170

Hoffmann E, Mahmood K, Yang CF, Webster RG, Greenberg HB, Kemble G. (2002b) Rescue of influenza B virus from eight plasmids. Proc Natl Acad Sci USA 99(17):11411–11416

Hoffmann E, Mahmood K, Chen Z, Yang CF, Spaete J, Greenberg HB, Herlocher ML, Jin H, Kemble G. (2005) Multiple gene segments control the temperature sensitivity and attenuation phenotypes of ca B/Ann Arbor/1/66. J Virol 79(17):11014–11021

Horimoto T, Kawaoka Y. (1994) Reverse genetics provides direct evidence for a correlation of hemagglutinin cleavability and virulence of an avian influenza A virus. J Virol 68(5):3120–3128

Horimoto T, Kawaoka Y. (2001) Pandemic threat posed by avian influenza A viruses. Clin Microbiol Rev 14(1):129–149

Izurieta HS, Haber P, Wise RP, Iskander J, Pratt D, Mink C, Chang S, Braun MM, Ball R. (2005) Adverse events reported following live, cold-adapted, intranasal influenza vaccine. JAMA 294(21):2720–2725

Jin H, Lu B, Zhou H, Ma C, Zhao J, Yang CF, Kemble G, Greenberg H. (2003) Multiple amino acid residues confer temperature sensitivity to human influenza virus vaccine strains derived from cold-adapted A/Ann Arbor/6/60. Virology 306(1):18–24

Johnson PR, Feldman S, Thompson JM, Mahoney JD, Wright PF. (1986) Immunity to influenza A virus infection in young children: a comparison of natural infection, live cold-adapted vaccine, and inactivated vaccine. J Infect Dis 154(1):121–127

Karron RA, Steinhoff MC, Subbarao EK, Wilson MH, Macleod K, Clements ML, Fries LF, Murphy BR. (1995) Safety and immunogenicity of a cold-adapted influenza A (H1N1) reassortant virus vaccine administered to infants less than six months of age. Pediatr Infect Dis J 14(1):10–16

Keitel WA, Piedra PA. (1998) Live cold-adapted, reassortant influenza vaccines. In: Nicholson K, Webster RG, Hay AJ (eds) Textbook of influenza. Blackwell, Malden

Kendal AP. (1997) Cold-adapted live attenuated influenza vaccines developed in Russia: can they contribute to meeting the needs for influenza control in other countries? Eur J Epidemiol 13(5):591–609

Kermode-Scott B. (2004) WHO confirms avian flu infections in Canada. BMJ 328(7445):913

Klenk HD, Garten W. (1994) Host cell proteases controlling virus pathogenicity. Trends Microbiol 2(2):39–34

Klimov AI, Cox NJ, Yotov WV, Rocha E, Alexandrova GI, Kendal AP. (1992) Sequence changes in the live attenuated, cold-adapted variants of influenza A/Leningrad/134/57 (H2N2) virus. Virology 186(2):795–797

Klimov AI, Egorov AY, Gushchina MI, Medvedeva TE, Gamble WC, Rudenko LG, Alexandrova GI, Cox NJ. (1995) Genetic stability of cold-adapted A/Leningrad/134/47/57 (H2N2) influenza virus: sequence analysis of live cold-adapted reassortant vaccine strains before and after replication in children. J Gen Virol 76(Pt 6):1521–1525

Klimov A, Rudenko LG, Egorov AY, Romanova JR, Polezhaev FI, Alexandrova GI, Cox NJ. (1996) Genetic stability of Russian cold-adapted live attenuated reassortant influenza vaccines. In: Brown L, Hampson A, Webster R. (eds) Options for the control of influenza III. Elsevier, Amsterdam, pp 129–136

Koopmans M, Wilbrink B, Conyn M, Natrop G, van der Nat H, Vennema H, Meijer A, van Steenbergen J, Fouchier R, Osterhaus A, Bosman A. (2004) Transmission of H7N7 avian influenza A virus to human beings during a large outbreak in commercial poultry farms in the Netherlands. Lancet 363(9409):587–593

Kurtz J, Manvell RJ, Banks J. (1996) Avian influenza virus isolated from a woman with conjunctivitis. Lancet 348:901–902

Li S, Liu C, Klimov A, Subbarao K, Perdue ML, Mo D, Ji Y, Woods L, Hia S, Bryant M. (1999) Recombinant influenza A virus vaccines for the pathogenic human A/Hong Kong/97 (H5N1) viruses. J Infect Dis 179(5):1132–1138

Lin YP, Shaw M, Gregory V, Cameron K, Lim W, Klimov A, Subbarao K, Guan Y, Krauss S, Shortridge K, Webster R, Cox N, Hay A. (2000) Avian-to-human transmission of H9N2 subtype influenza A viruses: relationship between H9N2 and H5N1 human isolates. Proc Natl Acad Sci USA 97(17):9654–9658

Luke CJ, Subbarao K. (2006) Vaccines for pandemic influenza. Emerg Infect Dis 12 (1):66–72

Massab HF. (1967) Adaptation and growth characteristics of influenza virus at 25 degrees C. Nature 213(5076):612–614

Massab HF. (1969) Biologic and immunologic characteristics of cold-adapted influenza virus. J Immunol 102(3):728–732

Massab HF, Bryant ML. (1999) The development of live attenuated cold-adapted influenza virus vaccine for humans. Rev Med Virol 9(4):237–244

Massab HF, Kendal AP, Abrams GD, Monto AS. (1982) Evaluation of a cold-recombinant influenza virus vaccine in ferrets. J Infect Dis 146(6):780–790

Mendelman PM, Rappaport R, Cho I, Block S, Gruber W, August M, Dawson D, Cordova J, Kemble G, Mahmood K, Palladino G, Lee MS, Razmpour A, Stoddard J, Forrest BD. (2004) Live attenuated influenza vaccine induces cross-reactive antibody responses in children against an a/Fujian/411/2002–like H3N2 antigenic variant strain. Pediatr Infect Dis J (11):1053–1055

Murphy B, Coelingh K. (2002) Principles underlying the development and use of live attenuated cold-adapted influenza A and B virus vaccines. Viral Immunol 15(2):295–323

Nichol KL, Mendelman PM, Mallon KP, Jackson LA, Gorse GJ, Belshe RB, Glezen WP, Wittes J. (1999) Effectiveness of live attenuated intranasal influenza virus vaccine in healthy, working adults: a randomized controlled trial. JAMA 282(2):137–144

Neumann G, Watanabe T, Ito H, Watanabe S, Goto H, Gao P, Hughes M, Perez D, Donis R, Hoffmann E, Hobom G, Kawaoka Y. (1999) Generation of influenza A viruses entirely from cloned cDNAs. Proc Natl Acad Sci USA 96:9345–9350

Ohmit SE, Victor JC, Rotthoff JR, Teich ER, Truscon RK, Baum LL, Rangarajan B, Newton DW, Boulton ML, Monto AS. (2006) Prevention of antigenically drifted influenza by inactivated and live attenuated vaccines. N Engl J Med 355(24):2513–2512

Peiris M, Yuen KY, Leung CW, Chan KH, Ip PL, Lai RW, Orr WK, Shortridge KF. (1999) Human infection with influenza H9N2. Lancet 354(9182):916–917

Perdue ML, Garcia M, Senne D, Fraire M. (1997) Virulence-associated sequence duplication at the hemagglutinin cleavage site of avian influenza viruses. Virus Res (1997) 49(2):173–86

Powell TJ, Strutt T, Reome J, Hollenbaugh JA, Roberts AD, Woodland DL, Swain SL, Dutton RW. (2007) Priming with cold-adapted influenza A does not prevent infection but elicits long-lived protection against supralethal challenge with heterosubtypic virus. J Immunol 178(2):1030–1038

Powers DC, Murphy BR, Fries LF, Adler WH, Clements ML. (1992) Reduced infectivity of cold-adapted influenza A H1N1 viruses in the elderly: correlation with serum and local antibodies. J Am Geriatr Soc 40(2):163–167

Puzelli S, DiTrani L, Fabiani C, Campitelli L, De Marco MA, Capua I, Aguilera JF, Zambon M, Donatelli I. (2005) Serological analysis of serum samples from humans exposed to avian H7 influenza viruses in Italy between 1999 and 2003. J Infect Dis 192(8):1318–1322

Richt JA, Lekcharoensuk P, Lager KM, Vincent AL, Loiacono CM, Janke BH, Wu WH, Yoon KJ, Webby RJ, Solórzano A, García-Sastre A. (2006) Vaccination of pigs against swine influenza viruses by using an NS1-truncated modified live-virus vaccine. J Virol 80(22):11009–11018

Rudenko LG, Lonskaya NI, Klimov AI, Vasilieva RI, Ramirez A. (1996) Clinical and epidemio-logical evaluation of a live, cold-adapted influenza vaccine for 3–14-year-olds. Bull World Health Organ 74(1):77–84

Subbarao K, Joseph T. (2007) Scientific barriers to developing vaccines against avian influenza viruses. Nat Rev Immunol 7(4):267–278

Subbarao K, Katz JM. (2004) Influenza vaccines generated by reverse genetics. Curr Top Microbiol Immunol 283:313–342

Suguitan AL Jr, McAuliffe J, Mills KL, Jin H, Duke G, Lu B, Luke CJ, Murphy B, Swayne DE, Kemble G, Subbarao K. (2006) Live attenuated influenza A H5N1 candidate vaccines provide broad cross-protection in mice and ferrets. PLoS Med 3(9):e360

Talon J, Salvatore M, O'Neill RE, Nakaya Y, Zheng H, Muster T, García-Sastre A, Palese P. (2000) Influenza A and B viruses expressing altered NS1 proteins: a vaccine approach. Proc Natl Acad Sci USA 97(8):4309–4314

Treanor JJ, Roth FK, Betts RF. (1990) Use of live cold-adapted influenza A H1N1 and H3N2 virus vaccines in seropositive adults. J Clin Microbiol 28(3):596–599

Treanor JJ, Kotloff K, Betts RF, Belshe R, Newman F, Iacuzio D, Wittes J, Bryant M. (1999) Evaluation of trivalent, live, cold-adapted and inactivated influenza vaccines in prevention of virus infection and illness following challenge of adults with wild-type influenza A (H1N1), A (H3N2), and B viruses. Vaccine 18(9–10):899–906

Tweed SA, Skowronski DM, David ST, Larder A, Petric M, Lees W, Li Y, Katz J, Kraiden M, Tellier R, Halpert C, Hirst M, Astell C, Lawrence D, Mak A. (2004) Human illness from avian influenza H7N3, British Columbia. Emerg Infect Dis 10(12):2196–2199

Vesikari T, Karvonen A, Korhonen T, Edelman K, Vainionpää R, Salmi A, Saville MK, Cho I, Razmpour A, Rappaport R, O'Neill R, Georgiu A, Gruber W, Mendelman PM, Forrest B. (2006) A randomized, double-blind study of the safety, transmissibility and phenotypic and genotypic stability of cold-adapted influenza virus vaccine. Pediatr Infect Dis J 25(7):590–595

Vincent AL, Ma W, Lager KM, Janke BH, Webby RJ, García-Sastre A, Richt JA. (2007) Efficacy of intranasal administration of a truncated NS1 modified live influenza virus vaccine in swine. Vaccine 25(47):7999–8009

Watanabe T, Watanabe S, Kim JH, Hatta M, Kawaoka Y. (2007) A novel approach to the development of effective H5N1 influenza A virus vaccines: the use of M2 cytoplasmic tail mutants. J Virol 82(5):2486–2492

Webster RG, Geraci J, Petursson G, Skirnisson K. (1981) Conjunctivitis in human beings caused by influenza A virus of seals. N Engl J Med 304:911

Wright PF, Neumann G, Kawaoka Y. (2007) Orthomyxoviruses. In: Howley PM, Knipe DM (eds) Fields virology, 5th edn. Lippincott Williams & Wilkins, Philadelphia, pp 1691–1740

Youngner JS, Treanor JJ, Betts RF, Whitaker-Dowling P. (1994) Effect of simultaneous administration of cold-adapted and wild-type influenza A viruses on experimental wild-type influenza infection in humans. J Clin Microbiol 32(3):750–4

# Influenza Vaccines for Avian Species

Darrell R. Kapczynski and David E. Swayne

## Contents

**Abstract**  Beginning in Southeast Asia in 2003, a multinational epizootic outbreak of H5N1 highly pathogenic avian influenza (HPAI) was identified in commercial poultry and wild bird species. This lineage, originally identified in Southern China in 1996 and then Hong Kong in 1997, caused severe morbidity and mortality in many bird species, was responsible for considerable economic losses via trade restrictions, and crossed species barriers (including its recovery from human cases). To date, these H5N1 HPAI viruses have been isolated in European, Middle Eastern, and African countries, and are considered endemic in many areas where regulatory control and different production sectors face substantial hurdles in controlling the spread of this disease. While control of avian influenza (AI) virus infections in wild bird populations may not be feasible at this point, control and eradiation of AI from

D.R. Kapczynski (✉) and D.E. Swayne
Exotic and Emerging Avian Viral Diseases Research Unit, Southeast Poultry Research
Laboratory, USDA–Agricultural Research Service–South Atlantic Area,
934 College Station Road, Athens, GA 30605, USA
e-mail: darrell.kapczynski@ars.usda.gov

R.W. Compans and W.A. Orenstein (eds.), *Vaccines for Pandemic Influenza*,
Current Topics in Microbiology and Immunology 333,
DOI 10.1007/978-3-540-92165-3_6, © Springer-Verlag Berlin Heidelberg 2009

commercial, semicommercial, zoo, pet, and village/backyard birds will be critical
to preventing events that could lead to the emergence of epizootic influenza virus.
Efficacious vaccines can help reduce disease, viral shedding, and transmission to
susceptible cohorts. However, only when vaccines are used in a comprehensive
program including biosecurity, education, culling, diagnostics and surveillance
can control and eradication be considered achievable goals. In humans, protection
against influenza is provided by vaccines that are chosen based on molecular,
epidemiologic, and antigenic data. In poultry and other birds, AI vaccines are
produced against a specific hemagglutinin subtype of AI, and use is decided by
government and state agricultural authorities based on risk and economic considerations, including the potential for trade restrictions. In the current H5N1 HPAI
epizootic, vaccines have been used in a variety of avian species as a part of an
overall control program to aid in disease management and control.

# 1   Introduction

Avian influenza (AI) is a viral disease of birds that remains an economic threat
to poultry throughout the world. AI viruses are classified in the family Orthomyxoviridae, genus Influenza A (Type A), and contain a segmented, negative-sense RNA
genome (Swayne and Halvorson 2003). Type A influenza virions are enveloped,
spherical to pleomorphic in shape, and approximately 80–120 nm in size (Swayne
and Halvorson 2003). Type A influenza virus genomes consist of eight segments,
which code for at least ten viral proteins, including the hemagglutinin (HA),
matrix (M1), membrane-bound ion channel-like protein (M2), neuraminidase (NA),
nucleocapsid (NP), nonstructural protein 1 (NS1), nuclear export protein (NS2), and
three proteins associated with polymerase activity (PA, PB1, and PB2). The lack of
proofreading mechanisms of the viral RNA polymerase results in a viral genome
that is highly variable, with mutations readily occurring in the HA and NA genes.
Antigenically, 16 HA (H1–H16) and 9 NA (N1–N9) subtypes have been described
(Ada and Jones 1986; Fouchier et al. 2005; Lambkin and Dimmock 1996). The
segmented nature of the virus genome allows for reassortment of genes when a
susceptible host is coinfected with different influenza viruses. Thus, influenza
viruses with up to 144 possible combinations of HA and NA subtypes can be isolated
from birds.

Two clinical forms of the AI virus can be seen, including the less severe low
pathogenic (LP) form that causes mucosal infections with variable morbidity
and mortality, and the high pathogenic (HP) form, which causes high morbidity and
mortality in gallinaceous poultry. A major virulence factor associated with the different clinical forms of AI is the presence of basic amino acids in at the HA fusion
cleavage site. LPAI isolates do not contain multiple basic amino acids at this site or
long insertions of extraneous nucleic acids, thus limiting their replication to

mucosal tissues. The presence of multiple basic amino acids found in HPAI isolates allows the virus to replicate in deeper tissues, resulting in increased virulence.

Wild aquatic birds are a primary natural reservoir for the LPAI viruses. The LPAI virus replicates in the intestinal tracts of these birds and is spread primarily by fecal contamination of the water environment. Although wild aquatic birds do not normally get sick from AI, they have on occasion transmitted the virus to domesticated birds, including chickens, ducks, and turkeys (Halvorson et al. 1985). Within the last several years, outbreaks of H5N1 HPAI in Southeast Asian, European, African, and Middle Eastern countries have resulted in one of the most severe animal disease outbreaks in recent history, with hundreds of millions of wild birds and poultry dead or depopulated (Capua and Marangon 2007a).

In the poultry industry, vaccines have been used for decades to combat production losses due to various infectious organisms. Some of these pathogenic organisms are reportable to the World Organization for Animal Health (OIE), including H5 and H7 HP and LPAI viruses, and thus their isolation in poultry has been linked to significant economic trade implications. Although infection with HPAI viruses can result in devastating morbidity and mortality of infected flocks, the use of vaccines to control AI in commercial poultry operations has been uncommon in most developed countries. In fact, extraordinary circumstances are generally required before approval is given to use H5 or H7 AI vaccines in the field. In the USA, the legal authority to conduct and implement H5 and H7 AI vaccination programs requires both state and federal approval. AI eradication programs, which include vaccination, have been previously used to control outbreaks of H5 or H7 AI in the USA (H7N3 and H7N2 LPAI), Mexico (H5N2 LPAI and HPAI), Italy (H7 LPAI), Hong Kong (H5N1 HPAI), Indonesia (H5N1 HPAI), Pakistan (H7N3 HPAI), Russia (H5N1 HPAI) and Vietnam (H5N1 HPAI) (Capua and Marangon 2007b; Sims 2007; Swayne 2003; Swayne and Halvorson 2008; Swayne and Kapczynski 2008; Villareal 2006).

## 1.1   History of Avian Influenza Vaccines

Avian influenza vaccines have their origins back in the late 1920s, when chickens infected with fowl plague virus (now known as HPAI virus) recovered from the disease and were resistant to re-exposure (Beaudette et al. 1932; Todd 1928). Initially, experimental vaccines against fowl plague were based on the experiences of Pasteur in rabies, who used spinal cords from infected animals as vaccines to protect against virulent rabies virus. Although many AI vaccine failures were observed, either from lack of immune response or inefficient viral inactivation, efficacious AI vaccines were eventually produced that protected birds from disease. However, by that point, the policy of "stamping out" HPAI virus infections of poultry through culling and depopulation in order to control the spread of disease had gained support throughout Europe and vaccines were generally not used as part of any control strategy.

More recently, vaccines have been developed and approved for use against LPAI virus infections of poultry. Beginning after the mid-1960s, when the economic impact of LPAI virus infections in poultry was realized, control strategies were implemented based on economic need. Early field management strategies included controlled exposure of pullets to LPAI viruses in order to produce immunity prior to egg lay (D. Halvorson, personal communication). In the USA beginning in 1979, AI vaccines were primarily used to prevent production losses in turkeys and egg-laying chickens (breeders and table-egg production). In the past decade, H1N1 and H3N2 swine influenza virus infections of turkeys have resulted in significant decreases in egg production and quality (McCapes and Bankowski 1987; Price 1981) (E. Gonder, personal communications). However, because these are low pathogenic isolates, limited conditional-use inactivated vaccines have been used in such turkey flocks as a management strategy. The first LPAI vaccine, a H4 and H6, was used in 1979 in the USA (Swayne and Kapczynski 2008). A multivalent inactivated AI vaccine containing H5N2, H6N2, and H10N2 along with Newcastle disease virus (NDV) was reported to have been used in Italy in 1980 to control multiple subtypes of LPAI virus (Zanella et al. 1981). Finally, following an outbreak of H7N2 LPAI in an isolated commercial chicken facility containing laying hens in Connecticut, agreements between state, federal, and industry representatives allowed the use of an inactivated LPAI vaccine as part of a comprehensive strategy to control the virus as an alternative to immediate depopulation (Adriatico 2005). As a part of the control strategy, vaccinated flocks were intensively monitored for virus shedding through dead bird testing, and serological surveillance using non-vaccinated sentinels and a neuraminidase (NA)-based DIVA (differentiation of infected from vaccinated animal) approach to detect infections in vaccinated birds. Taken together, vaccination can now be considered a valuable component in comprehensive AI control strategies. While situational or local outbreaks of HPAI may always require stamping out in nonendemic areas, in the face of an epizootic event, vaccines can be formulated and used based on field isolates recovered.

## 2  Avian Influenza Vaccines

Many different types of experimental AI vaccines have been described, and some have been licensed for commercial use (Swayne and Halvorson 2003; Swayne and Kapczynski 2008). Categories of vaccines include the following: inactivated, live, subunit, recombinant vectors expressing AI genes, and DNA vaccines. While many of these vaccines have been shown to induce protective immunity in the laboratory under optimal conditions, the final proof of protection and efficacy is still derived from field application. For field use, the overwhelming majority of AI vaccines produced and sold have been oil emulsion inactivated whole AI virus vaccines delivered via the parenteral route (subcutaneous or intramuscular) and, less frequently, recombinant vectored vaccines.

## 2.1 Protective Antigenic Component of Avian Influenza Vaccines

Protective immunity in poultry against AI viruses is primarily the result of an antibody response directed against the hemagglutinin (HA), of which there are 16 different subtypes (Swayne and Halvorson 2008). Antibodies produced against the HA are neutralizing and thus prevent attachment of the virus to host cells. When bursectomized chickens (i.e., unable to produce antibody responses) were vaccinated and challenged, no protection was provided, indicating the role of antibodies in protection from AI (Chambers et al. 1988). A minor contribution to immunity is provided against the NA protein, of which there are nine different subtypes (Swayne et al. 2000). Our current knowledge of protective immunity against AI is derived from experimental HPAI challenge studies (Alexander and Parsons 1980; Allan et al. 1971; McNulty et al. 1986; Rott et al. 1974), in which the NA in a whole AI virus vaccine provided mostly partial protection from mortality. In addition, immunization with NA protein alone produced only partial protection following 2–3 vaccinations (Sylte et al. 2007). AI vaccines are generally custom-made against the specific HA subtype and/or NA subtypes of the current field virus. Because protection is provided through an immune response to the HA, the more efficacious vaccines target the specific phylogenetic lineages of the virus within a HA subtype (Fig. 1). Tumpey et al. (2004) showed that antigenically related H7 AI viruses provide protection in turkeys through decreased challenge virus shedding (Table 1). In addition, H5 vaccines have been demonstrated to provide broad protection against diverse H5 viruses when the vaccine and challenge virus vary by up to 13% in HA-deduced amino acid sequence (Swayne et al. 1999).

Vaccines composed of internal AI proteins, including the matrix or nucleoprotein, did not provide significant protection in poultry, although a measurable immune response occurred (Brown et al. 1992; Webster et al. 1991). Because many different serotypes of AI viruses exist through different combinations of the HA and NA (potentially 144), there is no one universal vaccine which will protect against all AI viruses. However, because the M2 transmembrane protein is highly conserved between both avian and human influenza isolates, current research on human influenza is targeting this protein in the hope of producing such a vaccine (Huleatt et al. 2008; Watanabe et al. 2008). In mice, vaccination with the ectodomain of M2 (M2e) and hepatitis B virus core protein (HBc) resulted in protection against homologous and heterologous human influenza A virus challenge (De Filette et al. 2005, 2006). In a limited study, partial protection was provided following two vaccinations of chickens with a recombinant salmonella vaccine expressing the M2e against LPAI virus (D. Kapczynski, unpublished data). The mode of protection appeared to be humoral; purportedly the antibodies interfere with the ion-transport function of the protein required for virus uncoating. However, this type of vaccine is likely years away from any potential field application, and so, before adoption, there must be a demonstration of consistent protection that is comparable to HA-based vaccines.

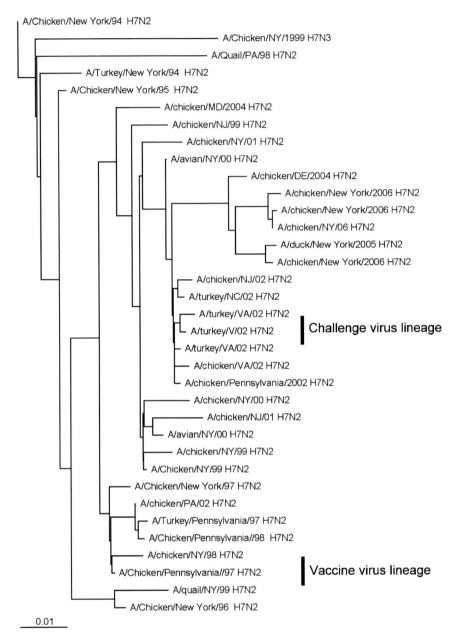

**Fig. 1** Phylogenetic analysis of the HA gene of H7N2 subtype AIVs isolated in North America. The tree was generated by the neighbor-joining method. The names of the states where the viruses were isolated are represented by their standard two-letter postal codes

Limited data exist on the contribution of cellular immunity to protection against AI in birds. In mice, live influenza virus vaccines induce both humoral and cellular immunity, but the overall contribution of the cellular aspect to protection remains

**Table 1** AI vaccine protection, as measured by the decrease in viral shedding following vaccination of eight-week-old turkeys with a commercial AI H7N2 oil emulsion vaccine

| Inactivated vaccine dose | Viral titer from oropharyngeal swabs | |
| --- | --- | --- |
| | 2 days postchallenge[a] | 4 days postchallenge |
| Sham control | 6.4[A] | 5.6[A] |
| H7N2 1X | 4.0[B] | 2.5[B] |
| H7N2 2X | 2.8[B] | 2.3[B] |

[a] Virus end-point titers are expressed as mean $\log_{10}$ $EID_{50}$ per ml. Numbers with different uppercase letters indicate significant differences between the H7N2 vaccinated groups and the sham-control group by analysis of variance

Turkeys received a single (at four weeks of age) or double (boost at six weeks of age) injection with a commercial AI vaccine H7N2 (A/chicken/PA/97), applied according to the manufacturers recommendations. Birds were challenged intranasally with $10^7$ $EID_{50}$ at eight weeks of age with a current H7N2 field isolate (A/turkey/VA/02). Taken from Tumpey et al. (2004)

unknown, since it takes 5–7 days for virus-specific CD8+ thymocyte-derived T cells to migrate and localize in lungs (Lawrence et al. 2005). However, the benefit of a secondary cellular response against influenza virus resulted in decreased duration and amount of viral shedding, which could reduce transmission potential to susceptible cohorts and which decreases severity of disease (Flynn et al. 1998). In chickens, Seo and Webster reported that adaptive transfer of AI-primed CD4+ and CD8+ T cells protected birds against lethal H5N1 challenge (A/chicken/Hong Kong/97), suggesting that cellular immunity alone could protect against HPAI virus challenge (Seo and Webster 2001). However, because recent H5N1 HPAI viruses caused rapid, sudden mortality (mean death times of 2–3 days), and mobilization of memory T cells to target organs may take 3–10 days, protection from disease may not be reasonable without a supporting innate and/or humoral immune response. Vaccine-induced cellular immunity requires either active viral replication in host cells for major-histocompatibility complex (MHC)-class I presentation or a recombinant-vectored or DNA vaccine expressing AI genes in host cells. However, the use of live AI virus vaccines in poultry is universally discouraged because of the potential for reassortment with genes from other AI viruses (Swayne and Halvorson 2008). While inducing both humoral and cellular immunity represents the best overall protective qualities in a vaccine, limitations on vaccine formulations capable of inducing cellular immunity and licensure for field application make this an achievable goal only when using non-AI virus vaccines, such as recombinant virus vectors.

## 2.2  Inactivated Virus Vaccines

Inactivated AI vaccines are prepared by first propagating the virus in embryonated chicken eggs, followed by chemical or physical inactivation (Swayne and Kapczynski 2008). The most widely used inactivation chemical is formalin, which crosslinks the viral proteins such that viral replication cannot occur. Other chemicals such as B-propiolactone or binary ethyleneimine can be used as inactivants (King 1991). Inactivated human influenza vaccines are usually detergent treated, which allows for partial purification of the viral surface proteins, HA and NA, into a subunit vaccine. In contrast, AI vaccines for poultry use nonpurified allantoic fluid (i.e., crude preparation), which keeps the production costs lower; however, as well as inducing a protective immune response against HA and NA, it also induces non-neutralizing antibodies against internal viral proteins.

To increase the immunogenicity of inactivated AI vaccines, most are blended with oil to form an emulsion. The emulsions are prepared by mixing a water-based antigen phase with an oil phase, normally containing surfactant for stabilization (e.g., sorbitan monoleate and Tween 80) to produce a water-in-oil emulsion. Without other components, oil-based adjuvants stimulate mainly antibody responses, although under some circumstances water-in-oil emulsions may be able to activate CTLs (Aucouturier et al. 2001; Beck et al. 2003; Folitse et al. 1998; Stone 1986, 1989; Xie and Stone 1990, http://www.seppic.com). The type of oil used in vaccine production can affect the overall immune response to the vaccine such that nonmetabolizable oils (e.g., mineral oil) enhance antibody responses over biodegradable oils (e.g., vegetable) (Stone 1993, 1997). Besides having adjuvant properties, oil emulsion vaccines slowly release antigen over time, resulting in higher immune responses than would be produced by the antigen alone.

Besides oil, different types of adjuvants have been added to vaccine formulations to enhance immune reactions (Fatunmbi et al. 1992; Jansen et al. 2006; Katz et al. 1993; Stephenson et al. 2005). Vaccine adjuvants are chemicals, microbial components, or mammalian proteins that enhance immune responses to vaccine antigens (Vogel 2000). In general, the antigen itself is an inert element which must be taken up by host cells for presentation to the immune system. The adjuvant, which represents the stimulating signal, is responsible for the activation and mobilization of professional lymphocytes to the vaccine antigen. Many different adjuvants have been described in experimental and commercial vaccines, including aluminum salts, bacterial derivates, cytokines, emulsions, ISCOMS, liposomes, nonionic block copolymers, polysaccharides, and saponins (Aucouturier et al. 2001; He et al. 2002; Hu et al. 2001; Roth 1999; Sambhara et al. 2001; Schijns 2003). Despite their extreme relevance and the demand for novel vaccine adjuvants, the various processes explaining how the immune system "knows" when to become activated by the adjuvant remain unknown (Schijns and Tangeras 2005).

Both homologous and heterologous inactivated AI vaccines have been produced. Homologous vaccines contain the same HA and NA subtypes as the field virus, whereas heterologous vaccines contain a matched HA with a different NA subtype. Heterologous vaccines can be used in a DIVA strategy (differentiating infected from vaccinated animals) that enables serological screening of flocks to determine if vaccinated flocks become infected from field virus exposure (Capua et al. 2003).

Because of their safety and immunogenicity, inactivated AI vaccines have been used in multiple avian species, including commercial poultry (e.g., chickens, turkeys, ducks, and geese), exotic (e.g., zoo and pet birds) and endangered species (Oh et al. 2005; Philippa et al. 2007). A minimum of two vaccinations, which can be adjusted based on weight, may be necessary to induce protective immunity, although subsequent boosting is needed in long-lived birds (Bertelsen et al. 2007).

## 2.3   Live Virus Vaccines

No live AI virus vaccines have been approved for use in poultry, although the use of live LPAI vaccines has been studied experimentally in poultry species (Beard and Easterday 1973; McNulty et al. 1986). One example of the latter is a laboratory-passaged, attenuated AI virus that contains a truncation of the NS1 gene in A/turkey/Oregon/1971 (H7N3). In chickens, this virus exhibited decreased replication and attenuation of infectivity (Cauthen et al. 2007). Although this virus has the potential for vaccine use, its safety would still need to be examined in the field because of the potential to reassort with NS1 genes from another AI virus, resulting in a field virus with increased infectivity. Because of this reassortant potential, live LPAI virus strains are not recommended for use as poultry vaccines, especially H5 and H7 subtypes, which have shown the ability to mutate into HP forms. LPAI vaccines may also be associ-ated with increased losses due to respiratory disease caused by viral replication and may spread to surrounding farms (Beard 1981; Lee and Suarez 2005).

## 2.4   AI Subunit Vaccines

Most AI-expressed subunit vaccines are based on expression of the HA gene in animal or plant cells, bacteria, viruses, or yeast (Chambers et al. 1988; Crawford et al. 1999; De et al. 1988; Schultz-Cherry et al. 2000). The resulting protein is then purified and quantified prior to application. Following purification, the antigen can be emulsified in oil with or without adjuvant as described for inactivated vaccines. In studies with mice, yeast-expressed influenza HA and NA protected against lethal challenge (Martinet et al. 1997; Saelens et al. 1999). Other vectors, such as baculovirus, have been used to express HA in cell-culture supernatants, which was utilized for experimental vaccine preparation in poultry (Crawford et al. 1999). One advantage of this vaccine platform is that it can be utilized in a DIVA strategy. However, because of the time and cost involved with production, none of these have yet been commercialized.

## 2.5   Recombinant Virus-Vectored Vaccines

Using recombinant DNA technologies, these vaccines utilize noninfluenza virus vectors to express influenza genes in vivo so as to induce protective immunity in the host (Beard et al. 1992; Boyle and Coupar 1988; Boyle et al. 2000; Bublot et al. 2006, 2007;

Jia et al. 2003; Qiao et al. 2003, 2006; Swayne et al. 2000; Webster et al. 1991). Because these are live viruses, they can provide the benefits of live virus vaccines, including the stimulation of humoral, mucosal, and cellular immunity. Unlike live AI virus vaccines, recombinant vectored vaccines do not have the potential for AI virus gene reassortment, making them safer to apply in the field. However, pre-exposure of the host to the vector may render the vaccine ineffective, as the vaccine virus may not be capable of sufficient replication and thus adequate immunological stimulation. The use of Newcastle disease virus (NDV) as a vectored vaccine for AI has shown promising results (Swayne et al. 2003). However, as most commercial birds are vaccinated multiple times during the course of their lives with NDV vaccines, and progeny contain maternally derived antibodies in ovo, the timing, route of administration and antibody levels for the NDV vector must be considered before the vaccine can be used in the field (Ge et al. 2007; Park et al. 2006; Veits et al. 2006). In China, a commercially licensed NDV-vectored H5 AI vaccine has been used in the field since 2006. Recombinant poxvirus-based AI vaccines have also been licensed and used in the field, most notably in China, Mexico, Guatemala, and El Salvador, against H5 subtype viruses (Bublot et al. 2006; Ge et al. 2007; Swayne et al. 1997, 2000). As with other AI vaccines, matching the HA sequence in the vaccine construct to the field-circulating virus resulted in decreased morbidity and morality, decreased duration of virus shedding, and decreased titer levels shed of challenge virus. A major advantage of recombinant AI vectored vaccines is the ability to rapidly change the AI gene insert into the recombinant backbone to compensate for the drift of the field virus. The production capacity and size of the vaccine market are limiting factors that a manufacturer uses before deciding to make and sell an AI vaccine. Other vectored viruses have been constructed to express AI genes as experimental vaccines, and include infectious laryngotracheitis virus (*Herpesviridae*), vaccinia virus (*Poxviridae*), human adenovirus 5 (*Adenoviridae*), Venezuelan equine encephalitis virus (*Togaviridae*), and retrovirus (*Retroviridae*) (Brown et al. 1992; Chambers et al. 1988; De et al. 1988; Gao et al. 2006; Hunt et al. 1988; Luschow et al. 2001; Schultz-Cherry et al. 2000; Toro et al. 2007; Veits et al. 2003). Finally, the ability to mass administer vaccines to commercial poultry, including spray and drinking water, offer significant cost savings to poultry producers and may improve field usage.

## 2.6 DNA Vaccines

DNA vaccines containing *E. coli*-derived plasmid DNA expressing viral gene(s), under the control of a eukaryotic promoter, have demonstrated protective immunity in chickens against AI viruses (Fynan et al. 1993; Le Gall-Recule et al. 2007; Robinson et al. 1993; Suarez and Schultz-Cherry 2000). Following vaccine application and cellular uptake, the DNA vaccine is transcribed into RNA and exported to the cytoplasm for translation. The in situ expressed protein antigen can either be processed intracellularly with the resulting peptides bound to MHC class I for cytotoxic T lymphocyte stimulation, or concomitantly released from the cell and taken up by antigen-presenting cells for MHC class II presentation to stimulate humoral immunity. The ability to elicit both humoral and cellular adaptive immune responses is a major advantage of DNA vaccines, and mimics an active viral infection. Although different viral genes have been tested, the HA appears most protective, since it contains epitopes for inducing virus-neutralizing antibodies. Recently, a small number of DNA vaccines have been licensed for veterinary use, including West Nile virus for use in horses (see http://www.aphis.usda.gov/lpa/news/2005/07/wnvdna_vs.html) and infectious hematopoietic necrosis virus in salmon (see http://www.ah.novartis.com).

## 2.7   Efficacy of Avian Influenza Vaccines in Birds

Under ideal conditions, AI vaccination of commercial poultry would stimulate a protective immune response that would prevent infection and disease following a single dose. Vaccine-induced protection from AI virus challenge is measurable in vivo using various avian models, including those bird species most likely to experience AI virus infection during an epizootic, such as chickens, turkeys, and other avian species. Because initial infection occurs at respiratory or alimentary mucosal surfaces, and parenteral vaccines primarily induce a humoral immune response, sterilizing immunity with complete inhibition of infection is not realistic in the field.

Generally, protection from mortality and morbidity are the easiest measurable (qualitative) outcomes following challenge. In addition, antibody titers (quantitative) following vaccination, but prior to challenge, can be used to predict protection from lethal challenge. It is generally believed that HI antibody titers of 1:32–40 or greater will protect birds against mortality following challenge. Other parameters of protection include decreasing virus shed titers from the respiratory and intestinal tracts following challenge, as well as decreases in the duration of shedding. These parameters can be indirect measures of the efficacy of the vaccine in limiting environmental contamination and thus decreasing transmission to susceptible cohorts (Capua et al. 2004; Swayne 2003; Swayne et al. 1997, 1999). A hundredfold reduction in challenge virus shedding in vaccinated vs. nonvaccinated animals has been recommended as the minimum standard of vaccine-induced immune potency (Suarez et al. 2006; Swayne et al. 1997). In some cases, protection from decreases in egg production and quality is associated with vaccine efficacy (D. Kapczynski, personnel observation) (Fig. 2).

Most vaccine studies involving the protection of birds against both LPAI and HPAI viruses have been performed in chickens and to a lesser extent turkeys and ducks (Cristalli and Capua 2007; Middleton et al. 2007; Philippa et al. 2005; Steensels et al. 2007; Swayne 2006; van der Goot et al. 2005, 2007). Generally, wild waterfowl are less susceptible to mortality following HPAI virus infection. However, the most recent H5N1 isolates of Southeast Asian origin have demonstrated high virulence for young domestic waterfowl, which have been identified as contributors to the spread of the disease (Hulse-Post et al. 2005). For this reason, it is important to establish

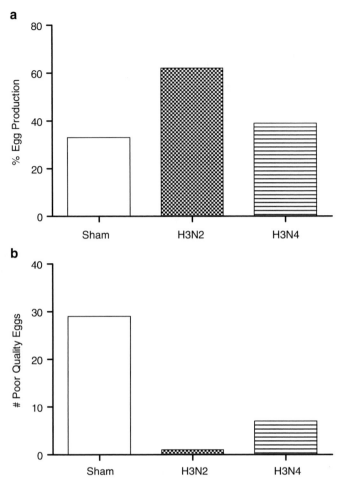

**Fig. 2a–b** AI vaccine protection against egg production losses in vaccinated turkey breeder hens following AI H3N2 challenge. Commercial egg-laying turkey hens received two doses of autogenous killed oil emulsion vaccine containing either H3N2 (A/turkey/North Carolina/05) or H3N4 (A/duck/Minnesota/79/79) at 26 and 30 weeks of age, followed by a challenge with a recent H3N2 field virus (A/turkey/NC/16108/03). Hens receiving phylogenetically matched H3N2 vaccine displayed increased production (**a**) and egg quality (**b**) over sham and H3N4-vaccinated hens

vaccine potency and efficacy for domestic ducks and other minor poultry species. Although the immune system of chickens is the best studied of that of any bird species, ducks are a distant second in terms of anatomical and immunological understanding and knowledge. While many features of their immune systems are similar, one dramatic anatomical difference is the presence of lymph nodes in ducks, which are absent in chickens. The overall impact of lymph nodes in vaccine-induced protection is unclear, although at least one study suggests that ducks may require twice the antigen content to produce a protective immune response when compared to chickens

(Middleton et al. 2007). All vaccinated ducks were protected from disease, and very few (2/15) shed virus at day 4 postchallenge. However these differences (and likely others) in other bird species underscore the need to evaluate a vaccine's potency and efficacy in each host species before licensing. Extrapolating the data generated in one species and applying it to another should be pursued with caution. Each species has distinct differences in immune response that impact on the protection induced by vaccines.

In the event of an avian influenza epizootic, zoo birds would constitute a potential source of infection and transmission among wild birds and poultry, and would likely need prophylactic vaccination. Only a few studies have examined the use of AI vaccination in zoo birds. Within the European Union (EU), following an outbreak of H7N7 HPAI in poultry, vaccination with an inactivated H7N1 virus was undertaken at a number of zoological parks (Philippa et al. 2005). Of the 211 birds vaccinated, 81.5% seroconverted (HI $\geq$ 1:40) after two doses of a commercial inactivated H7 AI poultry vaccine, with birds of the orders Anseriformes, Galliformes, and Phoenicopteriformes exhibiting better responses than birds in other orders. In Singapore during 2005, inactivated H5N2 vaccine was applied to zoo birds following a H5N1 HPAI outbreak, with 84% of birds exhibiting seroconversion after two doses of commercial inactivated H5N2 AI vaccine (Oh et al. 2005). A similar study utilizing commercial inactivated H5N2 virus vaccine in 334 zoo birds resulted in the seroconversion of 80.4% of birds (Philippa et al. 2007). Finally, a third research study with commercial inactivated H5N9 in zoos produced an 84% seroconversion rate among all bird orders tested, with black winged stilts, parrots, Congo peafowl, rheas, ibis, flamingos, and kookaburras displaying greater seroconversion rates than penguins, pelicans, ducks, geese, herons, Guinea fowl, and cranes (Bertelsen et al. 2007). In that study, the dose of commercial vaccine was applied based on body weight (1–99 g: 0.1 ml; 100–2,999 g: 0.3 ml; 3–20 kg: 0.6 ml; >20 kg: 1 ml), which may have influenced some of the differences observed in seroconversion; however, it underscores the need to adjust the antigenic load based on bird weight.

## 2.8   Scenarios for Avian Influenza Vaccine Application

Several situations lend themselves to the application of vaccines for AI control. Although not every outbreak requires the use of vaccine, in the event of an epizootic, vaccines must be safely and effectively applied to curb the spread of the virus in susceptible birds. When vaccination campaigns begin, an increased level of biosecurity must be practiced to reduce the risk of personnel mechanically spreading the virus from farm to farm on their shoes and clothing or by moving the virus as fomites on equipment and supplies. In addition, a planned eradication program must be developed, as well as surveillance and monitoring of all vaccinated flocks. An exit strategy for withdrawal of vaccine use should be developed to avoid overdependence on vaccination for control and to allow true eradication of the virus from poultry sectors. Vaccination can be used in three different ways as part of an

AI control strategy, including prophylactic or preventive vaccination, endemic (routine) vaccination, and emergency (outbreak) vaccination (Marangon et al. 2008).

Prophylactic vaccination is applied when an area has high risk for the introduction of H5 or H7 virus to susceptible bird species. Under this scenario, a country, geographical zone or production compartment that is free of AI may be bordered by an AI outbreak zone. The use of preventive vaccine increases the resistance of poultry to infection and reduces the risk of introduction within the country, zone, or compartment. Although vaccination can decrease the potential for virus introduction, this is highly dependent upon the population immunity within the area. Appropriate surveillance measures, including either sentinel bird placement, DIVA testing, and/ or appropriate diagnostics, are needed to assess the effectiveness of vaccination and to detect any potential virus introductions, because vaccination could mask the clinical detection of infection and allow the potential spread of AI virus. In the event of virus introduction, prophylactic vaccination will decrease the duration and level of virus shedding, resulting in decreased risk of secondary outbreaks, but biosecurity is necessary in all vaccinated flocks to make vaccination an effective tool in AI control (Capua and Marangon 2007a; Swayne and Kapczynski 2008).

When HPAI is endemic in a country, vaccination can be applied as an overarching control measure to reduce environmental load and spread, in preparation for future stamping out of the disease. In this situation, all birds that can be accessed should be vaccinated, with multiple doses applied throughout the period of potential virus exposure. The goal is to increase flock immunity to the level of 60–80%, which should prevent the spread of the virus between well-vaccinated flocks and allow the eradication of the virus within the area. However, it may also be necessary to apply stamping-out procedures to infected flocks to reduce the spread of the virus and prevent the development of a reservoir. Surveillance and increased biosecurity of all bird species and premises, including poultry flocks, waterfowl, and at-risk zoo birds, are critical to stamping out the disease from endemic areas. Constant veterinary medical vigilance, supported by government, commercial, and local officials, is needed to achieve this goal.

Emergency vaccination may be needed when a HPAI epizootic is spreading rapidly, causing large numbers of outbreaks on farms with susceptible poultry within a defined area or country, and may be necessary to contain the disease and reduce the risk of further virus spread. An initial step would be to establish a movement control zone around the index case and all noninfected birds; within the ring, all birds would be vaccinated or euthanized. Outside the primary ring, a secondary ring would be established, and all birds within that ring would be vaccinated and monitored for presence of virus. Because immunity may require 10–14 days to develop postvaccination, a rapid decision among government officials, along with input from local and commercial entities, is required for the emergency release of vaccines. Preplanning among officials is critical for the successful implementation of emergency vaccination due to inherent delays in vaccine application. Such planning should include (1) assessments of the number of vaccine doses on hand, (2) vaccine production, (3) mobilization of vaccination crews, (4) establishing routes of vaccination crews, (5) surveillance measures and reagents, and (6) development of the administrative control structure. While emergency vaccination may be considered a worst-case scenario for any country, the decision to implement the procedure must be developed based on cost–benefit ratios for each situation.

# 3 Conclusions

AI vaccines provide protection by stimulating host immunity, which is largely based on antibody production against the HA glycoprotein. These antibodies are capable of neutralizing the virus, thereby preventing infection, and reducing disease and virus transmission. A successful vaccine campaign will use a vaccine of the same HA subtype as that of the field virus. The establishment of neutralizing antibodies prior to HPAI virus infection is critical for disease control because these viruses induce rapid mortality within days of exposure and the immune system of the nonimmune bird does not have time to mobilize an efficient defense against the virus. Although many AI vaccine constructs have been described, only two are licensed for use in the field: inactivated AI virus and recombinant vectored vaccines. Until recently, inactivated vaccines were prepared from historically available LPAI virus isolates which grew to high titers in embryonating chickens eggs and required low biocontainment during manufacture because of the low escape risk. However, reverse genetic techniques are creating vaccine viruses with HA glycoproteins that are genetically a closer match and that have high growth properties in eggs but none of the safety concerns associated with of HPAI viruses. To standardize vaccine potency, quantification of HA per dose or seroconversion postvaccination can be used. Minimal protective titers described for birds should be >1:32–40. However, new research is needed to determine if these levels are adequate for all bird species, as only limited information is available on vaccine efficacy and protection outside of commercial chickens.

Avian vaccines applied during an AI event can be utilized as an integrated component of a comprehensive control strategy to limit the spread of the virus within animal and human populations and provide a safe, economical food source for the world's population. The strategic objectives should also include surveillance, biosecurity and quarantine, and education. While no single strategy exists to cover all outbreak situations for all countries and all susceptible bird species, appropriate programs should be developed by government officials, veterinary researchers, commercial industry and local entities to provide the leadership and science-based knowledge needed to limit the impact of AI influenza.

# References

Ada GL, Jones PD (1986) The immune response to influenza infection. Curr Top Microbiol Immunol 128:1–54

Adriatico N (2005) Controlling AI by vaccination: the Connecticut experience. North Central Avian Disease Conference, pp 25–28

Alexander DJ, Parsons G (1980) Protection of chickens against challenge with virulent influenza A viruses of Hav5 subtype conferred by prior infection with influenza A viruses of Hsw1 subtype. Arch Virol 66(3):265–269

Allan WH, Madeley CR, Kendal AP (1971) Studies with avian influenza A viruses: cross protection experiments in chickens. J Gen Virol 12(2):79–84

Aucouturier J, Dupuis L, Ganne V (2001) Adjuvants designed for veterinary and human vaccines. Vaccine 19(17–19):2666–2672

Beard CW (1981) Turkey influenza vaccination. Vet Rec 108(25):545

Beard CW, Easterday BC (1973) A-Turkey-Oregon-71, an avirulent influenza isolate with the hemagglutinin of fowl plague virus. Avian Dis 17(1):173–181

Beard CW, Schnitzlein WM, Tripathy DN (1992) Effect of route of administration on the efficacy of a recombinant fowlpox virus against H5N2 avian influenza. Avian Dis 36(4): 1052–1055

Beaudette FR, Hudson CB, Saxe AH (1932) An outbreak of fowl plague in New Jersey in 1929. J Agric Res 49:83–92

Beck I, Gerlach H, Burkhardt E, Kaleta EF (2003) Investigation of several selected adjuvants regarding their efficacy and side effects for the production of a vaccine for parakeets to prevent a disease caused by a paramyxovirus type 3. Vaccine 21(9–10):1006–1022

Bertelsen MF, Klausen J, Holm E, Grondahl C, Jorgensen PH (2007) Serological response to vaccination against avian influenza in zoo-birds using an inactivated H5N9 vaccine. Vaccine 25(22):4345–4349

Boyle DB, Coupar BE (1988) Construction of recombinant fowlpox viruses as vectors for poultry vaccines. Virus Res 10(4):343–356

Boyle DB, Selleck P, Heine HG (2000) Vaccinating chickens against avian influenza with fowlpox recombinants expressing the H7 haemagglutinin. Aust Vet J 78(1):44–48

Brown DW, Kawaoka Y, Webster RG, Robinson HL (1992) Assessment of retrovirus-expressed nucleoprotein as a vaccine against lethal influenza virus infections of chickens. Avian Dis 36(3):515–520

Bublot M, Pritchard N, Swayne DE, Selleck P, Karaca K, Suarez DL, Audonnet JC, Mickle TR (2006) Development and use of fowlpox vectored vaccines for avian influenza. Ann N Y Acad Sci 1081:193–201

Bublot M, Pritchard N, Cruz JS, Mickle TR, Selleck P, Swayne DE (2007) Efficacy of a fowlpox-vectored avian influenza H5 vaccine against Asian H5N1 highly pathogenic avian influenza virus challenge. Avian Dis 51(suppl 1):498–500

Capua I, Marangon S (2007a) Control and prevention of avian influenza in an evolving scenario. Vaccine 25(30):5645–5652

Capua I, Marangon S (2007b) The use of vaccination to combat multiple introductions of notifiable avian influenza viruses of the H5 and H7 subtypes between 2000 and 2006 in Italy. Vaccine 25(27):4987–4995

Capua I, Terregino C, Cattoli G, Mutinelli F, Rodriguez JF (2003) Development of a DIVA (Differentiating Infected from Vaccinated Animals) strategy using a vaccine containing a heterologous neuraminidase for the control of avian influenza. Avian Pathol 32(1):47–55

Capua I, Terregino C, Cattoli G, Toffan A (2004) Increased resistance of vaccinated turkeys to experimental infection with an H7N3 low-pathogenicity avian influenza virus. Avian Pathol 33(2):158–163

Cauthen AN, Swayne DE, Sekellick MJ, Marcus PI, Suarez DL (2007) Amelioration of influenza virus pathogenesis in chickens attributed to the enhanced interferon-inducing capacity of a virus with a truncated NS1 gene. J Virol 81(4):1838–1847

Chambers TM, Kawaoka Y, Webster RG (1988) Protection of chickens from lethal influenza infection by vaccinia-expressed hemagglutinin. Virology 167(2):414–421

Crawford J, Wilkinson B, Vosnesensky A, Smith G, Garcia M, Stone H, Perdue ML (1999) Baculovirus-derived hemagglutinin vaccines protect against lethal influenza infections by avian H5 and H7 subtypes. Vaccine 17(18):2265–2274

Cristalli A, Capua I (2007) Practical problems in controlling H5N1 high pathogenicity avian influenza at village level in Vietnam and introduction of biosecurity measures. Avian Dis 51(suppl 1):461–462

De BK, Shaw MW, Rota PA, Harmon MW, Esposito JJ, Rott R, Cox NJ, Kendal AP (1988) Protection against virulent H5 avian influenza virus infection in chickens by an inactivated vaccine produced with recombinant vaccinia virus. Vaccine 6(3):257–261

De Filette M, Min Jou W, Birkett A, Lyons K, Schultz B, Tonkyro A, Resch S, Fiers W (2005) Universal influenza A vaccine: optimization of M2-based constructs. Virology 337(1):149–161

De Filette M, Ramne A, Birkett A, Lycke N, Lowenadler B, Min Jou W, Saelens X, and Fiers W (2006) The universal influenza vaccine M2e-HBc administered intranasally in combination with the adjuvant CTA1-DD provides complete protection. Vaccine 24(5):544–551

Fatunmbi OO, Newman JA, Sivanandan V, Halvorson DA (1992) Enhancement of antibody response of turkeys to trivalent avian influenza vaccine by positively charged liposomal avridine adjuvant. Vaccine 10(9):623–626

Flynn KJ, Belz GT, Altman JD, Ahmed R, Woodland DL, Doherty PC (1998) Virus-specific CD8+ T cells in primary and secondary influenza pneumonia. Immunity 8(6):683–691

Folitse R, Halvorson DA, Sivanandan V (1998) Efficacy of combined killed-in-oil emulsion and live Newcastle disease vaccines in chickens. Avian Dis 42(1):173–178

Fouchier RA, Munster V, Wallensten A, Bestebroer TM, Herfst S, Smith D, Rimmelzwaan GF, Olsen B, Osterhaus AD (2005) Characterization of a novel influenza A virus hemagglutinin subtype (H16) obtained from black-headed gulls. J Virol 79(5):2814–2822

Fynan EF, Robinson HL, Webster RG (1993) Use of DNA encoding influenza hemagglutinin as an avian influenza vaccine. DNA Cell Biol 12(9):785–789

Gao W, Soloff AC, Lu X, Montecalvo A, Nguyen DC, Matsuoka Y, Robbins PD, Swayne DE, Donis RO, Katz JM, Barratt-Boyes SM, Gambotto A (2006) Protection of mice and poultry from lethal H5N1 avian influenza virus through adenovirus-based immunization. J Virol 80(4):1959–1964

Ge J, Deng G, Wen Z, Tian G, Wang Y, Shi J, Wang X, Li Y, Hu S, Jiang Y, Yang C, Yu K, Bu Z, Chen H (2007) Newcastle disease virus-based live attenuated vaccine completely protects chickens and mice from lethal challenge of homologous and heterologous H5N1 avian influenza viruses. J Virol 81(1):150–158

Halvorson DA, Kelleher CJ, Senne DA (1985) Epizootiology of avian influenza: effect of season on incidence in sentinel ducks and domestic turkeys in Minnesota. Appl Environ Microbiol 49(4):914–919

He Q, Mitchell A, Morcol T, Bell SJ (2002) Calcium phosphate nanoparticles induce mucosal immunity and protection against herpes simplex virus type 2. Clin Diagn Lab Immunol 9(5):1021–1024

Hu KF, Lovgren-Bengtsson K, Morein B (2001) Immunostimulating complexes (ISCOMs) for nasal vaccination. Adv Drug Deliv Rev 51(1–3):149–159

Huleatt JW, Nakaar V, Desai P, Huang Y, Hewitt D, Jacobs A, Tang J, McDonald W, Song L, Evans RK, Umlauf S, Tussey L, Powell TJ (2008) Potent immunogenicity and efficacy of a universal influenza vaccine candidate comprising a recombinant fusion protein linking influenza M2e to the TLR5 ligand flagellin. Vaccine 26(2):201–214

Hulse-Post DJ, Sturm-Ramirez KM, Humberd J, Seiler P, Govorkova EA, Krauss S, Scholtissek C, Puthavathana P, Buranathai C, Nguyen TD, Long HT, Naipospos TS, Chen H, Ellis TM, Guan Y, Peiris JS, Webster RG (2005) Role of domestic ducks in the propagation and biological evolution of highly pathogenic H5N1 influenza viruses in Asia. Proc Natl Acad Sci USA 102(30):10682–10687

Hunt LA, Brown DW, Robinson HL, Naeve CW, Webster RG (1988) Retrovirus-expressed hemagglutinin protects against lethal influenza virus infections. J Virol 62(8):3014–3019

Jansen T, Hofmans MP, Theelen MJ, Manders F, Schijns VE (2006) Structure- and oil type-based efficacy of emulsion adjuvants. Vaccine 24(26):5400–5405

Jia L, Peng D, Zhang Y, Liu H, Liu X (2003) Construction, genetic stability and protective efficacy of recombinant fowlpox virus expressing hemagglutinin gene of H5N1 subtype avian influenza virus. Wei Sheng Wu Xue Bao 43(6):722–727

Katz D, Inbar I, Samina I, Peleg BA, Heller DE (1993) Comparison of dimethyl dioctadecyl ammonium bromide, Freund's complete adjuvant and mineral oil for induction of humoral antibodies, cellular immunity and resistance to Newcastle disease virus in chickens. FEMS Immunol Med Microbiol 7(4):303–313

King DJ (1991) Evaluation of different methods of inactivation of Newcastle disease virus and avian influenza virus in egg fluids and serum. Avian Dis 35(3):505–514

Lambkin R, Dimmock NJ (1996) Longitudinal study of an epitope-biased serum haemagglutination-inhibition antibody response in rabbits immunized with type A influenza virions. Vaccine 14(3):212–218

Lawrence CW, Ream RM, Braciale TJ (2005) Frequency, specificity, and sites of expansion of CD8+ T cells during primary pulmonary influenza virus infection. J Immunol 174(9):5332–5340

Le Gall-Recule G, Cherbonnel M, Pelotte N, Blanchard P, Morin Y, Jestin V (2007) Importance of a prime-boost DNA/protein vaccination to protect chickens against low-pathogenic H7 avian influenza infection. Avian Dis 51(suppl 1):490–494

Lee CW, Suarez DL (2005) Avian influenza virus: prospects for prevention and control by vaccination. Anim Health Res Rev 6(1):1–15

Luschow D, Werner O, Mettenleiter TC, Fuchs W (2001) Protection of chickens from lethal avian influenza A virus infection by live-virus vaccination with infectious laryngotracheitis virus recombinants expressing the hemagglutinin (H5) gene. Vaccine 19(30):4249–4259

Marangon S, Cecchinato M, Capua I (2008) Use of vaccination in avian influenza control and eradication. Zoonoses Public Health 55(1):65–72

Martinet W, Saelens X, Deroo T, Neirynck S, Contreras R, Min Jou W, Fiers W (1997) Protection of mice against a lethal influenza challenge by immunization with yeast-derived recombinant influenza neuraminidase. Eur J Biochem 247(1):332–338

McCapes RH, Bankowski RA (1987) Use of avian influenza vaccines in California turkey breeders. In: Proceedings of the Second International Symposium on Avian Influenza. US Animal Health Association, Richmond, VA, pp 271–278

McNulty MS, Allan GM, Adair BM (1986) Efficacy of avian influenza neuraminidase-specific vaccines in chickens. Avian Pathology 15:107–115

Middleton D, Bingham J, Selleck P, Lowther S, Gleeson L, Lehrbach P, Robinson S, Rodenberg J, Kumar M, Andrew M (2007) Efficacy of inactivated vaccines against H5N1 avian influenza infection in ducks. Virology 359(1):66–71

Oh S, Martelli P, Hock OS, Luz S, Furley C, Chiek EJ, Wee LC, Keun NM (2005) Field study on the use of inactivated H5N2 vaccine in avian species. Vet Rec 157(10):299–300

Park MS, Steel J, Garcia-Sastre A, Swayne D, Palese P (2006) Engineered viral vaccine constructs with dual specificity: avian influenza and Newcastle disease. Proc Natl Acad Sci USA 103(21):8203–8208

Philippa JD, Munster VJ, Bolhuis H, Bestebroer TM, Schaftenaar W, Beyer WE, Fouchier RA, Kuiken T, Osterhaus AD (2005) Highly pathogenic avian influenza (H7N7): vaccination of zoo birds and transmission to non-poultry species. Vaccine 23(50):5743–5750

Philippa J, Baas C, Beyer W, Bestebroer T, Fouchier R, Smith D, Schaftenaar W, Osterhaus A (2007) Vaccination against highly pathogenic avian influenza H5N1 virus in zoos using an adjuvanted inactivated H5N2 vaccine. Vaccine 25(19):3800–3808

Price RJ (1981) Commercial avian influenza vaccines. In: Proceedings of the First International Symposium on Avian Influenza. US Animal Health Association, Richmond, VA, pp 178–179

Qiao CL, Yu KZ, Jiang YP, Jia YQ, Tian GB, Liu M, Deng GH, Wang XR, Meng QW, Tang XY (2003) Protection of chickens against highly lethal H5N1 and H7N1 avian influenza viruses with a recombinant fowlpox virus co-expressing H5 haemagglutinin and N1 neuraminidase genes. Avian Pathol 32(1):25–32

Qiao C, Yu K, Jiang Y, Li C, Tian G, Wang X, Chen H (2006) Development of a recombinant fowlpox virus vector-based vaccine of H5N1 subtype avian influenza. Dev Biol (Basel) 124:127–132

Robinson HL, Hunt LA, Webster RG (1993) Protection against a lethal influenza virus challenge by immunization with a haemagglutinin-expressing plasmid DNA. Vaccine 11(9):957–960

Roth JA (1999) Mechanistic bases for adverse vaccine reactions and vaccine failures. Adv Vet Med 41:681–700

Rott R, Becht H, Orlich M (1974) The significance of influenza virus neuraminidase in immunity. J Gen Virol 22(1):35–41

Saelens X, Vanlandschoot P, Martinet W, Maras M, Neirynck S, Contreras R, Fiers W, Jou WM (1999) Protection of mice against a lethal influenza virus challenge after immunization with yeast-derived secreted influenza virus hemagglutinin. Eur J Biochem 260(1):166–175

Sambhara S, Kurichh A, Miranda R, Tumpey T, Rowe T, Renshaw M, Arpino R, Tamane A, Kandil A, James O, Underdown B, Klein M, Katz J, Burt D (2001) Heterosubtypic immunity against human influenza A viruses, including recently emerged avian H5 and H9 viruses, induced by FLU-ISCOM vaccine in mice requires both cytotoxic T-lymphocyte and macrophage function. Cell Immunol 211(2):143–153

Schijns VE (2003) Mechanisms of vaccine adjuvant activity: initiation and regulation of immune responses by vaccine adjuvants. Vaccine 21(9–10):829–831

Schijns VE, Tangeras A (2005) Vaccine adjuvant technology: from theoretical mechanisms to practical approaches. Dev Biol (Basel) 121:127–134

Schultz-Cherry S, Dybing JK, Davis NL, Williamson C, Suarez DL, Johnston R, Perdue ML (2000) Influenza virus (A/HK/156/97) hemagglutinin expressed by an alphavirus replicon system protects chickens against lethal infection with Hong Kong-origin H5N1 viruses. Virology 278(1):55–59

Seo SH, Webster RG (2001) Cross-reactive, cell-mediated immunity and protection of chickens from lethal H5N1 influenza virus infection in Hong Kong poultry markets. J Virol 75(6): 2516–2525

Sims LD (2007) Lessons learned from Asian H5N1 outbreak control. Avian Dis 51(suppl 1): 174–181

Steensels M, Van Borm S, Lambrecht B, De Vriese J, Le Gros FX, Bublot M, van den Berg T (2007) Efficacy of an inactivated and a fowlpox-vectored vaccine in Muscovy ducks against an Asian H5N1 highly pathogenic avian influenza viral challenge. Avian Dis 51(suppl 1): 325–331

Stephenson I, Bugarini R, Nicholson KG, Podda A, Wood JM, Zambon MC, Katz JM (2005) Cross-reactivity to highly pathogenic avian influenza H5N1 viruses after vaccination with nonadjuvanted and MF59-adjuvanted influenza A/Duck/Singapore/97 (H5N3) vaccine: a potential priming strategy. J Infect Dis 191(8):1210–1215

Stone HD (1986) Efficacy of avian influenza oil-emulsion vaccines in chickens of various ages. Avian Dis 31:483–490

Stone HD (1989) Efficacy of oil-emulsion vaccines prepared with pigeon paramyxovirus-1, Ulster, and La Sota Newcastle Disease virus. Avian Dis 33:157–162

Stone HD (1993) Efficacy of experimental animal and vegetable oil-emulsion vaccines for Newcastle disease and avian influenza. Avian Dis 37(2):399–405

Stone HD (1997) Newcastle disease oil emulsion vaccines prepared with animal, vegetable, and synthetic oils. Avian Dis 41(3):591–597

Suarez DL, Schultz-Cherry S (2000) The effect of eukaryotic expression vectors and adjuvants on DNA vaccines in chickens using an avian influenza model. Avian Dis 44(4):861–868

Suarez DL, Lee CW, Swayne DE (2006) Avian influenza vaccination in North America: strategies and difficulties. Dev Biol (Basel) 124:117–124

Swayne DE (2003) Vaccines for List A poultry diseases: emphasis on avian influenza. Dev Biol (Basel) 114:201–212

Swayne DE (2006) Principles for vaccine protection in chickens and domestic waterfowl against avian influenza: emphasis on Asian H5N1 high pathogenicity avian influenza. Ann N Y Acad Sci 1081:174–181

Swayne DE, Halvorson DA (2003) Influenza. In: Saif YM, Barnes HJ, Fadly AM, Glisson JR (eds) Diseases of poultry, 11th edn. Iowa State University Press, Ames, IA, pp 135–160

Swayne DE, Halvorson DA (2008) Influenza. In: Saif YM, Barnes HJ, Fadly A, Glisson JR, McDougald LR, Nolan L, Swayne DE (eds) Diseases of poultry, vol 12. Iowa State University Press, Ames, IA (in press)

Swayne DE, Kapczynski DR (2008) Vaccines, vaccination, and immunology for avian influenza viruses in poultry. In: Swayne DE (ed) Avian influenza. Blackwell, Ames, IA, pp 407–451

Swayne DE, Beck JR, Mickle TR (1997) Efficacy of recombinant fowl poxvirus vaccine in protecting chickens against a highly pathogenic Mexican-origin H5N2 avian influenza virus. Avian Dis 41(4):910–922

Swayne DE, Beck JR, Marcia M, Stone HD (1999) Influence of virus strain and antigen mass on efficacy of H5 avian influenza inactivated vaccines. Avian Pathology 28:245–255

Swayne DE, Garcia M, Beck JR, Kinney N, Suarez DL (2000) Protection against diverse highly pathogenic H5 avian influenza viruses in chickens immunized with a recombinant fowlpox vaccine containing an H5 avian influenza hemagglutinin gene insert. Vaccine 18(11–12):1088–1095

Swayne DE, Suarez DL, Schultz-Cherry S, Tumpey TM, King DJ, Nakaya T, Palese P, Garcia-Sastre A (2003) Recombinant paramyxovirus type 1-avian influenza-H7 virus as a vaccine for protection of chickens against influenza and Newcastle disease. Avian Dis 47(suppl 3):1047–1050

Sylte MJ, Hubby B, Suarez DL (2007) Influenza neuraminidase antibodies provide partial protection for chickens against high pathogenic avian influenza infection. Vaccine 25(19):3763–3772

Todd C (1928) Experiments on the virus of fowl plague. Br J Exp Pathol 9:101–106

Toro H, Tang DC, Suarez DL, Sylte MJ, Pfeiffer J, Van Kampen KR (2007) Protective avian influenza in ovo vaccination with non-replicating human adenovirus vector. Vaccine 25(15):2886–2891

Tumpey TM, Kapazynsk DR, Swayne DE (2004) Comparative susceptibility of chickens and turkeys to avian influenza A H7N2 vins infection and protective efficacy of a commercial avian influenza H7N2 vins vaccine. Avian Dis 48(1):167–176.

van der Goot JA, Koch G, de Jong MC, van Boven M (2005) Quantification of the effect of vaccination on transmission of avian influenza (H7N7) in chickens. Proc Natl Acad Sci USA 102(50):18141–18146

van der Goot JA, van Boven M, Koch G, de Jong MC (2007) Variable effect of vaccination against highly pathogenic avian influenza (H7N7) virus on disease and transmission in pheasants and teals. Vaccine 25(49):8318–8325

Veits J, Luschow D, Kindermann K, Werner O, Teifke JP, Mettenleiter TC, Fuchs W (2003) Deletion of the non-essential UL0 gene of infectious laryngotracheitis (ILT) virus leads to attenuation in chickens, and UL0 mutants expressing influenza virus haemagglutinin (H7) protect against ILT and fowl plague. J Gen Virol 84(Pt 12):3343–3352

Veits J, Wiesner D, Fuchs W, Hoffmann B, Granzow H, Starick E, Mundt E, Schirrmeier H, Mebatsion T, Mettenleiter TC, Romer-Oberdorfer A (2006) Newcastle disease virus expressing H5 hemagglutinin gene protects chickens against Newcastle disease and avian influenza. Proc Natl Acad Sci USA 103(21):8197–8202

Villareal CL (2006) Control and eradication strategies of avian influenza in Mexico. Dev Biol 124:125–126

Vogel FR (2000) Improving vaccine performance with adjuvants. Clin Infect Dis 30(suppl 3): S266–S270

Watanabe T, Watanabe S, Kim JH, Hatta M, Kawaoka Y (2008) Novel approach to the development of effective H5N1 influenza A virus vaccines: use of M2 cytoplasmic tail mutants. J Virol 82(5):2486–2492

Webster RG, Kawaoka Y, Taylor J, Weinberg R, Paoletti E (1991) Efficacy of nucleoprotein and haemagglutinin antigens expressed in fowlpox virus as vaccine for influenza in chickens. Vaccine 9(5):303–308

Xie ZX, Stone HD (1990) Immune response to oil-emulsion vaccines with single or mixed antigens of Newcastle disease, avian influenza, and infectious bronchitis. Avian Dis 34(1):154–162

Zanella A, Poli G, Bigami M (1981) Avian influenza: approaches in the control of disease with inactivated vaccines in oil emulsion. In: Proceedings of the First International Symposium on Avian Influenza. US Animal Health Association, Richmond, VA, pp 180–183

# Development and Application of Avian Influenza Vaccines in China

**Hualan Chen and Zhigao Bu**

## Contents

**Abstract** Following the first detection of the highly pathogenic H5N1 avian influenza virus in sick geese in Guangdong Province in China in 1996, scientists began to develop vaccines in preparation for an avian influenza pandemic. An inactivated H5N2 vaccine was produced from a low pathogenic virus, A/turkey/England/ N-28/73, and was used for buffer zone vaccination during H5N1 outbreaks in 2004 in China. We also generated a low pathogenic H5N1 reassortant virus (Re-1) that derives its HA and NA genes from the GS/GD/96 virus and six internal genes from the high-growth A/Puerto Rico/8/34 (PR8) virus using plasmid-based reverse genetics. The inactivated vaccine derived from the Re-1 strain could induce more than ten months of protective immunity in chickens after one-dose inoculation; most importantly, this vaccine is immunogenic for geese and ducks.

H. Chen (✉) and Z. Bu
Harbin Veterinary Research Institute, Chinese Academy of Agricultural Sciences,
427 Maduan Street, Harbin 150001, People's Republic of China
e-mail: hlchen1@yahoo.com

R.W. Compans and W.A. Orenstein (eds.), *Vaccines for Pandemic Influenza*, 153
Current Topics in Microbiology and Immunology 333,
DOI 10.1007/978-3-540-92165-3_7, © Springer-Verlag Berlin Heidelberg 2009

We recently developed a Newcastle virus-vectored live vaccine that exhibits great promise for use in the field to prevent highly pathogenic avian influenza and Newcastle disease in chickens. Over 30 billion doses of these vaccines have been used in China and other countries, including Vietnam, Mongolia, and Egypt, and have played an important role in H5N1 avian influenza control in these countries.

# 1 Introduction

China is one of the largest producers of poultry in the world, with a production of domestic poultry totaling 15.2 billion in 2005—accounting for 20% of the total global production. Among the 15.2 billion poultry, over 60% are bred in small-scale farms or in backyards. China is home to an even larger population of waterfowl—approximately 70% of the world's total. The majority of these waterfowl are ducks that are distributed in the provinces of southern China, and these ducks are raised in the open field, which is rich in lakes and rivers. During breeding season, the ducks may migrate from one province to another, over hundreds of miles. This special breeding style brings domestic waterfowl into contact with both wild waterfowl and other domestic animals, such as chickens and pigs, allowing the waterfowl to play an important role as intermediate hosts in the transfer of influenza viruses from wild birds to domestic animals. Waterfowl migration also serves to spread influenza from one place to another, which poses huge difficulties in the control of avian influenza in China.

The highly pathogenic H5N1 avian influenza virus was first detected in a goose in the Guangdong Province of China in 1996. Multiple genotypes of H5N1 viruses have been identified from apparently healthy waterfowl since 1999 (Chen et al. 2004). In the years 2004, 2005, 2006, and 2007, there were 50, 31, 10, and 4 outbreaks, respectively, in domestic poultry and wild birds in China (Chen et al. 2006; Liu et al. 2005; Wan et al. 2005). These outbreaks occurred in 23 provinces and caused severe economic damage to the poultry industry in China. Since 2004, over 35,000,000 poultry have been depopulated in order to control the disease. Meanwhile, vaccines have also been used in the poultry in the buffer areas of disease outbreak sites.

The development of vaccines for H5 avian influenza has been supported by the government since the detection of the highly pathogenic H5N1 virus GS/GD/96 in 1996. During the last ten years, we have successfully developed several inactivated vaccines using naturally isolated low pathogenic H5N2 virus or artificially generated, low-pathogenic, high-growth reassortant virus in reverse genetics as seed viruses (Tian et al. 2005). Two kinds of live virus-vectored vaccines using the fowlpox virus and Newcastle disease virus (NDV) as backbones were also developed in China (Ge et al. 2007; Qiao et al. 2003). Here, we will present the development and application of vaccines to control H5N1 avian influenza in China, and we will also summarize the development of our two new vaccines.

## 2 Inactivated Vaccines

### 2.1 Inactivated H5N2 Vaccine

An inactivated oil-emulsified vaccine has been developed using an low pathogenic H5N2 virus, A/Turkey/England/N-28/73 (kindly provided by Dr. Danis Alexander), as a seed virus. The vaccine was approved for use in August of 2003 in Guangdong Province in chickens that were exported to Hong Kong and Macau. This vaccine was fully evaluated by the Chinese Veterinary Drug Evaluation Committee and was certified by the end of 2003. After the H5N1 outbreak in 2004, this vaccine was licensed to nine companies that have Good Manufacture Practice (GMP) facilities and the experience to produce egg-cultured vaccines. In total, 2.5 billion doses of inactivated H5N2 vaccine were used in the districts containing H5N1 outbreaks in 2004 (Table 1).

### 2.2 Inactivated H5N1 Vaccine

The use of the H5N2 vaccines, along with other measures, facilitated rapid control over H5N1 outbreaks in China in 2004. However, this vaccine is not an ideal one. First, the vaccine seed virus exhibited antigenic diversity from the H5N1 strains prevalent in China at the time. Second, the seed virus could not grow to high titers in eggs, which severely impaired vaccine production. To solve these problems, we used plasmid-based reverse genetics (Fodor et al. 1999; Hoffmann et al. 2000; Neumann et al. 1999) to generate several reassortant viruses that contained the internal genes from the high-growth A/Puerto Rico/8/34 (PR8) virus and the HA and NA genes from the H5N1 viruses GS/GD/1/96, A/bar-headed goose/Qinghai/3/2005 and A/duck/Anhui/1/2006 (Table 1). The multiple basic amino acids (–RRRKKR–) in the cleavage site of the HA protein that are associated with virulence in H5 avian influenza viruses were changed into –RETR– (Li et al. 1999; Subbarao et al. 2003), a characteristic of low pathogenic avian influenza viruses (Perdue et al. 1997; Senne et al. 1996). The reassortant virus, Re-1, that bears the HA and NA genes from GS/GD/1/96 was investigated extensively. The virus is completely attenuated in chicken embryos and chickens (Tian et al. 2005). It does not kill eggs within 72 h after inoculation and achieves a titer of more than 11 (log2). Most importantly, the Re-1 virus contains the HA and NA genes of the GS/GD/1/96 virus, which antigenically matches well with the H5N1 viruses that circulated in China (Chen et al. 2004). This inactivated H5N1 vaccine induced higher HI antibody responses and longer lasting protective immunity in chickens than the H5N2 vaccines, and was shown to be effective in ducks and geese (Tian et al. 2005). The vaccine was approved for use in the field by the end of 2004, and over 20 billion doses of the Re-1 vaccine have been used in China (18.04 billion doses) (Table 1), Vietnam, Mongolia, and Egypt so far.

**Table 1** Vaccines developed and used for H5N1 avian influenza control in China from 2004–2007

| Vaccine | | Seed virus generated | | Doses used in the year (billions) | | | | |
|---|---|---|---|---|---|---|---|---|
| | | Seed name | HA and/or NA gene donor virus | 2004 | 2005 | 2006 | 2007 | Total |
| Inactivated vaccine | H5N2 subtype | A/Turkey/England/N-28/73 (H5N2) (N-28) | – | 2.5 | 4.08 | 3.6 | – | 10.18 |
| | H5N1 subtype | H5N1/PR8(H5N1) (Re-1) | A/goose/Guangdong/1/1996 | 0.57 | 3.3 | 4.57 | 9.6 | 18.04 |
| | | H5N1/PR8(H5N1) (Re-3) | A/bar-headed goose/Qinghai/3/2005 | – | – | – | – | – |
| | | H5N1/PR8 (H5N1) (Re-4) | A/chicken/Shanxi/2/2006 | – | – | 0.84 | 0.42 | 1.26 |
| | | H5N1/PR8 (H5N1) (Re-5) | A/duck/Anhui/1/2006 | – | – | – | – | – |
| | | H5N1/PR8-5B19 | A/goose/Guangdong/1/1996 | – | – | – | – | – |
| | | Re-1/Re-4 | – | – | – | – | 2.2 | 2.2 |
| Live virus vector vaccine | Recombinant fowlpox vaccine | rFPF-HA-NA | A/goose/Guangdong/1/1996 | – | 0.615 | – | – | 0.615 |
| | Recombinant NDV vaccine | rLH5-1 | A/goose/Guangdong/1/1996 | – | – | 2.6 | 1.3 | 3.9 |
| | | rLH5-3 | A/bar-headed goose/Qinghai/3/2005 | – | – | – | – | – |
| | | rLH5-4 | A/chicken/Shanxi/2/2006 | – | – | – | – | – |
| | | rLH5-5 | A/duck/Anhui/1/2006 | – | – | – | – | – |
| DNA | – | pCAGGoptiHA | A/goose/Guangdong/1/1996 | – | – | – | – | – |

In early 2006, an H5N1 avian influenza virus was isolated from a chicken flock that had been vaccinated with the inactivated H5 vaccines. The disease in those flocks was recorded as a decrease in egg production and a mortality range of 10–20%. The virus, denoted CK/SX/06, exhibited huge antigenic drift from the viruses that were isolated in China previously. Though 187,000 poultry were depopulated to control the spread of this new virus after its first detection in February, the virus was reisolated in June from the Shanxi and Ningxia provinces. We found that the inactivated H5 vaccines used in China provided only 80% protection against the variant strain in a laboratory challenge study in specific pathogen-free (SPF) chickens, which was quite different from the protective efficacy we had reported previously (Tian et al. 2005). We therefore developed a new reassortant virus, designated Re-4, which contained the cleavage site-modified HA and NA genes from CK/SX/06 and six internal genes from the PR8 virus. This new vaccine was approved for use in Shanxi, Ningxia, and several of their neighboring provinces in northern China in August. A total of 1.26 billion doses were used in 2006 and 2007 (Table 1).

In some areas, cocirculation of both the GS/GD/96-like virus and CK/SX/06-like viruses was detected; therefore, an H5N1 vaccine that was produced from the combined antigens of Re-1 and Re-4 was also approved for use in a limited area in northern China. A total of 2.2 billion doses of this vaccine were used in 2007 (Table 1).

## 2.3   Inactivated H5N1 Marker Vaccine

The current commercially used inactivated H5N1 vaccine is safe and effective, providing complete protection from highly pathogenic H5N1 influenza viruses (Tian et al. 2005). However, the current inactivated H5N1 vaccine does not allow for serological distinction between vaccination and field infection. Recently, intensive vaccination with marker vaccines and stamping-out strategies have been gaining popularity in veterinary medicine for eradicating specific diseases of national or international interest. A marker vaccine is defined as one that can be used in conjunction with a diagnostic test to differentiate a vaccinated animal from a naturally infected animal (Babiuk 1999). A genetically marked H5N1 influenza vaccine that could readily be distinguished from wild-type strains would therefore be of great value in the eradication plan, allowing for vaccination programs that would not interfere with the serological surveillance of influenza viruses circulating in the wild.

We recently generated an attenuated H5N1 influenza marker vaccine seed virus, denoted H5N1/PR8-5B19, which derives its internal genes from the PR8 virus and modified H5 HA and N1 NA genes from the GS/GD/1/96 virus (Li et al. 2008). H5N1/PR8-5B19 encodes an attenuated HA molecule to allow for low pathogenicity in poultry and an NA molecule bearing the foreign 5B19 epitope of the S2 glycoprotein of murine hepatitis virus (MHV). H5N1/PR8-5B19 grew to high titers in embryonated eggs and in chickens without leading to sickness. When chickens were inoculated with one dose of the inactivated vaccine generated from the H5N1/PR8-5B19 virus, 70% of the chickens were positive for anti-5B19 antibody postvaccination. The chickens that received a booster dose at the end of two weeks after the first immunization were 100% positive for anti-5B19 antibody. In contrast, sera obtained from chickens vaccinated with the inactivated H5N1/PR8 vaccine or infected with the H5N1/PR8 virus showed no reactivity against the 5B19 epitope in a peptide-ELISA diagnostic test. In the vaccine trial, at 21 days after inoculation with one dose of the vaccine, chickens were challenged intranasally with different H5N1 viruses. No signs of disease associated with H5N1 infection were observed in any of the chickens immunized with inactivated H5N1/PR8-5B19 vaccine. No virus was recovered from the tracheal and cloacal swabs on day 3 after the challenge with homologous or heterologous H5N1 viruses. In contrast, the challenge virus was detected in the tracheal and cloacal swabs from all of the control chickens, and killed all of the control chickens in the observation period. These data show that the H5N1/PR8-5B19 marker vaccine elicited strong antibody responses to influenza HA and to the MHV 5B19 epitope, and provided complete immunity to H5N1 HPAIV challenge.

H5N1/PR8-5B19 is the first H5N1 vaccine candidate with the desired properties of efficient replication in eggs, safe use in birds, and the ability to serologically discriminate between infected and vaccinated chickens. Although additional experiments are necessary, the results we got show that the recombinant H5N1/ RP8-5B19 virus should be considered a potential candidate for an H5N1 influenza marker vaccine for chickens that may eventually be used in the field to control the spread of H5N1 influenza virus infection in poultry.

# 3 Live Virus-Vectored Vaccine

## 3.1 Recombinant Fowlpox Vaccine

Whole-virus inactivated vaccines and fowlpox virus-based recombinant vaccines have been used as control strategies for highly pathogenic avian influenza in the laboratory and in poultry farms located in different geographic regions in the world (Capua et al. 2003; Ellis et al. 2004; Swayne et al. 2000; van der Goot et al. 2005). In addition to the inactivated vaccines, we also developed two kinds of recombinant vaccines using fowlpox virus and NDV as vectors (Ge et al. 2007). After the detection of the GS/GD/96 virus, we began developing a recombinant fowlpox virus expressing the HA and NA genes of H5N1 virus as a live virus-vectored vaccine. The vaccine efficacy of this recombinant virus was proven in both laboratory and field tests (Qiao et al. 2003). About 0.615 billion doses of the recombinant fowlpox vaccine have been used in poultry in China since 2005.

## 3.2   Recombinant Newcastle Disease Virus Vaccine

The NDV live virus-vectored vaccine against influenza has several advantages, including ease of production, high production yield, ease of widespread administration to animals in the field, and its ability to serve as a bivalent vaccine against two viruses that can decimate bird populations. The use of NDV as the vaccine backbone should prevent confusion between vaccinated birds and infected birds for surveillance purposes, which is a problematic issue with the use of whole-virus influenza vaccines. Highly pathogenic Newcastle disease has been endemic, and more than thirty billion doses of live vaccines are used in chickens every year in China.

In 2005, we established a reverse genetics system for NDVs (LaSota) and generated several recombinant NDVs expressing the avian influenza virus HA genes from several H5N1 viruses representing different phylogenetic lineages of the viruses isolated in China (our unpublished data). These viruses included GS/GD/96, A/duck/Anhui/1/06, and A/bar-headed goose/Qinghai/3/05. Recently, we also generated a recombinant NDV expressing the HA gene of the CK/SX/06 virus. We have demonstrated that the recombinant NDVs expressing the various HA genes induce strong HI antibody responses to NDV and to H5 avian influenza viruses in chickens. The recombinant NDV-vaccinated chickens were protected from disease signs and death from challenge with highly pathogenic NDV. Most importantly, the vaccinated chickens were completely protected from homologous and heterologous H5N1 virus challenges and displayed no virus shedding, signs of disease, or death (Ge et al. 2007).

At the beginning of 2006, a recombinant NDV virus that expressed the HA gene of GS/GD/96 was approved for use in chickens as a bivalent, live attenuated vaccine for controlling the H5N1 avian influenza and highly pathogenic Newcastle disease. By the end of 2007, a total of four billion doses of this vaccine had been applied in chickens (Table 1), which dramatically increased the vaccination coverage.

## 4   DNA Vaccine

Although inactivated whole-virus vaccine (Tian et al. 2005), recombinant fowlpox vaccines (Qiao et al. 2003), and the recombinant NDV vaccine (Ge et al. 2007) have been used in China and some other countries, DNA vaccines may offer a number of advantages over these vaccine strategies for avian influenza virus control and infection prevention. First, DNA immunization can achieve both humoral and cell-mediated immune responses, similar to an attenuated live virus vaccine, and it has the safety of a killed or subunit vaccine (Donnelly et al. 2000; Garmory et al. 2003; Liu et al. 1998; Webster 1999). Second, DNA vaccines are easier to manufacture and store than inactivated whole-virus vaccines. Third, immune responses are generated against the expressed gene product and not the DNA vaccine vector.

Although previous studies (Chen et al. 2001; Fynan et al. 1993; Sharma et al. 2000; Suarez and Schultz-Cherry 2000) have confirmed the efficacy of the HA gene-based DNA vaccine against highly pathogenic avian influenza virus challenge in chickens, the high dosage (200–400 μg plasmid DNA) needed is a major obstacle to the field application of such vaccines. We therefore explored a strategy to decrease the dosage of DNA vaccine required by improving the expression of the target gene.

We constructed an H5 HA gene, optiHA, containing chicken-biased codons based on the HA amino acid sequence of the highly pathogenic H5N1 virus GS/GD/96. The optiHA was inserted to the plasmid pCAGGS under the control of chicken β-actin, and designated pCAGGoptiHA. We evaluated the vaccine efficacy of pCAGGoptiHA by intramuscular injection with different dosages (100, 10, or 1 μg) of the plasmid. All of the vaccinated chickens developed detectable HI and NT antibodies, with the titers of the antibodies correlating with the dosage of plasmid inoculated. When the chickens were challenged with a lethal dose of highly pathogenic H5N1 avian influenza virus, all of the 100- and 10-μg plasmid inoculated chickens were completely protected from disease signs and death, while 1-μg plasmid inoculated chickens were only partially protected. The vaccine efficacy of the low-dosage plasmid inoculation (1 μg) could be improved by a second immunization (Jiang et al. 2007). Further investigation demonstrated that two doses of 10 μg of the pCAGGoptiHA inoculation could induce protection lasting more than a year (Jiang et al. 2007), which covers the entire time period of the layers kept at the farm, thus eliminating the need for costly serial vaccinations. The clinical field trial of this DNA vaccine is currently ongoing.

# 5    Conclusion

In this chapter, we have briefly summarized the development and application of vaccines for the control of highly pathogenic avian influenza in China. The epidemic of H5N1 avian influenza in China resulted in the deaths of over 35,000,000 poultry through either infection or depopulation during 2004–2007, and led to severe economic damage to the poultry industry. China employs the culling plus vaccination strategy to control H5N1 avian influenza, and financial support from the government ensures the implementation of this strategy. Billions of doses of the vaccines have been used in the field, and the vaccines are antigenically well matched to the circulating strains. Though the government has required 100% vaccine coverage in domestic poultry since the end of 2005, it is impossible to give every single bird one or two doses of the vaccine in practice, as over 70% of the birds are reared in small-scale or backyard farms, often in the open field with ducks and geese. It is clear that the increased vaccination coverage results in decreased disease epidemics. There is no doubt that vaccination has played an important role in protecting poultry from H5N1 virus infection, reducing the virus load in the environment, and preventing the transmission of the H5N1 virus from poultry to humans. However, it is worth noting that the complete control and eradication of highly pathogenic

H5N1 avian influenza viruses can ultimately only be achieved through a combination of vaccination, improved biosecurity, extensive surveillance and an effective monitoring program.

**Acknowledgment** The authors are supported by the Animal Infectious Disease Control Program of the Ministry of Agriculture, and the Chinese National S&T Plan Grants 2004BA519A-57 and 2006BAD06A05.

# References

Babiuk LA (1999) Broadening the approaches to developing more effective vaccines. Vaccine 17:1587–1595

Capua I, Marangon S, dalla Pozza M, Terregino C, Cattoli G (2003) Avian influenza in Italy 1997–2001. Avian Dis 47:839–843

Chen H, Yu K, Jiang Y, Tang X (2001). DNA immunization elicits high HI antibody and protects chicken from AIV challenge. Options for the control of influenza IV, Crete, Greece, 23–28 Sept 2000, 1219:917–921

Chen H, Deng G, Li Z, Tian G, Li Y, Jiao P, Zhang L, Liu Z, Webster RG, Yu K (2004) The evolution of H5N1 influenza viruses in ducks in southern China. Proc Natl Acad Sci USA 101: 10452–10457

Chen H, Li Y, Li Z, Shi J, Shinya K, Deng G, Qi Q, Tian G, Fan S, Zhao H, Sun Y, Kawaoka Y (2006) Properties and dissemination of H5N1 viruses isolated during an influenza outbreak in migratory waterfowl in western China. J Virol 80:5976–5983

Donnelly JJ, Liu MA, Ulmer JB (2000) Antigen presentation and DNA vaccines. Am J Respir Crit Care Med 162:S190–S193

Ellis TM, Leung CY, Chow MK, Bissett LA, Wong W, Guan Y, Malik Peiris JS (2004) Vaccination of chickens against H5N1 avian influenza in the face of an outbreak interrupts virus transmission. Avian Pathol 33:405–412

Fodor E, Devenish L, Engelhardt OG, Palese P, Brownlee GG, Garcia-Sastre A (1999) Rescue of influenza A virus from recombinant DNA. J Virol 73:9679–9682

Fynan EF, Webster RG, Fuller DH, Haynes JR, Santoro JC, Robinson HL (1993) DNA vaccines: protective immunizations by parenteral, mucosal, and gene-gun inoculations. Proc Natl Acad Sci USA 90:11478–11482

Garmory HS, Brown KA, Titball RW (2003) DNA vaccines: improving expression of antigens. Genet Vaccines Ther 1:2

Ge J, Deng G, Wen Z, Tian G, Wang Y, Shi J, Wang X, Li Y, Hu S, Jiang Y, Yang C, Yu K, Bu Z, Chen H (2007) Newcastle disease virus-based live attenuated vaccine completely protects chickens and mice from lethal challenge of homologous and heterologous H5N1 avian influenza viruses. J Virol 81:150–158

Hoffmann E, Neumann G, Kawaoka Y, Hobom G, Webster RG (2000) A DNA transfection system for generation of influenza A virus from eight plasmids. Proc Natl Acad Sci USA 97:6108–6113

Jiang Y, Yu K, Zhang H, Zhang P, Li C, Tian G, Li Y, Wang X, Ge J, Bu Z, Chen H (2007) Enhanced protective efficacy of H5 subtype avian influenza DNA vaccine with codon optimized HA gene in a pCAGGS plasmid vector. Antiviral Res 75:234–241

Li S, Liu C, Klimov A, Subbarao K, Perdue ML, Mo D, Ji Y, Woods L, Hia S, BryantMetal (1999) Recombinant influenza A virus vaccines for the pathogenic human A/Hong Kong/97 (H5N1) viruses. J Infect Dis 179:1132–1138

Li C, Ping J, Jing B, Deng G, Jiang Y, Li Y, Tian G, Yu K, Bu Z, Chen H (2008) H5N1 influenza marker vaccine for serological differentiation between vaccinated and infected chickens. Biochem Biophys Res Commun 3722:293–297

Liu MA, Fu TM, Donnelly JJ, Caulfield MJ, Ulmer JB (1998) DNA vaccines. Mechanisms for generation of immune responses. Adv Exp Med Biol 452:187–191

Liu J, Xiao H, Lei F, Zhu Q, Qin K, Zhang XW, Zhang XL, Zhao D, Wang G, Feng Y, Ma J, Liu W, Wang J, Gao GF (2005) Highly pathogenic H5N1 influenza virus infection in migratory birds. Science 309:1206

Neumann G, Watanabe T, Ito H, Watanabe S, Goto H, Gao P, Hughes M, Perez DR, Donis R, Hoffmann E, Hobom G, Kawaoka Y (1999) Generation of influenza A viruses entirely from cloned cDNAs. Proc Natl Acad Sci USA 96:9345–9350

Perdue ML, Garcia M, Senne D, Fraire M (1997) Virulence-associated sequence duplication at the hemagglutinin cleavage site of avian influenza viruses. Virus Res 49:173–186

Qiao CL, Yu KZ, Jiang YP, Jia YQ, Tian GB, Liu M, Deng GH, Wang XR, Meng QW, Tang XY (2003) Protection of chickens against highly lethal H5N1 and H7N1 avian influenza viruses with a recombinant fowlpox virus co-expressing H5 haemagglutinin and N1 neuraminidase genes. Avian Pathol 32:25–32

Senne DA, Panigrahy B, Kawaoka Y, Pearson JE, Suss J, Lipkind M, Kida H, Webster RG (1996) Survey of the hemagglutinin (HA) cleavage site sequence of H5 and H7 avian influenza viruses: amino acid sequence at the HA cleavage site as a marker of pathogenicity potential. Avian Dis 40:425–437

Sharma S, Mamane Y, Grandvaux N, Bartlett J, Petropoulos L, Lin R, Hiscott J (2000) Activation and regulation of interferon regulatory factor 4 in HTLV type 1-infected T lymphocytes. AIDS Res Hum Retroviruses 16:1613–1622

Suarez DL, Schultz-Cherry S (2000) The effect of eukaryotic expression vectors and adjuvants on DNA vaccines in chickens using an avian influenza model. Avian Dis 44:861–868

Subbarao K, Chen H, Swayne D, Mingay L, Fodor E, Brownlee G, Xu X, Lu X, Katz J, Cox N, Matsuoka Y (2003) Evaluation of a genetically modified reassortant H5N1 influenza A virus vaccine candidate generated by plasmid-based reverse genetics. Virology 305:192–200

Swayne DE, Garcia M, Beck JR, Kinney N, Suarez DL (2000) Protection against diverse highly pathogenic H5 avian influenza viruses in chickens immunized with a recombinant fowlpox vaccine containing an H5 avian influenza hemagglutinin gene insert. Vaccine 18:1088–1095

Tian G, Zhang S, Li Y, Bu Z, Liu P, Zhou J, Li C, Shi J, Yu K, Chen H (2005) Protective efficacy in chickens, geese and ducks of an H5N1-inactivated vaccine developed by reverse genetics. Virology 341:153–162

van der Goot JA, Koch G, de Jong MC, van Boven M (2005) Quantification of the effect of vaccination on transmission of avian influenza (H7N7) in chickens. Proc Natl Acad Sci USA 102:18141–18146

Wan XF, Ren T, Luo KJ, Liao M, Zhang GH, Chen JD, Cao WS, Li Y, Jin NY, Xu D, Xin CA (2005) Genetic characterization of H5N1 avian influenza viruses isolated in southern China during the 2003–04 avian influenza outbreaks. Arch Virol 150:1257–1266

Webster RG (1999) Potential advantages of DNA immunization for influenza epidemic and pandemic planning. Clin Infect Dis 28:225–229

# Part III
# Novel Vaccine Approaches

# Designing Vaccines for Pandemic Influenza

**Taisuke Horimoto and Yoshihiro Kawaoka**

**Contents**

**Abstract** Recent outbreaks of highly pathogenic avian influenza A virus infections (including those of the H5N1 subtype) in poultry and in humans (through contact with infected birds) have raised concerns that a new influenza pandemic will soon occur. Effective vaccines against H5N1 virus are therefore urgently needed. Reverse genetics-based inactivated vaccines have been prepared according to WHO recommendations and licensed in several countries following their assessment in clinical trials. However, the effectiveness of these vaccines in a pandemic is not guaranteed.

T. Horimoto and Y. Kawaoka (✉)
Division of Virology, Department of Microbiology and Immunology, Institute of Medical Science, University of Tokyo, 4-6-1 Shirokanedai, Minato-ku, Tokyo 108-8639, Japan
e-mail: kawaoka@ims.u-tokyo.ac.jp

Y. Kawaoka
International Research Center for Infectious Diseases, Institute of Medical Science, University of Tokyo, 4-6-1 Shirokanedai, Minato-ku, Tokyo 108–8639, Japan;
Department of Pathological Sciences, School of Veterinary Medicine, University of Wisconsin-Madison, Madison, WI 53706, USA

R.W. Compans and W.A. Orenstein (eds.), *Vaccines for Pandemic Influenza,*
Current Topics in Microbiology and Immunology 333,
DOI 10.1007/978-3-540-92165-3_8, © Springer-Verlag Berlin Heidelberg 2009

We must therefore continue to develop alternative pandemic vaccine strategies. Here, we review the current strategies for the development of H5N1 influenza vaccines, as well as some future directions for vaccine development.


# 1 Introduction

Only type A influenza virus exhibits pandemic potential due to antigenic variation. Currently, 16 hemagglutinin (HA) and nine neuraminidase (NA) subtypes have been identified among type A viruses (Wright et al. 2007). Three influenza pandemics emerged during the twentieth century, the most devastating of which was the Spanish influenza, which was caused by an H1N1 virus and was responsible for the deaths of at least 40 million people in 1918–1919 (Johnson and Mueller 2002). Sequence information from resurrected lung tissue samples suggest that the 1918 virus was derived from "an unusual avian precursor" (Reid et al. 2004). The HA of the 1918 virus retained the residues in the host–receptor-binding site that are characteristic of an avian precursor HA, but could bind human cell-surface receptors containing $\alpha$-2,6-linked sialic acid (Gamblin et al. 2004; Stevens et al. 2004). Reverse genetics studies, including the reconstitution of the 1918 virus itself and experimental infection of macaques with the reconstituted virus, suggest that, unlike other human influenzas, the 1918 virus induced dysregulation of the antiviral response, causing acute respiratory distress and a fatal outcome in the nonhuman primate model (Kobasa et al. 2007). The other two, less serious, pandemics of the twentieth century occurred in 1957 (Asian influenza [H2N2]), and 1968 (Hong Kong influenza [H3N2]) (Wright et al. 2007). The 1957 virus consisted of HA (H2), NA (N2), and PB1 gene segments from an avian virus, with the other gene segments derived from a previously circulating human virus. The 1968 virus had avian HA (H3) and PB1 segments in a background of human viral genes. The acquisition of novel surface antigens allowed these viruses to circumvent the human immune response, resulting in these pandemics. The HAs of these two pandemic strains also bind preferentially to human-type receptors, although they originated from avian viruses (Matrosovich et al. 2000). Thus, for a virus to become a pandemic strain, it appears to require a novel HA subtype to which humans are immunologically naive, that efficiently binds to human-type receptors. It must also possess internal proteins to promote efficient growth in human upper respiratory cells and thereby facilitate its human-to-human transmission (Horimoto and Kawaoka 2005).

Vaccination is considered one of the most effective preventive measures for the control of influenza pandemics. Recent direct transmissions of avian viruses to humans suggest that avian viruses of HA subtypes other than H1 and H3 have pandemic potential, emphasizing the need for vaccines against these viruses. In particular, the widespread circulation of H5N1 viruses has focused current research on the development of H5N1 vaccines.

# 2  Developing H5N1 Vaccines for Humans

Although antivirals against H5N1 influenza viruses such as NA inhibitors (oseltamivir and zanamivir) may be effective for pandemic control, the possible emergence of drug-resistant viruses highlights the need for vaccination (Le et al. 2005; Gupta et al. 2006). Vaccination is considered the most effective preventive measure to combat an influenza pandemic. Currently, inactivated vaccines are typically used for influenza prophylaxis. They are usually prepared from virus that is grown in embryonated chicken eggs, purified from the allantoic fluids of the inoculated eggs, and inactivated with formaldehyde or β-propiolactone for "whole virus" vaccine formulation. Alternatively, the purified virus is treated with detergent for "split" or "subunit" vaccine formulation. These inactivated vaccines are then inoculated intramuscularly or subcutaneously into individuals. However, the high pathogenicity of the currently circulating H5N1 viruses presents difficulties for this type of vaccine preparation. Highly pathogenic avian influenza (HPAI) H5N1 viruses cannot be used as seed viruses for inactivated vaccine production because their virulence threatens the lives of vaccine producers and it is difficult to obtain high-quality allantoic fluid with acceptable virus titers from embryonated eggs.

## 2.1  The Conventional Approach

Seed viruses for inactivated vaccines must be antigenically similar to the circulating viruses and grow efficiently in eggs. Faced with a pandemic threat posed by the Hong Kong H5N1 outbreak in 1997, the low pathogenic avian influenza (LPAI) virus A/duck/Singapore/F119-3/97 (H5N3) was selected as a vaccine seed virus. Vaccine prepared with this virus was assessed in a randomized phase I clinical trial (Nicholson et al. 2001; Stephenson et al. 2004), the results of which showed that although antibody responses indicative of protection were achieved by administrating the vaccine with the oil-in-water MF59™ adjuvant, this strain was not suitable for large-scale vaccine production due to its inefficient growth in eggs. However, this adjuvanted vaccine candidate induced cross-reactive neutralizing antibody responses in humans to heterologous H5N1 viruses, including 2004 isolates (Stephenson et al. 2005), demonstrating its potential for use until an antigenically matched vaccine becomes available. Since no such antigenically matched natural avirulent isolates have been found for recent H5N1 viruses, an alternative approach is needed to produce safe vaccine seed viruses to protect humans from this virus infection.

## 2.2  The Practical Approach

New vaccines are being developed that exploit reverse genetics technology (Neumann et al. 1999; Fodor et al. 1999) and the knowledge that the pathogenicity of avian influenza viruses is primarily determined by HA cleavability (Kawaoka and Webster 1988; Horimoto and Kawaoka 1994). Conversion of the HA cleavage site sequence of HPAI viruses to that of avian influenza with low pathogenic (LPAI) viruses attenuates virulence but does not affect antigenicity. Several researchers have used reverse genetics to produce candidate H5N1 vaccine strains, whose HA and NA were derived from a human H5N1 virus and the remainder of their genes from a virus (termed a backbone virus) that grows well in eggs (Takada et al. 1999; Subbarao et al. 2003; Webby et al. 2004; Wood and Robertson 2004; Lipatov et al. 2005; Horimoto et al. 2006; Govorkova et al. 2006). For these vaccine strains, the HA cleavage site sequence was modified from virulent- to avirulent-type sequences. The WHO has recommended the use of A/Puerto Rico/8/34 (H1N1; PR8) as a backbone virus. PR8, originally a human isolate, has been passaged extensively in eggs and has proven to be attenuated for humans. Indeed, the PR8 backbone has been used to produce annual vaccines against human H1N1 and H3N2 virus infections. Several high-growth reassortant viruses (PR8/H5N1 6:2 reassortant virus) have been developed by reverse genetics, including the NIBRG-14 reference strain (produced by the National Institute for Biological Standards and Control, UK). Following extensive clinical testing, these viruses have now been licensed as vaccine seed viruses for inactivated vaccine in several countries (Treanor et al. 2006; Bresson et al. 2006). Inactivated H5N1 vaccines require adjuvants in addition to the "whole virus" formulation for adequate immunogenicity in humans, unlike seasonal vaccines. To compensate for the low immunogenicity of H5N1 vaccines in humans, antigenic matching between vaccine seeds and circulating strains should be considered. A panel of vaccine seed viruses with antigenic variations that reflect the genetic diversity of H5N1 viruses is required.

## 2.3   A Promising Approach

### 2.3.1   A High-Growth Seed for an Egg-Based Vaccine

To increase the total number of vaccine doses from the limited production capacity of current vaccine manufacturers, seed viruses with high growth properties in eggs are required. Given that the NIBRG-14 seed for inactivated H5N1 vaccine grows less efficiently in eggs than the seeds used for seasonal vaccines, the selection of other seeds with higher growth potential is a germane strategy for H5N1 vaccine production and stockpiling. The PR8(UW) strain maintained in our laboratory is a superior donor virus for H5N1 vaccine production compared to the PR8(Cambridge) used to produce the NIBRG-14 seed virus with respect to in ovo growth (Horimoto et al. 2007). PR8 strains differ in their growth properties depending on their passage histories; PR8(UW) may be more highly adapted in eggs than PR8(Cambridge). The high growth property of PR8(UW) in eggs was determined via several muta-tions in polymerases and NP.

Inclusion of an alterative NA protein in PR8(UW) further enhances its growth in eggs. The HA–NA functional balance affects the growth in eggs of influenza viruses

(Castrucci and Kawaoka 1993) and of seed strains for influenza seasonal vaccines (Lu et al. 2005). We found that HA–NA functional balance also determines the in ovo growth of H5N1 vaccine seed viruses; 7:1 reassortant viruses containing only modified HA from H5N1 viruses (and PR8 NA) grow significantly better than standard 6:2 reassortant viruses (Horimoto et al. 2007) (Fig. 1). One might argue that reassortants that lack NA from an H5N1 isolate would induce a less protective immune response than recombinant viruses with H5N1 NA because of antigenic differences in these proteins (even though the NA of PR8 is of the N1 subtype). However, since HA is the major protective antigen in inactivated vaccines, the enhanced growth potential conferred by the PR8 NA should offset the limited antigenic mismatch in this minor protective antigen. We propose that, in addition to the 6:2 reassortant viruses recommended by the WHO, 7:1 reassortant viruses (containing only a modified H5 derived from circulating strains) in the background of the PR8(UW) strain should be considered as vaccine seeds for inactivated H5N1 vaccine

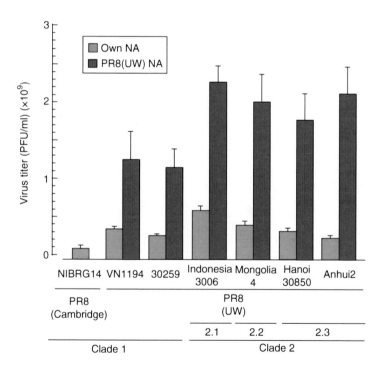

**Fig. 1** Growth enhancement of vaccine seed viruses in embryonated chicken eggs. PR8/H5N1 6:2 reassortants were prepared using H5N1 viruses of different clades with PR8(UW) donor virus; A/Vietnam/1194/04 (clade 1), A/Vietnam/30259 (clade 1), A/Indonesia/3006/05 (clade 2.1), A/ whooper swan/Mongolia/4/05 (clade 2.2), A/Vietnam/30850/05 (clade 2.3), and A/Anhui/2/05 (clade 2.3). All reassortants replicated significantly better in eggs than did the reference seed NIBRG-14 with PR8(Cambridge). PR8/H5 7:1 reassortants containing PR8(UW) NA and HA from H5N1 viruses of different clades replicated significantly better than the corresponding PR8/ H5N1 6:2 reassortants. Virus titers of the allantoic fluids were determined 48 h postinoculation after incubation at 33°C

production. This approach would increase the available doses of prepandemic or pandemic H5N1 vaccines in a timely, cost-efficient manner.

### 2.3.2 Cell Culture-Based Vaccines

Given that embryonated chicken eggs, which are currently used for inactivated vaccine production, would be in a short supply during a pandemic, the development of cell culture-based H5 vaccines is an attractive alternative approach. In fact, inactivated influenza vaccines produced with Madin–Darby canine kidney (MDCK) and African green monkey Vero cells have been licensed in The Netherlands (Medema et al. 2006). Important considerations for this approach include the selection of background viruses that grow well in these cell cultures and monitoring for antigenic changes during the propagation of the virus in the cell culture. The safety of the vaccine product for human use is also important with respect to tumorigenicity.

We found that our PR8(UW) strain also supports better growth in MDCK cells than does the PR8(Cambridge) strain due to its enhanced polymerase activity in this cell line. Interestingly, the NS gene of PR8(Cambridge) possesses higher interferon-antagonized activity in MDCK cells compared with that of PR8(UW). Accordingly, we believe that a chimeric PR8 construct whose NS gene is derived from the Cambridge strain and its remaining five internal genes from the UW strain would be the optimal donor for H5N1 vaccine production in MDCK cells. In addition, we found that inclusion of an HK213 (A/Hong Kong/213/2003) NA, whose stalk region does not have the deletion observed in most other H5N1 virus NAs, enhances the viral titers of 6:2 reassortants, indicating that the HA–NA functional balance is a determinant for viral growth in MDCK cells, as it is in eggs (Murakami et al. unpublished).

### 2.3.3 Live Attenuated Vaccines

To overcome the potential low immunogenicity of inactivated H5 vaccines for humans, live H5N1 attenuated vaccines with HA that has been altered to a nonpathogenic form have been developed. These vaccines are based on the recently licensed product FluMist®, possess a cold-adapted backbone virus, and are nonpathogenic in mammalian and chicken models (Li et al. 1999; Suguitan et al. 2006). Live influenza vaccines elicit systemic and local mucosal immune responses that include stimulating secretory IgA (sIgA) in the respiratory tract, a portal for the virus. They also elicit cellular immunity, which may provide better protection than that afforded by inactivated vaccines (Beyer et al. 2002). Live attenuated vaccines may also offer wider protection by protecting against viruses that have undergone antigenic drift. However, live H5N1 vaccines will not be used until the H5N1 virus has become widespread among humans, so as not to introduce new influenza viral HA and NA genes into the human population. In addition, they may not be used for the major high-risk groups of infants and the elderly due to safety considerations, as is the case with FluMist®. The use of live influenza vaccines for a pandemic is also currently limited by production capacity.

### 2.3.4 Mucosal Inactivated Vaccines

The primary target tissue of influenza virus infection is the respiratory mucosa. Therefore, virus-specific sIgA antibody induced in this organ by the mucosal immune system would efficiently protect against virus infection. sIgA antibody also exhibits cross-protection to antigenically drifted viruses, since its response to subsequent infection is driven by nonantigen-specific bystander help (Sangster et al. 2003). In this context, the protective efficacy of the licensed parenteral inactivated H5N1 vaccines, which induce mainly serum IgG antibodies, may be less satisfactory in pandemic vaccines.

Mucosal inactivated vaccines can induce sIgA antibody in respiratory organs and are safer for the vaccinees compared with live vaccines. Therefore, they can be used for people of all ages, including high-risk patients. In clinical trials with seasonal vaccine, both inactivated whole-virus particles and split vaccines are effective in preventing live virus infection when administrated intranasally, although stronger immunogenicity is seen with the whole-virus vaccine, probably due to the stimulation of innate immunity by single-strand RNA via toll-like receptor (TLR)-7 (Hasegawa et al. 2007).

To enhance the immunogenicity of nasal inactivated vaccines, mucosal adjuvants should be combined with these vaccines. Commercially available double-strand RNA, poly(I:$C_{12}$U) (Ampligen®) is effective as an adjuvant for H5N1 nasal inactivated vaccines. In addition, NKT cell-specific glycolipid ligand ($\alpha$-galactosylceramide), bacterial toxin-derived forms of cholera toxin B subunits (CTB) and *Escherichia coli* heat-labile enterotoxin (LT), as well as physiological complement component C3d all possess adjuvant effects with nasal vaccines, although neurological side effects were associated with toxin-derived LT in clinical use (Hasegawa et al. 2007).

The sublingual mucosal route is an attractive alternative to mucosal immunization routes for administering inactivated vaccines (Song et al. 2008). Studies in a mouse model revealed appreciably high levels of virus-specific IgG in serum and sIgA antibodies in mucosal secretions, even in the absence of adjuvants. Coadministration of a toxin-derived mucosal adjuvant enhanced these immune responses.

## 2.4  An Improved Method for Reverse Genetics

Currently, prepandemic H5N1 vaccines are being stockpiled in many countries. These inactivated vaccines were produced from viruses propagated in embryonated chicken eggs following inoculation of the vaccine seed virus, generated by reverse genetics in an African green monkey Vero cell line that is approved for human vaccine production (Nicolson et al. 2005). However, the generation of the H5N1 vaccine seed

viruses in Vero cells is not optimal due to the low plasmid transfection efficiency of
these cells for reverse genetics. In a pandemic situation, vaccines whose antigenicities
match the circulating strain(s) need to be rapidly produced. Therefore, a more robust
reverse genetics system is desirable for pandemic vaccine preparedness. Although
twelve- or eight-plasmid reverse genetics systems may prove useful for the production
of pandemic and interpandemic vaccines, there is the possibility that the transfection
efficiency of sets of plasmids is so low that the rapid and robust generation of vaccine
seed viruses is impeded. To overcome this possibility, we have reduced the number
of plasmids required to generate virus by reverse genetics (Fig. 2) (Neumann et al.
2005). In this system, one plasmid synthesizes the six gene segments for the internal
proteins (PB2, PB1, PA, NP, M, and NS), and a second plasmid synthesizes the HA
and NA segments. Two viral protein-expressing plasmids (one expressing NP and the
other expressing PB2, PB1, and PA) complete this system. Thus, only four plasmids
are transfected into Vero cells, which results in the generation of virus with significantly

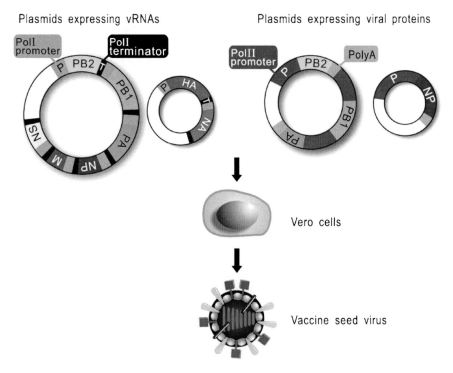

**Fig. 2** Schematic diagram of a four-plasmid-based reverse genetic approach to generating vaccine
seed viruses. A Pol I plasmid that synthesizes modified HA and NA gene segments is prepared from
the circulating wild-type strains. Another Pol I plasmid that synthesizes the other six gene segments
is prepared from a backbone virus that grows well in eggs or in cell culture. Two plasmids expressing
polymerase proteins or NP, respectively, are also prepared. Thus, a total of four plasmids are used
to generate the vaccine seed viruses with higher efficiency than other current systems, which can
contain more than eight plasmids

higher efficiency than that achieved with traditional twelve-plasmid systems. This four-plasmid system could, therefore, be valuable in the future generation of pandemic vaccine seed viruses.

Besides Vero cells, a limited number of other cells are approved for human vaccine production, such as MDCK cells and chicken embryonic fibroblasts (CEF). A modified reverse genetics system that uses the chicken RNA polymerase I (PolI) promoter also supports the generation of influenza virus in CEF (Massin et al. 2005) with an efficiency of virus generation comparable to that of the human PolI system in Vero cells. MDCK cells also support the efficient growth of influenza virus and are used as a substrate for the production of seasonal influenza vaccines (Brands et al. 1999, Govorkova et al. 1999, Halperin et al. 2002). In MDCK cells, however, reverse genetics with the human PolI promoter does not work well due to the host specificity of the PolI promoter. Recently, another reverse genetics system with T7 RNA polymerase II was shown to support influenza virus generation in MDCK cells (de Wit et al. 2007), although the efficiency of virus generation was inconsistent. We and others also established an alternative reverse genetics system driven by canine PolI (Murakami et al. 2008, Wang and Duke 2007) and generated recommended H5N1 vaccine seed viruses in MDCK cells with high efficiency.

## 3 Concluding Remarks

Research on H5N1 vaccine development has revealed that in clinical trials, the immune responses in humans to inactivated H5N1 vaccines are lower than those to annual vaccines, and therefore multiple doses of the vaccines with adjuvants would be required. Reverse genetics-based inactivated H5N1 vaccines with adjuvants have been licensed in several countries; however, whether these vaccines will be effective against antigenically different strains in humans is unknown. Thus, H5N1 vaccine libraries that reflect the different antigenicities of currently circulating strains are needed, as we cannot predict which strain will cause a pandemic. Furthermore, the adverse effects of adjuvants might also become apparent upon large-scale vaccination.

In the event of a pandemic caused by an HPAI virus, chicken eggs will likely be in short supply. Under such conditions, a reassortant vaccine seed virus with higher growth properties in eggs than the current seed strains is needed to produce sufficient vaccine. For this reason, we propose the reselection of a background virus for the production of seed viruses. As an alternative approach, cell-culture-based vaccines are currently being developed. Such egg-free vaccines may also be useful for those who have egg allergies.

Another consideration is the development of vaccines against influenza viruses of other subtypes, such as H2N2, H9N2 and H7N7 viruses, which also possess pandemic potential (Hehme et al. 2002, de Wit et al. 2005, Stephenson et al. 2003). Essentially, the same strategies as those used for H5N1 vaccines can be employed for the development of vaccines to these virus subtypes.

Lastly, there may be concerns regarding production capacity and global accessibility of vaccines, manufacturing costs, and the cooperation of international governments, which must be addressed. It is also essential that we continue to promote alterative approaches to the development of cross-reactive and long-lasting pandemic vaccines that are egg- and adjuvant-independent, although it will likely take years to achieve this objective.

**Acknowledgments** We thank Susan Watson for editing the manuscript. We also thank those in our laboratories who contributed to the data cited in this review. Our original research was supported by National Institute of Allergy and Infectious Diseases Public Health Service research grants; by CREST (Japan Science and Technology Agency); by Grants-in-Aid for Specially Promoted Research and for Scientific Research (B); by a Contract Research Fund for Program of Founding Research Centers for Emerging and Reemerging Infectious Diseases; and by the Special Coordination Funds for Promoting Science and Technology from the Ministry of Education, Culture, Sports, Science, and Technology of Japan.

# References

Beyer WEP, Palache AM, de Jong JC, Osterhaus ADME (2002) Cold-adapted live influenza vaccine versus inactivated vaccine: systemic vaccine reactions, local and systemic antibody response, and vaccine efficacy. A meta-analysis. Vaccine 20:1340–1353

Brands R, Visser J, Medema J, Palache AM, van Scharrenburg GJ (1999) Influvac: a safe Madin Darby Canine Kidney (MDCK) cell culture-based influenza vaccine. Dev Biol Stand 98:93–100; discussion 111

Bresson JL, Perronne C, Launay O, Gerdil C, Saville M, Wood J, Houml;schler K, Zambon MC (2006) Safety and immunogenicity of an inactivated split-virion influenza A/Vietnam/1194/2004 (H5N1) vaccine: phase I randomised trial. Lancet 367:1657–1664

Castrucci MR, Kawaoka Y (1993) Biological importance of neuraminidase stalk length in influenza A virus. J Virol 67:759–764

de Wit E, Munster VJ, Spronken MI, Bestebroer TM, Baas C, Beyer WE, Rimmelzwaan GF, Osterhaus AD, Fouchier RA (2005) Protection of mice against lethal infection with highly pathogenic H7N7 influenza A virus by using a recombinant low-pathogenicity vaccine strain. J Virol 79:12401–12407

de Wit E, Spronken MI, Vervaet G, Rimmelzwaan GF, Osterhaus AD, Fouchier RA (2007) A reverse-genetics system for Influenza A virus using T7 RNA polymerase. J Gen Virol 88:1281–1287

Fodor E, Devenish L, Engelhardt OG, Palese P, Brownlee GG, Garcia-Sastre A (1999) Rescue of influenza A virus from recombinant DNA. J Virol 73:9679–8296

Gamblin SJ, Haire LF, Russell RJ, Stevens DJ, Xiao B, Ha Y, Vasisht N, Steinhauer DA, Daniels RS, Elliot A, Wiley DC, Skehel JJ (2004) The structure and receptor binding properties of the 1918 influenza hemagglutinin. Science 303:1838–1842

Govorkova EA, Kodihalli S, Alymova IV, Fanget B, Webster RG (1999) Growth and immunogenicity of influenza viruses cultivated in Vero or MDCK cells and in embryonated chicken eggs. Dev Biol Stand 98:39–51; discussion 73–74

Govorkova EA, Webby RJ, Humberd J, Seiler JP, Webster RG (2006) Immunization with reverse-genetics-produced H5N1 influenza vaccine protects ferrets against homologous and heterologous challenge. J Infect Dis 194:159–167

Gupta RK, Nguyen-Van-Tam JS (2006) Oseltamivir resistance in influenza A (H5N1) infection. N Eng J Med 354:1423–1424

Halperin SA, Smith B, Mabrouk T, Germain M, Trepanier P, Hassell T, Treanor J, Gauthier R, Mills EL (2002) Safety and immunogenicity of a trivalent, inactivated, mammalian cell culture-derived influenza vaccine in healthy adults, seniors, and children. Vaccine 20:1240–1247

Hasegawa H, Ichinohe T, Tamura S, Kurata T (2007) Development of a mucosal vaccine for influenza viruses: preparation for a potential influenza pandemic. Exp Rev Vaccines 6:193–201

Hehme N, Engelmann H, Kuumlnzel W, Neumeier E, Saumlnger R (2002) Pandemic preparedness: lessons learnt from H2N2 and H9N2 candidate vaccines. Med Microbiol Immunol 191:203–208

Horimoto T, Kawaoka Y (1994) Reverse genetics provides direct evidence for a correlation of hemagglutinin cleavability and virulence of an avian influenza A virus. J Virol 68:3120–3128

Horimoto T, Kawaoka Y (2005) Influenza: lessons from past pandemics, warnings from current incidents. Nat Rev Microbiol 3:591–600

Horimoto T, Takada A, Fujii K, Goto H, Hatta M, Watanabe S, Iwatsuki-Horimoto K, Ito M, Tagawa-Sakai Y, Yamada S, Ito H, Ito T, Imai M, Itamura S, Odagiri T, Tashiro M, Lim W, Guan Y, Peiris M, Kawaoka Y (2006) The development and characterization of H5 influenza virus vaccines derived from a 2003 human isolate. Vaccine 24:3669–3676

Horimoto T, Murakami S, Muramoto Y, Yamada S, Fujii K, Kiso M, Iwatsuki-Horimoto K, Kino Y, Kawaoka Y (2007) Enhanced growth of seed viruses for H5N1 influenza vaccines. Virology 366:23–27

Johnson NP, Mueller J (2002) Updating the accounts: global mortality of the 1918–1920 "Spanish" influenza pandemic. Bull Hist Med 76:105–115

Kawaoka Y, Webster RG (1988) Sequence requirement for cleavage activation of influenza virus hemagglutinin expressed in mammalian cells. Proc Natl Acad Sci USA 85:324–328

Kobasa D, Jones SM, Shinya K, Kash JC, Copps J, Ebihara H, Hatta Y, Kim JH, Halfmann P, Hatta M, Feldmann F, Alimonti JB, Fernando L, Li Y, Katze MG, Feldmann H, Kawaoka Y (2007) Aberrant innate immune response in lethal infection of macaques with the 1918 influenza virus. Nature 445:319–323

Le QM, Kiso M, Someya K, Sakai YT, Nguyen TH, Nguyen KH, Pham ND, Ngyen HH, Yamada S, Muramoto Y, Horimoto T, Takada A, Goto H, Suzuki T, Suzuki Y, Kawaoka Y (2005) Avian flu: isolation of drug-resistant H5N1 virus. Nature 437:1108

Li S, Liu C, Klimov A, Subbarao K, Perdue ML, Mo D, Ji Y, Woods L, Hietala S, Bryant M (1999) Recombinant influenza A virus vaccines for the pathogenic human A/Hong Kong/97 (H5N1) viruses. J Infect Dis 179:1132–1138

Lipatov AS, Webby RJ, Govorkova EA, Krauss S, Webster RG (2005) Efficacy of H5 influenza vaccines produced by reverse genetics in a lethal mouse model. J Infect Dis 191:1210–1215

Lu B, Zhou H, Ye D, Kemble G, Jin H (2005) Improvement of influenza A/Fujian/411/02 (H3N2) virus growth in embryonated chicken eggs by balancing the hemagglutinin and neuraminidase activities, using reverse genetics. J Virol 79:6763–6771

Massin P, Rodrigues P, Marasescu M, van der Werf S, Naffakh N (2005) Cloning of the chicken RNA polymerase I promoter and use for reverse genetics of influenza A viruses in avian cells. J Virol 79:13811–13816

Matrosovich M, Tuzikov A, Bovin N, Gambaryan A, Klimov A, Castrucci MR, Donatelli I, Kawaoka Y (2000) Early alterations of the receptor-binding properties of H1, H2, and H3 avian influenza virus hemagglutinins after their introduction into mammals. J Virol 74:8502–8512

Medema JK, Meijer J, Kersten AJ, Horton R (2006) Safety assessment of Madin Darby canine kidney cells as vaccine substrate. Dev Biol (Basel) 123:243–250

Murakami S, Horimoto T, Yamada S, Kakugawa S, Goto H, Kawaoka Y (2008) Establishment of canine RNA polymerase I-driven reverse genetics for influenza A virus: its application for H5N1 vaccine production. J Virol 82:1605–1609

Neumann G, Watanabe T, Ito H, Watanabe S, Goto H, Gao P, Hughes M, Perez DR, Donis R, Hoffmann E, Hobom G, Kawaoka Y (1999) Generation of influenza A viruses entirely from cloned cDNAs. Proc Natl Acad Sci USA 96:9345–5093

Neumann G, Fujii K, Kino Y, Kawaoka Y (2005) An improved reverse genetics system for influenza A virus generation and its implications for vaccine production. Proc Natl Acad Sci USA 102:16825–16829

Nicholson KG, Colegate AE, Podda A, Stephenson I, Wood J, Ypma E, Zambon MC (2001) Safety and antigenicity of non-adjuvanted and MF59-adjuvanted influenza A/Duck/Singapore/97 (H5N3) vaccine: a randomised trial of two potential vaccines against H5N1 influenza. Lancet 357:1937–1943

Nicolson C, Major D, Wood JM, Robertson JS (2005) Generation of influenza vaccine viruses on Vero cells by reverse genetics: an H5N1 candidate vaccine strain produced under a quality system. Vaccine 23:2943–2952

Reid AH, Taubenberger JK, Fanning TG (2004) Evidence of an absence: the genetic origins of the 1918 pandemic influenza virus. Nat Rev Microbiol 2:909–914

Sangster MY, Riberdy JM, Gonzalez M, Topham DJ, Baumgarth N, Doherty PC (2003) An early CD4+T cell-dependent immunoglobulin A response to influenza infection in the absence of key cognate T–B interactions. J Exp Med 198:1011–1021

Song J-H, Hguyen HH, Cuburu N, Horimoto T, Ko S-Y, Park S-H, Czerkinsky C, Kweon M-N (2008) Sublingual vaccination with influenza virus protects mice against lethal viral infection. Proc Natl Acad Sci USA 105:1644–1649

Stephenson I, Nicholson KG, Glück R, Mischler R, Newman RW, Palache AM, Verlander NQ, Warburton F, Wood JM, Zambon MC (2003) Safety and antigenicity of whole virus and subunit influenza A/Hong Kong/1073/99 (H9N2) vaccine in healthy adults: phase I randomised trial. Lancet 362:1959–1966

Stephenson I, Nicholson KG, Wood JM, Zambon MC, Katz JM (2004) Confronting the avian influenza threat: vaccine development for a potential pandemic. Lancet Infect Dis 4:499–509

Stephenson I, Bugarini R, Nicholson KG, Podda A, Wood JM, Zambon MC, Katz JM (2005) Cross-reactivity to highly pathogenic avian influenza H5N1 viruses after vaccination with nonadjuvanted and MF59-adjuvanted influenza A/duck/Singapore/97 (H5N3) vaccine: a potential priming strategy. J Infect Dis 191:1210–1215

Stevens J, Corper AL, Basler CF, Taubenberger JK, Palese P, Wilson IA (2004) Structure of the uncleaved human H1 hemagglutinin from the extinct 1918 influenza virus. Science 303:1866–1870

Subbarao K, Chen H, Swayne D, Mingay L, Fodor E, Brownlee G, Xu X, Lu X, Katz J, Cox N, Matsuoka Y (2003) Evaluation of a genetically modified reassortant H5N1 influenza A virus vaccine candidate generated by plasmid-based reverse genetics. Virology 305:192–200

Suguitan AL Jr, McAuliffe J, Mills KL, Jin H, Duke G, Lu B, Luke CJ, Murphy B, Swayne DE, Kemble G, Subbarao K (2006) Live, attenuated influenza A H5N1 candidate vaccines provide broad cross-protection in mice and ferrets. PLoS Med 3:e360

Takada A, Kuboki N, Okazaki K, Ninomiya A, Tanaka H, Ozaki H, Itamura S, Nishimura H, Enami M, Tashiro M, Shortridge KF, Kida H (1999) Avirulent avian influenza virus as a vaccine strain against a potential human pandemic. J Virol 73:8303–8307

Treanor JJ, Campbell JD, Zangwill KM, Rowe T, Wolff M (2006) Safety and immunogenicity of an inactivated subvirion influenza A (H5N1) vaccine. N Eng J Med 354:1343–1351

Wang Z, Duke GM (2007) Cloning of the canine RNA polymerase I promoter and establishment of reverse genetics for influenza A and B in MDCK cells. Virology J 4:102–xx

Webby RJ, Perez DR, Coleman JS, Guan Y, Knight JH, Govorkova EA, McClain-Moss LR, Peiris JS, Rehg JE, Tuomanen EI, Webster RG (2004) Responsiveness to a pandemic alert: use of reverse genetics for rapid development of influenza vaccines. Lancet 363:1099–1103

Wood JM, Robertson JS (2004) From lethal virus to life-saving vaccine: the development of inactivated influenza vaccines for pandemic influenza. Nat Rev Microbiol 2:842–847

Wright PF, Neumann G, Kawaoka Y (2007) Orthomyxoviruses. In: Knipe et al. (eds) Field's virology, 5th edn. Lippincott Williams & Wilkins, Philadelphia, pp1691–1740

# Attenuated Influenza Virus Vaccines with Modified NS1 Proteins

Jüergen A. Richt and Adolfo García-Sastre

**Contents**

**Abstract** The development of reverse genetics techniques allowing the rescue of influenza virus from plasmid DNA has opened up the possibility of inserting mutations into the genome of this virus for the generation of novel live attenuated influenza virus vaccines. Modifications introduced into the viral NS1 gene via reverse genetics have resulted in attenuated influenza viruses with promising vaccine potential. One of the main functions of the NS1 protein of influenza virus is the inhibition of the innate host type I interferon-mediated antiviral response. Upon viral infection, influenza viruses with modified NS1 genes induce a robust local type I interferon response that limits their replication, resulting in disease attenuation in different animal models. Nevertheless, these viruses can be grown to high

J.A. Richt
Department of Diagnostic Medicine and Pathobiology, College of Veterinary Medicine,
Kansas State University, Manhattan , KS 66506, USA

A. García-Sastre (✉)
Department of Microbiology; Department of Medicine, Division of Infectious Diseases;
and Global Health and Emerging Pathogens Institute, Mount Sinai School of Medicine,
New York, NY 10029, USA
e-mail: adolfo.garcia-sastre@mssm.edu

R.W. Compans and W.A. Orenstein (eds.), *Vaccines for Pandemic Influenza*,     177
Current Topics in Microbiology and Immunology 333,
DOI 10.1007/978-3-540-92165-3_9, © Springer-Verlag Berlin Heidelberg 2009

titers in cell- and egg-based substrates with deficiencies in the type I IFN system. Intranasal inoculation of mice, pigs, horses, and macaques with NS1-modified influenza virus strains induced robust humoral and cellular immune responses, and generated immune protection against challenge with wild-type virus. This protective response was not limited to homologous strains of influenza viruses, as reduced replication of heterologous strains was also demonstrated in animals vaccinated with NS1-modified viruses, indicating the induction of a broad cross-neutralizing response by these vaccine candidates. The immunogenicity of NS1-modified viruses correlated with enhanced activation of antigen-presenting cells. While further studies on their safety and efficacy are still needed, the results obtained so far indicate that NS1-modified viruses could represent a new generation of improved influenza virus vaccines, and they suggest that modifying viral interferon antagonists in other virus families is a promising strategy for the generation of live attenuated virus vaccines.

# 1    Introduction

The segmented negative-strand RNA influenza A virus contains eight RNA segments encoding eleven viral proteins. Of all these proteins, only two are exclusively expressed in virus-infected cells and not present in virus particles, PB1-F2 and nonstructural protein 1 or NS1. The development of reverse genetics techniques to rescue influenza viruses has allowed the generation of recombinant viruses lacking PB1-F2 and NS1 genes (Chen et al. 2001; García-Sastre et al. 1998). The study of these recombinant viruses has revealed significant information on the functional roles of these two nonstructural genes during viral infection. The NS1 protein increases viral replication in the host by evading innate immune responses through the attenuation of the host type I interferon (IFN) response. The PB1-F2 protein has proapoptotic effects in vivo, but the functional consequences of these effects are still not fully elucidated. Both NS1 and PB1-F2 are virulence factors, and their deletion results in attenuated disease in animal models. However, viruses deleted in their NS1 genes are more attenuated than those deleted in PB1-F2. These observations led to NS1-modified influenza viruses being considered as potential live attenuated influenza virus vaccines (Talon et al. 2000b).

Influenza viruses with modified NS1 proteins induce high levels of type I interferon locally upon infection, which in turn inhibits viral replication through the induction of an antiviral state (García-Sastre et al. 1998). This provides the molecular basis for the attenuation of NS1-modified influenza viruses. These modified viruses can be propagated in systems devoid of type I interferon responses, and can therefore be manufactured as vaccines (Talon et al. 2000b). Due to the nonstructural nature of the NS1 protein, NS1-modified virions have an identical antigenic composition and structure to wild-type virions.

In general, live attenuated virus vaccines induce a more robust and broad protective immune response than inactivated virus vaccines, and this has proven to be the case when the efficacy of a live cold-adapted influenza virus vaccine was compared with an inactivated influenza virus vaccine in young children (Belshe et al. 2007). However, one of the challenges in designing live virus vaccines is to achieve the right balance between safety (attenuation) and immunogenicity. The ideal live virus vaccine should be attenuated for replication to the extent that does not induce disease, but not to the extent that viral antigen production in vivo is so limited that no robust immune responses are elicited. Interestingly, it is possible to generate a panel of NS1-modified viruses with different degrees of attenuation. By inserting different truncations into the NS1 gene, the ability of the NS1 protein to inhibit the type I IFN response becomes compromised to different degrees, and this translates into different degrees of attenuation (Quinlivan et al. 2005; Solórzano et al. 2005; Talon et al. 2000b). This interesting property of the NS1 protein should permit the selection of a live virus vaccine candidate with an optimal balance between attenuation and immunogenicity among a panel of NS1-modified viruses.

The IFN antagonist properties of the NS1 protein of influenza viruses also have consequences for the activation of antigen-presenting cell function, and NS1-modified viruses are in fact more potent activators of dendritic cell function than wild-type virus, which could explain their success as immunogens in different animal models (Fernandez-Sesma et al. 2006; López et al. 2003). In this chapter, we will review how NS1 modulates the type I IFN response and the vaccine properties of modified NS1 viruses in different animal models.

# 2 NS1 Functions

It has been recognized since the early 1990s that the NS1 protein of influenza A virus is a multifunctional protein that regulates several cellular and viral processes during influenza virus infection. Many of the functions of NS1 appear to be focused on the inhibition of type I IFN response. The NS1 protein intersects at different levels with the type I IFN system and in doing so inhibits early, intermediate, and late stages of this important arm of the host antiviral innate response (Fig. 1).

Cytoplasmic recognition of viral RNA products by cellular sensors appears to be one of the first steps of the IFN system inhibited by the NS1 protein of influenza A virus. The RNA helicase RIG-I has emerged as the critical sensor that initiates induction of type I IFN in most cell types upon influenza virus infection (Guo et al. 2007; Kato et al. 2006; Le Goffic et al. 2007; Loo et al. 2008; Mibayashi et al. 2007; Opitz et al. 2007; Pichlmair et al. 2006). RIG-I is known to recognize both cytoplasmic dsRNA and 5′-triphosphate-containing RNA (Hornung et al. 2006; Pichlmair et al. 2006), products that are generated during infection with several viruses, including influenza viruses. The ability of NS1 to bind dsRNA through an unconventional RNA-binding domain located within its first 73 amino acids at its amino terminal has been proposed to be responsible for sequestering these molecules from cellular sensors (Wang et al. 2002). Consistent with this, mutations that abrogate dsRNA binding by NS1 result in recombinant influenza viruses that more readily induce

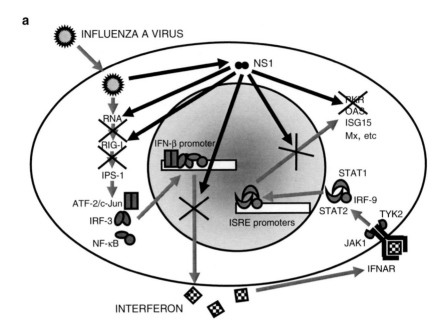

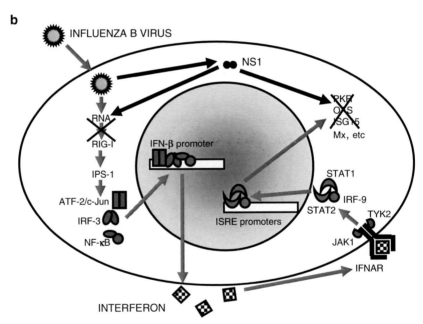

**Fig. 1a–b** Induction and inhibition of the type I IFN response by influenza viruses. **a** Influenza A virus. Viral RNA produced in virus-infected cells is recognized by the cellular protein RIG-I, which becomes activated and interacts with the adaptor molecule IPS-1 to activate transcription factors ATF-2/c-Jun, NF-κB, and IRF-3. The concerted actions of these transcription factors in the IFN-β promoter results in IFN synthesis and secretion. Secreted IFN interacts with its receptor

type I IFN (Donelan et al. 2004). The NS1 protein also prevents downstream signaling events mediated by RIG-I in a dsRNA-binding-independent manner (Donelan et al. 2004), through a process that involves interactions between NS1 and RIG-I (Mibayashi et al. 2007; Opitz et al. 2007; Pichlmair et al. 2006), but the nature of these interactions remains to be elucidated. In any case, as a consequence of the inhibition of RIG-I function by NS1, the downstream signaling events mediated by this sensor and its mitochondrial adaptor molecule IPS-1 are attenuated during influenza virus infection. This results in low levels of activation of transcription factors involved in the induction of type I IFN during influenza virus infection, including IRF-3 (Talon et al. 2000a), IRF-7 (Smith et al. 2001), NF-κB (Wang et al. 2000) and AP-1 (Ludwig et al. 2002).

While RIG-I inhibition by NS1 in the cytoplasm reduces the transcriptional activation of type I IFN genes, the influenza A virus NS1 also inhibits cellular mRNA processes at multiple posttranscriptional levels within the nucleus, preventing host responses that depend on de novo protein synthesis, including the synthesis of IFN- and IFN-stimulated genes. This second level of inhibition of cellular processes by NS1 probably reflects the need for a second blockade in IFN-stimulated gene expression during viral infection that cooperates with the first blockade at the level of transcription factor activation for the optimal inhibition of the type I IFN system. Posttranscriptional inhibition of cellular processes by the NS1 protein is at least partly mediated by a domain located downstream of the dsRNA binding domain that mediates NS1 binding to the cellular factor CPSF (Kochs et al. 2007a; Nemeroff et al. 1998; Twu et al. 2007). CPSF is a cellular factor required for the polyadenylation of cellular mRNA. As a result of NS1 binding to CPSF, cellular mRNA polyadenylation is downregulated, resulting in reduced cellular gene expression. However, the CPSF-binding domain is not completely conserved among influenza virus strains, suggesting that a loss of CPSF inhibitory function by the NS1 could be compensated for by other mechanisms, at least in some strains (Kochs et al. 2007a; Twu et al. 2007). In addition to CPSF, the NS1 proteins of some influenza virus strains bind to and inhibit PABP2, a cellular protein that is also involved in cellular mRNA polyadenylation (Chen et al. 1999). Other posttranscriptional processes inhibited by NS1 include mRNA splicing (Fortes et al. 1994; Lu et al. 1994) and mRNA nucleocytoplasmic export, the latter through

---

(IFNAR), resulting in the activation of the kinases JAK1 and TYK2 and formation of the STAT1/STAT2/IRF9 transcription factor. This factor promotes transcription from interferon stimulated response elements (ISRE), resulting in the expression of antiviral IFN inducible proteins such as PKR, OAS, ISG15, Mx, and others. Expression of the NS1 protein during viral infection inhibits this cascade by sequestering activating viral RNA, by preventing the action of RIG-I, by inhibiting cellular mRNA maturation and export, and by preventing the activation of PKR and OAS. **b** Influenza B virus. Infection triggers the same processes as described above for influenza A virus-infected cells, except that it is not known whether the NS1 of influenza B virus specifically inhibits RIG-I-mediated downstream signaling, it is believed not to affect cellular mRNAs, and in addition to inhibiting PKR and OAS by sequestering dsRNA, it also inhibits the IFN-induced protein ISG15 through a binding interaction

specific interactions with components of the nuclear pore complex (Satterly et al. 2007). Through these multiple processes, cytoplasmic expression of cellular RNAs is reduced in virus-infected cells.

Finally, a third level of downmodulation by the NS1 protein of influenza A virus of the type I IFN response involves the inhibition of the antiviral activity of at least two IFN-inducible gene products, PKR and OAS. PKR is a kinase that inhibits protein synthesis by phosphorylating the translation initiation factor eIF2α, while OAS catalyzes the synthesis of 2′,5′-oligoadenylates, which in turn activate a latent RNAse, RNAseL, promoting RNA degradation, translational arrest and apoptosis in virus-infected cells (Samuel 2001). Both PKR and OAS use dsRNA as a cofactor in their activation, and sequestering of dsRNA by the NS1 protein appears to prevent their activation (Lu et al. 1995; Min and Krug 2006). Direct binding of NS1 to PKR further enhances the inhibition of the kinase activity of this IFN-induced antiviral enzyme (Li et al. 2006a).

Other activities of the NS1 protein of influenza A viruses have also been reported, and these include the activation of PI3 kinase (Ehrhardt et al. 2007; Hale et al. 2006), the stimulation of viral mRNA translation (Aragón et al. 2000), the inhibition of RNA silencing (Li et al. 2004), and the binding, at least in the case of NS1 proteins derived from avian influenza viruses, to proteins that are involved in multiple cellular processes, such as PDZ (Obenauer et al. 2006) and Crk/CrkL proteins (Heikkinen et al. 2008). Although several of these activities, if not all, are likely to modulate virulence (Jackson et al. 2008), the physiological consequences of these NS1-mediated activities are not well understood.

Structurally, the NS1 protein requires dimerization in order to inhibit the type I IFN system. Multiple regions of the NS1 are involved in optimal dimer formation (Nemeroff et al. 1995), and therefore mutations that affect efficient dimerization result in NS1 proteins with impaired IFN antagonistic functions (Wang et al. 2002). Consistent with the multifunctional properties of the NS1, which requires both nuclear and cytoplasmic localizations, trafficking of NS1 within the cell is governed by a complex regulation of multiple nuclear import and export signals (Qian et al. 1994), and the precise mechanisms that regulate NS1 localization are still to be elucidated.

While the NS1 protein of influenza A virus has been extensively studied, such a detailed analysis is still lacking for the NS1 protein of influenza B virus. However, there are several striking similarities between these two proteins (Fig. 1b). The NS1 of influenza B virus, similarly to the NS1 of influenza A virus, has a dsRNA binding domain in its amino terminal, and inhibits both PKR (Wang and Krug 1996) and the activation of transcription factors involved in type I IFN induction (Dauber et al. 2004, 2006). However, the NS1 protein of influenza B virus does not seem to inhibit polyadenylation of cellular mRNAs (Wang and Krug 1996). Moreover, influenza B virus NS1 and not influenza A virus NS1 binds to ISG15, an IFN-induced ubiquitin-like molecule, preventing its conjugation (ISGylation) to substrates (Yuan and Krug 2001). Although the functional consequences of ISGylation are still unknown, this inhibitory activity of the NS1 of influenza B virus is likely to contribute to immune evasion, as it has recently been found that ISG15 has antiviral properties against both influenza A and B viruses in mice (Lenschow et al. 2007).

## 2.1 Biology of DelNS1 Influenza Viruses

The generation of recombinant influenza A and B viruses lacking the NS1 gene allowed a more clear understanding of the biological role of this viral gene. The NS gene of influenza virus encodes two viral proteins from two alternatively spliced viral mRNAs derived from this gene. Unspliced NS mRNA codes for the NS1 protein, while spliced NS mRNA codes for the NEP protein, involved in viral ribonucleoprotein trafficking within the cell (Fig. 2). A delNS1 influenza A virus was first generated in 1998 by eliminating the intron of the NS mRNA through the use of reverse genetics techniques (García-Sastre et al. 1998). This recombinant virus demonstrated a striking replication host range, and was able to grow to titers close to wild-type viruses in IFN-deficient cells, such as Vero cells, but replicated poorly in cells with an intact IFN response, such as MDCK cells (García-Sastre et al. 1998). DelNS1 virus grew well in young, six- to eight-day-old chicken embryonated eggs known to lack a robust IFN response; it develops later on during embryo development (Sekellick and Marcus 1985). DelNS1 virus was impaired for replication in older embryonated genes and in the respiratory tracts of mice (García-Sastre et al. 1998; Talon et al. 2000b). By contrast, mice genetically deficient in key genes of the type I IFN response, such as STAT1–/– mice, which lack a transcription factor required for IFN action, and Mx1–/– PKR–/– double-deficient mice, which lack two important IFN inducible genes with influenza virus antiviral activity, supported the replication of delNS1 virus in their respiratory tracts and developed severe disease upon infection with this virus, demonstrating that the main function of the NS1

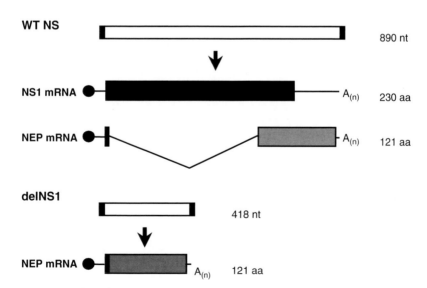

**Fig. 2** Schematic representation of NS RNA segments from wild-type influenza virus and from recombinant delNS1 virus. Numbers correspond to nt and aa length based on the influenza A/PR/8/34 virus

gene is to promote viral replication by attenuating the IFN antiviral response (Bergmann et al. 2000; García-Sastre et al. 1998). Thus, in the absence of a functional NS1 protein, delNS1 virus became an efficient inducer of type I IFN through an unimpeded activation of transcription factors IRF3 and NF-κB involved in type I IFN transcriptional activation, and of the IFN-inducible protein PKR (Bergmann et al. 2000; Talon et al. 2000a; Wang et al. 2000). This cellular activation in turn inhibited viral replication, and prevented disease after intranasal infection of mice with delNS1 virus, opening up the possibility of developing NS1-deficient influenza viruses as live attenuated influenza vaccines.

Likewise, it has been possible to generate an influenza B virus lacking the NS1 gene by reverse genetics. Similarly to the delNS1 influenza A virus, the delNS1 influenza B virus is highly restricted in terms of replication in type I IFN competent systems, including the respiratory tracts of mice, and is a potent activator of IRF3 and type I IFN expression in infected cells (Dauber et al. 2004; Donelan et al. 2004).

In addition to an unimpaired type I IFN response in cells infected with delNS1 viruses, an absence of NS1 function during influenza virus infection results in enhanced proapoptotic responses, which are also likely to contribute to reduced replication of delNS1 viruses (Stasakova et al. 2005; Zhirnov et al. 2002). Inhibition of apoptosis and of IFN induction by the NS1 might be functionally linked to RIG-I-mediated inhibition by NS1, since RIG-I activation has recently be demonstrated to result not only in type I IFN induction but also in caspase 1 and 3 activation (Rintahaka et al. 2008).

# 3    Interactions of NS1 Mutant Viruses with Dendritic Cells

Cellular sensors of viral infection are not only implicated in the induction of type I IFN but they are also responsible for the activation of antigen-presenting cells during the first stages of viral infection, providing a functional link between innate and adaptive immune responses (García-Sastre and Biron 2006). Consistent with this notion, the ability of a virus to induce activation of type I IFN was found to correlate with its ability to induce dendritic cell maturation (López et al. 2003). This concept led to an investigation of the functional consequences of NS1 expression during influenza virus infection of human dendritic cells (Fernandez-Sesma 2007). Expression of NS1 during influenza virus infection was found to attenuate not only type I IFN production by dendritic cells but also many other genes involved in dendritic cell maturation, migration, and T cell stimulation, such as costimulatory molecules, proinflammatory proteins, chemokines, and chemokine receptors. As a result, dendritic cells infected with delNS1 influenza virus were more potent stimulators of T cells than those infected with wild-type virus, suggesting that, at equal levels of virus infection, impairment of NS1 function translates into increased immunogenicity (Fernandez-Sesma et al. 2006). Thus, modifications in the NS1 protein would be predicted to result in viruses not only with an attenuated phenotype in vivo, but with increased immunogenicity too, due to the known adjuvant properties

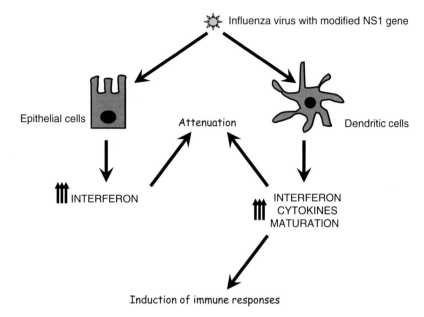

**Fig. 3** Impact of NS1 modification on influenza virus virulence and immunogenicity. In the absence of NS1 function, influenza virus-infected cells produce high levels of type I IFN that inhibits further viral replication. The lack of NS1 also results in a more robust activation of infected dendritic cells, where NS1-deleted virus induces not only type I IFN but also proinflammatory cytokines and maturation processes associated with enhanced activation of T cells

of type I IFN (the expression of which is increased locally during delNS1 virus infection) and to enhanced dendritic cell activation (Fig. 3). This is a novel and powerful paradigm in the rational design of live attenuated viral vaccines.

# 4 Studies of NS1-Modified Viruses in Mice

First studies on the vaccine efficacies of NS1-modified influenza viruses were conducted in mice using a mouse model of influenza virus infection based on intranasal infection with a mouse-adapted influenza A virus (influenza A/PR/8/33 virus, PR8) (Talon et al. 2000b). In these studies, two different NS1-modified PR8 viruses were used, the first one containing a full deletion of the NS1 gene (or delNS1 virus), and the second one (NS1-99) expressing a truncated NS1 protein of only 99 amino acids as opposed to the 230-amino-acid wild-type NS1 protein. The NS1-99 truncated protein contains the dsRNA binding domain of the NS1 but lacks the carboxy-terminal region required for optimal NS1 dimerization and other NS1 functions, and so it was expected to have an attenuated phenotype that is intermediate between the wild-type virus and the delNS1 virus. The rationale for using this NS1 mutant virus in addition to delNS1 virus was the possibility that delNS1 virus was overattenuated in replication, resulting in poor expression of viral antigens in vivo and therefore in poor immunogenicity. Both delNS1 and NS1-99 viruses proved attenuated in mice after intranasal administration, and they induced influenza virus-specific humoral and cellular immune responses, with NS1-99 being the most potent inducer of humoral responses of the two. Intranasal administration of a single dose of $10^6$ pfu of any of the two viruses resulted in 100% protection against challenge with a high lethal dose of wild-type influenza virus. However, when doses of $3.3 \times 10^4$ pfu were administered, delNS1 immunized mice were not protected against wild-type challenge, while NS1-99 mice still demonstrated high levels of protection (Talon et al. 2000b). This experiment illustrated the concept of modulating the levels of NS1 function in order to determine the most optimal live vaccine virus candidate with the right balance between attenuation and immunogenicity.

The same principle was applied to influenza B viruses. Two influenza B viruses containing partial deletions in their NS1 genes that arose spontaneously during passage in the laboratory were also tested for vaccine efficacy after intranasal infections in mice, and they proved to be (similar to the NS1-99 influenza A virus) attenuated and immunogenic, providing protection against wild-type influenza B virus challenge (Talon et al. 2000b). The establishment of reverse genetics techniques for the influenza B virus has more recently allowed the generation of recombinant influenza B viruses with defined NS1 mutations (Dauber et al. 2004, 2006), and this will facilitate the rational design of NS1 mutant influenza B virus for use as potential influenza virus vaccines in combination with NS1-modified influenza A viruses.

Further studies in mice have validated the use of NS1-modified influenza viruses as potential vaccines (Falcón et al. 2005; Ferko et al. 2004). Interestingly, the addition of foreign epitopes into NS1-mutant viruses have resulted in immune responses against these novel epitopes, indicating the potential of these viruses as not only influenza virus vaccines but also as vector vaccines against other diseases, such as HIV-1 (Ferko et al. 2001), cancer (Efferson et al. 2003, 2006), and tuberculosis (Sereinig et al. 2006; Stukova et al. 2006). It might also be possible to further improve the immunogenicity of NS1-modified influenza viruses by expressing cytokines such as IL-2 (Ferko et al. 2006).

# 5   Studies of NS1-Modified Viruses in Pigs

Although the studies in mice described above are suggestive, they may not faithfully represent the phenotypes of the NS1-modified influenza viruses in their natural hosts. More recently, studies with NS1-modified viruses have been performed using a naturally occurring H3N2 swine influenza A virus strain, influenza Sw/A/TX/98 virus (TX98), and pigs. Swine influenza is a respiratory disease that is very similar to human influenza, and swine influenza A virus strains are closely related to human influenza A virus strains. Although inactivated swine influenza virus vaccines are available, the diversity of the antigenic strains that have been circulating in recent years in pigs makes it desirable to develop live attenuated or vectored subunit vaccines for swine that could induce a balanced immune response, including humoral and cell-mediated mechanisms, in order to improve homotypic and hetero-subtypic protection.

With this idea in mind, we generated a panel of recombinant swine influenza TX98 viruses expressing NS1-truncated proteins of 73 (TX98 NS1Δ73), 99 (TX98 NS1Δ99) and 126 (TX98 NS1Δ126) amino acids (Solórzano et al. 2005). All NS1-truncated viruses had decreased replication as compared to the wild-type virus in pig kidney cells and in the respiratory tracts of pigs, with the TX98 NS1Δ126 virus being the most restricted, followed by the TX98 NS1Δ99 and by the TX98 NS1Δ73 viruses. Remarkably, the degree of attenuation both in vitro and in vivo correlated with the levels of type I IFN induction by these viruses in pig cells (Solórzano et al. 2005). Therefore, by using different truncations in the NS1 gene, we obtained a panel of recombinant viruses displaying different degrees of attenuation in vivo according to the remaining levels of IFN antagonism mediated by their truncated NS1 proteins. The length of the NS1-truncated region did not correlate with the degree of attenuation; rather, different truncations had different impacts on NS1 expression, and the most attenuated virus, TX98 NS1Δ126, displayed the lowest level of NS1 expression. It appears then that truncations in NS1 have multiple effects on NS1 function. First, they eliminate specific domains required for optimal IFN antagonism and viral replication in the host. Second, they affect NS1 dimerization, indirectly reducing NS1 function (Wang et al. 2002). Third, they differentially affect levels of NS1 expression.

Based on the good safety profile of experimentally infected naïve pigs with TX98 NS1Δ126, we further characterized this NS1-modified virus in swine. Intratracheal infection in pigs with this virus resulted in minimal lung lesions, both macroscopically as well as by immunohistochemical examination of lung tissue. The virus did not shed from the noses of infected animals but was capable of stimulating an immune response, as measured by the presence of virus-neutralizing antibodies in sera of inoculated pigs (Solórzano et al. 2005). To further evaluate the vaccine efficacy of the TX98 NS1Δ126 virus, four-week old pigs were vaccinated and boosted with this virus via the intratracheal route. Pigs were subsequently challenged with the wild-type homologous H3N2 virus or with a heterosubtypic classical H1N1 swine influenza virus (Richt et al. 2006). Administration of TX98 NS1Δ126 virus completely protected against challenge with the homologous TX98 virus. Vaccinated pigs challenged with the heterosubtypic H1N1 virus demonstrated pathologic lung changes similar to the unvaccinated H1N1 control pigs, but had significantly reduced virus shedding from the respiratory tract when compared to unvaccinated, H1N1 challenged pigs. All vaccinated pigs developed a significant level of hemagglutinin inhibitory titers, serum IgG, and mucosal IgG and IgA antibodies against parental H3N2 antigens (Richt et al. 2006).

A follow-up study evaluated the vaccine efficacy of the TX98 NS1Δ126 virus when used through practical routes of immunization, intranasal and intramuscular (Vincent et al. 2007). The intranasal route was more efficient than the intramuscular route at inducing mucosal anti-influenza virus antibodies. A single dose of TX98

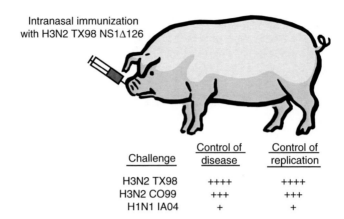

Intranasal immunization
with H3N2 TX98 NS1Δ126

| Challenge | Control of disease | Control of replication |
|---|---|---|
| H3N2 TX98 | ++++ | ++++ |
| H3N2 CO99 | +++ | +++ |
| H1N1 IA04 | + | + |

**Fig. 4** Intranasal vaccination of pigs with a single dose of swine influenza H3N2 TX98 NS1Δ126 virus provides protection against challenge with homologous and heterologous swine H3N2 viruses and partial protection against challenge with heterosubtypic swine H1N1 viruses. Adapted from Vincent et al. (2007)

NS1Δ126 virus administered intranasally conferred complete protection against homologous TX98 wild-type virus challenge and nearly complete protection against challenge with an antigenically distant H3N2 virus (CO99 virus) from a second lineage also circulating in pigs (Fig. 4). In vaccinated pigs, there was substantial cross-reactivity in antibodies at the mucosal level with the TX98 and CO99 H3N2 viruses. Moreover, when vaccinated animals were challenged with an H1N1 swine virus isolated from pigs in 2004 (IA04 virus), they displayed reduced fever and virus titers despite minimal reduction in lung lesions (Fig. 4). Thus, it would seem that a complex host response involving both cellular and humoral mechanisms contributes to the broad vaccine efficacy of the TX98 NS1Δ126 virus after intranasal delivery, and this efficacy appears to be superior to that induced by inactivated influenza vaccines (Vincent et al. 2007).

# 6   Studies of NS1-Modified Viruses in Horses

Another natural animal host of influenza virus is the horse. Equine influenza is one of the most economically important respiratory diseases of horses in countries with substantial breeding and racing industries. The most widely used influenza vaccines in horses are based on inactivated virus preparations of the circulating H3N8 strains, but the efficacy of these vaccines is limited (Morley et al. 1999). A cold-adapted live attenuated intranasal equine influenza vaccine is also available and appears to be more efficient than the inactivated vaccine (Townsend et al. 2001).

Similarly to the studies performed with NS1-modified swine influenza viruses, Quinlivan et al. (2005) generated a panel of recombinant equine H3N8 viruses

(KY02 strain) bearing truncations in the NS1 gene, resulting in the expression of only the first 126, 99, and 73 amino acids of the NS1 protein. These viruses displayed a phenotype similar to that of the swine influenza viruses, and demonstrated different degrees of attenuation in equine tissue culture cells and in mice that correlated with their levels of IFN induction, with the KY02 NS1Δ126 virus being the most attenuated of the three, followed by KY02 NS1Δ99 and by KY02 NS1Δ73 (Quinlivan et al. 2005). Intranasal inoculation of horses with the three recombinant NS1 viruses produced no adverse effects (Chambers et al., in press). Intranasal vaccination of horses with KY02 NS1Δ126 virus resulted in protection against fever and clinical symptoms and in reduced viral replication following wild-type virus challenge (Chambers et al., in press). Thus, studies in horses with NS1-modified equine influenza viruses appear to mirror those in pigs, indicating the value of this vaccination strategy across different mammalian species infected by influenza viruses.

# 7  Studies of NS1-Modified Viruses in Birds

The NS1 protein of avian influenza A virus has also been found to inhibit the type I IFN system and contribute to virulence (Jiao et al. 2008; Li et al. 2006b; Long et al. 2008; Seo et al. 2002; Zhu et al. 2008). Cauthen et al. (2007) have recently evaluated the phenotype of an avian H7N3 influenza virus originally isolated from turkeys in 1971 (OR71 virus), and a variant of this virus that was spontaneously generated upon passage, and that expresses a carboxy-terminally truncated NS1 protein of 124 amino acids (OR71 NS1Δ124). As expected, OR71 NS1Δ124 virus induced more IFN, replicated to lower levels, and induced less severe lesions in chicken tissue than its wild-type counterpart. Similar results were obtained by Kochs et al. using recombinant H7N7 virus lacking the NS1 gene (Kochs et al. 2007b). Although more studies are still needed in birds using NS1-modified viruses to demonstrate their vaccine potential in these hosts, the data available so far indicate that regulation of attenuation and IFN induction of avian influenza virus strains is also possible through the modification of their NS1 genes.

# 8  Studies of NS1-Modified Viruses in Macaques

We are still lacking published studies describing the safety and vaccine efficacy in humans of NS1-modified influenza viruses. Nevertheless, Baskin et al. (2007) have conducted a vaccination study using an NS1 mutant human influenza virus in macaques, which are among the closest genetic relatives to humans after great apes. First, a recombinant human A/Texas/91 H1N1 virus expressing a truncated NS1 protein of 126 amino acids, TX91 NS1Δ126, was generated. One group of macaques was vaccinated by the respiratory route with $6 \times 10^7$ pfu of live TX91

NS1Δ126 virus, while another group was vaccinated intramuscularly with 45 μg of HA of a whole inactivated TX91 virus vaccine preparation. Vaccination with the NS1-modified virus was safely tolerated and associated with a robust induction of influenza virus-specific CD4[+] T cell proliferation and virus neutralizing antibody in sera, and with transcriptional induction of T- and B-cell pathways in lung tissue (Baskin et al. 2007). By contrast, no neutralizing antibodies were detected in animals immunized with a single dose of inactivated vaccine, consistent with the requirement of two doses of inactivated vaccine in naïve humans to elicit protective responses. Importantly, macaques vaccinated with the live virus, but not with the inactivated virus, were protected against challenge with wild-type TX91 virus. These results illustrate the superior vaccine efficacy of live attenuated NS1-modified influenza viruses as compared to inactivated vaccines in a nonhuman primate model.

# 9   Conclusions and Future Perspectives of NS1-Modified Viruses as Epidemic and Pandemic Influenza Virus Vaccines

Current inactivated influenza virus vaccines, while providing significant levels of protection against human epidemic influenza, are not yet optimal. Importantly, vaccine efficacy is reduced in the elderly, who are at high risk of severe disease and death upon influenza virus infection. In addition, vaccine efficacy is diminished during years when antigenic mismatches between the vaccine and the current circulating viruses occur. Thus, alternative vaccination approaches with increased efficacy are desirable for the prevention of epidemic influenza. One of these approaches is the use of live attenuated vaccines administered intranasally. Live attenuated vaccines have the potential to replicate and provide exposure to large amounts of antigens despite a low starting dose. Another advantage is their potential to induce broader and longer-lasting protection based on the induction of mucosal humoral immunity and cross-reactive cellular immune responses. The available live influenza virus vaccines, based on influenza virus cold adaptation, while having some of these advantages, are not yet optimal. Thus, they have not been approved for use in the elderly, and they still appear to require large amounts of viruses for efficacy, which would suggest that they are overattenuated. The use of NS1-modified influenza viruses as live influenza virus vaccines provides an alternative strategy. While attenuation of the cold-adapted viruses relies on their thermosensitivity, which restricts their replication to the cooler upper respiratory tract, attenuation of NS1-modified viruses is due to their IFN inducing capabilities, which also increase their immunogenic properties (Fig. 3). This provides a new paradigm in vaccine development that could be applied to other viruses encoding IFN antagonistic proteins.

In the event of an influenza pandemic, two high doses of a conventional inactivated vaccine would be needed to protect an immunologically naïve human population, and the delay associated with this regimen could result in many deaths (Treanor et al. 2006). Alternative approaches are therefore needed for antigen sparing of influenza vaccines, allowing for rapid production and/or the wide use of stockpiles of pre-existing vaccines in the case of a pandemic. A live attenuated vaccine might represent a viable solution, since it is likely to require a single low dose of antigen to generate protection.

The proof-of-concept of the initial safety and vaccine efficacy of NS1-modified influenza viruses has now been obtained in mice, pigs, horses, and macaques. Pigs and horses are natural hosts of influenza viruses, and further studies in these animals will indicate whether the use of NS1-modified viruses as vaccines against swine and equine influenza represent an improvement over the existing approaches. The ability to modulate NS1 function by choosing specific truncations has the added advantage of allowing the vaccine that demonstrates the best balance of safety and immunogenicity to be determined in field trials. Nevertheless, in order to use NS1-modified viruses as veterinary influenza vaccines, additional safety and immunogenicity studies, including studies to determine the influence of maternal antibody responses on vaccine efficacy, are still needed.

While NS1-modified viruses are attenuated with respect to disease induction and viral shedding, additional safety considerations include their potential to revert to a virulent phenotype through mutation and/or reassortment processes. Reversion by mutation is unlikely due to the inclusion of extensive truncations in the NS1 genes of the vaccine viruses. Reassortment processes are mainly of concern if a new antigenic type could be introduced into the host through the use of the live modified virus vaccine. This will be only the case when considering potential prepandemic vaccination approaches. In any case, it may be possible in the future to generate reassortment-impaired influenza viruses through a better understanding of the mechanisms of viral RNA packaging.

In order to provide optimal protection, a human vaccine based on NS1-modified viruses will still need to be trivalent and updated yearly for the H1, H3, and B virus components to antigenically match circulating strains. Ideally, this vaccine could be generated by reverse genetics using the six internal conserved viral genes, including the modified NS1 gene, from master NS1-modified influenza A and B virus strains, and the HA and NA genes of the circulating strains. The results in animals point to the induction of broad cross-protection responses induced by NS1-modified viruses that might translate into protection against antigenically mismatched strains. Nevertheless, field studies in animals and in humans are still needed to ascertain whether NS1-modified viruses are safe and to demonstrate improved efficacy with respect to current existing influenza virus vaccination approaches.

**Acknowledgments** We thank Thomas Chambers, Ana Fernández-Sesma, Kelly M. Lager, Wenjun Ma and Amy L. Vincent for helpful discussions. Work on influenza in the authors' laboratories has been funded in part by the National Institute of Allergy and Infectious Diseases, the National Institutes of Health, and the Department of Health and Human Services, under contracts HHSN266200700005C (JAR) and HHSN2662007000010C (AG-S) and under grants R01AI46954 (AG-S), P01AI58113 (AG-S), U19AI62623 (Center for Investigating Viral Immunity and Antagonism, AG-S), U54AI57158 (Northeast Biodefense Center, AG-S), U01AI70469 (AG-S) and U01AI174539 (AG-S), as well as by the Centers for Disease Control and Prevention grant U01CI000357-01 (JAR), the US Department of Agriculture grant 2006-35204-17437 (AG-S), and the W. M. Keck Foundation (AG-S).

# References

Aragón T, de La Luna S, Novoa I, Carrasco L, Ortín J, Nieto A (2000) Eukaryotic translation initiation factor 4GI is a cellular target for NS1 protein, a translational activator of influenza virus. Mol Cell Biol 20:6259–6268

Baskin CR, Bielefeldt-Ohmann H, García-Sastre A, Tumpey TM, Van Hoeven N, Carter VS, Thomas MJ, Proll S, Solorzano A, Billharz R, Fornek JL, Thomas S, Chen CH, Clark EA, Murali-Krishna K, Katze MG (2007) Functional genomic and serological analysis of the protective immune response resulting from vaccination of macaques with an NS1-truncated influenza virus. J Virol 81:11817–11827

Belshe RB, Edwards KM, Vesikari T, Black SV, Walker RE, Hultquist M, Kemble G, Connor EM (2007) Live attenuated versus inactivated influenza vaccine in infants and young children. N Engl J Med 356:685–696

Bergmann M, García-Sastre A, Carnero E, Pehamberger H, Wolff K, Palese P, Muster T (2000) Influenza virus NS1 protein counteracts PKR-mediated inhibition of replication. J Virol 74:6203–6206

Cauthen AN, Swayne DE, Sekellick MJ, Marcus PI, Suarez DL (2007) Amelioration of influenza virus pathogenesis in chickens attributed to the enhanced interferon-inducing capacity of a virus with a truncated NS1 gene. J Virol 81:1838–1847

Chen Z, Li Y, Krug RM (1999) Influenza A virus NS1 protein targets poly(A)-binding protein II of the cellular 3′-end processing machinery. EMBO J 18:2273–2283

Chen W, Calvo PA, Malide D, Gibbs J, Schubert U, Bacik I, Basta S, O'Neill R, Schickli J, Palese P, Henklein P, Bennink JR, Yewdell JW (2001) A novel influenza A virus mitochondrial protein that induces cell death. Nat Med 7:1306–1312

Dauber B, Heins G, Wolff T (2004) The influenza B virus nonstructural NS1 protein is essential for efficient viral growth and antagonizes beta interferon induction. J Virol 78:1865–1872

Dauber B, Schneider J, Wolff T (2006) Double-stranded RNA binding of influenza B virus nonstructural NS1 protein inhibits protein kinase R but is not essential to antagonize production of alpha/beta interferon. J Virol 80:11667–11677

Donelan NR, Dauber B, Wang X, Basler CF, Wolff T, García-Sastre A (2004) The N- and C-terminal domains of the NS1 protein of influenza B virus can independently inhibit IRF-3 and beta interferon promoter activation. J Virol 78:11574–11582

Efferson CL, Schickli J, Ko BK, Kawano K, Mouzi S, Palese P, García-Sastre A, Ioannides CG (2003) Activation of tumor antigen-specific cytotoxic T lymphocytes (CTLs) by human dendritic cells infected with an attenuated influenza A virus expressing a CTL epitope derived from the HER-2/neu proto-oncogene. J Virol 77:7411–7424

Efferson CL, Tsuda N, Kawano K, Nistal-Villan E, Sellappan S, Yu D, Murray JL, García-Sastre A, Ioannides CG (2006) Prostate tumor cells infected with a recombinant influenza virus expressing a truncated NS1 protein activate cytolytic CD8+ cells to recognize noninfected tumor cells. J Virol 80:383–394

Ehrhardt C, Wolff T, Pleschka S, Planz O, Beermann W, Bode JG, Schmolke M, Ludwig S (2007) Influenza A virus NS1 protein activates the PI3K/Akt pathway to mediate antiapoptotic signaling responses. J Virol 81:3058–3067

Falcón AM, Fernández-Sesma A, Nakaya Y, Moran TM, Ortín J, García-Sastre A (2005) Attenuation and immunogenicity in mice of temperature-sensitive influenza viruses expressing truncated NS1 proteins. J Gen Virol 86:2817–2821

Ferko B, Stasakova J, Sereinig S, Romanova J, Katinger D, Niebler B, Katinger H, Egorov A (2001) Hyperattenuated recombinant influenza A virus nonstructural-protein- encoding vectors induce human immunodeficiency virus type 1 Nef- specific systemic and mucosal immune responses in mice. J Virol 75:8899–8908

Ferko B, Stasakova J, Romanova J, Kittel C, Sereinig S, Katinger H, Egorov A (2004) Immunogenicity and protection efficacy of replication-deficient influenza A viruses with altered NS1 genes. J Virol 78:13037–13045

Ferko B, Kittel C, Romanova J, Sereinig S, Katinger H, Egorov A (2006) Live attenuated influenza virus expressing human interleukin-2 reveals increased immunogenic potential in young and aged hosts. J Virol 80:11621–11627

Fernandez-Sesma A (2007) The influenza virus NS1 protein: inhibitor of innate and adaptive immunity. Infect Disord Drug Targets 7:336–343

Fernandez-Sesma A, Marukian S, Ebersole BJ, Kaminski D, Park MS, Yuen T, Sealfon SC, García-Sastre A, Moran TM (2006) Influenza virus evades innate and adaptive immunity via the NS1 protein. J Virol 80:6295–6304

Fortes P, Beloso A, Ortín J (1994) Influenza virus NS1 protein inhibits pre-mRNA splicing and blocks mRNA nucleocytoplasmic transport. EMBO J 13:704–712

García-Sastre A, Biron CA (2006) Type 1 interferons and the virus–host relationship: a lesson in détente. Science 312:879–882

García-Sastre A, Egorov A, Matassov D, Brandt S, Levy DE, Durbin JE, Palese P, Muster T (1998) Influenza A virus lacking the NS1 gene replicates in interferon-deficient systems. Virology 252:324–330

Guo Z, Chen LM, Zeng H, Gomez JA, Plowden J, Fujita T, Katz JM, Donis RO, Sambhara S (2007) NS1 protein of influenza A virus inhibits the function of intracytoplasmic pathogen sensor, RIG-I. Am J Respir Cell Mol Biol 36:263–269

Hale BG, Jackson D, Chen YH, Lamb RA, Randall RE (2006) Influenza A virus NS1 protein binds p85beta and activates phosphatidylinositol-3-kinase signaling. Proc Natl Acad Sci USA 103:14194–14199

Heikkinen LS, Kazlauskas A, Melen K, Wagner R, Ziegler T, Julkunen I, Saksela K (2008) Avian and 1918 Spanish influenza a virus NS1 proteins bind to Crk/CrkL Src homology 3 domains to activate host cell signaling. J Biol Chem 283:5719–5727

Hornung V, Ellegast J, Kim S, Brzozka K, Jung A, Kato H, Poeck H, Akira S, Conzelmann KK, Schlee M, Endres S, Hartmann G (2006) 5′-Triphosphate RNA is the ligand for RIG-I. Science 314:994–997

Jackson D, Hossain MJ, Hickman D, Perez DR, Lamb RA (2008) A new influenza virus virulence determinant: the NS1 protein four C-terminal residues modulate pathogenicity. Proc Natl Acad Sci USA 105:4381–4386

Jiao P, Tian G, Li Y, Deng G, Jiang Y, Liu C, Liu W, Bu Z, Kawaoka Y, Chen H (2008) A single-amino-acid substitution in the NS1 protein changes the pathogenicity of H5N1 avian influenza viruses in mice. J Virol 82:1146–1154

Kato H, Takeuchi O, Sato S, Yoneyama M, Yamamoto M, Matsui K, Uematsu S, Jung A, Kawai T, Ishii KJ, Yamaguchi O, Otsu K, Tsujimura T, Koh CS, Reis e Sousa C, Matsuura Y, Fujita T, Akira S (2006) Differential roles of MDA5 and RIG-I helicases in the recognition of RNA viruses. Nature 441:101–105

Kochs G, García-Sastre A, Martinez-Sobrido L (2007a) Multiple anti-interferon actions of the influenza A virus NS1 protein. J Virol 81:7011–7021

Kochs G, Koerner I, Thiel L, Kothlow S, Kaspers B, Ruggli N, Summerfield A, Pavlovic J, Stech J, Staeheli P (2007b) Properties of H7N7 influenza A virus strain SC35M lacking interferon antagonist NS1 in mice and chickens. J Gen Virol 88:1403–1409

Le Goffic R, Pothlichet J, Vitour D, Fujita T, Meurs E, Chignard M, Si-Tahar M (2007) Influenza A virus activates TLR3-dependent inflammatory and RIG-I-dependent antiviral responses in human lung epithelial cells. J Immunol 178:3368–3372

Lenschow DJ, Lai C, Frias-Staheli N, Giannakopoulos NV, Lutz A, Wolff T, Osiak A, Levine B, Schmidt RE, García-Sastre A, Leib DA, Pekosz A, Knobeloch KP, Horak I, Virgin HWt (2007) IFN-stimulated gene 15 functions as a critical antiviral molecule against influenza herpes, and, Sindbis viruses. Proc Natl Acad Sci USA 104:1371–1376

Li WX, Li H, Lu R, Li F, Dus M, Atkinson P, Brydon EW, Johnson KL, García-Sastre A, Ball LA, Palese P, Ding SW (2004) Interferon antagonist proteins of influenza and vaccinia viruses are suppressors of RNA silencing. Proc Natl Acad Sci USA 101:1350–13555

Li S, Min JY, Krug RM, Sen GC (2006a) Binding of the influenza A virus NS1 protein to PKR mediates the inhibition of its activation by either PACT or double-stranded RNA. Virology 349:13–21

Li Z, Jiang Y, Jiao P, Wang A, Zhao F, Tian G, Wang X, Yu K, Bu Z, Chen H (2006b) The NS1 gene contributes to the virulence of H5N1 avian influenza viruses. J Virol 80:11115–11123

Long JX, Peng DX, Liu YL, Wu YT, Liu XF (2008) Virulence of H5N1 avian influenza virus enhanced by a 15-nucleotide deletion in the viral nonstructural gene. Virus Genes 36:471–478

Loo YM, Fornek J, Crochet N, Bajwa G, Perwitasari O, Martinez-Sobrido L, Akira S, Gill MA, García-Sastre A, Katze MG, Gale M Jr (2008) Distinct RIG-I and MDA5 signaling by RNA viruses in innate immunity. J Virol 82:335–345

López CB, García-Sastre A, Williams BRG, Moran TM (2003) Type I interferon induction pathway, but not released interferon, participates in the maturation of dendritic cells induced by negative-strand RNA viruses. J Infect Dis 187:1126–1136

Lu Y, Qian XY, Krug RM (1994) The influenza virus NS1 protein: a novel inhibitor of pre-mRNA splicing. Genes Dev 8:1817–1828

Lu Y, Wambach M, Katze MG, Krug RM (1995) Binding of the influenza virus NS1 protein to double-stranded RNA inhibits the activation of the protein kinase that phosphorylates the eIF-2 translation initiation factor. Virology 214:222–228

Ludwig S, Wang X, Ehrhardt C, Zheng H, Donelan N, Planz O, Pleschka S, García-Sastre A, Heins G, Wolff T (2002) The influenza A virus NS1 protein inhibits activation of Jun N-terminal Kinase and AP-1 transcription factors. J Virol 76:11166–11171

Mibayashi M, Martinez-Sobrido L, Loo YM, Cardenas WB, Gale M Jr, García-Sastre A (2007) Inhibition of retinoic acid-inducible gene I-mediated induction of beta interferon by the NS1 protein of influenza A virus. J Virol 81:514–524

Min JY, Krug RM (2006) The primary function of RNA binding by the influenza A virus NS1 protein in infected cells: Inhibiting the 2′-5′ oligo (A) synthetase/RNase L pathway. Proc Natl Acad Sci USA 103:7100–7105

Morley PS, Townsend HG, Bogdan JR, Haines DM (1999) Efficacy of a commercial vaccine for preventing disease caused by influenza virus infection in horses. J Am Vet Med Assoc 215:61–66

Nemeroff ME, Qian XY, Krug RM (1995) The influenza virus NS1 protein forms multimers in vitro and in vivo. Virology 212:422–428

Nemeroff ME, Barabino SM, Li Y, Keller W, Krug RM (1998) Influenza virus NS1 protein interacts with the cellular 30 kDa subunit of CPSF and inhibits 3′end formation of cellular pre-mRNAs. Mol Cell 1:991–1000

Obenauer JC, Denson J, Mehta PK, Su X, Mukatira S, Finkelstein DB, Xu X, Wang J, Ma J, Fan Y, Rakestraw KM, Webster RG, Hoffmann E, Krauss S, Zheng J, Zhang Z, Naeve CW (2006) Large-scale sequence analysis of avian influenza isolates. Science 311:1576–1580

Opitz B, Rejaibi A, Dauber B, Eckhard J, Vinzing M, Schmeck B, Hippenstiel S, Suttorp N, Wolff T (2007) IFNbeta induction by influenza A virus is mediated by RIG-I which is regulated by the viral NS1 protein. Cell Microbiol 9:930–938

Pichlmair A, Schulz O, Tan CP, Naslund TI, Liljestrom P, Weber F, Reis e Sousa C (2006) RIG-I-mediated antiviral responses to single-stranded RNA bearing 5′-phosphates. Science 314:997–1001

Qian XY, Alonso-Caplen F, Krug RM (1994) Two functional domains of the influenza virus NS1 protein are required for regulation of nuclear export of mRNA. J Virol 68:2433–2441

Quinlivan M, Zamarin D, García-Sastre A, Cullinane A, Chambers T, Palese P (2005) Attenuation of equine influenza viruses through truncations of the NS1 protein. J Virol 79:8431–8439

Richt JA, Lekcharoensuk P, Lager KM, Vincent AL, Loiacono CM, Janke BH, Wu WH, Yoon KJ, Webby RJ, Solorzano A, García-Sastre A (2006) Vaccination of pigs against swine influenza viruses by using an NS1-truncated modified live-virus vaccine. J Virol 80:11009–11018

Rintahaka J, Wiik D, Kovanen PE, Alenius H, Matikainen S (2008) Cytosolic antiviral RNA recognition pathway activates caspases 1 and 3. J Immunol 180:1749–1757

Samuel CE (2001) Antiviral actions of interferons. Clin Microbiol Rev 14:778–809

Satterly N, Tsai PL, van Deursen J, Nussenzveig DR, Wang Y, Faria PA, Levay A, Levy DE, Fontoura BM (2007) Influenza virus targets the mRNA export machinery and the nuclear pore complex. Proc Natl Acad Sci USA 104:1853–1858

Sekellick MJ, Marcus PI (1985) Interferon induction by viruses. XIV. Development of interferon inducibility and its inhibition in chick embryo cells "aged" in vitro. J Interferon Res 5:651–667

Seo SH, Hoffmann E, Webster RG (2002) Lethal H5N1 influenza viruses escape host anti-viral cytokine responses. Nat Med 8:950–954

Sereinig S, Stukova M, Zabolotnyh N, Ferko B, Kittel C, Romanova J, Vinogradova T, Katinger H, Kiselev O, Egorov A (2006) Influenza virus NS vectors expressing the mycobacterium tuberculosis ESAT-6 protein induce CD4+ Th1 immune response and protect animals against tuberculosis challenge. Clin Vaccine Immunol 13:898–904

Smith EJ, Marié I, Prakash A, García-Sastre A, Levy DE (2001) IRF3 and IRF7 phosphorylation in virus-infected cells does not require double-stranded RNA-dependent protein kinase R or IκB kinase but is blocked by vaccinia virus E3L protein. J Biol Chem 276:8951–8957

Solórzano A, Webby RJ, Lager KM, Janke BH, García-Sastre A, Richt JA (2005) Mutations in the NS1 protein of swine influenza virus impair anti-interferon activity and confer attenuation in pigs. J Virol 79:7535–7543

Stasakova J, Ferko B, Kittel C, Sereinig S, Romanova J, Katinger H, Egorov A (2005) Influenza A mutant viruses with altered NS1 protein function provoke caspase-1 activation in primary human macrophages, resulting in fast apoptosis and release of high levels of interleukins 1beta and 18. J Gen Virol 86:185–195

Stukova MA, Sereinig S, Zabolotnyh NV, Ferko B, Kittel C, Romanova J, Vinogradova TI, Katinger H, Kiselev OI, Egorov A (2006) Vaccine potential of influenza vectors expressing *Mycobacterium tuberculosis* ESAT-6 protein. Tuberculosis (Edinb) 86:236–246

Talon J, Horvath CM, Polley R, Basler CF, Muster T, Palese P, García-Sastre A (2000a) Activation of interferon regulatory factor 3 is inhibited by the influenza A virus NS1 protein. J Virol 74:7989–7996

Talon J, Salvatore M, O'Neill RE, Nakaya Y, Zheng H, Muster T, García-Sastre A, Palese P (2000b) Influenza A and B viruses expressing altered NS1 proteins: a vaccine approach. Proc Natl Acad Sci USA 97:4309–4314

Townsend HG, Penner SJ, Watts TC, Cook A, Bogdan J, Haines DM, Griffin S, Chambers T, Holland RE, Whitaker-Dowling P, Youngner JS, Sebring RW (2001) Efficacy of a cold-adapted, intranasal, equine influenza vaccine: challenge trials. Equine Vet J 33:637–643

Treanor JJ, Campbell JD, Zangwill KM, Rowe T, Wolff M (2006) Safety and immunogenicity of an inactivated subvirion influenza A (H5N1) vaccine. N Engl J Med 354:1343–1351

Twu KY, Kuo RL, Marklund J, Krug RM (2007) The H5N1 influenza virus NS genes selected after 1998 enhance virus replication in mammalian cells. J Virol 81:8112–8121

Vincent AL, Ma W, Lager KM, Janke BH, Webby RJ, García-Sastre A, Richt JA (2007) Efficacy of intranasal administration of a truncated NS1 modified live influenza virus vaccine in swine. Vaccine 25:7999–8009

Wang W, Krug RM (1996) The RNA-binding and effector domains of the viral NS1 protein are conserved to different extents among influenza A and B viruses. Virology 223:41–50

Wang X, Li M, Zheng H, Muster T, Palese P, Beg AA, García-Sastre A (2000) Influenza A virus NS1 protein prevents the activation of NF-κB and induction of type I IFN. J Virol 74:11566–11573

Wang X, Basler CF, Williams BRG, Silverman RH, Palese P, García-Sastre A (2002) Functional replacement of the carboxy-terminal two thirds of the influenza A virus NS1 protein with short heterologous dimerization domains. J Virol 76:12951–12962

Yuan W, Krug RM (2001) Influenza B virus NS1 protein inhibits conjugation of the interferon (IFN)-induced ubiquitin-like ISG15 protein. EMBO J 20:362–371

Zhirnov OP, Konakova TE, Wolff T, Klenk HD (2002) NS1 protein of influenza A virus down-regulates apoptosis. J Virol 76:1617–1625

Zhu Q, Yang H, Chen W, Cao W, Zhong G, Jiao P, Deng G, Yu K, Yang C, Bu Z, Kawaoka Y, Chen H (2008) A naturally occurring deletion in its NS gene contributes to the attenuation of an H5N1 swine influenza virus in chickens. J Virol 82:220–228

# DNA Vaccines Against Influenza Viruses

Jin Hyang Kim and Joshy Jacob

**Contents**

**Abstract** As an attractive alternative to conventional vaccines, DNA vaccines play a critical role in inducing protection against several infectious diseases. In this review, we discuss the advantages that DNA vaccines offer in comparison to conventional protein-based vaccines. We discuss strategies to improve the potency and efficacy of DNA vaccines. Specifically, we focus on the potential use of DNA-based vaccines to elicit broad-spectrum humoral and cellular immunity against influenza virus. Finally, we discuss the advances made in the use of DNA vaccines to prevent avian H5N1 influenza.

## 1  Introduction

DNA vaccines represent a novel and powerful alternative to conventional vaccine approaches. Their novelty and usefulness stems from the fact that they are noninfectious, nonreplicating, extremely stable, and can be produced en masse at low cost. Most importantly, DNA vaccines against emerging pathogens or bioterrorism threats can be quickly constructed based solely upon the pathogen's genetic code.

J.H. Kim and J. Jacob (✉)

Department of Microbiology & Immunology and Emory Vaccine Center, School of Medicine, Emory University, 954 Gatewood Road, Atlanta, GA 30329, USA
e-mail: joshy.jacob@emory.edu

R.W. Compans and W.A. Orenstein (eds.), *Vaccines for Pandemic Influenza*,     197
Current Topics in Microbiology and Immunology 333,
DOI 10.1007/978-3-540-92165-3_10, © Springer-Verlag Berlin Heidelberg 2009

DNA vaccines have been tested for a variety of infectious agents, including bacteria, viruses, and parasites (Davis et al. 1996; Hoffman et al. 1994; Ulmer et al. 1998a; van Drunen Littel-van den Hurk et al. 2000). DNA vaccines require all elements needed for mass production of the plasmid DNA and proper protein expression in eukaryotic cells. Thus the prototypic plasmid DNA vaccine contains (1) an origin of replication, (2) a bacterial antibiotic resistance gene for selection in culture, (3) a strong promoter such as cytomegalovirus (CMV) promoter or simian virus 40 promoter that is active in eukaryotic cells, and (4) RNA transcripts stabilized by polyadenylation sequences (van Drunen Littel-van den Hurk et al. 2000). In addition to these structural features, bacterial plasmid DNA itself contains specific nucleotide sequences (e.g., unmethylated cytidine phosphate guanosine, CpG) that stimulate innate immune responses. This is characterized by the production of IL-6, tumor-necrosis factor (TNF)-$\alpha$, IFN-$\gamma$, and IFN-$\alpha$ (Halpern et al. 1996; Klinman et al. 1996; Sato et al. 1996), activation of natural killer cells (Cowdery et al. 1996), and antigen-presenting cells (APCs), including macrophages and dendritic cells (DCs) (Jakob et al. 1998; Stacey et al. 1996). CpG also links innate immunity to adaptive immunity by directly activating B cells (Krieg et al. 1995) and T cells (Bendigs et al. 1999). All of these factors of plasmid DNA contribute to the immunogenicity of DNA vaccination. At the site of immunization, cells transfected with plasmid DNA encode the protein of interest. This protein is then processed and presented to the immune system in the context of MHC class I and/or II to activate antigen-specific CD8+ and CD4+ T cells (see below).

DNA vaccines are commonly administered via two routes: intramuscular injection or bombardment of the skin using a gene gun. Following application, viral/bacterial DNA can be processed and presented to the immune system via at least three mechanisms: (1) direct priming by somatic cells such as myocytes and keratinocytes, (2) direct transfection of professional APCs, such as DCs, and (3) cross-priming, where secreted protein is taken up by professional APCs and presented to T cells via MHC class I-dependent pathways. With the direct priming mechanism, muscle cells are shown to be critical to the expression of the protein and the initiation of cellular immunity (Wolff et al. 1990). However, it remains unclear how muscle cells activate CD8+ T cells without costimulatory molecules B7-1 or B7-2. Nonetheless, the transfer of stably transfected myoblasts expressing influenza nucleoprotein (Schunemann et al. 2007) conferred protection against challenge (Ulmer et al. 1996), underscoring the role of muscle cells in cellular immunity following intramuscular DNA vaccination. Skin cells such as keratinocytes and Langerhans cells also contribute to antigen expression and generation of cellular immunity following skin DNA injection (Akbari et al. 1999; Klinman et al. 1998b). Many studies substantiate the second mechanism (direct transfection of DCs), showing that cytotoxic T lymphocyte (CTL) responses are initiated primarily by bone marrow DCs following DNA vaccination (Doe et al. 1996; Iwasaki et al. 1997). It is noteworthy that only a small proportion of DCs were directly transfected with plasmid DNA capable of presenting antigen to T cells in vitro (Casares et al. 1997). The majority of DCs remained untransfected, yet showed an activated phenotype and massively migrated following DNA vaccination. This indicates the alternative possibility of their cross-priming by a few transfected DCs. Most secreted or exogenous proteins are processed in the MHC class II pathway to stimulate CD4+ T cells, while endogenous proteins are presented to CD8+ T cells via the MHC class I pathway. While most exogenous proteins are excluded from the MHC class I pathway, evidence of exogenous proteins presented with MHC class I supports the phenomenon of cross-priming. Thus it is possible that despite the lack of costimulatory molecules, muscle cells initiate cellular immunity by secreting proteins that then are picked up by DCs to cross-prime CD8+ T cells at the site of DNA vaccination. Consistent with this idea, the transfer of myoblasts expressing influenza nucleoprotein into F1 hybrid mice induced MHC haplotype-restricted CTL responses (Fu et al. 1997). Taken together, DCs are the main target of DNA vaccination to initiate the immune response, while somatic cells like myocytes and keratinocytes transfer antigens for cross-priming.

## 2  Advantages of DNA Vaccines

Being essentially *E. coli*-derived plasmid DNA, DNA vaccines offer unique advantages over conventional protein-based vaccines/killed vaccines. The DNA is noninfectious and nonreplicating, thereby mitigating safety concerns associated with live attenuated vaccines. The manufacture of a DNA vaccine is cost-effective, enabling large-scale production with a high purity and stability (Hoare et al. 2005). In addition, it can be easily stored without the need for a cold chain, which is expensive and difficult in developing countries. Upon administration, DNA vaccines encode only the protein of interest, not additional viral or bacterial antigens. This minimizes undesirable side effects and enables multiple vaccinations to be administered to individuals without inducing immune-dampening vector-specific responses. More importantly, an advantage of DNA vaccination is its ability to induce broad immunity, including CTL and humoral responses. For most infectious diseases, primary protection is mediated by existing antibodies, whereas for intracellular pathogens (*Mycobacterium tuberculosis, Leishmania major*), protection is mediated by MHC class I-restricted CD8 T cell responses. Most protein-based vaccines/killed vaccines promote a good humoral immune response but fail to induce significant CD8+ CTL responses. This is because killed vaccines or proteins are taken up by professional APCs and presented to CD4+ T cells via the MHC class II pathway, which then aids in the production of high-affinity antibodies. On the other hand, live attenuated vaccines induce both antibody and strong CTL responses. However, some live vaccines may be associated with unwanted properties such as virus shedding and genetic mutation, causing reversion back to the wild-type phenotype. (Cinatl et al. 2007). DNA vaccines are also capable of inducing long-term memory, a requirement for a vaccine. Vaccination of mice with a plasmid encoding influenza hemagglutinin (HA) generated persisting anti-HA antibodies for more than a year (Deck et al. 1997). In addition, DNA vaccine encoding a specific leishmanial antigen was shown to induce and maintain strong antigen-specific Th1 cells to control *L. major* infection (Gurunathan et al. 1998). Of note, antigen-specific CD4[+] T cells persist in spleen or lymph nodes for up to 40 weeks post DNA vaccination without detectable antigen (Akbari et al. 1999). These immunological aspects make plasmid DNA an attractive vaccine candidate.

# 3  Disadvantages of DNA Vaccines

There are several safety concerns associated with DNA vaccines. First, the DNA vaccines possibly integrate into host genomes, increasing the risk of malignancy. Although there is no clear evidence as yet that plasmids integrate into host genome, this possibility has not been formally excluded either. Second, the plasmid DNA encoding small amounts of protein may induce autoimmunity. This concern is raised because the CpG motifs within plasmid DNA activate B cells specific for double-stranded DNA. In normal mice, bacterial plasmid DNA induced the production of anti-double stranded DNA autoantibodies, and in lupus-proned NZB/NZW mice, DNA vaccination accelerated the development of autoimmunity (Gilkeson et al. 1993, 1995). In addition, CpG motifs activate polyclonal B cells while stimulating IL-6 production and conferring B cell resistance to apoptosis. These observations may link persistence of activated autoreactive B cells to exacerbation of disease (Klinman 1990; Klinman and Steinberg 1987; Krieg 1995; Linker-Israeli et al. 1991; Watanabe-Fukunaga et al. 1992; Yi et al. 1996). Although these concerns are legitimate, evidence of autoimmunity induced by CpG motifs is most prominent in autoimmune-predisposed animals. In normal mice, serum anti-DNA IgG was transiently elevated following multiple DNA vaccinations but failed to develop disease (Mor et al. 1997). Even in NZB/NZW mice, DNA vaccination did not alter the onset or change the course of disease (Mor et al. 1997). The third issue is that DNA vaccines may cause tolerance rather than immunity due to the persistent production of small amounts of antigen. This is especially relevant in infants, children, and the elderly. Due to immature development of the immune system, newborns are susceptible to developing tolerance to foreign antigens. This is a concern because protein encoded by DNA vaccines is endogenously produced and expressed in the context of MHC class I, and it may be recognized by the immune system as self, leading to tolerance to this protein. Consistent with this idea, a DNA vaccine encoding the circumsporozoite protein of malaria induced tolerance in newborn mice, as demonstrated by the failure to generate cellular and humoral immune responses upon challenge with circumsporozoite protein as adults (Mor et al. 1996). Induction of tolerance is dependent on the age at vaccination; only mice younger than eight days of age are susceptible to tolerance. Mice older than two years also showed diminished protection, suggesting that DNA vaccines are less effective in the elderly as well as newborns (Klinman et al. 1998a,b). Other factors that influence neonatal tolerance are the nature and concentration of the antigen, route of administration, and the mode of antigen presentation to the immune system (Marodon and Rocha 1994; Sarzotti et al. 1996). Lastly, DNA vaccines encoding antigen from certain infectious agents generate a suboptimal response for protection. In the case of HIV, DNA vaccination did not generate persistent high titers of neutralizing antibody against HIV envelope protein (Sedegah et al. 2000). For proper protection, boosting with Env-expressing recombinant vaccinia virus or recombinant Env protein was required (Richmond et al. 1997, 1998). However, suboptimal potency with DNA vaccines is dependent on antigen, as immunization with influenza virus HA-encoding DNA did not require a protein boost (Richmond et al. 1998).

# 4   Improving DNA Vaccines

Protection against infectious diseases is the ultimate goal of vaccines. Although DNA vaccines generate a high titer of neutralizing antibodies against some antigens (e.g., influenza HA, NP proteins), they fail to elicit optimal responses against others (e.g., HIV, malaria). In addition, DNA vaccination is often good at priming small animals (e.g., mice) but less effective in larger animals (Turnes et al. 1999; Ugen et al. 1998). Many strategies have been devised to improve the potency of DNA vaccines. These include (1) the use of better promoters/enhancers (Garg et al. 2004; Harms and Splitter 1995), (2) increased availability of proteins in the cytosol, (3) coadministration of immunomodulatory cytokines (Chow et al. 1997; James et al. 2007; Manickan et al. 1997; Prince et al. 1997; Sailaja et al. 2003; Sarzotti et al. 1997; Wang et al. 1997), (4) optimization of vaccine administration and delivery (Babiuk et al. 2002; Sharpe et al. 2007), (5) protein boosting following DNA vaccination (Epstein et al. 2005; Jones et al. 2001; Richmond et al. 1998), (6) use of adjuvants (Ozaki et al. 2005), (7) direct targeting of DNA vaccines to APCs (Deliyannis et al. 2000; Lew et al. 2000; Tachedjian et al. 2003), and (8) vectors encoding antigens fused to molecules that facilitate antigen spread and cross-priming (Hung et al. 2001, 2002; Ross et al. 2000; Wills et al. 2001).

One aspect of the reduced efficacy of DNA vaccines is insufficient expression of the protein in vivo. At the level of protein production, gene expression driven by most common promoters used in DNA plasmid constructs is inhibited by the cytokine IFN-$\gamma$, which often accompanies cellular immune responses (Harms and Splitter 1995). Thus, the SV40 or CMV promoter may not be a suitable choice for a DNA vaccine where elevated IFN-$\gamma$ is expected. In this case, the MHC I promoter/enhancer represents an alternative promoter for driving protein expression (Harms and Splitter 1995). Efficacy of DNA vaccines can also be enhanced by modifying the promoter/enhancer to increase protein expression. We have developed a DNA vaccine encoding HA driven by a hybrid CMV enhancer/chicken beta-actin promoter and/or the mRNA-stabilizing posttranscriptional regulatory element from the woodchuck hepatitis virus (WPRE). We have shown that this modified DNA vaccine effectively lowered the immunization dose tenfold while providing complete protection from a lethal challenge (Garg et al. 2004). The potency of DNA vaccines can be posttranscriptionally augmented by manipulating protein localization.

An example of this strategy is influenza nucleoprotein (Schunemann et al. 2007), which localizes in the nucleus, resulting in lower vaccine efficacy (Neumann et al. 1997). To overcome the low availability of protein in the cytoplasm, Ohba and colleagues developed a DNA vaccine with an N-terminal mutant NP that preferentially localized in the cytoplasm and demonstrated higher immunogenicity and cross-reactivity with the mutant vaccine than the wild-type NP DNA vaccine (Ohba et al. 2007). Similarly, fusion of the NP to herpes simplex virus genes (VP22) augmented the immunogenicity of NP-based DNA vaccines (Saha et al. 2006). Another strategy for improving the immunogenicity of DNA vaccines is to increase protein production in vivo by enhancing transfection efficiency. This has been shown to be effective in large animals, which show poor vaccine efficacy. Transfection by electroporation, called intramuscular electroporation technology (imEPT), significantly enhanced gene expression in pigs (Babiuk et al. 2002), and the amount of plasmid required to induce an optimal response was reduced by 20-fold (Saha et al. 2006). Likewise, administering a plasmid encoding HA from H5N1 influenza virus by particle-mediated epidermal delivery (PMED) led to potent anti-HA response and conferred protection from a lethal challenge (Sharpe et al. 2007).

Quality and magnitude of an immune response can be influenced by local cytokine milieu and activation of innate immunity. Consistent with this idea, the immunogenicity of DNA vaccines is improved by coadministering DNA plasmids expressing cytokines or costimulatory molecules. In neonates, this approach showed strong immunity and protection (Manickan et al. 1997; Prince et al. 1997; Sarzotti et al. 1997; Wang et al. 1997). Inclusion of the cytokine IL-2, GM-CSF, or Flt-3 ligand-encoding plasmid in DNA vaccine regimen not only effectively reduced the immunization dose but also generated strong CTL response and memory (Chow et al. 1997; Sailaja et al. 2003; Sedegah et al. 2000). The efficacy of DNA vaccines is affected by the choice of immunomodulatory cytokine. Among type I IFN multigene family members, the subtype IFNA6 was shown to be most effective while IFNA1 was the least effective at reducing lung virus replication during influenza challenge (James et al. 2007). Adjuvants have been used in conjunction with protein-based vaccines to improve the quality of immune response. They activate innate immunity to antigen by augmenting the activities of DCs and macrophages, thereby establishing local inflammatory surroundings. Use of adjuvants has been applied to DNA vaccination, such that a plasmid encoding influenza virus matrix (M) protein significantly increased protection against a lethal challenge with heterologous strains of virus when adjuvanted with cholera toxin and CpG motifs (Ozaki et al. 2005).

## 5   DNA Vaccines Against H1N1 Influenza

The efficacy of DNA vaccines that confer protection against H1N1 influenza virus has been demonstrated in many animal models. Protection against influenza infection is mediated predominantly by antibodies. These antibodies are primarily directed against surface glycoproteins of influenza virus, in other words hemagglutinin (HA) and neuraminidase (NA). Antibodies to HA neutralize the infectivity of the virus while antibodies to NA efficiently prevent the release of the virus from the infected cells. Many studies have aimed at eliciting strong anti-HA or anti-NA antibody responses. Indeed, DNA vaccines encoding HA induced long-lasting serum hemagglutination inhibiting antibodies and conferred protection against challenge with homologous strains of influenza virus in mice, chickens, and ferrets (Liu et al. 1997; Pertmer et al. 1995; Robinson et al. 1993; Ulmer et al. 1998a,b). However, because of antigenic drift of the virus, virus-specific neutralizing antibodies may fail to offer protection against secondary infection with heterologous strain. Further, HA vaccines are unlikely to provide protection against different subtypes. NA protein, on the other hand, exhibits slower antigenic mutation in comparison to HA (Kilbourne et al. 1990). Consistent with this idea, a DNA vaccine encoding NA from A/Guizhou/54/89 (H3N2) conferred complete protection against homologous virus challenge and significant cross-protection against a variant (drift) virus (A/Aichi/2/68; H3N2) (Chen et al. 2000). However, it failed to provide protection against infection with a heterosubtypic virus (H1N1). This indicates that HA or NA-based vaccines are not suitable for broad protection.

While the HA-based vaccine remains effective at preventing infection, cellular immunity to influenza has been widely investigated. Cellular immunity provides protection from the morbidity and mortality associated with pathogenic influenza virus, although it may not prevent infection per se (Hogan et al. 2001). A DNA vaccine encoding the conserved internal protein NP of influenza virus generated a cytotoxic T cell response and conferred protection from heterosubtypic virus challenge in mice (Ulmer et al. 1993). In addition, a NP-encoding DNA vaccine elicited CD4[+] helper T cells of the Th1 phenotype producing IFN-γ and IL-2 (Ulmer et al. 1998a,b). Strategies to improve the efficacy of NP-DNA vaccines have been devised, and they include deleting the nuclear localization signal of NP (Neumann et al. 1997), N-terminal mutation (Ohba et al. 2007), and fusion of the NP into the tegument protein VP22 of herpes simplex virus 1 (Kim et al. 2004). In addition to the NP protein, other conserved internal proteins (e.g., matrix protein M) of influenza virus have been explored for DNA vaccination (Ozaki et al. 2005). In an effort to induce broad protection against heterologaus subtypes of influenza virus, the approach of combined immunization with DNA vaccines encoding surface protein (e.g., HA) and internal protein (e.g., NP or M1) has been studied. In ferrets, combined immunization with HA-encoding DNA and NP-encoding DNA vaccines conferred protection against challenge with the antigenic drift variants A/Georgia/03/93 and A/Johannesburg/33/94 (Donnelly et al. 1995, 1997). In mice, vaccination with a mixture of HA, NA-encoding DNAs or HA, NA and M1-encoding DNAs prepared from A/PR/8/34 (H1N1) provided complete protection against not only homologous virus infection but also an antigenically drifted strain (A/Yamagata/120/86; H1N1). The degree of protection afforded in combined vaccine groups was significantly higher than that in mice given each DNA alone (Chen et al. 1999).

## 6   DNA Vaccines Against H5N1 Influenza

The 1997 outbreak of H5N1 avian influenza in humans in Hong Kong and frequent subsequent outbreaks in China and Eastern Europe have resulted in great concern in the world health community (Subbarao et al. 1998; Yuen et al. 1998). This is because the outbreaks were caused by highly pathogenic strains of an influenza subtype to which humans lack immunity, and hence poses the potential to cause an influenza pandemic, as seen in 1918. Our current understanding of avian influenza virus, including the mutations that allow its transformation into a human-transmittable virus, remains limited. Considering the molecular basis for the virulence of H5N1 viruses (Hatta et al. 2001), the most promising method of controlling a pandemic is the use antiviral drugs (Laver and Garman 2001). However, these drugs only partially reduce the symptoms and duration of the disease, and drug resistance has been found in multiple isolates (Beigel et al. 2005; Nicholson et al. 2000; Schunemann et al. 2007). Genomic and antigenic analyses of H5N1 viruses isolated since 2004 have revealed at least two distinct sublineages with different geographic distributions, designated clade 1 and clade 2 (Webster and Govorkova 2006). This raises the concern that it is not possible to predict which strain may emerge in a future pandemic; therefore, a vaccine generated from a single selected strain may not be able to protect against a diverse set of viruses. In addition, there are several inherent difficulties associated with H5N1 vaccines (Stephenson et al. 2006; Subbarao and Luke 2007; Subbarao et al. 2006). H5 viruses are highly pathogenic, yet H5-HA is poorly immunogenic for unknown reasons. Little is known about the antigenic sites on avian HAs and immune correlates of protection from avian influenza infections. In comparison with human-adapted influenza viruses, the yield of candidate vaccines of H5N1 in embryonic chicken eggs is reduced, and limited manufacturing capacity represents an additional obstacle in the development of H5N1 vaccines. Thus, the major focus in H5N1 vaccine development is on testing vaccine candidates for priming, cross-reactivity and cross-protection against infection with viruses from different clades and subclades.

Studies have explored the possibility of developing DNA vaccines against H5N1 influenza viruses (Epstein et al. 2002; Kodihalli et al. 1999, 2000; Laddy et al. 2007). As in human influenza viruses, the protective ability of HA-based DNA vaccines of H5N1 virus is limited to homologous strains of virus. Kodihalli and colleagues showed that a DNA vaccine encoding HA from the index human influenza isolate A/HongKong/156/97 provided immunity against homologous H5N1 infection of mice (Kodihalli et al. 1999). However, a DNA vaccine encoding the HA from A/ Ty/Ir/1/83 (H5N8), which differs from A/HK/156/97 (H5N1) by 12% in HA1, prevented death but not H5N1 infection (Kodihalli et al. 1999). The possibility of protection conferred by NA-based DNA vaccines has also been explored (Sandbulte et al. 2007). Based on the idea that the NA of H5N1 viruses (avN1) and of endemic human H1N1 viruses (huN1) are classified in the same serotype, Sandbulte's group tested whether an immune response to huN1 could mediate cross-protection against H5N1 influenza virus infection (Sandbulte et al. 2007). A DNA vaccine encoding huN1 from A/PR/8/34 (H1N1) partially protected mice from lethal challenge with

H5N1 virus or recombinant PR8-avN1. These findings suggest that a portion of the human population could have some degree of resistance to H5N1 influenza (Sandbulte et al. 2007). More promising results regarding cross-protection are found in studies where internal protein-based DNA vaccines were applied. Epstein and colleagues showed that DNA vaccination encoding the PR8 (H1N1)-NP and M1 proteins reduced replication of A/HongKong/486/97, a nonlethal H5N1 strain in mice, and completely protected and minimized morbidity upon lethal challenge with the more virulent A/HongKong/156/97. Upon challenge with a highly virulent strain, HK/483, half of the vaccinated mice survived (Epstein et al. 2002). The strategy of combined DNA vaccination against H5N1 virus was also shown to be effective against other viruses, such that H5- and H7-encoding DNA vaccines protected chickens against lethal infection by both A/Ck/Vic/1/85 (H7N7) and A/Ty/Ir/1/83 (H5N8) (Kodihalli et al. 2000). However, in this study, chickens immunized with NP-encoding DNA from TyIr83 showed signs of infection, and approximately 50% of chickens survived lethal challenge by both viruses. These observations point out the need for improvements in H5N1 DNA vaccines that target cell-mediated immunity. A recent study that employed novel consensus DNA vaccines represents a promising alternative (Laddy et al. 2007). By analyzing a large number of circulating avian influenza viruses, Laddy and colleagues generated constructs containing the most common amino acids with the potential to induce highly cross-reactive cellular and humoral immune responses. Sequences were chosen from HA proteins from avian flu viruses isolated between 1997 and 2005, and NA and M1 proteins from 40 and 45 primary sequences of both H1N1 and H5N1 in multiple countries. This vaccine elicited strong CTL and humoral immune responses in mice, and the recombinant protein of the H5-HA construct protected animals against lethal challenge with a highly pathogenic H5N1 strain (A/Hanoi/30408/05). While more improvement is needed, these studies highlight the effectiveness of DNA vaccines for protection against a possible pandemic by avian influenza virus.

# References

Akbari O, Panjwani N, Garcia S, Tascon R, Lowrie D, Stockinger B (1999) DNA vaccination: transfection and activation of dendritic cells as key events for immunity. J Exp Med 189:169–178

Babiuk S, Baca-Estrada ME, Foldvari M, Storms M, Rabussay D, Widera G, Babiuk LA (2002) Electroporation improves the efficacy of DNA vaccines in large animals. Vaccine 20:3399–3408

Beigel JH, Farrar J, Han AM, Hayden FG, Hyer R, de Jong MD, Lochindarat S, Nguyen TK, Nguyen TH, Tran TH, Nicoll A, Touch S, Yuen KY (2005) Avian influenza A (H5N1) infection in humans. N Engl J Med 353:1374–1385

Bendigs S, Salzer U, Lipford GB, Wagner H, Heeg K (1999) CpG-oligodeoxynucleotides co-stimulate primary T cells in the absence of antigen-presenting cells. Eur J Immunol 29:1209–1218

Casares S, Inaba K, Brumeanu TD, Steinman RM, Bona CA (1997) Antigen presentation by dendritic cells after immunization with DNA encoding a major histocompatibility complex class II-restricted viral epitope. J Exp Med 186:1481–1486

Chen Z, Matsuo K, Asanuma H, Takahashi H, Iwasaki T, Suzuki Y, Aizawa C, Kurata T, Tamura S (1999) Enhanced protection against a lethal influenza virus challenge by immunization with both hemagglutinin- and neuraminidase-expressing DNAs. Vaccine 17:653–659

Chen Z, Kadowaki S, Hagiwara Y, Yoshikawa T, Matsuo K, Kurata T, Tamura S (2000) Cross-protection against a lethal influenza virus infection by DNA vaccine to neuraminidase. Vaccine 18:3214–3222

Chow YH, Huang WL, Chi WK, Chu YD, Tao MH (1997) Improvement of hepatitis B virus DNA vaccines by plasmids coexpressing hepatitis B surface antigen and interleukin-2. J Virol 71:169–178

Cinatl J Jr, Michaelis M, Doerr HW (2007) The threat of avian influenza A (H5N1). Part IV: Development of vaccines. Med Microbiol Immunol 196:213–225

Cowdery JS, Chace JH, Yi AK, Krieg AM (1996) Bacterial DNA induces NK cells to produce IFN-gamma in vivo and increases the toxicity of lipopolysaccharides. J Immunol 156:4570–4575

Davis HL, McCluskie MJ, Gerin JL, Purcell RH (1996) DNA vaccine for hepatitis B: evidence for immunogenicity in chimpanzees and comparison with other vaccines. Proc Natl Acad Sci USA 93:7213–7218

Deck RR, DeWitt CM, Donnelly JJ, Liu MA, Ulmer JB (1997) Characterization of humoral immune responses induced by an influenza hemagglutinin DNA vaccine. Vaccine 15:71–78

Deliyannis G, Boyle JS, Brady JL, Brown LE, Lew AM (2000) A fusion DNA vaccine that targets antigen-presenting cells increases protection from viral challenge. Proc Natl Acad Sci USA 97:6676–6680

Doe B, Selby M, Barnett S, Baenziger J, Walker CM (1996) Induction of cytotoxic T lymphocytes by intramuscular immunization with plasmid DNA is facilitated by bone marrow-derived cells. Proc Natl Acad Sci USA 93:8578–8583

Donnelly JJ, Friedman A, Martinez D, Montgomery DL, Shiver JW, Motzel SL, Ulmer JB, Liu MA (1995) Preclinical efficacy of a prototype DNA vaccine: enhanced protection against antigenic drift in influenza virus. Nat Med 1:583–587

Donnelly JJ, Friedman A, Ulmer JB, Liu MA (1997) Further protection against antigenic drift of influenza virus in a ferret model by DNA vaccination. Vaccine 15:865–868

Epstein SL, Tumpey TM, Misplon JA, Lo CY, Cooper LA, Subbarao K, Renshaw M, Sambhara S, Katz JM (2002) DNA vaccine expressing conserved influenza virus proteins protective against H5N1 challenge infection in mice. Emerg Infect Dis 8:796–801

Epstein SL, Kong WP, Misplon JA, Lo CY, Tumpey TM, Xu L, Nabel GJ (2005) Protection against multiple influenza A subtypes by vaccination with highly conserved nucleoprotein. Vaccine 23:5404–5410

Fu TM, Ulmer JB, Caulfield MJ, Deck RR, Friedman A, Wang S, Liu X, Donnelly JJ, Liu MA (1997) Priming of cytotoxic T lymphocytes by DNA vaccines: requirement for professional antigen presenting cells and evidence for antigen transfer from myocytes. Mol Med 3:362–371

Garg S, Oran AE, Hon H, Jacob J (2004) The hybrid cytomegalovirus enhancer/chicken beta-actin promoter along with woodchuck hepatitis virus posttranscriptional regulatory element enhances the protective efficacy of DNA vaccines. J Immunol 173:550–558

Gilkeson GS, Ruiz P, Howell D, Lefkowith JB, Pisetsky DS (1993) Induction of immune-mediated glomerulonephritis in normal mice immunized with bacterial DNA. Clin Immunol Immunopathol 68:283–292

Gilkeson GS, Pippen AM, Pisetsky DS (1995) Induction of cross-reactive anti-dsDNA antibodies in preautoimmune NZB/NZW mice by immunization with bacterial DNA. J Clin Invest 95:1398–1402

Gurunathan S, Prussin C, Sacks DL, Seder RA (1998) Vaccine requirements for sustained cellular immunity to an intracellular parasitic infection. Nat Med 4:1409–1415

Halpern MD, Kurlander RJ, Pisetsky DS (1996) Bacterial DNA induces murine interferon-gamma production by stimulation of interleukin-12 and tumor necrosis factor-alpha. Cell Immunol 167:72–78

Harms JS, Splitter GA (1995) Interferon-gamma inhibits transgene expression driven by SV40 or CMV promoters but augments expression driven by the mammalian MHC I promoter. Hum Gene Ther 6:1291–1297

Hatta M, Gao P, Halfmann P, Kawaoka Y (2001) Molecular basis for high virulence of Hong Kong H5N1 influenza A viruses. Science 293:1840–1842

Hoare M, Levy MS, Bracewell DG, Doig SD, Kong S, Titchener-Hooker N, Ward JM, Dunnill P (2005) Bioprocess engineering issues that would be faced in producing a DNA vaccine at up to 100 m3 fermentation scale for an influenza pandemic. Biotechnol Prog 21:1577–1592

Hoffman SL, Sedegah M, Hedstrom RC (1994) Protection against malaria by immunization with a *Plasmodium yoelii* circumsporozoite protein nucleic acid vaccine. Vaccine 12:1529–1533

Hogan RJ, Usherwood EJ, Zhong W, Roberts AA, Dutton RW, Harmsen AG, Woodland DL (2001) Activated antigen-specific CD8+ T cells persist in the lungs following recovery from respiratory virus infections. J Immunol 166:1813–1822

Hung CF, Cheng WF, Chai CY, Hsu KF, He L, Ling M, Wu TC (2001) Improving vaccine potency through intercellular spreading and enhanced MHC class I presentation of antigen. J Immunol 166:5733–5740

Hung CF, He L, Juang J, Lin TJ, Ling M, Wu TC (2002) Improving DNA vaccine potency by linking Marek's disease virus type 1 VP22 to an antigen. J Virol 76:2676–2682

Iwasaki A, Torres CA, Ohashi PS, Robinson HL, Barber BH (1997) The dominant role of bone marrow-derived cells in CTL induction following plasmid DNA immunization at different sites. J Immunol 159:11–14

Jakob T, Walker PS, Krieg AM, Udey MC, Vogel JC (1998) Activation of cutaneous dendritic cells by CpG-containing oligodeoxynucleotides: a role for dendritic cells in the augmentation of Th1 responses by immunostimulatory DNA. J Immunol 161:3042–3049

James CM, Abdad MY, Mansfield JP, Jacobsen HK, Vind AR, Stumbles PA, Bartlett EJ (2007) Differential activities of alpha/beta IFN subtypes against influenza virus in vivo and enhancement of specific immune responses in DNA vaccinated mice expressing haemagglutinin and nucleoprotein. Vaccine 25:1856–1867

Jones TR, Narum DL, Gozalo AS, Aguiar J, Fuhrmann SR, Liang H, Haynes JD, Moch JK, Lucas C, Luu T, Magill AJ, Hoffman SL, Sim BK (2001) Protection of *Aotus* monkeys by *Plasmodium falciparum* EBA-175 region II DNA prime-protein boost immunization regimen. J Infect Dis 183:303–312

Kilbourne ED, Johansson BE, Grajower B (1990) Independent and disparate evolution in nature of influenza A virus hemagglutinin and neuraminidase glycoproteins. Proc Natl Acad Sci USA 87:786–790

Kim TW, Hung CF, Kim JW, Juang J, Chen PJ, He L, Boyd DA, Wu TC (2004) Vaccination with a DNA vaccine encoding herpes simplex virus type 1 VP22 linked to antigen generates long-term antigen-specific CD8-positive memory T cells and protective immunity. Hum Gene Ther 15:167–177

Klinman DM (1990) Polyclonal B cell activation in lupus-prone mice precedes and predicts the development of autoimmune disease. J Clin Invest 86:1249–1254

Klinman DM, Steinberg AD (1987) Systemic autoimmune disease arises from polyclonal B cell activation. J Exp Med 165:1755–1760

Klinman DM, Yi AK, Beaucage SL, Conover J, Krieg AM (1996) CpG motifs present in bacteria DNA rapidly induce lymphocytes to secrete interleukin 6, interleukin 12, and interferon gamma. Proc Natl Acad Sci USA 93:2879–2883

Klinman DM, Conover J, Bloom ET, Weiss W (1998a) Immunogenicity and efficacy of a DNA vaccine in aged mice. J Gerontol A Biol Sci Med Sci 53:B281–286

Klinman DM, Sechler JM, Conover J, Gu M, Rosenberg AS (1998b) Contribution of cells at the site of DNA vaccination to the generation of antigen-specific immunity and memory. J Immunol 160:2388–2392

Kodihalli S, Goto H, Kobasa DL, Krauss S, Kawaoka Y, Webster RG (1999) DNA vaccine encoding hemagglutinin provides protective immunity against H5N1 influenza virus infection in mice. J Virol 73:2094–2098

Kodihalli S, Kobasa DL, Webster RG (2000) Strategies for inducing protection against avian influenza A virus subtypes with DNA vaccines. Vaccine 18:2592–2599

Krieg AM (1995) CpG DNA: a pathogenic factor in systemic lupus erythematosus. J Clin Immunol 15:284–292

Krieg AM, Yi AK, Matson S, Waldschmidt TJ, Bishop GA, Teasdale R, Koretzky GA, Klinman DM (1995) CpG motifs in bacterial DNA trigger direct B-cell activation. Nature 374:546–549

Laddy DJ, Yan J, Corbitt N, Kobasa D, Kobinger GP, Weiner DB (2007) Immunogenicity of novel consensus-based DNA vaccines against avian influenza. Vaccine 25:2984–2989

Laver G, Garman E (2001) Virology. The origin and control of pandemic influenza. Science 293:1776–1777

Lew AM, Brady BJ, Boyle BJ (2000) Site-directed immune responses in DNA vaccines encoding ligand–antigen fusions. Vaccine 18:1681–1685

Linker-Israeli M, Deans RJ, Wallace DJ, Prehn J, Ozeri-Chen T, Klinenberg JR (1991) Elevated levels of endogenous IL-6 in systemic lupus erythematosus. A putative role in pathogenesis. J Immunol 147:117–123

Liu MA, McClements W, Ulmer JB, Shiver J, Donnelly J (1997) Immunization of non-human primates with DNA vaccines. Vaccine 15:909–912

Manickan E, Yu Z, Rouse BT (1997) DNA immunization of neonates induces immunity despite the presence of maternal antibody. J Clin Invest 100:2371–2375

Marodon G, Rocha B (1994) Activation and 'deletion' of self-reactive mature and immature T cells during ontogeny of Mls-1a mice: implications for neonatal tolerance induction. Int Immunol 6:1899–1904

Mor G, Yamshchikov G, Sedegah M, Takeno M, Wang R, Houghten RA, Hoffman S, Klinman DM (1996) Induction of neonatal tolerance by plasmid DNA vaccination of mice. J Clin Invest 98:2700–2705

Mor G, Singla M, Steinberg AD, Hoffman SL, Okuda K, Klinman DM (1997) Do DNA vaccines induce autoimmune disease. Hum Gene Ther 8:293–300

Neumann G, Castrucci MR, Kawaoka Y (1997) Nuclear import and export of influenza virus nucleoprotein. J Virol 71:9690–9700

Nicholson KG, Aoki FY, Osterhaus AD, Trottier S, Carewicz O, Mercier CH, Rode A, Kinnersley N, Ward P (2000) Efficacy and safety of oseltamivir in treatment of acute influenza: a randomised controlled trial. Neuraminidase Inhibitor Flu Treatment Investigator Group. Lancet 355: 1845–1850

Ohba K, Yoshida S, Zahidunnabi Dewan M, Shimura H, Sakamaki N, Takeshita F, Yamamoto N, Okuda K (2007) Mutant influenza A virus nucleoprotein is preferentially localized in the cytoplasm and its immunization in mice shows higher immunogenicity and cross-reactivity. Vaccine 25:4291–4300

Ozaki T, Yauchi M, Xin KQ, Hirahara F, Okuda K (2005) Cross-reactive protection against influenza A virus by a topically applied DNA vaccine encoding M gene with adjuvant. Viral Immunol 18:373–380

Pertmer TM, Eisenbraun MD, McCabe D, Prayaga SK, Fuller DH, Haynes JR (1995) Gene gun-based nucleic acid immunization: elicitation of humoral and cytotoxic T lymphocyte responses following epidermal delivery of nanogram quantities of DNA. Vaccine 13:1427–1430

Prince AM, Whalen R, Brotman B (1997) Successful nucleic acid based immunization of newborn chimpanzees against hepatitis B virus. Vaccine 15:916–919

Richmond JF, Mustafa F, Lu S, Santoro JC, Weng J, O'Connell M, Fenyo EM, Hurwitz JL, Montefiori DC, Robinson HL (1997) Screening of HIV-1 Env glycoproteins for the ability to raise neutralizing antibody using DNA immunization and recombinant vaccinia virus boosting. Virology 230:265–274

Richmond JF, Lu S, Santoro JC, Weng J, Hu SL, Montefiori DC, Robinson HL (1998) Studies of the neutralizing activity and avidity of anti-human immunodeficiency virus type 1 Env antibody elicited by DNA priming and protein boosting. J Virol 72:9092–9100

Robinson HL, Hunt LA, Webster RG (1993) Protection against a lethal influenza virus challenge by immunization with a haemagglutinin-expressing plasmid DNA. Vaccine 11:957–960

Ross TM, Xu Y, Bright RA, Robinson HL (2000) C3d enhancement of antibodies to hemagglutinin accelerates protection against influenza virus challenge. Nat Immunol 1:127–131

Saha S, Yoshida S, Ohba K, Matsui K, Matsuda T, Takeshita F, Umeda K, Tamura Y, Okuda K, Klinman D, Xin KQ (2006) A fused gene of nucleoprotein (NP) and herpes simplex virus genes (VP22) induces highly protective immunity against different subtypes of influenza virus. Virology 354:48–57

Sailaja G, Husain S, Nayak BP, Jabbar AM (2003) Long-term maintenance of gp120-specific immune responses by genetic vaccination with the HIV-1 envelope genes linked to the gene encoding Flt-3 ligand. J Immunol 170:2496–2507

Sandbulte MR, Jimenez GS, Boon AC, Smith LR, Treanor JJ, Webby RJ (2007) Cross-reactive neuraminidase antibodies afford partial protection against H5N1 in mice and are present in unexposed humans. PLoS Med 4:e59

Sarzotti M, Robbins DS, Hoffman PM (1996) Induction of protective CTL responses in newborn mice by a murine retrovirus. Science 271:1726–1728

Sarzotti M, Dean TA, Remington MP, Ly CD, Furth PA, Robbins DS (1997) Induction of cytotoxic T cell responses in newborn mice by DNA immunization. Vaccine 15:795–797

Sato Y, Roman M, Tighe H, Lee D, Corr M, Nguyen MD, Silverman GJ, Lotz M, Carson DA, Raz E (1996) Immunostimulatory DNA sequences necessary for effective intradermal gene immunization. Science 273:352–354

Schunemann HJ, Hill SR, Kakad M, Bellamy R, Uyeki TM, Hayden FG, Yazdanpanah Y, Beigel J, Chotpitayasunondh T, Del Mar C, Farrar J, Tran TH, Ozbay B, Sugaya N, Fukuda K, Shindo N, Stockman L, Vist GE, Croisier A, Nagjdaliyev A, Roth C, Thomson G, Zucker H, Oxman AD (2007) WHO Rapid Advice Guidelines for pharmacological management of sporadic human infection with avian influenza A (H5N1) virus. Lancet Infect Dis 7:21–31

Sedegah M, Weiss W, Sacci JB Jr, Charoenvit Y, Hedstrom R, Gowda K, Majam VF, Tine J, Kumar S, Hobart P, Hoffman SL (2000) Improving protective immunity induced by DNA-based immunization: priming with antigen and GM-CSF-encoding plasmid DNA and boosting with antigen-expressing recombinant poxvirus. J Immunol 164:5905–5912

Sharpe M, Lynch D, Topham S, Major D, Wood J, Loudon P (2007) Protection of mice from H5N1 influenza challenge by prophylactic DNA vaccination using particle mediated epidermal delivery. Vaccine 25:6392–6398

Stacey KJ, Sweet MJ, Hume DA (1996) Macrophages ingest and are activated by bacterial DNA. J Immunol 157:2116–2122

Stephenson I, Gust I, Pervikov Y, Kieny MP (2006) Development of vaccines against influenza H5. Lancet Infect Dis 6:458–460

Subbarao K, Luke C (2007) H5N1 viruses and vaccines. PLoS Pathog 3:e40

Subbarao K, Klimov A, Katz J, Regnery H, Lim W, Hall H, Perdue M, Swayne D, Bender C, Huang J, Hemphill M, Rowe T, Shaw M, Xu X, Fukuda K, Cox N (1998) Characterization of an avian influenza A (H5N1) virus isolated from a child with a fatal respiratory illness. Science 279:393–396

Subbarao K, Murphy BR, Fauci AS (2006) Development of effective vaccines against pandemic influenza. Immunity 24:5–9

Tachedjian M, Boyle JS, Lew AM, Horvatic B, Scheerlinck JP, Tennent JM, Andrew ME (2003) Gene gun immunization in a preclinical model is enhanced by B7 targeting. Vaccine 21:2900–2905

Turnes CG, Aleixo JA, Monteiro AV, Dellagostin OA (1999) DNA inoculation with a plasmid vector carrying the faeG adhesin gene of Escherichia coli K88ab induced immune responses in mice and pigs. Vaccine 17:2089–2095

Ugen KE, Nyland SB, Boyer JD, Vidal C, Lera L, Rasheid S, Chattergoon M, Bagarazzi ML, Ciccarelli R, Higgins T, Baine Y, Ginsberg R, Macgregor RR, Weiner DB (1998) DNA vaccination with HIV-1 expressing constructs elicits immune responses in humans. Vaccine 16:1818–1821

Ulmer JB, Donnelly JJ, Parker SE, Rhodes GH, Felgner PL, Dwarki VJ, Gromkowski SH, Deck RR, DeWitt CM, Friedman A et al. (1993) Heterologous protection against influenza by injection of DNA encoding a viral protein. Science 259:1745–1749

Ulmer JB, Deck RR, Dewitt CM, Donnhly JI, Liu MA (1996) Generation of MHC class I-restricted cytotoxic T lymphocytes by expression of a viral protein in muscle cells: antigen presentation by non-muscle cells. Immunology 89:59–67

Ulmer JB, Fu TM, Deck RR, Friedman A, Guan L, DeWitt C, Liu X, Wang S, Liu MA, Donnelly JJ, Caulfield MJ (1998a) Protective CD4+ and CD8+ T cells against influenza virus induced by vaccination with nucleoprotein DNA. J Virol 72:5648–5653

Ulmer JB, Montgomery DL, Tang A, Zhu L, Deck RR, DeWitt C, Denis O, Orme I, Content J, Huygen K (1998b) DNA vaccines against tuberculosis. Novartis Found Symp 217:239–246; discussion 246–253

van Drunen Littel-van den Hurk S, Gerdts V, Loehr BI, Pontarollo R, Rankin R, Uwiera R, Babiuk LA (2000) Recent advances in the use of DNA vaccines for the treatment of diseases of farmed animals. Adv Drug Deliv Rev 43:13–28

Wang Y, Xiang Z, Pasquini S, Ertl HC (1997) Immune response to neonatal genetic immunization. Virology 228:278–284

Watanabe-Fukunaga R, Brannan CI, Copeland NG, Jenkins NA, Nagata S (1992) Lympho-proliferation disorder in mice explained by defects in Fas antigen that mediates apoptosis. Nature 356:314–317

Webster RG, Govorkova EA (2006) H5N1 influenza—continuing evolution and spread. N Engl J Med 355:2174–2177

Wills KN, Atencio IA, Avanzini JB, Neuteboom S, Phelan A, Philopena J, Sutjipto S, Vaillancourt MT, Wen SF, Ralston RO, Johnson DE (2001) Intratumoral spread and increased efficacy of a p53-VP22 fusion protein expressed by a recombinant adenovirus. J Virol 75:8733–8741

Wolff JA, Malone RW, Williams P, Chong W, Acsadi G, Jani A, Felgner PL (1990) Direct gene transfer into mouse muscle in vivo. Science 247:1465–1468

Yi AK, Hornbeck P, Lafrenz DE, Krieg AM (1996) CpG DNA rescue of murine B lymphoma cells from anti-IgM-induced growth arrest and programmed cell death is associated with increased expression of c-myc and bcl-xL. J Immunol 157:4918–4925

Yuen KY, Chan PK, Peiris M, Tsang DN, Que TL, Shortridge KF, Cheung PT, To WK, Ho ET, Sung R, Cheng AF (1998) Clinical features and rapid viral diagnosis of human disease associated with avian influenza A H5N1 virus. Lancet 351:467–471

# Recombinant Proteins Produced in Insect Cells

John Treanor

## Contents

**Abstract** Both purified expressed proteins and virus-like particles generated in insect cells by recombinant baculoviruses are being explored as potential vaccines for seasonal and pandemic influenza. Clinical trials have suggested that recombinant hemagglutinin vaccines are well tolerated in healthy and elderly adults, that they induce a functional antibody response, and that they provide protection against seasonal influenza in adults. In one trial, a pandemic formulation of H5 vaccine (rH5) induced neutralizing antibody in adults at rates roughly similar to that seen with egg-derived subvirion H5N1 vaccine. Preliminary data suggest that vaccination with the rH5 can also prime for booster responses on revaccination with drifted strains of H5. Recombinant approaches may be extremely valuable in combating future pandemics and further studies of recombinant pandemic vaccines in humans are needed.

## 1 Introduction

Most current influenza vaccines for both seasonal and pandemic influenza are generated in embryonated hen's eggs. Virions are harvested from the egg allantoic fluid, chemically inactivated and treated with detergent, and either a whole virion

J. Treanor
Infectious Diseases Division, Department of Medicine, University of Rochester
Medical Center, 601 Elmwood Avenue, Rochester, NY 14642, USA
e-mail: John_Treanor@urmc.rochester.edu

R.W. Compans and W.A. Orenstein (eds.), *Vaccines for Pandemic Influenza*,
Current Topics in Microbiology and Immunology 333,
DOI 10.1007/978-3-540-92165-3_11, © Springer-Verlag Berlin Heidelberg 2009

preparation is generated or the hemagglutinin (HA) and neuraminidase (NA) proteins are partially purified to produce split-product, subvirion, or subunit vaccines (Wood 1998). Although this system has served us well for over 50 years, there are several well-recognized disadvantages to the use of eggs as the substrate for vaccine production. Some of these include the potential vulnerability of the supply of embryonated eggs, especially in the context of an evolving avian pandemic, and the need to generate high-yielding reassortants of new antigenic variants. In addition, growth in eggs has been reported to result in selection of receptor variants that do not provide optimal protection against circulating strains (Katz and Webster 1989; Wood et al. 1989). Thus, there has been significant interest in other production techniques for both seasonal and pandemic vaccines.

The use of recombinant DNA techniques to generate vaccine antigens expressed in cell culture is an alternative that avoids the dependence on egg supply. One well-characterized system is the use of recombinant baculovirus to express foreign proteins in insect cells. An advantage of insect cells as an expression system is that these eukaryotic cells use the same N-glycosylation sites as mammalian cells, although they do not add terminal sialic acid or galactose residues (Kost et al. 2005). Baculovirus-derived proteins have been used in the production of several pharmaceutical products, most notably including a safe and highly effective vaccine for human papillomavirus (HPV) (Harper et al. 2004, 2006).

Baculovirus expression of influenza proteins in insect cells provides several advantages for the production of influenza vaccines for both seasonal and pandemic use. Because the process of gene cloning is more rapid than the process of adapting new variants for growth in eggs, selection of appropriate antigenic variants can be delayed until more epidemiologic information becomes available. The recombinant proteins are highly purified and may be less reactogenic, and the vaccine does not contain egg protein, which may produce hypersensitivity reactions in a small number of individuals. Because of the efficiency of the expression system, higher dose vaccines can be effectively produced, which may be associated with more frequent or more vigorous immune responses. The use of recombinant techniques also allows control of the sequence of the final product, potentially obviating concerns regarding selection of incorrect receptor phenotypes. Finally, because the use of the recombinant approach does not require handling the live virus, this approach also avoids biocontainment and biohazard issues that might be particularly relevant to pandemic vaccines.

## 2   Preclinical Studies

Two general approaches to generating influenza vaccine in insect cells have been described. One approach is to simply express the HA antigen and/or NA antigen and purify the product to make a subunit vaccine. For use as a vaccine in humans, the HA and/or NA are generated as full-length, nonsecreted proteins and do not contain antibody tags for purification purposes. The hemagglutinin proteins expressed under these conditions form trimeric structures under electron microscopy and are not cleaved in insect cells in the absence of exogenously added proteases (with the exception of HAs containing the highly cleavable sequence of basic amino acids at the cleavage site). Therefore, they are sometimes referred to as rHA0. Since the cleavage site is not involved in the immune response, there should be no significant difference between the immune response to cleaved or uncleaved HA. The proteins are typically further purified by a combination of anion-exchange and lectin chromatography (Wang et al. 2006).

Studies with rHA in chickens have demonstrated protection against lethal infection with highly pathogenic avian influenza viruses (HPAIV) of both the H5 (Swayne et al. 2000) and H7 (Crawford et al. 1999) subtypes. Recombinant HA has also been shown to be highly protective against influenza in murine models (Johansson 1999; Brett and Johansson 2005), including protection against lethal infection with H5N1 influenza (Katz et al. 2000). Intranasal administration of rHA formulated with *N. meningitidis* outer membrane proteins induces mucosal antibodies and also provides protection against challenge (Jones et al. 2003). Baculovirus-expressed recombinant NA is also protective in animal models (Kilbourne et al. 2004), and studies in mice have suggested that inclusion of NA in the vaccine increases the level of protection, particularly against antigenically drifted strains within the same subtype (Brett and Johansson 2005). Recombinant M2 protein expressed in insect cells has also been evaluated in the mouse model and shown to provide both homo-subtypic and heterosubtypic protection (Slepushkin et al. 1995).

A second approach is to coexpress multiple influenza proteins in insect cells, leading to the generation of a virus-like particle (VLP). Coexpression of the influenza HA and M proteins (Quan et al. 2007) or of the HA, NA, and M proteins (Pushko et al. 2005) results in VLP formation. Influenza VLPs generated in insect cells are morphologically similar to authentic influenza virions (Fig. 1), with a predominantly spherical rather than filamentous morphology (Pushko et al. 2005; Quan et al. 2007), and could be considered to be the equivalent of whole-virion vaccines, as compared to the subunit vaccines generated by expression of individual proteins. In mouse models, insect cell-derived VLPs are immunogenic, and provide protection against both seasonal influenza viruses (Quan et al. 2007) as well as H9N2 influenza virus (Pushko et al. 2005). In some previous studies of pandemic vaccination in man, whole-virion vaccines have been more immunogenic than subvirion vaccines, and for similar reasons it is possible that recombinant VLP vaccines would be more immunogenic than rHA vaccines in naïve hosts. In one study in mice, an H3N2 VLP vaccine appeared to be more immunogenic than the corresponding H3 rHA vaccine (Bright et al. 2007). Immunogenicity of a candidate H9N2 VLP in ferrets was substantially enhanced by formulation with a liposomal adjuvant (Pushko et al. 2007). Studies with VLP vaccines are discussed in more detail in a subsequent chapter in this volume (Kang et al. 2008).

**A / Brisbane / 10 / 07**     **A / Brisbane / 59 / 07**     **B / Florida / 4 / 06**

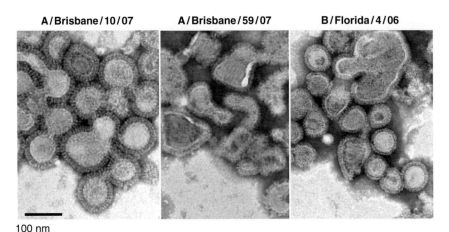

100 nm

**Fig. 1** Negative staining electron microcopy of virus-like particles (VLPs) from the 2008–2009 influenza strains recommended for the Northern Hemisphere. The hemagglutinin (HA) and neu-raminidase (NA) genes were cloned from non-egg-adapted influenza virus strains A/Brisbane/10/07 (H3N2), A/Brisbane/59/07 (H1N1), and B/Florida/4/06. HA and NA genes from each strain were cloned along with an influenza matrix protein (M1) in tandem into baculovirus expression vectors. Recombinant VLPs were secreted from infected Sf9 insect cells then purified by a tangent flow filtration–chromatography process which removes baculovirus and other host cell contaminants. Purified VLPs were adsorbed on freshly discharged plastic-carbon coated grids and stained with 2% sodium phosphotungstate, pH 6.5. Stained VLPs were observed by transmission electron microscopy at magnifications from ×6,000 to ×120,000. The *bar* represents 100 nm. VLPs were produced and the figure was provided by Novavax, Inc., Rockville, MD

## 3  Clinical Studies: Seasonal Influenza

Currently available, published clinical data for influenza vaccines generated in insect cells are restricted to results obtained using subunit vaccines. Four clinical studies of monovalent or bivalent baculovirus-generated recombinant HA (HA0) for seasonal influenza have been reported in healthy adults (summarized in Treanor et al. 1996). In the first three of these studies, the recombinant HA vaccines used contained full-length HA0 from the influenza A/Beijing/32/92 (H3) and A/Texas/31/91 (H1) viruses. In the first study, adults aged 18–45 were randomly assigned to receive A/Beijing/92 (H3) rHA0 vaccine at doses of 15 µg with alum, 15, 45, or 90 µg without alum, trivalent subvirion vaccine, or placebo (sterile saline) (Powers et al. 1995). In the second study, an additional cohort of adults 18–45 years old were randomly assigned to receive A/Beijing/92 rHA0 at 15, 45, or 135 µg, subvirion vaccine, or placebo (Treanor et al. 1996). In the third study, adults aged 18–45 were randomly assigned to receive A/Beijing/92 rHA0 45 µg, A/Texas/91 (H1) rHA0 15, 45, or 135 µg, bivalent H1 + H3 vaccine at 45 µg each, or trivalent subvirion vaccine (Lakey et al. 1996).

All of these doses of rHA0 vaccines were well tolerated, and there were no significant adverse events in any group. Administration of rHA0 vaccine stimulated both functional (neutralizing and HAI) and binding (IgG ELISA) HA-specific serum antibody in young adults. The results of HAI antibody assays in young adults pooled across all three studies are summarized in Fig. 2. Taken as a whole, there was a significant dose–response effect for both rHA0s, which was more pronounced for the H3 vaccine. For both H1 and H3 vaccines, the responses to doses of 15–45 µg rHA0 were similar to the responses to licensed subvirion inactivated vaccine. Antibody responses to doses of 90 or 135 µg of H3 vaccine occurred with greater frequency and resulted in higher titers of antibody than did subvirion vaccine. Antibody responses to both the H1 and H3 vaccine were similar when the vaccines were administered as a bivalent vaccine containing 45 µg of each antigen to when the vaccines were administered separately at doses of 45 µg each. Similar results were seen for neutralizing and HA-specific IgG ELISA antibody. Antibody responses were not significantly enhanced by the addition of alum, similar to previous studies using conventionally generated subvirion influenza vaccines.

In one of these studies of monovalent H3 rHA0 vaccine in healthy adults, subjects were followed during the subsequent influenza season to ascertain influenza-like illnesses (Powers et al. 1995). Four individuals (three placebo recipients and one recipient of 15 µg of rHA0 vaccine) had influenza-like illness associated with the

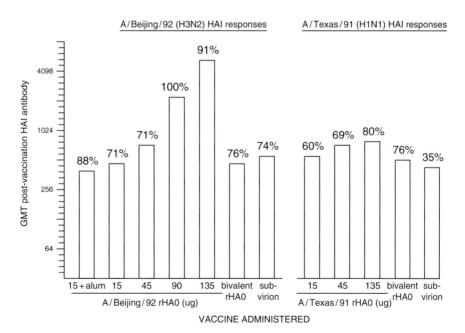

**Fig. 2** Pooled results of studies of monovalent and bivalent rHA0 vaccines in healthy adult volunteers, showing the GMT of postvaccination serum HAI antibody and the percent of subjects in each group manifesting a fourfold or greater increase in antibody titer between prevaccination and postvaccination sera

isolation of H3 influenza virus resembling the A/Beijing/92 virus from nasal secretions. The combined rate of laboratory-documented influenza in this study was therefore 1/77 in recipients of rHA0 vaccine at any dose (26 received 15 µg, 25 received 15 µg plus alum, and 26 received 90 µg) compared to 3/24 in placebo recipients ($P < 0.05$, Fisher's exact test).

In a preliminary evaluation in the elderly, adults 65 years of age or older were randomly assigned to receive monovalent A/Beijing/92 (H3) rHA0 at 15, 45, or 135 µg, subvirion vaccine, or placebo (Treanor et al. 1996). Both the frequency and level of serum antibody responses in elderly subjects given doses of 45–135 µg of H3 rHA0 were comparable to or in excess of those seen in this age group with licensed subvirion vaccine. The geometric mean serum HAI titer in elderly subjects after 135 µg of H3 rHA0 was 630 (76% responding), compared to 315 (59% responding) following subvirion vaccine, but these differences were not statistically significant in this small study.

This study was followed by a larger phase II study in which 399 healthy adults aged 65 and over (mean age: 72 years) were randomly assigned to receive a trivalent rHA0 preparation containing the rHAs of A/Panama/2007/99 (H3); A/New Caledonia/20/99 (H1); and B/Hong Kong/330/2001 at doses of 15, 45, or 135 µg per rHA0, or licensed subvirion vaccine (Treanor et al. 2006b). The rHA0 vaccine was highly immunogenic at both the 135 µg/HA and the 45 µg/HA doses in these elderly subjects. In this study, the predefined primary efficacy endpoint was the proportion of subjects achieving a postvaccination serum HI titer of 1:128 or greater against the H3 component of the vaccine. This endpoint was chosen because H3 influenza viruses have been repeatedly shown to cause the bulk of the increased hospitalizations and deaths seen in the elderly, and because some studies have suggested that the levels of antibody required to protect elderly individuals may be somewhat higher than the traditional 1:40 titers often used to predict protective efficacy in young adults (Arden et al. 1986; Betts et al. 1993). The proportion of subjects achieving this endpoint against the H3 component of the vaccine were 62%, 76%, and 88% in those receiving rHA0 at 15, 45, or 135 µg, while the rate in those receiving TIV was 33% (Treanor et al. 2006b).

In addition, recipients of rHA at either 135 or 45 µg/HA had significantly higher postvaccination GMT against the H3 component of the vaccine than did recipients of licensed subvirion vaccine. There were no significant differences in the postvaccination GMT of antibody to either the H1 or the B component of the vaccine when comparing either the 135 or the 45 µg/HA dose with subvirion vaccine.

Based on these results, a subsequent small field trial evaluation was undertaken in healthy adults 18–49 years to refine estimates of the dose–response to trivalent vaccine, and to obtain more definitive evidence of protective efficacy (Treanor et al. 2007). A total of 460 subjects were enrolled in this study and randomized to receive either 75 µg rHA0 (containing 45 µg of the H3 component and 15 µg each of the H1 and B components), 135 µg rHA0 (containing 45 mcg of each component), or placebo. Injection of rHA0 vaccine was associated with local injection site pain that was significantly more frequent than after saline placebo and that was dose dependent. However, 97% of all complaints of pain after rHA0 vaccine were rated as mild.

The frequencies of HAI antibody responses to vaccination ranged from 51% to 92%. The frequency of responses to both the A/New Caledonia/99 (H1) and B/Jiangsu/03 viruses were significantly higher in the group receiving the 135 μg dose, consistent with the higher dose of H1 and B components in the 135 μg vaccine, while (as expected) the frequency of HAI antibody responses to the A/Wyoming/03 (H3N2) virus was not different between the two doses. Similarly, there were significant differences in the day 28 geometric mean titer (GMT) of HAI antibody between the 75- and 135-μg doses for both the H1 and B components, but not the H3 component. Based on this analysis, a trivalent formulation containing 45 μg of each component has been selected as the lead candidate for licensure trials, which are in progress.

Subjects in this study were followed throughout the subsequent influenza season with weekly phone calls and instructed to return to the study clinics for any acute respiratory illness, at which time a nasopharyngeal swab for viral culture was obtained. A total of 116 such cultures were obtained, 43 in the placebo group, 39 in the 75 μg group, and 34 in the 135 μg group. The primary efficacy endpoint for this study was the development of culture-confirmed influenza illness meeting the influenza-like illness case definition of the Centers for Disease Control (CDC-ILI), i.e., the presence of fever >99.8°F and either sore throat or cough, or both.

There were a total of 13 positive cultures for influenza in the study population, of which ten were influenza A (all of which were confirmed as influenza H3) and three were influenza B. Of these 13 cases, nine (69%) occurred in individuals meeting the CDC influenza illness case definition. The rates of culture-positive influenza illness, the prespecified primary efficacy endpoint, were 7/153 (4.6%) in placebo recipients, 2/150 (1.4%) in recipients of the 75 μg dose, and 0/151 in recipients of the 135 μg dose (Table 1). When considering both vaccine groups combined, the cumulative incidence of culture-positive CDC-ILI was reduced by 86%. For comparison, in a recently reported study conducted in the same influenza season, the efficacy of TIV in healthy adults against culture-confirmed influenza meeting a similar case definition was 77% (95% CI 37%, 92%) (Ohmit et al. 2006).

The rHA0 vaccine evaluated in this study also had a positive effect on a number of interrelated secondary endpoints, including laboratory-confirmed influenza (CDC-ILI with either a positive culture or a serologic response), the total number of respiratory illnesses associated with a positive culture (any positive culture), the number of subjects with any evidence of influenza infection, and the overall rate of respiratory and influenza-like illness regardless of culture results (Table 1).

The majority of cases in this study were due to influenza A, and all of the influenza A viruses isolated in this study that were further subtyped were of the H3N2 subtype, consistent with the report that 98.5% of all influenza A viruses typed in the USA during the 2004–2005 season were H3N2 viruses (CDC 2005). In addition, all of the influenza A H3N2 viruses isolated from subjects in this trial were genetically similar to A/California/7/2004, a significant antigenic variant which reacts poorly with antisera from persons who received the 2004–2005 formulation of TIV (Anonymous 2005), as were 75% of H3N2 isolates in the 2004–2005 season (CDC 2005). These results suggest that it is possible to generate a substantial amount of

**Table 1** Rates of study endpoints and the percent protective efficacy (95% confidence interval) upon comparing vaccine to placebo recipients

| Study outcome | No. cases (%) in those receiving | | | Percent protective efficacy (95% CI) in those receiving | | |
|---|---|---|---|---|---|---|
| | Placebo (n = 153) | 75 mcg (n = 150) | 135 mcg (n = 151) | 75 mcg | 135 mcg | Any rHA0 |
| Culture-documented influenza[a] | 7 (4.6) | 2 (1.3) | 0 | 70.9 (−38, 94) | 100 | 85.5 (31, 97) |
| Laboratory-documented influenza[b] | 8 (5.2) | 2 (1.3) | 1 (0.7) | 74.5 (−18, 95) | 87.3 (0, 98) | 80.1 (29, 95) |
| Any positive culture | 8 (5.2) | 4 (2.7) | 1 (0.7) | 49.0 (−66, 84) | 87.3 (0, 98) | 68.2 (5, 89) |
| Influenza infection[c] | 29 (19) | 10 (7) | 17 (11) | 65 (30, 82) | 41 (−3, 66) | 53 (23, 71) |
| Influenza-like illness[d] | 20 (13.1) | 14 (9.3) | 9 (6.0) | 28.6 (−36, 63) | 54.4 (3, 79) | 41.5 (−3, 67) |
| Any respiratory illness | 43 (28.1) | 39 (26.0) | 34 (22.5) | | | |

[a] Illness meeting the CDC-ILI case definition (fever >38°C plus either cough or sore throat) with a positive NP culture for influenza
[b] Any individual with CDC-ILI plus either positive culture or serologic evidence of infection
[c] Any individual with either positive culture or serologic evidence of infection, regardless of illness history
[d] Cases meeting the CDC-ILI definition regardless of evidence of influenza infection

protection in an immunologically primed population against influenza with a pure hemagglutinin vaccine, even in the presence of antigenic drift.

## 4 Clinical Studies: Pandemic Formulations

As described above, recombinant proteins are an excellent option for the development of pandemic vaccines because of the rapidity with which new vaccines could potentially be generated as well as the relatively less stringent biocontainment issues associated with the use of a recombinant protein approach. The first known emergence of H5N1 viruses in man took place in late 1997 with human infections with influenza A/Hong Kong/97 (H5N1) virus in a small group of individuals living on the island of Hong Kong (Yuen et al. 1998). Two main subgroups of H5 viruses were identified among the cases, as represented by the prototype A/Hong Kong/156 (group A) and A/Hong Kong/483 (group B) viruses (Bender et al. 1999). Subsequently, these viruses have been characterized as belonging to clade 3 among the H5 avian influenza viruses.

Among the earliest attempts to develop and test human vaccines for H5 influenza viruses was the use of recombinant A/Hong Kong/97 virus expressed in insect cells, one of the few approaches available at the time that could feasibly deal with the highly pathogenic nature of H5 viruses (Fig. 3). The study was conducted in healthy adults aged 18–40 who did not have agricultural or occupational exposures to H5 viruses, and was designed to determine the effects of both dose and schedule on the serum antibody response. All subjects received a two-dose schedule, but subjects were randomly assigned to one of three intervals between doses (three, four, and six weeks), and within each interval were randomly assigned to receive either two doses of placebo, two doses of 25 µg rH5, two doses of 45 µg rH5, two doses of 90 µg rH5, or one dose of 90 µg followed by one dose of 10 µg of rH5, in double-blind fashion (Treanor et al. 2001). There were thus a total of 15 groups (five doses × three intervals), with approximately ten subjects per group.

Vaccine was well tolerated, and there were no serious adverse events. Increases in neutralizing antibody were seen after the first dose of vaccine, but were restricted to individuals who received 90 µg (Fig. 4). Administration of a second dose of vaccine tended to result in further increases in antibody titer, especially in those who initially received 90 µg. However, the mean neutralizing antibody titer met or exceeded 1:80 only in those who received two doses of 90 µg, and then only at two weeks after the second dose of vaccine. In retrospect, the results of this study were remarkably similar to those seen subsequently in healthy adults who received egg-derived, subvirion A/Vietnam/1203/04 vaccine (Treanor et al. 2006a).

The rate and magnitude of the serum antibody response was dose related, with the response depending on the total dose of vaccine received (Fig. 4a). That is, greater responses were seen in those who received two doses of 90 µg (total 180 µg) than in those who received one dose of 90 µg followed by one dose of 10 µg (total 100 µg), which in turn were greater than seen in those receiving two

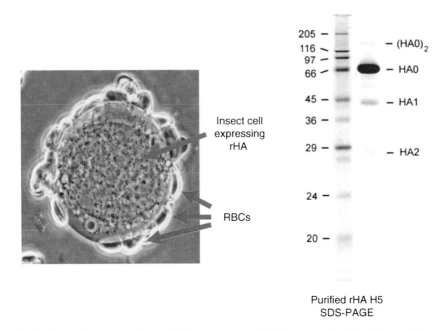

Purified rHA H5
SDS-PAGE

**Fig. 3** Expression of recombinant H5 hemagglutinin (HA) in insect cells. The *left panel* shows the expression of functional H5 HA in insect cells, which are agglutinating chick red blood cells (RBC). The *right panel* shows a Coomassie-stained protein gel of the expressed protein. Because the H5 HA contains basic amino acids in the cleavage site associated with cleavage by ubiquitous proteases, the HA is partially cleaved into HA1 and HA2. Conventional HAs are not cleaved in insect cells. Both types of HAs are produced as fully functional trimers in the cells

doses of 45 μg (total 90 μg). In contrast, there was relatively little variation in the response rate depending on the interval between doses within the relatively narrow range of dosing schedules employed in this study (Fig. 4b). Overall, rates of neutralizing response were relatively low, even at the highest doses of vaccine employed, with slightly more than half of the subjects responding to two doses of 90 μg of vaccine. Sera were also tested by ELISA for antibody against the 483 virus HA, and the responses were qualitatively similar to the response to the 156 virus.

Approximately eight years after this priming vaccination, a subset of subjects in this study were vaccinated in open label fashion with a single 90 μg dose of the egg-derived subvirion A/Vietnam/1203/04 (clade 1) vaccine in order to determine whether the rH5 (clade 3) vaccine had resulted in significant priming for the antigenically variant clade 1 vaccine (Goji et al. 2008). Sera for assessing hemagglutination-inhibition (HAI) and neutralizing (MN) antibody responses were obtained prior to vaccination and on days 28 and 56 following vaccination, and the results were compared to those in naïve subjects receiving two 90 μg doses of the A/Vietnam/1203/04 vaccine (Treanor et al. 2006a). In total, 37 subjects who had previously received the baculovirus-derived A/Hong Kong/156/97 vaccine received a single dose of 90 mcg of A/Vietnam/1203/04. Vaccination of primed subjects was not associated

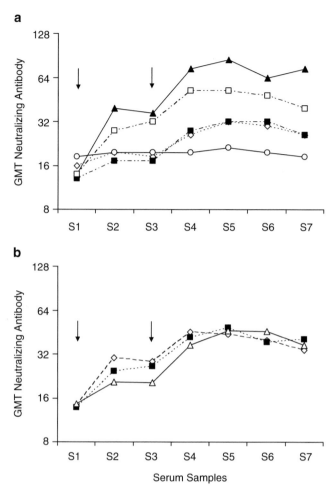

**Fig. 4a–b** Effect of dose and interval on the antibody response to rH5 in humans (Treanor et al. 2001). Subjects were randomized by dose and interval into 15 groups of approximately ten subjects each. To evaluate the effect of dose (**a**), subjects in each interval were pooled into five groups of approximately 30 subjects each. Placebo × 2 (*open circle*), 25 µg × 2 (*diamond*), 45 × 2 (*filled square*), 90 µg/10 µg (*open square*), 90 µg × 2 (*filled triangle*). To evaluate the effect of interval (**b**), the dose groups (other than placebo) were pooled into three groups of approximately 40 subjects each. 21 days (*diamond*), 28 days (*filled square*), 42 days (*filled triangle*)

with increased local or systemic reactogenicity when compared to naïve subjects. As expected, the primed subjects did not have detectable antibody to either the A/Vietnam/1203/04 virus or to the A/Hong Kong/156/97 virus at the beginning of the study. However, a single 90 mcg dose resulted in serum HAI and MN antibody titers that were significantly higher than seen after one dose in vaccine naïve subjects, and actually higher than seen after two doses in these subjects (Fig. 5). Primed subjects had more frequent responses to a single dose of subvirion vaccine than did naïve

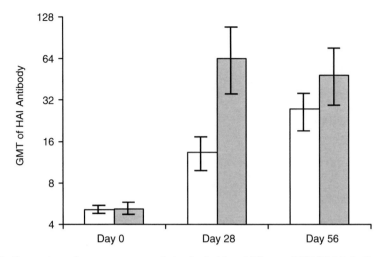

**Fig. 5** Comparison of responses to egg-derived subvirion A/Vietnam/1203/04 (clade 1) vaccine in naïve subjects receiving two 90 µg doses on days 0 and 28 (*white bars*), and in subjects previously primed with rH5 A/Hong Kong/97 (clade 3) vaccine and receiving a single 90 µg dose of subvirion A/Vietnam/1203/04 vaccine (*gray bars*). *Error bars* show 95% confidence intervals for titer. Data from Goji et al. (2008)

subjects, and were more likely to achieve HAI and MN titers of 1:40 or greater at day 28 than did naïve subjects after either one or two doses of vaccine. Although the numbers of subjects are small, both the frequency of HAI and MN responses as well as the GMT achieved at day 28 were higher in those who had responded to the vaccine in 1998 than in those who did not. However, there was not a clear relationship between the dose of vaccine received in 1998 and the ability to respond in 2006, or in the interval between doses in 1998 and the response in 2006.

In order to characterize the effect of prior vaccination on the breadth of the antibody response, sera from the primed subjects as well as titer- and age-matched H5 naïve subjects were retested in HAI assays against the wild-type A/Vietnam/1203/04 (Clade 1), A/Indonesia/05/05 (Clade 2), and A/Hong Kong/156/97 (Clade 3) viruses. Because of the matching process, titers against the homologous virus were similar in the two groups. However, subjects primed with A/Hong Kong/156/97 receiving a single dose of A/Vietnam/1203/04 had slightly higher titers and were more likely to have a titer of ≥1:40 against both A/Indonesia/05/05 and A/Hong Kong/156/97 viruses than were H5 vaccine-naïve subjects.

## 5  Summary and Conclusions

The use of recombinant influenza proteins (such as HA) expressed in insect cells meets several of the desirable properties of a pandemic influenza vaccine. Because live influenza virus is not used during the production of the vaccine, elaborate biocontainment procedures are not needed in vaccine manufacture. In addition, the use of insect cells has the same types of advantages as other cell culture systems used for vaccine production, and does not rely on a supply of embryonated eggs during an emerging pandemic. In addition, global bioreactor capacity could theoretically be harnessed to quickly scale up production if needed (Fedson and Dunnill 2007). Very importantly in the case of rHA vaccine, there is already substantial relevant safety and immunogenicity data in humans with seasonal preparations, including convincing evidence of protective efficacy. However, immune responses to the pandemic H5 formulation were relatively poor, similar to the findings with conventional egg-derived subvirion H5 vaccines in humans. It is however likely that immunogenicity of these vaccines could be substantially improved by the same sorts of adjuvants that have shown substantial dose sparing and improved immunogenicity for conventional subvirion pandemic candidates (Atmar et al. 2006; Leroux-Roels et al. 2007).

# References

Anonymous (2005) Recommended composition of influenza virus vaccines for use in the 2005–2006 influenza season. Wkly Epidemiol Rec 80:71–75

Arden NH, Patriarca PA, Kendal AP. (1986) Experiences in the use and efficacy of inactivated influenza vaccine in nursing homes. Options for the control of influenza. Alan R Liss, Keystone, CO

Atmar RL, Keitel WA, Patel SM, Katz JM, She D, El Sahly H, Pompey J, Cate TR, Couch RB. (2006) Safety and immunogenicity of nonadjuvanted and MF59-adjuvanted influenza A/H9N2 vaccine preparations. Clin Infect Dis 43:1135–1142

Bender C, Hall H, Huang J, Klimov A, Cox N, Hay A, Gregory V, Cameron K, Lim W, Subbarao K. (1999) Characterization of the surface proteins of influenza A (H5N1) viruses isolated from humans in 1997–98. Virology 254:115–123

Betts RF, O'Brien D, Menegus M, Falsey AR, Kouides RW, Yuen JB, Barker WH. (1993) A comparison of the protective benefit of influenza (FLU) vaccine in reducing hospitalization of patients infected with FLU A or FLU B. Clin Infect Dis 17(3):573 (A257)

Brett EC, Johansson BE. (2005) Immunization against influenza A virus: comparison of conventional inactivated, live-attenuated and recombinant baculovirus produced purified hemagglutinin and neuraminidase vaccines in a murine model system. Virology 339:273–280

Bright RA, Carter DM, Daniluk S, Toapanta FR, Ahmad A, Gavrilov V, Massare M, Pushko P, Mytle N, Rowe T, Smith G, Ross TM. (2007) Influenza virus-like particles elicit broader immune responses than whole virion inactivated influenza virus or recombinant hemagglutinin. Vaccine 25(19):3871–3878

CDC. (2005) Update: influenza activity—United States and worldwide, 2004–05 season. MMWR 54:631–634

Crawford J, Wilkinson B, Vosnesensky A, Smith G, Garcia M, Stone H, Perdue ML. (1999) Baculovirus-derived hemagglutinin vaccines protect against lethal influenza infections by avian H5 and H7 subtypes. Vaccine 17(18):2265–2274

Fedson DS, Dunnill P. (2007) From scarcity to abundance: pandemic vaccines and other agents for "have not" countries. J Public Health Policy 28:322–340

Goji NA, Nolan C, Hill H, Wolf M, Rowe T, Treanor JJ. (2008) Immune responses of healthy subjects to a single dose of intramuscular inactivated influenza A/Vietman/1203/04 (H5N1 vaccine after priming with an antigenic variant. J Infect Dis 198(5):635–641

Harper DM, Franco EL, Wheeler CM, Ferris DG, Jenkins D, Schuind A, Zahaf T, Innis BL, Naud P, De Carvalho NS, Rotelli-Martins CM, Teixeira J, Blatter MM, Korn AP, Quint W, Dubin G. (2004)

Efficacy of a bivalent L1 virus-like particle vaccine in prevention of infection with human papillomavirus types 16 and 18 in young women: a randomised controlled trial. Lancet 364:1757–1765

Harper DM, Franco EL, Wheeler CM, Moscicki A-B, Romanowski B, Roteli-Martins CM, Jenkins D, Schuind A, Clemens SAC, Dubin G. (2006) Sustained efficacy up to 4.5 years of a bivalent L1 virus-like particle vaccine against human papillomavirus types 16 and 18: follow-up from a randomized control trial. Lancet 367:1247–1255

Johansson BE. (1999) Immunization with influenza A virus hemagglutinin and neuraminidase produced in recombinant baculovirus results in a balanced and broadened immune response superior to conventional vaccine. Vaccine 17(15–16):2073–2080

Jones T, Allard F, Cyr SL, Tran SP, Plante M, Gauthier J, Bellerose N, Lowell GH, Burt DS. (2003) A nasal proteosome influenza vaccine containing baculovirus-derived hemagglutinin induces protective mucosal and systemic immunity. Vaccine 21(25–26):3706–3712

Katz JM, Webster RG. (1989) Efficacy of inactivated influenza A virus (H3N2) vaccines grown in mammalian cells or embryonated eggs. J Infect Dis 160(2):191–198

Katz JM, Lu X, Frace AM, Morken T, Zaki SR, Tumpey TM. (2000) Pathogenesis of and immunity to avian influenza A H5 viruses. Biomed Pharmacother 54(4):178–187

Kilbourne ED, Pokorny BA, Johansson B, Brett I, Milev Y, Matthews JT. (2004) Protection of mice with recombinant influenza virus neuraminidase. J Infect Dis 189(3):459–461

Kost TA, Condreay JP, Jarvis DL. (2005) Baculovirus as versatile vectors for protein expression in insect and mammalian cells. Nat Biotechnol 32:567–575

Lakey DL, Treanor JJ, Betts RF, Smith GE, Thompson J, Sannella E, Reed G, Wilkinson BE, Wright PF. (1996) Recombinant baculovirus influenza A hemagglutinin vaccines are well tolerated and immunogenic in healthy adults. J Infect Dis 174:838–841

Leroux-Roels I, Borkowski A, Vanwolleghem T, Drame M, Clement F, Hons E, Devaster J-M, Leroux-Roels G. (2007) Antigen sparing and cross-reactive immunity with an adjuvanted rH5N1 prototype pandemic influenza vaccine: a randomised controlled trial. Lancet 370:580–589

Ohmit SE, Victor JC, Rotthoff JR, Teich ER, Truscon RK, Baum LL, Rangarajan B, Newton DW, Boulton ML, Monto AS. (2006) Prevention of antigenically drifted influenza by inactivated and live attenuated vaccines. New Engl J Med 355(24):2513–2522

Powers DC, Smith GE, Anderson EL, Kennedy DJ, Hackett CS, Wilkinson BE, Volvovitz F, Belshe RB, Treanor JJ. (1995) Influenza A virus vaccines containing purified recombinant H3 hemagglutinin are well-tolerated and induce protective immune responses in healthy adults. J Infect Dis 171(6):1595–1598

Pushko P, Tumpey TM, Bu F, Knell J, Robinson R, Smith G. (2005) Influenza virus-like particles comprised of the HA, NA, and M1 proteins of H9N2 influenza virus induce protective immune responses in BALB/c mice. Vaccine 23(50):5751–5759

Pushko P, Tumpey TM, Van Hoeven N, Belser JA, Robinson R, Nathan M, Smith G, Wright DC, Bright RA. (2007) Evaluation of influenza virus-like particles and Novasome adjuvant as candidate vaccine for avian influenza. Vaccine 25(21):4283–4290

Quan FS, Huang C, Compans RW, Kang SM. (2007) Virus-like particle vaccine induces protective immunity against homologous and heterologous strains of influenza virus. J Virol 81(7):3514–3524

Slepushkin VA, Katz JM, Black RA, Gamble WC, Rota PA, Cox NJ. (1995) Protection of mice against influenza A virus challenge by vaccination with baculovirus-expressed M2 protein. Vaccine 13(15):1399–1402

Swayne DE, Perdue ML, Beck JR, Garcia M, Suarez DL. (2000) Vaccines protect chickens against H5 highly pathogenic avian influenza in the face of genetic changes in field viruses over multiple years. Vet Microbiol 74(1–2):165–172

Treanor J, Betts R, Powers D, Belshe R, Anderson E, Lakey D, Wright P, Wilkinson B, Smith G. (1996) Evaluation of recombinant influenza virus hemagglutinins (HAs) expressed in insect cells as influenza vaccines in humans. In: Brown LE, Hampson AW, Webster RG. (eds.) Options for the control of influenza III. Elsevier, Amsterdam, pp 677–682

Treanor JJ, Wilkinson BE, Masseoud F, Hu-Primmer J, Battaglia R, O'Brien D, Wolff M, Rabinovich G, Blackwelder W, Katz JM. (2001) Safety and immunogenicity of a recombinant hemagglutinin vaccine for H5 influenza in humans. Vaccine 19:1732–1737

Treanor JJ, Campbell JD, Zangwill KM, Rowe T, Wolff M. (2006a) Safety and immunogenicity of an inactivated subvirion influenza A (H5N1) vaccine. New Engl J Med 354:1343–1351

Treanor JJ, Schiff GM, Couch RB, Cate TR, Brady RC, Hay CM, Wolff M, She D, Cox MMJ. (2006b) Dose-related safety and immunogenicity of a trivalent baculovirus-expressed influenza-virus hemagglutinin vaccine in elderly adults. J Infect Dis 193(9):1223–1228

Treanor JJ, Schiff GM, Hayden FG, Brady RC, Hay CM, Meyer AL, Holden-Wiltse J, Liang H, Gilbert A, Cox M. (2007) Safety and immunogenicity of a baculovirus-expressed hemagglutinin influenza vaccine: A randomized controlled trial. JAMA 297:1577–1582

Wang K, Holtz KM, Anderson K, Chubet R, Mahmoud W, Cox MMJ. (2006) Expression and purification of an influenza hemagglutinin – one step closer to a recombinant protein-based influenza vaccine. Vaccine 24(12):2176–2185

Wood JM. (1998) Standardization of inactivated influenza vaccine. In: Nicholson KG, Webster RG, Hay AJ (eds) Textbook of influenza. Blackwell, London, pp 333–345

Wood JM, Oxford JS, Dunleavy U, Newman RW, Major D, Robertson JS. (1989) Influenza A (H1N1) vaccine efficacy in animal models is influenced by two amino acid substitutions in the hemagglutinin molecule. Virology 171:214–221

Yuen KY, Chan PKS, Peiris M, Tsang DNC, Que TL, Shortridge KF, Cheung PT, To WK, Ho ETF, Sung R, Cheng AFB. (1998) Clinical features and rapid viral diagnosis of human disease associated with avian influenza A H5N1 virus. Lancet 351:467–471

# Influenza Neuraminidase as a Vaccine Antigen

Matthew J. Sylte and David L. Suarez

## Contents

**Abstract**   The neuraminidase protein of influenza viruses is a surface glycoprotein that shows enzymatic activity to remove sialic acid, the viral receptor, from both viral and host proteins. The removal of sialic acid from viral proteins plays a key role in the release of the virus from the cell by preventing the aggregation of the virus by the hemagglutinin protein binding to other viral proteins. Antibodies to the neuraminidase protein can be protective alone in animal challenge studies, but the neuraminidase antibodies appear to provide protection in a different manner than antibodies to the hemagglutinin protein. Neutralizing antibodies to the hemagglutinin protein can directly block virus entry, but protective antibodies to the neuraminidase protein are thought to primarily aggregate virus on the cell surface, effectively reducing the amount of virus released from infected cells. The neuraminidase protein can be divided into nine distinct antigenic subtypes, where there is little cross-protection of antibodies between subtypes. All nine subtypes of neuraminidase protein are commonly found in avian influenza viruses, but only selected subtypes are routinely found in mammalian influenza viruses; for example, only the N1 and N2 subtypes are commonly found in both humans and swine. Even within a subtype, the neuraminidase protein can have a high level of antigenic drift, and vaccination

M.J. Sylte
Department of Infectious Diseases, College of Veterinary Medicine,
University of Georgia, Athens, GA 30602, USA

D.L. Suarez (✉)
Southeast Poultry Research Laboratory, Agricultural Research Service,
US Dept. of Agriculture, 934 College Station Road, Athens, GA 30605, USA
e-mail: David.Suarez@ars.usda.gov

R.W. Compans and W.A. Orenstein (eds.), *Vaccines for Pandemic Influenza*,       227
Current Topics in Microbiology and Immunology 333,
DOI 10.1007/978-3-540-92165-3_12, © Springer-Verlag Berlin Heidelberg 2009

has to specifically be targeted to the circulating strain to give optimal protection. The levels of neuraminidase antibody also appear to be critical for protection, and there is concern that human influenza vaccines do not include enough neuraminidase protein to induce a strong protective antibody response. The neuraminidase protein has also become an important target for antiviral drugs that target sialic acid binding which blocks neuraminidase enzyme activity. Two different antiviral drugs are available and are widely used for the treatment of seasonal influenza in humans, but antiviral resistance appears to be a growing concern for this class of antivirals.

# 1 Influenza Neuraminidase: Structure, Function, and Subytpes

Type A influenza viruses are segmented, negative-sense, enveloped viruses that encode for at least ten separate proteins, including two surface glycoproteins, the hemagglutinin (HA) and neuraminidase (NA) proteins. The HA protein has two major functions, including the initial binding of the virus to the host cell by attachment to sialic acid, and to penetrate the endosome membrane to allow the release of viral RNA and polymerase complex into the cell interior to initiate viral replication. The NA protein exhibits enzymatic activity to remove $\alpha$-2,3- or $\alpha$-2,6-linked sialic acid moieties from host or viral glycoproteins, and is necessary for the efficient spread of the virus. As influenza viruses are being prepared for budding and release, viral HA may bind to host or viral glycoproteins, which may prevent viral release or cause aggregation of virions. This has been demonstrated by the clinical use of chemical inhibitors of NA such as zanamivir and oseltamivir, which are known to prevent influenza shedding from infected cells and ultrastructurally cause accumulation of influenza virions on the apical surface of infected cells (Moscona 2005). Recent evidence suggests that NA may also function in promoting viral adhesion to human respiratory epithelial cells (Matrosovich et al. 2004). Degradation of mucin in the respiratory tract (a sialic acid-containing glycoprotein) by viral NA may expose host receptors and promote influenza binding. The function of the NA protein, although critical to the efficient spread of the virus, can be replaced in cell culture with bacterial NAs (Liu et al. 1995).

Sialic acid is commonly conjugated to proteins throughout the body, but slight differences in the sialic acid linkages can be found between species and even in different locations of the host. In general, the HA protein of influenza viruses has a preference for binding to either $\alpha$-2,3 or $\alpha$-2,6 sialic acid-linked glycoproteins; the HA from avian influenza viruses typically binds to $\alpha$-2,3 sialic acid, whereas human influenza viruses preferentially bind to $\alpha$-2,6 sialic acid-containing glycoproteins (Suzuki 2005). The NA protein also has different affinities in substrate specificity, such as for $\alpha$-2,3- or $\alpha$-2,6-linked sialic acids. The differences in substrate specificity are not as defined for the NA protein as the HA binding, but the trend is for swine and human influenza viruses to have a greater specificity toward $\alpha$-2,6 sialic acid from more recent isolates (Baum and Paulson 1991; Xu et al. 1995).

At least one characterized amino acid difference that accounts for this difference in substrate specificity is the isoleucine to valine substitution at position 275 in the N2 protein that occurred in the early 1960s (Kobasa et al. 1999).

The influenza NA protein is a type II integral glycoprotein with enzymatic activity that is vital for the spread of virus from host cells. The gene encoding NA is approximately 1,413 nucleotides long and codes for a protein of around 470 amino acids which has at least five potential glycosylation sites (Varghese et al. 1983). The size of the protein differs between NA subtypes and even within viruses from the same subtype. The NA protein is normally found as a mushroom-shaped homotetrameric protein (Varghese et al. 1983) that possesses *N*-acetyl-neuraminosyl-glycohydrolase activity (Gottschalk 1957). The NA protein can be divided into four main regions, including a short hydrophilic amino terminal tail, a hydrophobic transmembrane domain, a stalk region, and a globular head that contains the enzymatic site for the protein (Colman et al. 1983). The hydrophilic tail consists of six amino acids, MNPNQK, and is highly conserved in most type A influenza viruses, with the exception of some swine origin N1 genes. The sequence of the transmembrane domain is extremely variable between subtypes, but a predicted stretch of hydrophobic amino acids is generally found between amino acids 8 and 37. The stalk region is also extremely variable in sequence between subtypes, but in general has a 30-amino-acid region that was predicted to be hydrophilic. However, amino acid deletions in the stalk region are common in poultry isolates, but stalk deletions do not appear to affect pathogenesis in poultry. For example, the highly pathogenic H5N2 avian influenza isolates from Pennsylvania have a 20 amino acid deletions in their stalk region, yet remain extremely virulent in chickens (Deshpande et al. 1985). The globular head has the largest number of conserved amino acids, including the enzymatic active site (Colman et al. 1983; Kobasa et al. 1999). Overall, the globular head region shows less amino acid variability than the stalk or transmembrane regions (Burmeister et al. 1993). The NA protein in the virion forms a noncovalently bound homotetramer, but the enzymatic activity is still present in individual units, even when the globular head is separated from the stalk and transmembrane regions of the protein. However, the stalk appears to play a major role in the budding function of the virus; viruses with stalk deletions have lower enzymatic activity, which, in severe cases, actually results in the aggregation of the virus on the cell surface, presumably affecting the efficient transmission of the virus. Influenza viruses with stalk deletions from several different NA subtypes are often associated with avian influenza isolated from poultry (Matrosovich et al. 1999). The putative amino acid sequence alignment for all nine subtypes of Type A influenza showed a size range of 453–471 amino acids, with most isolates being either 469 or 470 amino acids in length. When comparing all nine NA subtypes, a total of 102 amino acids were highly conserved (Fig. 1), but when the comparison included influenza type B isolates, only 76 amino acids were conserved (data not shown). Based on X-ray crystallography data and conservation among sequences available at the time, 21 amino acids were thought to form the NA active site (Colman 1989), and all 21 amino acids were strictly conserved in all the type A and B influenza viruses that were compared.

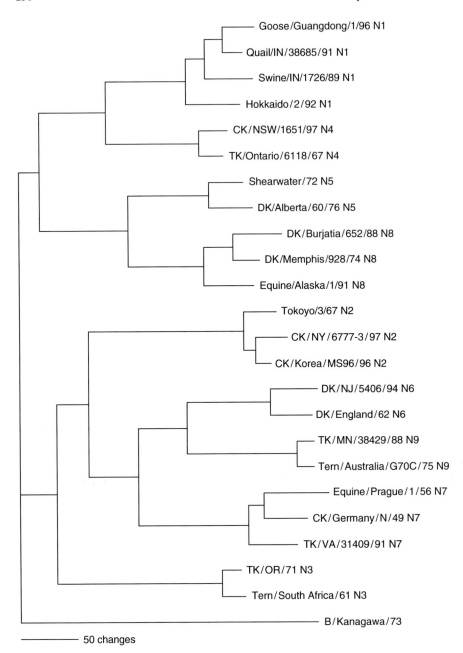

**Fig. 1** Phylogenetic tree of representative isolates from all nine neuraminidase subtypes. The amino acid sequence was aligned and a heuristic search was used to produce a phylogram. A Type B influenza neuraminidase gene was used as an outgroup

All nine NA subtypes have been isolated from mammalian and poultry species, with the N1, N2, N7, and N8 subtypes having become endemic in humans, swine and equine populations. However, the original source of all influenza type A viral genes is thought to be the wildlife reservoir; primarily waterfowl and shorebirds (Slemons et al. 1974; Kawaoka et al. 1988). These viral genes periodically cross over into poultry and mammalian populations, and on rare occasions become established in the new host species. The viral genes in the new hosts are thought to go through a rapid evolution and adaptation to the new host, and, particularly for the surface glycoproteins, immune pressure may also select for rapid evolutionary changes (Suarez and Schultz-Cherry 2000; Xu et al. 1996).

The NA gene has been shown to have a high rate of nucleotide substitutions in human influenza viruses (Yano et al. 2008), with an increased number of changes observed in the stalk region as compared to the globular head (Xu et al. 1996). However, the mutational rates for NA and HA occur independent of each other (Xu et al. 1996; Kilbourne et al. 1990; Abed et al. 2002), and NA undergoes a lower yearly rate of mutations than HA; 0.45–1.01% vs. 1–2%, respectively (Abed et al. 2002). This may be the result of decreased selective pressure by the immune response against the NA protein. However, because influenza viruses are segmented, the NA gene has the potential to reassort between different influenza viruses. For example in a study of swine H3N2 influenza viruses from Asia, evidence of reassortment of the N2 gene from multiple sources was observed (Nerome et al. 1995). Reassortment of human pandemic viruses has also been seen with both the HA and NA genes being replaced in 1957 with the emergence of the H2N2 virus, but the 1968 H3N2 outbreak was different because the HA, and not the NA gene was replaced in that virus.

The complete coding sequences of all nine NA subtypes from type A influenza viruses and a type B influenza virus were phylogenetically compared (Fig. 1). For each NA subtype, a representative viral gene sequence from established lineages was included, including a North American avian and a Eurasian avian influenza sequence for all nine subtypes. Mammalian influenza isolates for N1, N2, N7, and N8 subtypes including classical swine (N1), human (N1 and N2), and equine (N7 and N8) were also included. The phylogenetic tree, using B/Kanagawa/73 as the outgroup, showed two main clusters, with the N1, N4, N5 and N8 in one group, and N2, N3, N6, N7 and N9 in the other group (Fig. 1). Pairwise sequence comparisons of amino acids between the NA subtypes showed 66.8% similarity between the N5 and N8 subtypes, and the least conserved viruses had only 37.3% amino acid sequence similarity between the N5 and N9 subtypes. In contrast, almost 20% amino acid sequence divergence was seen within a subtype for both the N1 and N2 subtypes (data not shown).

# 2 Immune Response to NA and Vaccine Efficacy

The immune response to influenza surface glycoproteins, such as HA and NA, is mainly humoral (Webster et al. 1968), but NA-specific CD8+ T cell-mediated immunity is also produced (Oh et al. 2001; Stitz et al. 1985). The humoral immune response to NA was initially characterized in humans (Kilbourne et al. 1968a,b; Downie 1970; Murphy et al. 1972; Vonka et al. 1977), chickens (McNulty et al. 1986; Webster et al. 1988), mice (Schulman et al. 1968; Schulman 1969; Reichert and Mauler 1975; Bottex et al. 1981) and other species, and an inhibitory action of NA-specific antiserum was characterized in vitro using tissue culture neutralization tests, and in vivo (Kilbourne et al. 1968a; Murphy et al. 1972; Schulman et al. 1968, Schulman 1969; Kasel et al. 1973; Rott et al. 1974). Although the humoral response to HA is best characterized to protect against influenza infection (Webster et al. 1968), these aforementioned studies suggested that anti-NA antibodies afford some protection. The NA glycoprotein is clearly immunogenic, but not all antibodies generated against it can inhibit its function. Antibodies generated against the NA enzymatic site produce the greatest amount of neuraminidase inhibitory (NI) activity, as defined by the ability of antibodies to block NA enzymatic function. To test this hypothesis, chickens were treated with monoclonal antibodies mapped against different N2 epitopes and challenged with virus homologous to that against which the antibodies were raised. Results indicated that antibodies directed against the enzymatic site in the globular head region, as well as nonenzymatic globular head and stalk sites, produced NI activity and a reduction in disease mortality (Webster et al. 1988). It is not entirely clear why NI was produced from antibodies distant to the enzymatic site, but they may induce conformational changes in the tertiary structure of NA that affects its ability to bind substrate.

Several factors have limited the ability to study the role of NA-specific antibodies in the protection against influenza. First, the balance of humoral immune response to influenza glycoproteins is skewed toward a HA response because there is approximately four times more HA than NA protein expressed on the surface of an infectious influenza virion (Webster et al. 1968), and the HA immunologically outcompetes NA in the priming of B and T cell responses in mice (Johansson et al. 1987c). It is not clear if the effect seen in mice is conserved for all mammals and birds, but susceptible species tend to have higher serum levels of antibodies against HA compared to NA. The superiority of HA as an antigen occurs only when HA and NA are present in the context of a viral particle, and is abrogated if administered separately (Johansson and Kilbourne 1993). In order to determine if the low content of NA in whole or split virus vaccines was, in part, responsible for the lessened immune reaction, studies examining the effect of additional exogenous NA to standard vaccines were performed, and in mice a balanced humoral immune response to NA and HA proteins was seen (Johansson et al. 1998). The higher levels of NA antibody significantly reduced pulmonary influenza titers in mice as compared to mice receiving the standard vaccine (Johansson et al. 2002). Supplementation of human seasonal vaccines with exogenous NA may produce a balanced immune response resulting in higher NA antibody production and increased protection against disease. Second, the quantity of NA in whole or split licensed vaccines is not standardized (Gerentes et al. 1999), and may vary greatly from different manufacturers and production lots (Tanimoto et al. 2005; Aymard 2002). The opposite is true for HA, which is standardized for split vaccines.

Regulatory measures should be considered to standardize the HA and NA contents of whole or split virus influenza vaccines. Third, CD4+ T cells were shown to contribute to the competition between HA and NA antigens (Johansson et al. 1987a). Although the mechanism responsible for the disparate immune response to the surface glycoproteins is unknown, perhaps antigen-presenting cells load HA peptides more efficiently onto MHC II molecules than NA peptides, which results in an increased concentration of HA peptide presented to CD4+ T helper cells. Lastly, existing humoral immunity to HA subtypes appears to affect the immune response to the NA component of vaccines. Humans exposed to H7N2 virus produced significantly elevated anti-N2 serum levels compared to those exposed to H3N2 virus; most humans already have serum anti-H3 antibodies, which limits the priming of the immune response to N2 (Johansson et al. 1987b). These factors must be overcome to successfully develop efficacious NA vaccines.

Although antibodies are generated against HA and NA, they appear to function differently in protecting against influenza (Johansson et al. 1989). Antibodies against HA are neutralizing and block infection, whereas antibodies against NA produce "permissive immunity" (Johansson et al. 1989, 1993). The permissive effect of NA antibodies blocks the release of infectious virions from the apical surface of infected cells, and lessens viral spread. The concept of permissive immunity afforded by NA antibodies was initially characterized in laboratory animal models of disease (Webster et al. 1968; Schulman et al. 1968), and in humans recovering from disease (Kilbourne et al. 1968a,b; Downie 1970; Murphy et al. 1972; Schulman 1969; Kasel et al. 1973; Kendal et al. 1977). In these studies, addition of NA antisera from convalescent humans failed to block infection of cells in vitro (Webster et al. 1968; Kilbourne et al. 1968a,b; Kasel et al. 1973), but in mice NA antibody can reduce morbidity and mortality (Webster et al. 1968; Schulman et al. 1968). Antibodies directed against HA are known to prevent infection by influenza viruses because they block HA interaction with host receptors. Because NA is not directly involved with binding to host receptors, antibodies against NA probably play little role in preventing infection by influenza viruses (Kilbourne et al. 1968a). Thus, the immune response against NA antigen(s), whether from natural infection or vaccination, is expected to permit infection but limit viral spread within the host, reduce morbidity and mortality, and decrease viral shedding into the environment, reducing the opportunity for transmission to other susceptible individuals.

Numerous studies have used different approaches to examine the role of NA antibodies in protecting against influenza in mice, ferrets, chickens, and other species. Experimental strategies utilized include reassortment or recombinant viruses (Kilbourne et al. 1995, 2004; Lee et al. 2007), chromatography-purified protein from virus (Johansson et al. 1998), yeast-derived protein (Martinet et al. 1997), baculovirus-derived protein (Johansson et al. 1995, 2002, Deroo et al. 1996; Johansson 1999; Brett and Johansson 2005), DNA plasmid (Chen J et al. 2005; Chen Z et al. 1998, 1999a,b, 2000; Li et al. 2006; Zhang et al. 2005), and a variety of virus-vectored recombinant vaccines expressing the NA protein including vaccinia virus (Webster et al. 1988), replication-deficient alphavirus (Sylte et al. 2007) adenovirus (Gao 2006), and fowlpox virus (Qiao et al. 2003). Other studies have revealed important findings when considering utilizing NA as a vaccine antigen. First, that the concentration of divalent cations (e.g., $Ca^{2+}$ and $Mg^{2+}$) in the vaccine should be carefully considered when using vaccines containing N1 protein, because the stability and antigenicity of the molecule may be affected (Brett and Johansson 2006). Secondly, inclusion of the whole NA gene or specific segments of it appear to affect its antigenicity in experimental models. For example, removal of 60 and 66 nucleotides from the 5' and 3' ends of the coding region, respectively, significantly reduced the protective effect of N2 DNA against homologous viral challenge (Li et al. 2006). These results suggest that inclusion of these sequences is essential for optimal protection from NA DNA or vector-based vaccines.

In the aforementioned vaccine studies, infection by influenza was not affected by NA vaccines, but a reduction in morbidity and mortality was noted. Similarly, these studies revealed that NA vaccines provide poor heterosubtypic immunity. This was demonstrated in chickens vaccinated with N2, which were not protected against challenge with highly pathogenic H7N7 virus (Webster et al. 1988). Similar results demonstrated that mice vaccinated with N2 DNA failed to protect against heterologous H1N1, but protected against homologous H3N2 challenge (Chen Z et al. 2000). Less is known about the immunity conferred using divergent NA antigens within a subtype. Chickens vaccinated with a virus-vectored vaccine expressing N2 closely (98% amino acid homology) resembling the challenge virus N2 were 88% protected against mortality, whereas the chickens vaccinated with a genetically distant N2 (85% amino acid homology) were poorly protected from the challenge virus (Sylte et al. 2007). Recent evidence suggests that the N1 content of seasonal human influenza vaccines may provide some protection against H5N1 avian influenza infection, despite large sequence differences between the two lineages. Fifty percent of mice receiving serum from other mice vaccinated with a DNA vaccine expressing N1 from A/New Caledonia/20/99 H1N1 virus survived lethal challenge with A/Vietnam/1293/04 H5N1 virus (Sandbulte et al. 2006). Because most humans would be expected to have serum antibodies to N1 from vaccination or natural exposure to N1 viruses, this might provide some initial degree of resistance against pandemic H5N1 morbidity and mortality.

The role of NA antibodies in the protection of chickens from virulent challenge is not as well known. Rott et al. (1974) demonstrated a protective role for NA antibodies in chickens infected with avian influenza. Administration of killed H7N1 virus protected chickens against H5N1-associated mortality, but failed to completely protect against clinical disease because most vaccinated birds developed conjunctivitis postchallenge (McNulty et al. 1986). These data suggest that NA antibodies produce permissive immunity in avian species. The immune response to NA vaccines in chickens was further characterized to detect the production of serum NI activity after vaccinating with either inactivated virus, vaccinia virus encoding NA or purified NA (Webster et al. 1988). Vaccination with purified NA or killed virus completely protected against homologous challenge (H5N2) but did not protect against heterologous challenge (H7N7) (Webster et al. 1988). Qiao et al. (2003) demonstrated the production of NA-specific antibodies in chickens following administration of recombinant fowl-pox virus encoding H5 and N1, which were completely protected against H5N1 challenge. However, an NA-only control was not included in this study, which makes it difficult to determine a role of NA-antibodies using this vaccine. Administration of replication-deficient alphaviruses expressing N2 provided partial protection against highly pathogenic avian influenza challenge (Sylte et al. 2007). Chickens with a NI titer of ≥1:128 were significantly protected, suggesting that the amount of NA antigen should be carefully considered in avian vaccines.

The effecter mechanism of NI activity responsible for protecting mammals or birds from influenza is not clear, but appears to differ from antibodies directed against the HA protein. One likely mechanism is that the NI activity keeps influenza viruses trapped on the apical surfaces of infected cells (Kilbourne et al. 1968a), where they may be susceptible to cytotoxic T lymphocyte (CTL) or NK-mediated killing. Although CTLs readily kill influenza-infected cells in vitro (Stitz et al. 1985) and in the respiratory tracts of mice and humans (Flynn et al. 1999; Wiley et al. 2001), their contribution to influenza protection in mammals is poorly characterized and unknown in poultry. However, $\beta_2$-microglobulin-deficient mice, which lack MHC I molecule expression, that were vaccinated with NA and HA were protected against lethal influenza infection (Epstein et al. 1993), suggesting that antibody is the predominate effector molecule that confers protection. The use of vaccines known to generate CD8+-specific CTL responses may more clearly show a role CTL in protection. For example, replication-deficient alphavirus-based vaccines encoding tumor antigens yielded CTLs specific for tumors, and these cells were lysed by CTL in vitro (Vidalin et al. 2000; Colmenero et al. 1999). Replication-deficient alphaviruses expressing an avian N2 provided 88% protection against highly pathogenic avian influenza chickens in chickens (Sylte et al. 2007). Although not examined for in this study, the N2 alphavirus vaccine likely produced CTL specific for N2, which might function together with neuraminidase antibody to protect chickens against highly pathogenic avian influenza. Phylogenetic analysis of circulating human H1N1 and H3N2 viruses over the past six years in Denmark revealed significant mutations in CTL epitopes of NA genes (Bragstad et al. 2008). These data suggest that influenza evasion of CTL-mediated killing might already be occurring due to selective immune pressure. Serum NI activity may directly damage or kill influenza virus by activating the classical complement pathway, which may cause disruption of the viral envelope, lessen infectivity or kill the virus (Mozdzanowska et al. 2006). Likewise, neuraminidase antibody may kill influenza-infected cells via antibody-dependent cellular cytotoxicity (ADCC), a cytotoxic process where serum NI activity (e.g., IgG specific for NA) binds to NA on the surfaces of infected cells, and effecter cells (e.g., NK cells) recognize and kill these cells via apoptosis (Hashimoto et al. 1983). Kinetically, ADCC specific for NA was produced in humans before hemag-glutination-inhibiting antibodies were formed, and persisted for up to a year with broad reactivity within the NA subtype (Hashimoto et al. 1983). Finally, NI activity may indirectly kill influenza by acting as an opsonin to direct the virus toward phago-cytic cells (e.g., macrophages) with virucidal effects (Hartshorn et al. 1996).

## 3  Neuraminidase Inhibitor Resistance and Relevance to NA Vaccine Antigen

Chemical inhibitors of influenza NA (e.g., oseltamivir and zanamivir) were introduced as a first line of defense against influenza in humans from 1999 to 2002. Initially, all nine subtypes were sensitive to NA inhibitors, including N1 from H5N1 isolates, but these drugs must be administered within the first 6–48 h after initial clinical signs to block viral release from the respiratory tract, and significantly reduce morbidity (Moscona 2005; Aoki et al. 2003). Spontaneous resistance of NA to inhibitors is rare, and is induced by an increase in mutation rates in selected amino acids in the NA enzymaztic site (Yen et al. 2006). Resistance has been noted since their introduction into clinical use (1999–2002) (Monto et al. 2006), and viruses expressing mutations indicative of resistance to NA inhibitors were detected in 4% of children treated with oseltamivir (Whitley et al. 2001), and in 18% of a sampled population of Japanese youths (Kiso et al. 2004). Adults harboring influenza with NA resistant to chemical inhibitors is less prominent. These results are of interest because children represent a population that is highly susceptible to influenza. Different amino acid residues are involved in resistance to NA inhibitors among different NA subtypes (Ho et al. 2007), which indicates that each subtype should be individually assessed. Mutations in the NA protein that render the molecule resistant to oseltamivir may also reduce the overall fitness of the virus and affect its pathogenicity. However, ferrets challenged with NA inhibitor-resistant human isolates showed no decrease in viral replication in the respiratory tract (Herlocher et al. 2004). These results indicate that NA inhibitor-resistant influenza isolates might be able to circulate during epidemic or pandemic conditions.

Fear of pandemic H5N1 avian influenza has increased the frequency of NA inhibitor use. It is not certain whether emerging H5N1 isolates will remain sensitive to NA inhibitors, whose resistance would nullify a vital first line of influenza therapy. To assess this possibility, ferrets were challenged with 2004 avian H5N1 isolates and then treated with oseltamivir. They were protected from a lethal influenza infection, but more virulent isolates may require an increased dose to achieve protection (Govorkova et al. 2007). Additional evidence suggests that H5N1 isolated from humans may be increasingly resistant to oseltamivir (de Jong et al. 2005; Le et al. 2005), and computer modeling predicts that large-scale use of stockpiled NA inhibitors may impose strong selection for the evolution of NA inhibitor drug-resistant strains during an influenza pandemic (Regoes and Bonhoeffer 2006). Because resistance is mounting in H5N1 isolates, it is necessary to invest in the development of more efficacious NA subunit vaccines, or to produce a more balanced response to HA and NA in seasonal influenza vaccines (Johansson et al. 1998). Even though selective pressure drives an increase in NA mutation rates, the diversity and plasticity of the immune response (e.g., humoral and cell-mediated) is likely to neutralize NA. The immune response has great capacity to adapt and overcome multiple mutations in NA epitopes, whereas a single amino acid mutation may render NA inhibitors ineffective. The difficulty is to develop efficacious vaccines that produce an appropriate blend of HA- and NA-neutralizing antibodies at mucosal sites, where the primary wave of replication occurs, before exposure to influenza. Vaccine strategies with this goal are the future of NA vaccines. Ultimately, increased use of NA subunit vaccines may lead to increased selective pressure of NA to mutate to levels similar to that seen in HA following prolonged vaccination (Lee et al. 2004), which may lessen their efficacy and require more frequent changing of NA vaccine antigens to combat circulating influenza viruses.

# References

Abed Y et al. (2002) Divergent evolution of hemagglutinin and neuraminidase genes in recent influenza A:H3N2 viruses isolated in Canada. J Med Virol 67(4):589–595

Aoki FY et al. (2003) Early administration of oral oseltamivir increases the benefits of influenza treatment. J Antimicrob Chemother 51(1):123–129

Aymard M (2002) Quantification of neuramidase (NA) protein content. Vaccine 20(suppl 2): S59–S60

Baum LG, Paulson JC (1991) The N2 neuraminidase of human influenza virus has acquired a substrate specificity complementary to the hemagglutinin receptor specificity. Virology 180(1):10–15

Bottex C, Burckhart MF, Fontanges R (1981) Comparative immunogenicity of live influenza viruses and their solubilized neuraminidases: results of mouse protection experiments. Arch Virol 70(2):83–89

Bragstad K, Nielsen LP, Fomsgaard A (2008) The evolution of human influenza A viruses from 1999 to 2006: a complete genome study. Virol J 5:40

Brett IC, Johansson BE (2005) Immunization against influenza A virus: Comparison of conventional inactivated, live-attenuated and recombinant baculovirus produced purified hemagglutinin and neuraminidase vaccines in a murine model system. Virology 339(2):273–280

Brett IC, Johansson BE (2006) Variation in the divalent cation requirements of influenza A virus N1 neuraminidases. J Biochem 139(3):439–447

Burmeister WP et al. (1993) Comparison of structure and sequence of influenza B/Yamagata and B/Beijing neuraminidases shows a conserved "head" but much greater variability in the "stalk" and NB protein. Virology 192(2):683–686

Chen J et al. (2005) Protection against influenza virus infection in BALB/c mice immunized with a single dose of neuraminidase-expressing DNAs by electroporation. Vaccine 23(34):4322–4328

Chen Z et al. (1998) Comparison of the ability of viral protein-expressing plasmid DNAs to protect against influenza. Vaccine 16(16):1544–1549

Chen Z et al. (1999a) Enhanced protection against a lethal influenza virus challenge by immunization with both hemagglutinin- and neuraminidase-expressing DNAs. Vaccine 17(7–8):653–659

Chen Z et al. (1999b) Protection and antibody responses in different strains of mouse immunized with plasmid DNAs encoding influenza virus haemagglutinin, neuraminidase and nucleoprotein. J Gen Virol 80(Pt 10):2559–2564

Chen Z et al. (2000) Cross-protection against a lethal influenza virus infection by DNA vaccine to neuraminidase. Vaccine 18(28):3214–3222

Colman PM (1989) In: Krug RM (ed) Neuraminidase enzyme and antigen, in the influenza viruses. Plenum, New York, pp 175–218

Colman PM, Varghese JN, Laver WG (1983) Structure of the catalytic and antigenic sites in influenza virus neuraminidase. Nature 303(5912):41–44

Colmenero P, Liljestrom P, Jondal M (1999) Induction of P815 tumor immunity by recombinant Semliki Forest virus expressing the *P1A* gene. Gene Ther 6(10):1728–1733 de Jong MD et al. (2005) Oseltamivir resistance during treatment of influenza A (H5N1) infection. N Engl J Med 353(25):2667–2672

Deroo T, Jou WM, Fiers W (1996) Recombinant neuraminidase vaccine protects against lethal influenza. Vaccine 14(6):561–569

Deshpande KL, Naeve CW, Webster RG (1985) The neuraminidases of the virulent and avirulent A/Chicken/Pennsylvania/83 (H5N2) influenza A viruses: sequence and antigenic analyses. Virology 147(1):49–60

Downie JC (1970) Neuraminidase- and hemagglutinin-inhibiting antibodies in serum and nasal secretions of volunteers immunized with attenuated and inactivated influenza B-Eng-13–65 virus vaccines. J Immunol 105(3):620–626

Epstein SL et al. (1993) Beta 2-microglobulin-deficient mice can be protected against influenza A infection by vaccination with vaccinia-influenza recombinants expressing hemagglutinin and neuraminidase. J Immunol 150(12):5484–5493

Flynn KJ et al. (1999) In vivo proliferation of naive and memory influenza-specific CD8(+) T cells. Proc Natl Acad Sci USA 96(15):8597–8602

Gao W et al. (2006) Protection of mice and poultry from lethal H5N1 avian influenza virus through adenovirus-based immunization. J Virol 80(4):1959–1964

Gerentes L, Kessler N, Aymard M (1999) Difficulties in standardizing the neuraminidase content of influenza vaccines. Dev Biol Stand 98:189–196; discussion 197

Gottschalk A (1957) Neuraminidase: the specific enzyme of influenza virus and *Vibrio cholerae*. Biochim Biophys Acta 23(3):645–646

Govorkova EA et al. (2007) Efficacy of oseltamivir therapy in ferrets inoculated with different clades of H5N1 influenza virus. Antimicrob Agents Chemother 51(4):1414–1424

Hartshorn KL et al. (1996) Neutrophil deactivation by influenza A viruses: mechanisms of protection after viral opsonization with collectins and hemagglutination-inhibiting antibodies. Blood 87(8):3450–3461

Hashimoto G, Wright PF, Karzon DT (1983) Antibody-dependent cell-mediated cytotoxicity against influenza virus-infected cells. J Infect Dis 148(5):785–794

Herlocher ML et al. (2004) Influenza viruses resistant to the antiviral drug oseltamivir: transmission studies in ferrets. J Infect Dis 190(9):1627–1630

Ho HT et al. (2007) Neuraminidase inhibitor drug susceptibility differs between influenza N1 and N2 neuraminidase following mutagenesis of two conserved residues. Antiviral Res 76(3):263–266

Johansson BE (1999) Immunization with influenza A virus hemagglutinin and neuraminidase produced in recombinant baculovirus results in a balanced and broadened immune response superior to conventional vaccine. Vaccine 17(15–16):2073–2080

Johansson BE, Kilbourne ED (1993) Dissociation of influenza virus hemagglutinin and neuraminidase eliminates their intravirionic antigenic competition. J Virol 67(10):5721–5723

Johansson BE et al. (1987a) Immunologic response to influenza virus neuraminidase is influenced by prior experience with the associated viral hemagglutinin. III. Reduced generation of neuraminidase-specific helper T cells in hemagglutinin-primed mice. J Immunol 139(6):2015–2019

Johansson BE et al. (1987b) Immunologic response to influenza virus neuraminidase is influenced by prior experience with the associated viral hemagglutinin. II. Sequential infection of mice simulates human experience. J Immunol 139(6):2010–2014

Johansson BE, Moran TM, Kilbourne ED (1987c) Antigen-presenting B cells and helper T cells cooperatively mediate intravirionic antigenic competition between influenza A virus surface glycoproteins. Proc Natl Acad Sci USA 84(19):6869–6873

Johansson BE, Bucher DJ, Kilbourne ED (1989) Purified influenza virus hemagglutinin and neuraminidase are equivalent in stimulation of antibody response but induce contrasting types of immunity to infection. J Virol 63(3):1239–1246

Johansson BE, Grajower B, Kilbourne ED (1993) Infection-permissive immunization with influenza virus neuraminidase prevents weight loss in infected mice. Vaccine 11(10):1037–1039

Johansson BE, Price PM, Kilbourne ED (1995) Immunogenicity of influenza A virus N2 neuraminidase produced in insect larvae by baculovirus recombinants. Vaccine 13(9):841–845

Johansson BE, Matthews JT, Kilbourne ED (1998) Supplementation of conventional influenza A vaccine with purified viral neuraminidase results in a balanced and broadened immune response. Vaccine 16(9–10):1009–1015

Johansson BE, Pokorny BA, Tiso VA (2002) Supplementation of conventional trivalent influenza vaccine with purified viral N1 and N2 neuraminidases induces a balanced immune response without antigenic competition. Vaccine 20(11–12):1670–1674

Kasel JA et al. (1973) Effect of influenza anti-neuraminidase antibody on virus neutralization. Infect Immun 8(1):130–131

Kawaoka Y et al. (1988) Is the gene pool of influenza viruses in shorebirds and gulls different from that in wild ducks? Virology 163(1):247–250

Kendal AP, Noble GR, Dowdle WR (1977) Neuraminidase content of influenza vaccines and neuraminidase antibody responses after vaccination of immunologically primed and unprimed populations. J Infect Dis 136(suppl):S415–S424

Kilbourne ED et al. (1968a) Antiviral activity of antiserum specific for an influenza virus neuraminidase. J Virol 2(4):281–288

Kilbourne ED, Christenson WN, Sande M (1968b) Antibody response in man to influenza virus neuraminidase following influenza. J Virol 2(7):761–762

Kilbourne ED, Johansson BE, Grajower B (1990) Independent and disparate evolution in nature of influenza A virus hemagglutinin and neuraminidase glycoproteins. Proc Natl Acad Sci USA 87(2):786–790

Kilbourne ED et al. (1995) Purified influenza A virus N2 neuraminidase vaccine is immunogenic and non-toxic in humans. Vaccine 13(18):1799–1803

Kilbourne ED et al. (2004) Protection of mice with recombinant influenza virus neuraminidase. J Infect Dis 189(3):459–461

Kiso M et al. (2004) Resistant influenza A viruses in children treated with oseltamivir: descriptive study. Lancet 364(9436):759–765

Kobasa D et al. (1999) Amino acid residues contributing to the substrate specificity of the influenza A virus neuraminidase. J Virol 73(8):6743–6751

Le QM et al. (2005) Avian flu: isolation of drug-resistant H5N1 virus. Nature 437(7062):1108

Lee CW, Senne DA, Suarez DL (2004) Effect of vaccine use in the evolution of Mexican lineage H5N2 avian influenza virus. J Virol 78(15):8372–8381

Lee YJ et al. (2007) Effects of homologous and heterologous neuraminidase vaccines in chickens against H5N1 highly pathogenic avian influenza. Avian Dis 51(suppl 1):476–478

Li X et al. (2006) Essential sequence of influenza neuraminidase DNA to provide protection against lethal viral infection. DNA Cell Biol 25(4):197–205

Liu C et al. (1995) Influenza type A virus neuraminidase does not play a role in viral entry, replication, assembly, or budding. J Virol 69(2):1099–1106

Martinet W et al. (1997) Protection of mice against a lethal influenza challenge by immunization with yeast-derived recombinant influenza neuraminidase. Eur J Biochem 247(1):332–338

Matrosovich M et al. (1999) The surface glycoproteins of H5 influenza viruses isolated from humans, chickens, and wild aquatic birds have distinguishable properties. J Virol 73(2):1146–1155

Matrosovich MN et al. (2004) Neuraminidase is important for the initiation of influenza virus infection in human airway epithelium. J Virol 78(22):12665–12667

McNulty M, Allan G, McKracken R (1986) Efficacy of avian influenza neuraminidase-specific vaccines in chickens. Avian Pathol 15(1):107–115

Monto AS et al. (2006) Detection of influenza viruses resistant to neuraminidase inhibitors in global surveillance during the first 3 years of their use. Antimicrob Agents Chemother 50(7): 2395–2402

Moscona A (2005) Neuraminidase inhibitors for influenza. N Engl J Med 353(13):1363–1373

Mozdzanowska K et al. (2006) Enhancement of neutralizing activity of influenza virus-specific antibodies by serum components. Virology 352(2):418–426

Murphy BR, Kasel JA, Chanock RM (1972) Association of serum anti-neuraminidase antibody with resistance to influenza in man. N Engl J Med 286(25):1329–1332

Nerome K et al. (1995) Genetic analysis of porcine H3N2 viruses originating in southern China. J Gen Virol 76(Pt 3): 613–624

Oh S, Belz GT, Eichelberger MC (2001) Viral neuraminidase treatment of dendritic cells enhances antigen-specific CD8(+) T cell proliferation, but does not account for the CD4(+) T cell independence of the CD8(+) T cell response during influenza virus infection. Virology 286(2): 403–411

Qiao CL et al. (2003) Protection of chickens against highly lethal H5N1 and H7N1 avian influenza viruses with a recombinant fowlpox virus co-expressing H5 haemagglutinin and N1 neuraminidase genes. Avian Pathol 32(1):25–32

Regoes RR, Bonhoeffer S (2006) Emergence of drug-resistant influenza virus: population dynamical considerations. Science 312(5772):389–391

Reichert E, Mauler R (1975) Effect of neuraminidase on potency of inactivated influenza virus vaccines in mice. Dev Biol Stand 28:319–323

Rott R, Becht H, Orlich M (1974) The significance of influenza virus neuraminidase in immunity. J Gen Virol 22(1):35–41

Sandbulte MR et al. (2006) Cross-reactive neuraminidase antibodies afford partial protection against H5N1 in mice and are present in unexposed humans. PLoS Med 4(2):e59. doi:10.1371/journal.pmed.0040059

Schulman JL (1969) The role of antineuraminidase antibody in immunity to influenza virus infection. Bull World Health Organ 41(3):647–650

Schulman JL, Khakpour M, Kilbourne ED (1968) Protective effects of specific immunity to viral neuraminidase on influenza virus infection of mice. J Virol 2(8):778–786

Slemons RD et al. (1974) Type-A influenza viruses isolated from wild free-flying ducks in California. Avian Dis 18(1):119–124

Stitz L et al. (1985) Cytotoxic T cell lysis of target cells fused with liposomes containing influenza virus haemagglutinin and neuraminidase. J Gen Virol 66(Pt 6):1333–1339

Suarez DL, Schultz-Cherry S (2000) Immunology of avian influenza virus: a review. Dev Comp Immunol 24(2–3):269–283

Suzuki Y (2005) Sialobiology of influenza: molecular mechanism of host range variation of influenza viruses. Biol Pharm Bull 28(3):399–408

Sylte MJ, Hubby B, Suarez DL (2007) Influenza neuraminidase antibodies provide partial protection for chickens against high pathogenic avian influenza infection. Vaccine 25(19):3763–3772

Tanimoto T, et al. (2005) Estimation of the neuraminidase content of influenza viruses and split-product vaccines by immunochromatography. Vaccine 23(37):4598–4609

Varghese JN, Laver WG, Colman PM (1983) Structure of the influenza virus glycoprotein antigen neuraminidase at 2.9 A resolution. Nature 303(5912):35–40

Vidalin O et al. (2000) Use of conventional or replicating nucleic acid-based vaccines and recombinant Semliki forest virus-derived particles for the induction of immune responses against hepatitis C virus core and E2 antigens. Virology 276(2):259–270

Vonka V et al. (1977) Small-scale field trial with neuraminidase vaccine. Dev Biol Stand 39:337–339

Webster RG, Laver WG, Kilbourne ED (1968) Reactions of antibodies with surface antigens of influenza virus. J Gen Virol 3(3):315–326

Webster RG, Reay PA, Laver WG (1988) Protection against lethal influenza with neuraminidase. Virology 164(1):230–237

Whitley RJ et al. (2001) Oral oseltamivir treatment of influenza in children. Pediatr Infect Dis J 20(2):127–133

Wiley JA et al. (2001) Antigen-specific CD8(+) T cells persist in the upper respiratory tract following influenza virus infection. J Immunol 167(6):3293–3299

Xu G et al. (1995) Sialidase of swine influenza A viruses: variation of the recognition specificities for sialyl linkages and for the molecular species of sialic acid with the year of isolation. Glycoconj J 12(2):156–161

Xu X et al. (1996) Genetic variation in neuraminidase genes of influenza A (H3N2) viruses. Virology 224(1):175–183

Yano T et al. (2008) Effects of single-point amino acid substitutions on the structure and function neuraminidase proteins in influenza A virus. Microbiol Immunol 52(4):216–223

Yen HL et al. (2006) Importance of neuraminidase active-site residues to the neuraminidase inhibitor resistance of influenza viruses. J Virol 80(17):8787–8795

Zhang F et al. (2005) Maternal immunization with both hemagglutinin- and neuraminidase-expressing DNAs provides an enhanced protection against a lethal influenza virus challenge in infant and adult mice. DNA Cell Biol 24(11):758–765

# Recombinant Vectors as Influenza Vaccines

Sarah A. Kopecky-Bromberg and Peter Palese

## Contents

**Abstract** The antiquated system used to manufacture the currently licensed inactivated influenza virus vaccines would not be adequate during an influenza virus pandemic. There is currently a search for vaccines that can be developed faster and provide superior, long-lasting immunity to influenza virus as well as other highly pathogenic viruses and bacteria. Recombinant vectors provide a safe and effective method to elicit a strong immune response to a foreign protein or epitope. This review explores the advantages and limitations of several different vectors that are currently being tested, and highlights some of the newer viruses being used as recombinant vectors.

S.A. Kopecky-Bromberg and P. Palese
Department of Microbiology, Mount Sinai School of Medicine,
New York NY, 10029-6574, USA

P. Palese (✉)
Department of Medicine, Mount Sinai School of Medicine,
New York, NY 10029-6574, USA
e-mail: Peter.Palese@mssm.edu

R.W. Compans and W.A. Orenstein (eds.), *Vaccines for Pandemic Influenza*,
Current Topics in Microbiology and Immunology 333,
DOI 10.1007/978-3-540-92165-3_13, © Springer-Verlag Berlin Heidelberg 2009

# 1 Introduction

While the ideal vaccine would elicit the exact immune response that occurs during natural infection with highly pathogenic influenza virus, expression of influenza virus proteins from live replicating vectors can safely induce a strong humoral and cellular immune response comparable to natural infection (Souza et al. 2005). Recombinant vectors have been developed because it is considered to be too dangerous to vaccinate people with even vastly attenuated forms of dangerous viruses such as highly pathogenic influenza virus or Ebola virus. Vaccination with recombinant vectors offers several advantages over vaccination with inactivated influenza viruses. Inactivated vaccines induce short-lived antibody-mediated immunity, while recombinant vectors elicit a longer-lasting immune response that stimulates both memory B and T cells. Also, the manufacture of the recombinant vaccines entails substantially less risk than growing large quantities of influenza viruses expressing the highly pathogenic hemagglutinin (HA) and neuraminidase (NA) proteins required for inactivation, since the recombinant vectors cannot cause influenza. Many different viruses and bacteria are currently being tested for their ability to function as a good recombinant vector.

# 2 Qualities of Ideal Vectors

Good vectors should be easy to manipulate genetically, allowing the insertion of large foreign genes or epitopes. The vectors should grow well and be easy to produce in large-scale operations. The foreign proteins should be highly expressed from the vector to elicit the best immune response. Proteins from the vector should not elicit a strong immune response, as this may interfere with the induction of a response to the foreign proteins and also reduce the effectiveness of boosts after initial vaccination. The expression of foreign proteins in the host should be transient and the vector must be fully cleared from the host once the adaptive immune response has commenced. Integration of DNA from the vector into the host genome must not occur, as this can disrupt host genes and possibly lead to the development of cancer. Humans should not have pre-existing antibodies to the vector, as this could prevent replication of the vector and subsequently prevent the induction of an immune response to the foreign proteins. Ideally, the vector should not cause disease symptoms in humans and should be safe even for immunocompromised individuals and young children. Additionally, recombinant vectors that do not require refrigeration would facilitate the distribution of the vaccine to developing nations. Recombinant vectors that can be administered without needles would also aid distribution and enhance vaccination compliance (Babiuk and Tikoo 2000; Barouch and Nabel 2005; Souza et al. 2005; Barouch 2006; Li et al. 2007). Several viruses possess many, but not all, of the characteristics of a useful vector.

# 3   Newcastle Disease Virus Vectors

Newcastle disease virus (NDV) contains a nonsegmented, single-strand, negative-sense RNA genome and belongs to the family *Paramyxoviridae*. NDV contains six genes that encode seven proteins: nucleocapsid protein (NP), phosphoprotein and V protein (P/V), matrix (M) protein, fusion (F) protein, hemagglutinin-neuraminidase (HN), and large polymerase (L) protein. As the viral polymerase can disassociate from the viral genome after the transcription of each gene, expression levels of the proteins reduce in a sequential manner from the 3′ to the 5′ end of the genome. Thus, the expression level of a foreign protein can be controlled by its position on the viral genome (Huang et al. 2004). NDV naturally infects avian species, and it is a highly contagious virus with a pathogenicity ranging from avirulent to high levels of mortality (Huang et al. 2003). One determinant of NDV pathogenicity is the cleavage site of the F protein, which is necessary for the fusion of the viral envelope to the cell membrane. NDV strains containing an F protein cleavage site that has several basic amino acids is readily cleaved by numerous cellular proteases in a variety of tissues, leading to wide dissemination of the virus throughout the host organism and high virulence. NDV strains containing an F protein cleavage site that contains fewer basic amino acids is only cleaved by a secreted protease found in the lung, and thus its tropism is limited to the lung, leading to lower pathogenicity (Panda et al. 2004). The ability to easily adjust the pathogenicity of NDV is one of the reasons that NDV is an attractive recombinant vector.

NDV possesses many of the qualities of an ideal vector for use in humans, and several of its properties make it specifically suited as a recombinant vector for pandemic influenza. Since under natural conditions NDV infects only birds, humans do not have pre-existing immunity to NDV. Pre-existing antibodies to the recombinant vector drastically reduce or completely eliminate the formation of immunity to the foreign protein expressed from the vector, and this is one of the major reasons that recombinant vectors are not effective. The fact that birds are a major reservoir for both NDV and highly pathogenic avian influenza virus (HPAI) has led to the development of dual vaccines that can protect poultry against both diseases.

An NDV virus expressing an influenza virus HA, rNDV/B1-HA, was first rescued in 2001 (Nakaya et al. 2001). The influenza virus HA gene from the A/WSN/33 (H1N1) virus had been inserted between the P and M genes of the Hitchner B1 strain, which is avirulent, and this virus has been used as a live vaccine in birds (Russell and Ezeifeka 1995). The genomic structure of a recombinant NDV is illustrated in Fig. 1. The influenza virus HA was confirmed to be incorporated into the viral envelope and to be cleaved. The rNDV/B1-HA showed no pathogenicity in embryonated chicken eggs, which are used to grow large stocks of both influenza virus and NDV for vaccines. Most importantly, vaccination of mice with rNDV/B1-HA conferred complete protection against lethal challenge with A/WSN/33 influenza virus (Nakaya et al. 2001).

Recently, NDV recombinant vectors expressing HA genes from HPAI strains have also been generated. NDVs expressing HAs from H5 and H7 influenza virus

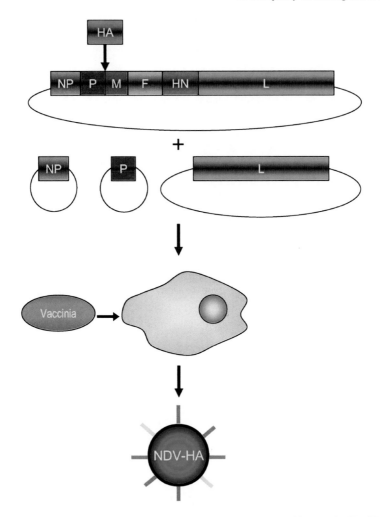

**Fig. 1** Rescue of a recombinant NDV vector expressing influenza virus HA protein. The HA gene was cloned between the P and the M gene in a plasmid containing the full-length NDV genome under the control of the T7 promoter that requires the T7 polymerase for expression. Cells were cotransfected with a plasmid containing the full-length NDV-HA genome as well as helper plasmids expressing NP, P, and L from a Pol II promoter. One hour prior to transfection, cells were infected with MVA-T7 vaccinia virus, which had been modified to express the T7 polymerase. The resulting NDV virus expressed NDV F and HN, as well as influenza HA on the virion surface. Conformation of the rescue of NDV-HA virus was determined by sequence analysis, as described in Nakaya et al. (2001)

strains, which have both caused illness and death in humans working with infected birds, were designed to be dual vaccines to protect birds from HPAI and NDV (Park et al. 2006; Veits et al. 2006; Ge et al. 2007). In order to enhance incorporation of H7 into the viral envelope, the ectodomain of H7 was fused to the transmembrane and cytoplasmic domains of the F gene and inserted into the NDV genome between the P and M genes. The virus was too attenuated to confer full protection after challenge with HPAI, so three basic amino acids were added to the F protein fusion site to enhance viral spread. The resulting virus was still highly attenuated compared to pathogenic strains of NDV, but it did induce 90% protection after stringent challenge with HPAI and 100% protection after challenge with NDV (Park et al. 2006). H5 HAs were cloned into a different nonpathogenic NDV strain, La Sota, and used to vaccinate chickens. Not only were the chickens fully protected after lethal challenge with HPAI and NDV, but the vaccine also prevented the chickens from shedding virus after challenge (Veits et al. 2006; Ge et al. 2007).

NDV has also been used as a recombinant vector for pathogens other than influenza virus. Infectious bursal disease virus (IBDV) causes immunosuppression in poultry, which reduces the effectiveness of vaccines and leaves the animals especially susceptible to other infections. A recombinant NDV expressing the IBDV VP2 protein provided protection against IBDV and NDV in chickens (Huang et al. 2004). Another recombinant NDV was designed to prevent respiratory syncytial virus (RSV), which causes severe respiratory disease in infants and the elderly. A recombinant NDV expressing RSV F protein protected mice against challenge with RSV (Martinez-Sobrido et al. 2006).

Recombinant NDV vaccines for pathogens such as severe acute respiratory syndrome virus (SARS-CoV) and human parainfluenza virus type 3 (HPIV3) have been tested in primates. SARS-CoV caused a worldwide outbreak in 2003 with a mortality rate of about 10%. Vaccination of African green monkeys with an NDV vector expressing SARS-CoV spike protein resulted in a dramatic reduction of viral replication after challenge with SARS-CoV (DiNapoli et al. 2007). An NDV vector expressing HPIV3 HN protein induced levels of antibody comparable to natural infection with HPIV3 in African green monkeys (Bukreyev et al. 2005).

One limitation of NDV as a recombinant vector is that it can be difficult to grow NDVs with long foreign genes or multiple foreign genes inserted into the NDV genome. This shortcoming has recently been overcome by rescuing a recombinant NDV with a bisegmented genome. One segment contains the genes for NP, P, and L, while the other segment contains the genes for M, F, and HN. A recombinant virus expressing GFP from the first segment and SARS-CoV spike protein from the second segment was rescued, demonstrating that NDV can be designed to express multiple foreign proteins and large proteins such as SARS-CoV spike protein (Fig. 2) (Gao et al. 2008).

**Fig. 2** Expression of two foreign proteins from a bisegmented NDV virus. The nonsegmented NDV genome was divided into two segments to allow the expression of two foreign proteins. Segment 1 contains the M, F, and HN genes as well as the SARS-CoV spike gene. The 3′ and 5′ noncoding regions were added onto the ends of segment 1. Segment 2 contains NP, P, and L genes as well as GFP inserted between the P and L genes. Figure adapted from Gao et al. (2008)

Interestingly, NDV is being tested as a recombinant vector for not only infectious diseases, but also for cancer. It had previously been observed that NDV replicates in human tumor cells much more readily than normal cells. This is believed to occur because cancer cells often have mutations in the interferon pathway, a key host antiviral immune response, while normal cells have an intact interferon pathway. Since NDV is very sensitive to the effects of interferon, it is rapidly eliminated from normal cells. The cancer cells, however, are killed by NDV since they cannot mount an antiviral response. An NDV vector expressing granulocyte/macrophage colony-stimulating factor (GM-CSF), which has been shown to enhance immunity to tumors, stimulates antitumor activity in human cells better than NDV vector alone (Janke et al. 2007). A recombinant NDV vector expressing IL-2 reduced tumor volume and caused a higher remission rate of colon carcinoma tumors in mice compared to NDV alone (Vigil et al. 2007). A new approach involving the expression of tumor-related antibodies from the NDV genome also has promise for cancer therapy (Puhler et al. 2008). Additionally, the safety of NDV has already been demonstrated in humans (Freeman et al. 2006; Laurie et al. 2006). Thus, NDV shows promise of being a valuable recombinant vector in both humans and birds.

# 4 Vesicular Stomatitis Virus Vectors

Vesicular stomatitis virus (VSV) contains a nonsegmented, single-strand, negative-sense RNA genome and belongs to the family *Rhabdoviridae*. The VSV genome is organized similarly to NDV except that VSV encodes a protein responsible for both fusion and attachment, glycoprotein (G), whereas NDV encodes two separate proteins for these functions, F and HN. VSV normally infects horses, cattle, and swine, where it causes vesicular lesions on the mouth, nose, teats, and hooves. Infection is usually cleared within two weeks without complications. In nature, VSV is spread primarily by arthropod vectors such as sand flies (*Lutzomyia shannoni*) and black flies (*Simulium vittatum*), though transmission by animal-to-animal contact has been reported (Letchworth et al. 1999; Stallknecht et al. 2001; Rodriguez 2002). VSV is considered a good vaccine vector candidate since humans do not have pre-existing antibodies that would interfere with the induction of an immune response to a foreign protein expressed from the virus.

A recombinant VSV expressing the HA of WSN influenza virus strain was rescued and subsequently shown to confer protection in mice after challenge with WSN virus (Kretzschmar et al. 1997; Roberts et al. 1998). VSV expressing HA from A/Hong Kong/156/97, a highly pathogenic H5N1 virus, was shown to elicit neutralizing antibodies in mice and confer protection after challenge with A/Hong Kong/156/97 virus (Fig. 3). Interestingly, this recombinant vector induced cross-reactive neutralizing antibodies to distantly related H5 viruses. Long-term protection was also achieved with this vector, as mice were fully protected after 7.5 months between vaccination and challenge (Schwartz et al. 2007). These experiments demonstrate that VSV may be a suitable recombinant vector for influenza virus HA proteins.

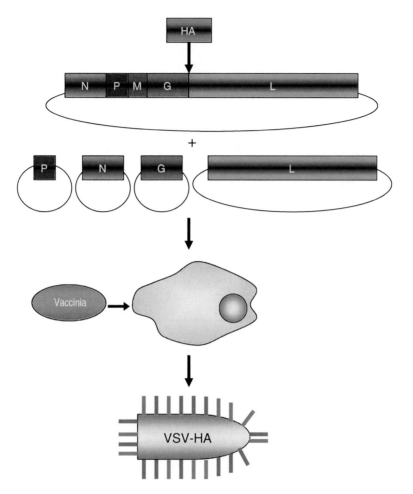

**Fig. 3** Rescue of a recombinant VSV vector expressing influenza virus HA protein. The HA gene was cloned between the G and the L gene in a plasmid containing the full-length VSV genome under the control of the T7 promoter. Cells were cotransfected with a plasmid containing the full-length VSV-HA genome as well as helper plasmids expressing NP, P, G, and L from a Pol II promoter. One hour prior to transfection, cells were infected with MVA-T7 vaccinia virus. The resulting VSV virus expressed VSV G, as well as influenza virus HA. Figure adapted from Schwartz et al. (2007)

Recombinant VSV vectors expressing various viral and bacterial proteins of pathogens have also been tested in animal models. VSV vectors expressing the *Yersinia pestis lcrV* gene provided protection in mice from lethal pulmonary challenge with *Yersinia pestis* (Palin et al. 2007). Vaccination with a recombinant VSV expressing SARS spike protein provided protection against challenge with SARS in both young and aged mice (Kapadia et al. 2005; Vogel et al. 2007). VSV vectors are also currently being tested as vaccines and therapeutic agents for HIV

(Johnson et al. 1997; Schnell et al. 1997; Haglund et al. 2000; Rose et al. 2001; Ramsburg et al. 2004; Publicover et al. 2005; Okuma et al. 2006; Cooper et al. 2008). Recombinant VSV vectors have been developed for many other viruses, such as herpes simplex virus type 2, Borna disease virus, Marburg virus, and papillomaviruses. (Daddario-DiCaprio et al. 2006; Natuk et al. 2006; Brandsma et al. 2007a,b; Perez et al. 2007).

One limitation of VSV as a recombinant vector is that it does appear to cause some pathogenicity in humans. Though humans are rarely infected, animal handlers have been exposed to VSV and show symptoms of disease ranging from asymptomatic to fever with myalgia that resolves within a week. Mouse studies of VSV infection demonstrate that the virus can replicate in the olfactory nerve soon after intranasal infection and can then cross the blood–brain barrier. The virus then spreads to many areas of the brain, resulting in neuropathology, hind-limb paralysis, and death (Huneycutt et al. 1994; Bi et al. 1995; Plakhov et al. 1995). VSV infection was also analyzed in nonhuman primates, and it is critical to determine the safety of vectors in this model prior to human use. Macaques inoculated with VSV intranasally shed virus in nasal washes for the first day after infection, but the virus did not cause viremia or enter the central nervous system. However, when macaques were injected with VSV directly into the brain with an intrathalamic injection, the virus spread and caused severe disease symptoms (Johnson et al. 2007).

In order for VSV to be used as a recombinant vector in humans, the virus must be attenuated so that disease symptoms are eliminated. Fortunately, much has already been discovered about the mechanisms of VSV pathogenicity, and so attenuated VSV vectors can be rationally designed. Insertion of the HIV Gag protein into the VSV genome attenuated the virus sufficiently so that it did not cause pathogenesis in macaques, though additional viral attenuation may be necessary for human trials (Johnson et al. 2007). Truncations of the cytoplasmic region of the G protein had been shown to attenuate VSV growth and pathogenesis in mice (Roberts et al. 1998). A recombinant vector containing a G protein deletion and expressing HIV Env protein elicited CD8[+] T cell responses comparable to wtVSV expressing Env protein (Publicover et al. 2004).

Another strategy for attenuating VSV expressing HIV proteins includes placing *gag* at the beginning of the genome, which results in a reduction of the expression of the VSV proteins and a reduction of viral replication. Also, the N gene was moved to a further downstream position, which reduces N protein expression and viral replication. The M protein, which has been shown to inhibit the interferon response and induce apoptosis, has also been mutated to reduce VSV pathogenesis (Clarke et al. 2007; Cooper et al. 2008). Applying a combination of these alterations to recombinant VSV vectors expressing Gag protein resulted in a drastic reduction in pathogenesis in mice but still induced a strong immune response to Gag (Cooper et al. 2008). VSV vector replication and pathogenesis can also be eliminated by using VSV mutants that can only complete one cycle of replication. This is achieved by eliminating the VSV G protein from the genome. Since VSV G protein, but not the G gene, is necessary for viral growth, the vector can be grown readily in cell lines that constitutively express VSV G so that G protein can be incorporated into the viral envelope. In animals, the single-cycle VSV vectors can enter and infect cells and express the viral and foreign proteins in its genome, but viral assembly cannot occur because the G protein is not synthesized. A VSV single-cycle vector expressing HIV Env was demonstrated to produce an immune response to Env that was similar to the response elicited by replicating VSV vectors expressing Env (Publicover et al. 2005). VSV vectors with attenuating mutations are being investigated carefully so that they can be safely administered to humans without side effects.

Similar to NDV, VSV is also being analyzed as a therapeutic cancer agent based on the observation that VSV replicates and induces apoptosis in cancer cells more readily than normal cells (Barber 2004). VSV is also very sensitive to the effects of interferon, and tumor cells without an intact interferon pathway are rapidly killed by VSV. VSV vectors expressing the cytokine IL-12 or the chemokine inhibitor equine herpes virus-1 glycoprotein were able to enhance tumor reduction of squamous cell carcinoma and hepatocellular carcinoma, respectively, in mice (Shin et al. 2007; Altomonte et al. 2008). Thus, VSV vectors show the potential to function as vaccines for infectious diseases as well as cancer.

# 5  Influenza Virus Vectors

There has been recent interest in using influenza virus itself as a recombinant vector to protect against highly virulent influenza virus strains as well as other pathogens. Influenza viruses need to be highly attenuated for use as a vector, and several strategies for attenuating influenza viruses have been successful. An attenuated cold-adapted strain was generated by growing influenza virus at 25°C in primary chick kidney cells, and is currently licensed for use in humans (Cox et al. 1988). An attenuated cold-adapted strain that was generated in embryonated chicken eggs grown at low temperatures is also being used in horses (Youngner et al. 2001). Reduction of virulence is also observed by influenza viruses containing deletions of the NS1 protein, the viral protein responsible for inhibiting the innate interferon response (Garcia-Sastre et al. 1998; Talon et al. 2000). Influenza virus vaccines containing deletions of the M2 gene, which are necessary for virus uncoating, are currently being tested (Watanabe et al. 2007). Thus, there are likely many ways to sufficiently attenuate influenza virus.

Several influenza virus vectors have shown promising results in animal models. A recombinant influenza virus containing the NDV HN ectodomain in place of the influenza virus NA ectodomain administered in ovo provided protection in chickens against both influenza virus and NDV after a lethal challenge (Fig. 4) (Steel et al. 2008). Millions of people die each year after infection with *Mycobacterium tuberculosis*, and an effective vaccine is urgently needed. Vaccination with influenza virus that expresses a truncated NS1 protein and the ESAT-6 protein of *Mycobacterium tuberculosis* from the NS gene segment provides protection in mice and guinea pigs from lethal challenge with the bacteria (Sereinig et al. 2006; Stukova et al. 2006). Recombinant influenza viruses expressing portions of *Bacillus anthracis* proteins

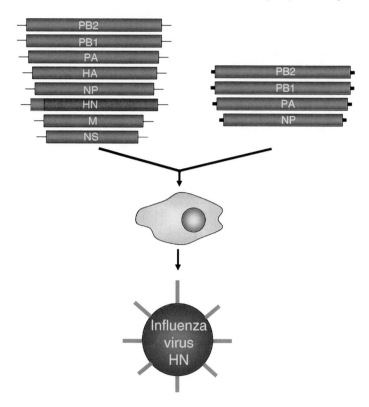

**Fig. 4** Rescue of a recombinant influenza virus expressing NDV HN. The ectodomain of influenza virus NA was replaced with the ectodomain of NDV HN. Eight of the plasmids contained a Pol I promoter (*left*) and four of the plasmids contained a Pol II promoter (*right*). Cells were transfected with the 12 plasmids and a recombinant virus expressing NDV HN was rescued, as described in Steel et al. (2008)

fused to the influenza virus HA protein elicited antibodies against both *Bacillus anthracis* and influenza virus (Li et al. 2005). Protection in mice was achieved after vaccination with an influenza virus expressing *Chlamydia trachomatis* epitopes in the NA protein (He et al. 2007). Mice vaccinated with an influenza virus expressing an epitope of *Pseudomonas aeruginosa*, which is the leading cause of mortality in cystic fibrosis patients, were fully protected after challenge (Gilleland et al. 2000). Chimeric influenza viruses have been developed to express HIV epitopes. After intranasal administration, these vectors induce a long-lasting mucosal antibody response in not only the respiratory tract, but also the genital tract (Li et al. 1993; Muster et al. 1994, 1995; Palese et al. 1997; Gonzalo et al. 1999; Gherardi et al. 2003; Nakaya et al. 2003). The encouraging results obtained using influenza virus vectors thus far demonstrate the need for further research in this field.

While the live cold-adapted influenza virus vaccine has been successfully administered to millions of individuals, safety must be carefully considered in the development of future live-attenuated influenza virus vaccines. Live-attenuated influenza virus vaccines must be designed so that a pathogenic virus could not result from reassortment of the vaccine with a circulating influenza virus strain. This issue was addressed for the cold-adapted vaccine by demonstrating that three internal genes contained attenuating mutations, making it unlikely that reassortment would lead to a virulent viral strain (Cox et al. 1986, 1988; Jin et al. 2003).

The issue of vector-mediated immunity is a concern for most live vaccines, but influenza virus rapidly evolves due to antigen drift. Live-attenuated influenza virus vaccines, like the cold-adapted vaccine, must be reformulated each year to reflect the newly emerged strains. Just as people can be infected multiple times with different strains of influenza virus, a recombinant vector based on influenza virus could be repeatedly administered if it was designed using different antigenic variants of influenza virus.

## 6   Adenovirus Vectors

Adenoviruses are nonenveloped DNA viruses that have been thoroughly explored for their potential use as recombinant vaccine vectors. Adenoviruses were originally identified as one of the causes of acute respiratory infections. Infection with adenoviruses has also been associated with conjunctivitis and gastroenteritis in infants. While adenovirus infection usually results in mild disease symptoms that are promptly resolved, adenovirus infection of immunocompromised individuals can result in severe disease symptoms, such as pneumonia, encephalitis, and even death (Krilov 2005). Adenovirus is often chosen as a recombinant vaccine vector to express foreign proteins because a live vaccine was administered to US military personal for over two decades with no incidence of significant side effects (Souza et al. 2005). This oral vaccine consisted of the two most prevalent strains of adenovirus among military personal, Ad4 and Ad7, contained in a capsule coated to prevent the release of the viruses until they reached the intestines (Howell et al. 1998; Lichtenstein and Wold 2004). Because of its potential to cause illness, many of the adenovirus vectors currently being developed and tested are replication defective and cannot spread cell-to-cell. Replication-defective vectors often have deletions of the E1 portion of the viral genome, since this region is necessary for the initiation of viral replication (Souza et al. 2005).

In addition to its extensive record as a military vaccine, adenovirus offers several advantages as a recombinant vaccine vector. The viral genome is relatively easy to manipulate and the virus grows to high titers. Adenovirus can by lyophilized, after which it does not need refrigeration (Souza et al. 2005). Because of these reasons, adenovirus has been one of the most popular recombinant vectors, and pharmaceutical companies have chosen to test adenovirus vectors in clinical trials.

Several groups have demonstrated that recombinant adenovirus vectors expressing influenza virus proteins can protect animals after challenge. Adenovirus vectors expressing HA and NP of an H3N2 swine influenza virus fully protected swine after a lethal challenge (Wesley et al. 2004). A recombinant adenovirus vector expressing HA of HPAI H5N1 induced both cellular and humoral immunity in mice, and the vaccine was completely protective in both mice and chickens after lethal challenge (Gao et al. 2006). An adenovirus vector expressing HA from a H9N2 strain that was used to vaccinate chickens in ovo provided complete protection against lethal challenge with an H5N2 strain and partial protection against an H5N1 strain (Toro et al. 2007). Long-lasting immunity after vaccination with adenovirus vector expressing HA from an H5N1 virus provides protection after lethal challenge for at least one year (Hoelscher et al. 2007).

Adenovirus vectors are currently being examined as possible vaccines for a variety of viruses. Adenovirus vectors, which have been shown to prevent disease after challenge, include those expressing herpes simplex virus and measles virus H, N, and F proteins (McDermott et al. 1989; Fooks et al. 1998; Sharpe et al. 2002). Perhaps the most famous adenovirus vector is the Merck-sponsored HIV vaccine V520 that recently went into clinical trials. The vaccine consisted of adenovirus vectors containing HIV *nef*, *gag*, and *pol* genes (Steinbrook 2007; Sekaly 2008). The vaccine was administered as three injections at zero, two, and six months. The clinical trials were halted early because it became clear that not only was the vaccine failing to prevent HIV infection, but the individuals given the vaccine also had a higher rate of HIV infection than those given the placebo (Sekaly 2008). This devastating failure necessitates a thorough analysis of what went wrong so that it will not be repeated in future trials. It is clear that a major problem is that many people have been exposed to the Ad5 strain used as the vaccine vector. About half of the individuals in western countries have antibodies to Ad5, and about 95% of people in developing countries have antibodies. The presence of pre-existing antibodies likely led to a rapid memory immune response that prevented the development of an immune response to the HIV proteins expressed from the vectors. What was unexpected and is not yet fully explained is that individuals that had been previously exposed to adenovirus before the vaccinations were more susceptible to HIV infection. Other researchers have been testing less common strains of adenovirus to use as vectors in hopes of circumventing the pre-existing immunity problems (Hofmann et al. 1999; Reddy et al. 1999; Farina et al. 2001).

Surprisingly, this was not the first time that adenovirus vectors have unexpectedly harmed clinical trial participants. Adenovirus vectors have been used in gene therapy trials as well as for therapy against cancer, and it was during a gene therapy trial that a participant died of an inflammatory response after receiving a high dose of vector ($3.8 \times 10^{13}$ virus particles) (Lehrman 1999; Marshall 1999). Another disadvantage of using this vector is that the adenovirus genome is DNA, and there is a risk that viral DNA may disrupt host genes and possibly cause cancer. Some adenovirus strains can cause cancer in laboratory animals (Trentin et al. 1962). Even though adenoviruses have not been shown to cause human cancer, it is possible that some cancer cases may arise after the vaccination of a large population. In light of the fiascos involving this vector, it is difficult to foresee high enthusiasm for adenovirus vectors in the future when other vectors seem more promising.

## 7   Venezuelan Equine Encephalitis Virus Vectors

Venezuelan equine encephalitis virus (VEE) is an RNA virus that primarily infects equines in Central and South America. In contrast to adenovirus, most humans do not have pre-existing antibodies to VEE that could interfere with the vaccine (Davis et al. 1996). Like VSV, VEE is transmitted by insects. Strains of VEE range in pathogenicity from avirulent to causing acute encephalitis and death in equines. Humans can be infected as well and usually only develop mild symptoms, but human deaths have occurred (Weaver et al. 2004). Thus, VEE must be highly attenuated in order to be used as a vaccine vector. Many of the VEE vaccines being tested use a viral replicon particle (VRP) that is capable of infecting cells but cannot spread throughout the host. Foreign proteins are expressed at high levels from the VRP vectors. One major advantage of using this vector it that the VEE targets antigen-presenting cells in the draining lymph node, so the foreign antigen is presented directly to the site where the adaptive immune response begins (Davis et al. 1996, 2002; Charles et al. 1997). Another advantage is that VEE vaccines can also induce an IgA mucosal immune response, even after subcutaneous injection of the vaccine (Charles et al. 1997). Since many pathogens, including HIV, initially invade mucosal surfaces, the induction of mucosal immunity by a vaccine is highly desirable.

VEE VRPs expressing HA from an H5N1 influenza virus were used to successfully protect two-week-old chickens from lethal challenge (Fig. 5) (Schultz-Cherry et al. 2000). VEE vaccines have also been developed for many other agents, including SIV, HIV, Lassa virus, Norwalk virus, *Borrelia burgdorferi* (the causative agent of Lyme disease), SARS-CoV, cowpox virus, dengue virus, and RSV (Caley et al. 1997, 1999; Pushko et al. 1997; Davis et al. 2000; Baric et al. 2002; Harrington et al. 2002; Gipson et al. 2003; Johnston et al. 2005; Deming et al. 2006; Cecil et al. 2007; Mok et al. 2007; Thornburg et al. 2007; White et al. 2007). Further testing will determine whether VEE vaccine vectors are safe and efficacious in humans.

## 8   Poxvirus Vectors

Poxviruses, DNA viruses with large genomes, have been studied as recombinant vectors after the successful eradication of smallpox using vaccinia virus. Vaccinia viruses possess several properties of an ideal vector (Panicali et al. 1983): they are easy and inexpensive to manufacture, can be lyophilized, can accommodate large inserts of foreign DNA, and can induce both mucosal and systemic immunity after oral administration (Gherardi and Esteban 2005; Souza et al. 2005). A major drawback of vaccinia vectors is that a large segment of the population has pre-existing immunity to vaccinia from the smallpox eradication program, which would interfere with the induction of an immune response to a foreign protein

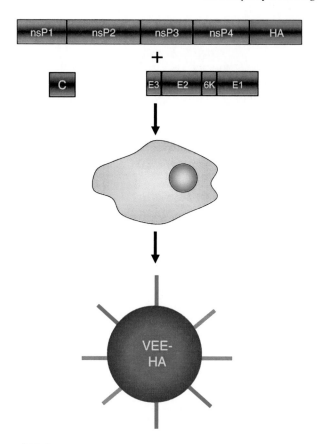

**Fig. 5** Rescue of VEE replicon expressing influenza virus HA from three messenger RNAs. Influenza virus HA was cloned into a plasmid containing the VEE nonstructural genes. Helper plasmids were prepared containing the capsid genes and glycoprotein genes. All plasmids were linearized and transcribed into mRNA. The mRNA was transfected into cells and a recombinant VEE replicon expressing influenza virus HA was rescued. Figure adapted from Pushko et al. (1997)

expressed from vaccinia virus. To overcome this, similar poxviruses from other species that do not cross-react with vaccinia virus, such as canarypox virus and fowlpox virus, are currently being tested (Johnson et al. 2005; Bublot et al. 2006). However, it appears that these recombinant vectors induce a weaker immune response to foreign proteins compared to recombinant vaccinia vectors (Zhang et al. 2007). A canarypox vaccine encoding HIV gp120 failed phase 2 clinical trials in humans since it failed to elicit a strong cellular immune response (Russell et al. 2007). Also, vaccinees can spread vaccinia virus to other individuals, which is especially dangerous for immunocompromised individuals. To address this, replication-defective attenuated vaccinia viruses, such as the Ankara strain, are being evaluated for recombinant vector potential, though these attenuated strains do cross-react with vaccinia virus, making it less likely that they will ultimately be successful (Souza et al. 2005).

## 9   Live Attenuated Measles Viruses as Recombinant Vectors

Another strategy involves altering currently licensed vaccines—such as the live attenuated measles virus vaccine—to express foreign proteins, in the hope that there would be a strong immune response to both measles virus and a foreign protein (Zuniga et al. 2007). The measles virus vaccine is highly efficacious in infants and has an excellent safety record. A measles virus vector expressing West Nile virus glycoprotein protected mice against a lethal challenge with West Nile virus (Despres et al. 2005). The only disadvantage of this vector is that most of the human population has already been vaccinated and has pre-existing immunity to measles virus. However, mice and macaques were vaccinated with the measles virus vaccine, and after 12 months were vaccinated with the measles virus vaccine expressing HIV gp140. The animals developed antibody titers to HIV that were similar to the antibody titers in naïve animals (Lorin et al. 2004). While more work is required to substantiate these results in order to recommend using this vector in humans with pre-existing measles virus immunity, at the very least this is a promising method for vaccinating naïve infants against both measles virus and another pathogen.

## 10   Other Recombinant Vectors

A current vaccine strategy under development is the use of bacteria as delivery vehicles of foreign antigens. Attenuated strains of intracellular bacteria such as *Salmonella enterica* serovar Typhimurium and *Listeria monocytogenes* are being engineered as recombinant vectors (Schoen et al. 2004; Cheminay and Hensel 2008; Schoen et al. 2008). While intracellular, the bacteria remain in a membrane-bound vesicle inside the host cell, which prevents foreign proteins expressed by bacteria from entering the host cytosol, a necessary step for antigens to be presented to the immune system. Two mechanisms of antigen delivery to combat this problem have been tested in bacterial vectors. One involves synthesis of the foreign protein inside the bacteria and release of the foreign protein into the human cell by the bacterial type III secretory pathway (Panthel et al. 2008). Because proteins must be unfolded prior to being secreted, foreign proteins with high stability cannot be completely unfolded and are unable to exit the bacterial cell. It has been demonstrated that the removal of small stabilizing domains in HIV proteins can allow these large foreign proteins to be secreted by the type III pathway (Chen et al. 2006). However, the complexity of the bacterial genome and the difficulty of secreting foreign proteins will limit the use of this system.

Another mechanism of antigen delivery by bacterial vectors involves the release of DNA encoding a foreign protein into the host cell, essentially a DNA vaccine delivered by a live bacterial organism. This strategy includes transforming bacteria with a naked DNA vector that encodes a foreign gene. Bacteria that target antigen-presenting cells, such as *Listeria monocytogenes* that targets dendritic cells, must be used for this technique. The chosen bacteria have to be highly attenuated and designed to lyse upon host cell entry. Once the bacteria are lysed, the DNA vector enters the cytosol and then transports to the nucleus, where it is transcribed. After being translated in the cytoplasm, the antigens can be processed to be presented on both MHC I and MHC II molecules in order to stimulate humoral and cellular immunity (Mollenkopf et al. 2001; Weiss 2003; Schoen et al. 2008). While this method shows promise, as with viral vectors, pre-existing immunity to bacterial vectors does appear to inhibit the production of an immune response to foreign proteins (Sevil Domenech et al. 2007).

A new area of recombinant vector research has been focusing on using transgenic plants as delivery vectors. Plants are safe and inexpensive vectors, can easily be grown in large quantities, are stable at room temperature, and can be designed to express many antigens (Webster et al. 2005). Expression of HIV antigens in plants has been reported, and these vaccine vectors are currently being evaluated for their efficacy (Yusibov et al. 1997; Marusic et al. 2001).

# 11 Conclusions

The growing interest in using recombinant vectors as vaccines for influenza virus and other dangerous pathogens reflects the reality that these vaccines have substantial advantages over most other types of vaccines. While some recombinant vectors appear to be more encouraging than others, ideally it is hoped that several different vectors will ultimately be used to vaccinate against different diseases. An important hurdle to overcome in the development of recombinant vectors is the problem of pre-existing immunity to many of the vectors being tested. The issue of pre-existing immunity must also be addressed for recombinant viruses that humans currently do not have immunity against. This is because people have to be vaccinated multiple times for influenza virus, as the viral HA protein mutates. If a strong immune response is generated against the vector after the first vaccination, the vector may not be able to replicate sufficiently after successive administrations. This would prevent the formation of an immune response to the mutated HAs, leaving the individual vulnerable to infection with the altered influenza viruses. This is being addressed for VSV by the generation of vectors that express different serotypes of VSV G protein that do not cross-react. Recombinant VSV vectors expressing HIV Env elicit a strong immune response to Env, and subsequent vaccination with different VSV vectors expressing HIV Env and other G proteins can be used to boost the initial immune response (Rose et al. 2000). Further work in this area is needed to overcome this limitation of recombinant vectors. A comparison of the viral vectors is shown in Table 1. Overall, recombinant vectors provide a

**Table 1** Comparison of the viral vectors

| Vector | Advantages | Disadvantages | References |
|---|---|---|---|
| NDV | RNA genome; humans do not have pre-existing immunity; multiple foreign genes can be expressed from bisegmented genome; can be constructed for avian and human use; phase I trials demonstrate safety in humans | Requires refrigeration | Freeman et al. (2006); Laurie et al. (2006); Park et al. (2006); Veits et al. (2006); Gao et al. (2008) |
| VSV | RNA genome; humans do not have pre-existing immunity | Vector safety in humans not yet established; requires refrigeration | Roberts and Rose (1999); Clarke et al. (2006) |
| Influenza virus | RNA genome; can be constructed for avian and human use | Segmented genome that could reassort; humans have pre-existing immunity; requires refrigeration | Garcia-Sastre and Palese (1995); Steel et al. (2008) |
| Adenovirus | Can be lyophilized; does not require refrigeration | DNA genome (may alter host genome); humans have pre-existing immunity; clinical trial failure | Lehrman (1999); Marshall (1999); Souza et al. (2005); Steinbrook (2007); Sekaly (2008) |
| VEE | RNA genome; humans do not have pre-existing immunity; targets APC cells; can induce mucosal immunity after s.c. injection | Vector safety in humans is currently being studied | Davis et al. (1996); Charles et al. (1997); Davis et al. (2002) |
| Poxviruses | Can be lyophilized; does not require refrigeration; can accommodate at least 30 KB of foreign DNA | DNA genome (may alter host genome); humans have pre-existing immunity | Gherardi and Esteban (2005); Souza et al. (2005); Vijaysri et al. (2008) |
| Measles virus | RNA genome; excellent safety record in humans | Humans have pre-existing immunity | Zuniga et al. (2007) |

safe and effective mechanism for eliciting humoral and cellular immunity to the most dangerous pathogens on the planet.

**Acknowledgements** Work done in the laboratory of the authors was funded by NIH grants U54 AI057158 (Northeast Biodefense Center) and U01 AI070469 (Live Attenuated Vaccines for Epidemic and Pandemic Flu), and with Bill and Melinda Gates Foundation grant 38648.

# References

Altomonte J, Wu L, Chen L, Meseck M, Ebert O, Garcia-Sastre A, Fallon J, Woo SL (2008) Exponential enhancement of oncolytic vesicular stomatitis virus potency by vector-mediated suppression of inflammatory responses in vivo. Mol Ther 16(1):146–153

Babiuk LA, Tikoo SK (2000) Adenoviruses as vectors for delivering vaccines to mucosal surfaces. J Biotechnol 83(1–2):105–113

Barber GN (2004) Vesicular stomatitis virus as an oncolytic vector. Viral Immunol 17(4):516–527

Baric RS, Yount B, Lindesmith L, Harrington PR, Greene SR, Tseng FC, Davis N, Johnston RE, Klapper DG, Moe CL (2002) Expression and self-assembly of Norwalk virus capsid protein from Venezuelan equine encephalitis virus replicons. J Virol 76(6):3023–3030

Barouch DH (2006) Rational design of gene-based vaccines. J Pathol 208(2):283–289

Barouch DH, Nabel GJ (2005) Adenovirus vector-based vaccines for human immunodeficiency virus type 1. Hum Gene Ther 16(2):149–156

Bi Z, Barna M, Komatsu T, Reiss CS (1995) Vesicular stomatitis virus infection of the central nervous system activates both innate and acquired immunity. J Virol 69(10):6466–6472

Brandsma JL, Shlyankevich M, Buonocore L, Roberts A, Becker SM, Rose JK (2007a) Therapeutic efficacy of vesicular stomatitis virus-based E6 vaccination in rabbits. Vaccine 25(4):751–762

Brandsma JL, Shylankevich M, Su Y, Roberts A, Rose JK, Zelterman D, Buonocore L (2007b) Vesicular stomatitis virus-based therapeutic vaccination targeted to the E1, E2, E6, and E7 proteins of cottontail rabbit papillomavirus. J Virol 81(11):5749–5758

Bublot M, Pritchard N, Swayne DE, Selleck P, Karaca K, Suarez DL, Audonnet JC, Mickle TR (2006) Development and use of fowlpox vectored vaccines for avian influenza. Ann N Y Acad Sci 1081:193–201

Bukreyev A, Huang Z, Yang L, Elankumaran S, St Claire M, Murphy BR, Samal SK, Collins PL (2005) Recombinant Newcastle disease virus expressing a foreign viral antigen is attenuated and highly immunogenic in primates. J Virol 79(21):13275–13284

Caley IJ, Betts MR, Irlbeck DM, Davis NL, Swanstrom R, Frelinger JA, Johnston RE (1997) Humoral, mucosal, and cellular immunity in response to a human immunodeficiency virus type 1 immunogen expressed by a Venezuelan equine encephalitis virus vaccine vector. J Virol 71(4):3031–3038

Caley IJ, Betts MR, Davis NL, Swanstrom R, Frelinger JA, Johnston RE (1999) Venezuelan equine encephalitis virus vectors expressing HIV-1 proteins: vector design strategies for improved vaccine efficacy. Vaccine 17(23–24):3124–3135

Cecil C, West A, Collier M, Jurgens C, Madden V, Whitmore A, Johnston R, Moore DT, Swanstrom R, Davis NL (2007) Structure and immunogenicity of alternative forms of the simian immunodeficiency virus gag protein expressed using Venezuelan equine encephalitis virus replicon particles. Virology 362(2):362–373

Charles PC, Brown KW, Davis NL, Hart MK, Johnston RE (1997) Mucosal immunity induced by parenteral immunization with a live attenuated Venezuelan equine encephalitis virus vaccine candidate. Virology 228(2):153–160

Cheminay C, Hensel M (2008) Rational design of Salmonella recombinant vaccines. Int J Med Microbiol 298(1–2):87–98

Chen LM, Briones G, Donis RO, Galan JE (2006) Optimization of the delivery of heterologous proteins by the *Salmonella enterica* serovar Typhimurium type III secretion system for vaccine development. Infect Immun 74(10):5826–5833

Clarke DK, Cooper D, Egan MA, Hendry RM, Parks CL, Udem SA (2006) Recombinant vesicular stomatitis virus as an HIV-1 vaccine vector. Springer Semin Immunopathol 28(3):239–253

Clarke DK, Nasar F, Lee M, Johnson JE, Wright K, Calderon P, Guo M, Natuk R, Cooper D, Hendry RM, Udem SA (2007) Synergistic attenuation of vesicular stomatitis virus by combination of specific G gene truncations and N gene translocations. J Virol 81(4):2056–2064

Cooper D, Wright KJ, Calderon PC, Guo M, Nasar F, Johnson JE, Coleman JW, Lee M, Kotash C, Yurgelonis I, Natuk RJ, Hendry RM, Udem SA, Clarke DK (2008) Attenuation of recombinant vesicular stomatitis virus-human immunodeficiency virus type 1 vaccine vectors by gene translocations and g gene truncation reduces neurovirulence and enhances immunogenicity in mice. J Virol 82(1):207–219

Cox NJ, Kitame F, Klimov A, Koennecke I, Kendal AP (1986) Comparative studies of wild-type and cold-mutant (temperature-sensitive) influenza virus: detection of mutations in all genes of the A/Ann Arbor/6/60 (H2N2) mutant vaccine donor strain. Microb Pathog 1(4):387–397

Cox NJ, Kitame F, Kendal AP, Maassab HF, Naeve C (1988) Identification of sequence changes in the cold-adapted, live attenuated influenza vaccine strain, A/Ann Arbor/6/60 (H2N2). Virology 167(2):554–567

Daddario-DiCaprio KM, Geisbert TW, Geisbert JB, Stroher U, Hensley LE, Grolla A, Fritz EA, Feldmann F, Feldmann H, Jones SM (2006) Cross-protection against Marburg virus strains by using a live, attenuated recombinant vaccine. J Virol 80(19):9659–9666

Davis NL, Brown KW, Johnston RE (1996) A viral vaccine vector that expresses foreign genes in lymph nodes and protects against mucosal challenge. J Virol 70(6):3781–3787

Davis NL, Caley IJ, Brown KW, Betts MR, Irlbeck DM, McGrath KM, Connell MJ, Montefiori DC, Frelinger JA, Swanstrom R, Johnson PR, Johnston RE (2000) Vaccination of macaques against pathogenic simian immunodeficiency virus with Venezuelan equine encephalitis virus replicon particles. J Virol 74(1):371–378

Davis NL, West A, Reap E, MacDonald G, Collier M, Dryga S, Maughan M, Connell M, Walker C, McGrath K, Cecil C, Ping LH, Frelinger J, Olmsted R, Keith P, Swanstrom R, Williamson C, Johnson P, Montefiori D, Johnston RE (2002) Alphavirus replicon particles as candidate HIV vaccines. IUBMB Life 53(4–5):209–211

Deming D, Sheahan T, Heise M, Yount B, Davis N, Sims A, Suthar M, Harkema J, Whitmore A, Pickles R, West A, Donaldson E, Curtis K, Johnston R, Baric R (2006) Vaccine efficacy in senescent mice challenged with recombinant SAR S-CoV bearing epidemic and zoonotic spike variants. PLoS Med 3(12):e525

Despres P, Combredet C, Frenkiel MP, Lorin C, Brahic M, Tangy F (2005) Live measles vaccine expressing the secreted form of the West Nile virus envelope glycoprotein protects against West Nile virus encephalitis. J Infect Dis 191(2):207–214

DiNapoli JM, Kotelkin A, Yang L, Elankumaran S, Murphy BR, Samal SK, Collins PL, Bukreyev A (2007) Newcastle disease virus, a host range-restricted virus, as a vaccine vector for intranasal immunization against emerging pathogens. Proc Natl Acad Sci USA 104(23):9788–9793

Farina SF, Gao GP, Xiang ZQ, Rux JJ, Burnett RM, Alvira MR, Marsh J, Ertl HC, Wilson JM (2001) Replication-defective vector based on a chimpanzee adenovirus. J Virol 75(23): 11603–11613

Fooks AR, Jeevarajah D, Lee J, Warnes A, Niewiesk S, ter Meulen V, Stephenson JR, Clegg JC (1998) Oral or parenteral administration of replication-deficient adenoviruses expressing the measles virus haemagglutinin and fusion proteins: protective immune responses in rodents. J Gen Virol 79(Pt 5):1027–1031

Freeman AI, Zakay-Rones Z, Gomori JM, Linetsky E, Rasooly L, Greenbaum E, Rozenman-Yair S, Panet A, Libson E, Irving CS, Galun E, Siegal T (2006) Phase I/II trial of intravenous NDV-HUJ oncolytic virus in recurrent glioblastoma multiforme. Mol Ther 13(1):221–228

Gao W, Soloff AC, Lu X, Montecalvo A, Nguyen DC, Matsuoka Y, Robbins PD, Swayne DE, Donis RO, Katz JM, Barratt-Boyes SM, Gambotto A (2006) Protection of mice and poultry from lethal H5N1 avian influenza virus through adenovirus-based immunization. J Virol 80(4):1959–1964

Gao Q, Park MS, Palese P (2008) Expression of transgenes from Newcastle disease virus with a segmented genome. J Virol 82:2692–2698

Garcia-Sastre A, Palese P (1995) Influenza virus vectors. Biologicals 23(2):171–178

Garcia-Sastre A, Egorov A, Matassov D, Brandt S, Levy DE, Durbin JE, Palese P, Muster T (1998) Influenza A virus lacking the NS1 gene replicates in interferon-deficient systems. Virology 252(2):324–330

Ge J, Deng G, Wen Z, Tian G, Wang Y, Shi J, Wang X, Li Y, Hu S, Jiang Y, Yang C, Yu K, Bu Z, Chen H (2007) Newcastle disease virus-based live attenuated vaccine completely protects chickens and mice from lethal challenge of homologous and heterologous H5N1 avian influenza viruses. J Virol 81(1):150–158

Gherardi MM, Esteban M (2005) Recombinant poxviruses as mucosal vaccine vectors. J Gen Virol 86(Pt 11):2925–2936

Gherardi MM, Najera JL, Perez-Jimenez E, Guerra S, Garcia-Sastre A, Esteban M (2003) Prime-boost immunization schedules based on influenza virus and vaccinia virus vectors potentiate cellular immune responses against human immunodeficiency virus Env protein systemically and in the genitorectal draining lymph nodes. J Virol 77(12):7048–7057

Gilleland HE, Gilleland LB, Staczek J, Harty RN, Garcia-Sastre A, Palese P, Brennan FR, Hamilton WD, Bendahmane M, Beachy RN (2000) Chimeric animal and plant viruses expressing epitopes of outer membrane protein F as a combined vaccine against *Pseudomonas aeruginosa* lung infection. FEMS Immunol Med Microbiol 27(4):291–297

Gipson CL, Davis NL, Johnston RE, de Silva AM (2003) Evaluation of Venezuelan equine encephalitis (VEE) replicon-based outer surface protein A (OspA) vaccines in a tick challenge mouse model of Lyme disease. Vaccine 21(25–26):3875–3884

Gonzalo RM, Rodriguez D, Garcia-Sastre A, Rodriguez JR, Palese P, Esteban M (1999) Enhanced CD8+ T cell response to HIV-1 env by combined immunization with influenza and vaccinia virus recombinants. Vaccine 17(7–8):887–892

Haglund K, Forman J, Krausslich HG, Rose JK (2000) Expression of human immunodeficiency virus type 1 Gag protein precursor and envelope proteins from a vesicular stomatitis virus recombinant: high-level production of virus-like particles containing HIV envelope. Virology 268(1):112–121

Harrington PR, Yount B, Johnston RE, Davis N, Moe C, Baric RS (2002) Systemic, mucosal, and heterotypic immune induction in mice inoculated with Venezuelan equine encephalitis replicons expressing Norwalk virus-like particles. J Virol 76(2):730–742

He Q, Martinez-Sobrido L, Eko FO, Palese P, Garcia-Sastre A, Lyn D, Okenu D, Bandea C, Ananaba GA, Black CM, Igietseme JU (2007) Live-attenuated influenza viruses as delivery vectors for Chlamydia vaccines. Immunology 122(1):28–37

Hoelscher MA, Jayashankar L, Garg S, Veguilla V, Lu X, Singh N, Katz JM, Mittal SK, Sambhara S (2007) New pre-pandemic influenza vaccines: an egg- and adjuvant-independent human adenoviral vector strategy induces long-lasting protective immune responses in mice. Clin Pharmacol Ther 82(6):665–671

Hofmann C, Loser P, Cichon G, Arnold W, Both GW, Strauss M (1999) Ovine adenovirus vectors overcome preexisting humoral immunity against human adenoviruses in vivo. J Virol 73(8):6930–6936

Howell MR, Nang RN, Gaydos CA, Gaydos JC (1998) Prevention of adenoviral acute respiratory disease in Army recruits: cost-effectiveness of a military vaccination policy. Am J Prev Med 14(3):168–175

Huang Z, Elankumaran S, Panda A, Samal SK (2003) Recombinant Newcastle disease virus as a vaccine vector. Poult Sci 82(6):899–906

Huang Z, Elankumaran S, Yunus AS, Samal SK (2004) A recombinant Newcastle disease virus (NDV) expressing VP2 protein of infectious bursal disease virus (IBDV) protects against NDV and IBDV. J Virol 78(18):10054–10063

Huneycutt BS, Plakhov IV, Shusterman Z, Bartido SM, Huang A, Reiss CS, Aoki C (1994) Distribution of vesicular stomatitis virus proteins in the brains of BALB/c mice following intranasal inoculation: an immunohistochemical analysis. Brain Res 635(1–2):81–95

Janke M, Peeters B, de Leeuw O, Moorman R, Arnold A, Fournier P, Schirrmacher V (2007) Recombinant Newcastle disease virus (NDV) with inserted gene coding for GM-CSF as a new vector for cancer immunogene therapy. Gene Ther 14(23):1639–1649

Jin H, Lu B, Zhou H, Ma C, Zhao J, Yang CF, Kemble G, Greenberg H (2003) Multiple amino acid residues confer temperature sensitivity to human influenza virus vaccine strains (FluMist) derived from cold-adapted A/Ann Arbor/6/60. Virology 306(1):18–24

Johnson JE, Schnell MJ, Buonocore L, Rose JK (1997) Specific targeting to CD4+ cells of recombinant vesicular stomatitis viruses encoding human immunodeficiency virus envelope proteins. J Virol 71(7):5060–5068

Johnson DC, McFarland EJ, Muresan P, Fenton T, McNamara J, Read JS, Hawkins E, Bouquin PL, Estep SG, Tomaras GD, Vincent CA, Rathore M, Melvin AJ, Gurunathan S, Lambert J (2005) Safety and immunogenicity of an HIV-1 recombinant canarypox vaccine in newborns and infants of HIV-1-infected women. J Infect Dis 192(12):2129–2133

Johnson JE, Nasar F, Coleman JW, Price RE, Javadian A, Draper K, Lee M, Reilly PA, Clarke DK, Hendry RM, Udem SA (2007) Neurovirulence properties of recombinant vesicular stomatitis virus vectors in non-human primates. Virology 360(1):36–49

Johnston RE, Johnson PR, Connell MJ, Montefiori DC, West A, Collier ML, Cecil C, Swanstrom R, Frelinger JA, Davis NL (2005) Vaccination of macaques with SIV immunogens delivered by Venezuelan equine encephalitis virus replicon particle vectors followed by a mucosal challenge with SIVsmE660. Vaccine 23(42):4969–4979

Kapadia SU, Rose JK, Lamirande E, Vogel L, Subbarao K, Roberts A (2005) Long-term protection from SARS coronavirus infection conferred by a single immunization with an attenuated VSV-based vaccine. Virology 340(2):174–182

Kretzschmar E, Buonocore L, Schnell MJ, Rose JK (1997) High-efficiency incorporation of functional influenza virus glycoproteins into recombinant vesicular stomatitis viruses. J Virol 71(8):5982–5989

Krilov LR (2005) Adenovirus infections in the immunocompromised host. Pediatr Infect Dis J 24(6):555–556

Laurie SA, Bell JC, Atkins HL, Roach J, Bamat MK, O'Neil JD, Roberts MS, Groene WS, Lorence RM (2006) A phase 1 clinical study of intravenous administration of PV701, an oncolytic virus, using two-step desensitization. Clin Cancer Res 12(8):2555–2562

Lehrman S (1999) Virus treatment questioned after gene therapy death. Nature 401(6753):517–518

Letchworth GJ, Rodriguez LL, Del cbarrera J (1999) Vesicular stomatitis. Vet J 157(3):239–260

Li S, Polonis V, Isobe H, Zaghouani H, Guinea R, Moran T, Bona C, Palese P (1993) Chimeric influenza virus induces neutralizing antibodies and cytotoxic T cells against human immunodeficiency virus type 1. J Virol 67(11):6659–6666

Li ZN, Mueller SN, Ye L, Bu Z, Yang C, Ahmed R, Steinhauer DA (2005) Chimeric influenza virus hemagglutinin proteins containing large domains of the *Bacillus anthracis* protective antigen: protein characterization, incorporation into infectious influenza viruses, and antigenicity. J Virol 79(15):10003–10012

Li S, Locke E, Bruder J, Clarke D, Doolan DL, Havenga MJ, Hill AV, Liljestrom P, Monath TP, Naim HY, Ockenhouse C, Tang DC, Van Kampen KR, Viret JF, Zavala F, Dubovsky F (2007) Viral vectors for malaria vaccine development. Vaccine 25(14):2567–2574

Lichtenstein DL, Wold WS (2004) Experimental infections of humans with wild-type adenoviruses and with replication-competent adenovirus vectors: replication, safety, and transmission. Cancer Gene Ther 11(12):819–829

Lorin C, Mollet L, Delebecque F, Combredet C, Hurtrel B, Charneau P, Brahic M, Tangy F (2004) A single injection of recombinant measles virus vaccines expressing human immunodeficiency virus (HIV) type 1 clade B envelope glycoproteins induces neutralizing antibodies and cellular immune responses to HIV. J Virol 78(1):146–157

Marshall E (1999) Gene therapy death prompts review of adenovirus vector. Science 286(5448): 2244–2245

Martinez-Sobrido L, Gitiban N, Fernandez-Sesma A, Cros J, Mertz SE, Jewell NA, Hammond S, Flano E, Durbin RK, Garcia-Sastre A, Durbin JE (2006) Protection against respiratory syncytial virus by a recombinant Newcastle disease virus vector. J Virol 80(3):1130–1139

Marusic C, Rizza P, Lattanzi L, Mancini C, Spada M, Belardelli F, Benvenuto E, Capone I (2001) Chimeric plant virus particles as immunogens for inducing murine and human immune responses against human immunodeficiency virus type 1. J Virol 75(18):8434–8439

McDermott MR, Graham FL, Hanke T, Johnson DC (1989) Protection of mice against lethal challenge with herpes simplex virus by vaccination with an adenovirus vector expressing HSV glycoprotein B. Virology 169(1):244–247

Mok H, Lee S, Utley TJ, Shepherd BE, Polosukhin VV, Collier ML, Davis NL, Johnston RE, Crowe Jr JE (2007) Venezuelan equine encephalitis virus replicon particles encoding respiratory syncytial virus surface glycoproteins induce protective mucosal responses in mice and cotton rats. J Virol 81(24):13710–13722

Mollenkopf H, Dietrich G, Kaufmann SH (2001) Intracellular bacteria as targets and carriers for vaccination. Biol Chem 382(4):521–532

Muster T, Guinea R, Trkola A, Purtscher M, Klima A, Steindl F, Palese P, Katinger H (1994) Cross-neutralizing activity against divergent human immunodeficiency virus type 1 isolates induced by the gp41 sequence ELDKWAS. J Virol 68(6):4031–4034

Muster T, Ferko B, Klima A, Purtscher M, Trkola A, Schulz P, Grassauer A, Engelhardt OG, Garcia-Sastre A, Palese P et al (1995) Mucosal model of immunization against human immunodeficiency virus type 1 with a chimeric influenza virus. J Virol 69(11):6678–6686

Nakaya T, Cros J, Park MS, Nakaya Y, Zheng H, Sagrera A, Villar E, Garcia-Sastre A, Palese P (2001) Recombinant Newcastle disease virus as a vaccine vector. J Virol 75(23):11868–11873

Nakaya Y, Zheng H, Garcia-Sastre A (2003) Enhanced cellular immune responses to SIV Gag by immunization with influenza and vaccinia virus recombinants. Vaccine 21(17–18):2097–2106

Natuk RJ, Cooper D, Guo M, Calderon P, Wright KJ, Nasar F, Witko S, Pawlyk D, Lee M, DeStefano J, Tummolo D, Abramovitz AS, Gangolli S, Kalyan N, Clarke DK, Hendry RM, Eldridge JH, Udem SA, Kowalski J (2006) Recombinant vesicular stomatitis virus vectors expressing herpes simplex virus type 2 gD elicit robust CD4+ Th1 immune responses and are protective in mouse and guinea pig models of vaginal challenge. J Virol 80(9):4447–4457

Okuma K, Boritz E, Walker J, Sarkar A, Alexander L, Rose JK (2006) Recombinant vesicular stomatitis viruses encoding simian immunodeficiency virus receptors target infected cells and control infection. Virology 346(1):86–97

Palese P, Zavala F, Muster T, Nussenzweig RS, Garcia-Sastre A (1997) Development of novel influenza virus vaccines and vectors. J Infect Dis 176 Suppl 1:S45–S49

Palin A, Chattopadhyay A, Park S, Delmas G, Suresh R, Senina S, Perlin DS, Rose JK (2007) An optimized vaccine vector based on recombinant vesicular stomatitis virus gives high-level, long-term protection against *Yersinia pestis* challenge. Vaccine 25(4):741–750

Panda A, Huang Z, Elankumaran S, Rockemann DD, Samal SK (2004) Role of fusion protein cleavage site in the virulence of Newcastle disease virus. Microb Pathog 36(1):1–10

Panicali D, Davis SW, Weinberg RL, Paoletti E (1983) Construction of live vaccines by using genetically engineered poxviruses: biological activity of recombinant vaccinia virus expressing influenza virus hemagglutinin. Proc Natl Acad Sci USA 80(17):5364–5368

Panthel K, Meinel KM, Sevil Domenech VE, Trulzsch K, Russmann H (2008) Salmonella type III-mediated heterologous antigen delivery: a versatile oral vaccination strategy to induce cellular immunity against infectious agents and tumors. Int J Med Microbiol 298(1–2):99–103

Park MS, Steel J, Garcia-Sastre A, Swayne D, Palese P (2006) Engineered viral vaccine constructs with dual specificity: avian influenza and Newcastle disease. Proc Natl Acad Sci USA 103(21):8203–8208

Perez M, Clemente R, Robison CS, Jeetendra E, Jayakar HR, Whitt MA, de la Torre JC (2007) Generation and characterization of a recombinant vesicular stomatitis virus expressing the glycoprotein of Borna disease virus. J Virol 81(11):5527–5536

Plakhov IV, Arlund EE, Aoki C, Reiss CS (1995) The earliest events in vesicular stomatitis virus infection of the murine olfactory neuroepithelium and entry of the central nervous system. Virology 209(1):257–262

Publicover J, Ramsburg E, Rose JK (2004) Characterization of nonpathogenic, live, viral vaccine vectors inducing potent cellular immune responses. J Virol 78(17):9317–9324

Publicover J, Ramsburg E, Rose JK (2005) A single-cycle vaccine vector based on vesicular stomatitis virus can induce immune responses comparable to those generated by a replication-competent vector. J Virol 79(21):13231–13238

Puhler F, Willuda J, Puhlmann J, Mumberg D, Romer-Oberdorfer A, Beier R (2008) Generation of a recombinant oncolytic Newcastle disease virus and expression of a full IgG antibody from two transgenes. Gene Ther 15(5):371–383

Pushko P, Parker M, Ludwig GV, Davis NL, Johnston RE, Smith JF (1997) Replicon-helper systems from attenuated Venezuelan equine encephalitis virus: expression of heterologous genes in vitro and immunization against heterologous pathogens in vivo. Virology 239(2):389–401

Ramsburg E, Rose NF, Marx PA, Mefford M, Nixon DF, Moretto WJ, Montefiori D, Earl P, Moss B, Rose JK (2004) Highly effective control of an AIDS virus challenge in macaques by using vesicular stomatitis virus and modified vaccinia virus Ankara vaccine vectors in a single-boost protocol. J Virol 78(8):3930–3940

Reddy PS, Idamakanti N, Chen Y, Whale T, Babiuk LA, Mehtali M, Tikoo SK (1999) Replication-defective bovine adenovirus type 3 as an expression vector. J Virol 73(11):9137–9144

Roberts A, Rose JK (1999) Redesign and genetic dissection of the rhabdoviruses. Adv Virus Res 53:301–319

Roberts A, Kretzschmar E, Perkins AS, Forman J, Price R, Buonocore L, Kawaoka Y, Rose JK (1998) Vaccination with a recombinant vesicular stomatitis virus expressing an influenza virus hemagglutinin provides complete protection from influenza virus challenge. J Virol 72(6):4704–4711

Rodriguez LL (2002) Emergence and re-emergence of vesicular stomatitis in the United States. Virus Res 85(2):211–219

Rose NF, Roberts A, Buonocore L, Rose JK (2000) Glycoprotein exchange vectors based on vesicular stomatitis virus allow effective boosting and generation of neutralizing antibodies to a primary isolate of human immunodeficiency virus type 1. J Virol 74(23):10903–10910

Rose NF, Marx PA, Luckay A, Nixon DF, Moretto WJ, Donahoe SM, Montefiori D, Roberts A, Buonocore L, Rose JK (2001) An effective AIDS vaccine based on live attenuated vesicular stomatitis virus recombinants. Cell 106(5):539–549

Russell PH, Ezeifeka GO (1995) The Hitchner B1 strain of Newcastle disease virus induces high levels of IgA, IgG and IgM in newly hatched chicks. Vaccine 13(1):61–66

Russell ND, Graham BS, Keefer MC, McElrath MJ, Self SG, Weinhold KJ, Montefiori DC, Ferrari G, Horton H, Tomaras GD, Gurunathan S, Baglyos L, Frey SE, Mulligan MJ, Harro CD, Buchbinder SP, Baden LR, Blattner WA, Koblin BA, Corey L (2007) Phase 2 study of an HIV-1 canarypox vaccine (vCP1452) alone and in combination with rgp120: negative results fail to trigger a phase 3 correlates trial. J Acquir Immune Defic Syndr 44(2):203–212

Schnell MJ, Johnson JE, Buonocore L, Rose JK (1997) Construction of a novel virus that targets HIV-1-infected cells and controls HIV-1 infection. Cell 90(5):849–857

Schoen C, Stritzker J, Goebel W, Pilgrim S (2004) Bacteria as DNA vaccine carriers for genetic immunization. Int J Med Microbiol 294(5):319–335

Schoen C, Loeffler DI, Frentzen A, Pilgrim S, Goebel W, Stritzker J (2008) Listeria monocytogenes as novel carrier system for the development of live vaccines. Int J Med Microbiol 298(1–2):45–58

Schultz-Cherry S, Dybing JK, Davis NL, Williamson C, Suarez DL, Johnston R, Perdue ML (2000) Influenza virus (A/HK/156/97) hemagglutinin expressed by an alphavirus replicon system protects chickens against lethal infection with Hong Kong-origin H5N1 viruses. Virology 278(1):55–59

Schwartz JA, Buonocore L, Roberts A, Suguitan A Jr, Kobasa D, Kobinger G, Feldmann H, Subbarao K, Rose JK (2007) Vesicular stomatitis virus vectors expressing avian influenza H5 HA induce cross-neutralizing antibodies and long-term protection. Virology 366(1):166–173

Sereinig S, Stukova M, Zabolotnyh N, Ferko B, Kittel C, Romanova J, Vinogradova T, Katinger
  H, Kiselev O, Egorov A (2006) Influenza virus NS vectors expressing the mycobacterium
  tuberculosis ESAT-6 protein induce CD4+ Th1 immune response and protect animals against
  tuberculosis challenge. Clin Vaccine Immunol 13(8):898–904
Sevil Domenech VE, Panthel K, Meinel KM, Winter SE, Russmann H (2007) Pre-existing anti-
  Salmonella vector immunity prevents the development of protective antigen-specific CD8 T
  cell frequencies against murine listeriosis. Microbes Infect 9(12–13):1447–1453
Sharpe S, Fooks A, Lee J, Hayes K, Clegg C, Cranage M (2002) Single oral immunization with
  replication deficient recombinant adenovirus elicits long-lived transgene-specific cellular and
  humoral immune responses. Virology 293(2):210–216
Shin EJ, Wanna GB, Choi B, Aguila D III, Ebert O, Genden EM, Woo SL (2007) Interleukin-12
  expression enhances vesicular stomatitis virus oncolytic therapy in murine squamous cell
  carcinoma. Laryngoscope 117(2):210–214
Souza AP, Haut L, Reyes-Sandoval A, Pinto AR (2005) Recombinant viruses as vaccines against
  viral diseases. Braz J Med Biol Res 38(4):509–522
Stallknecht DE, Perzak DE, Bauer LD, Murphy MD, Howerth EW (2001) Contact transmission
  of vesicular stomatitis virus New Jersey in pigs. Am J Vet Res 62(4):516–520
Steel J, Burmakina SV, Thomas C, Spackman E, Garcia-Sastre A, Swayne DE, Palese P (2008)
  A combination in-ovo vaccine for avian influenza virus and Newcastle disease virus. Vaccine
  26(4):522–531
Steinbrook R (2007) One step forward, two steps back—will there ever be an AIDS vaccine.
  N Engl J Med 357(26):2653–2655
Stukova MA, Sereinig S, Zabolotnyh NV, Ferko B, Kittel C, Romanova J, Vinogradova TI,
  Katinger H, Kiselev OI, Egorov A (2006) Vaccine potential of influenza vectors expressing
  Mycobacterium tuberculosis ESAT-6 protein. Tuberculosis (Edinb) 86(3–4):236–246
Talon J, Salvatore M, O'Neill RE, Nakaya Y, Zheng H, Muster T, Garcia-Sastre A, Palese P (2000)
  Influenza A and B viruses expressing altered NS1 proteins: a vaccine approach. Proc Natl
  Acad Sci USA 97(8):4309–4314
Thornburg NJ, Ray CA, Collier ML, Liao HX, Pickup DJ, Johnston RE (2007) Vaccination with
  Venezuelan equine encephalitis replicons encoding cowpox virus structural proteins protects
  mice from intranasal cowpox virus challenge. Virology 362(2):441–452
Toro H, Tang DC, Suarez DL, Sylte MJ, Pfeiffer J, Van Kampen KR (2007) Protective
  avian influenza in ovo vaccination with non-replicating human adenovirus vector. Vaccine
  25(15):2886–2891
Trentin JJ, Yabe Y, Taylor G (1962) The quest for human cancer viruses. Science 137:835–841
Veits J, Wiesner D, Fuchs W, Hoffmann B, Granzow H, Starick E, Mundt E, Schirrmeier H,
  Mebatsion T, Mettenleiter TC, Romer-Oberdorfer A (2006) Newcastle disease virus express-
  ing H5 hemagglutinin gene protects chickens against Newcastle disease and avian influenza.
  Proc Natl Acad Sci USA 103(21):8197–8202
Vigil A, Park MS, Martinez O, Chua MA, Xiao S, Cros JF, Martinez-Sobrido L, Woo SL,
  Garcia-Sastre A (2007) Use of reverse genetics to enhance the oncolytic properties of
  Newcastle disease virus. Cancer Res 67(17):8285–8292
Vijaysri S, Jentarra G, Heck MC, Mercer AA, McInnes CJ, Jacobs BL (2008) Vaccinia viruses
  with mutations in the E3L gene as potential replication-competent, attenuated vaccines: intra-
  nasal vaccination. Vaccine 26(5):664–676
Vogel LN, Roberts A, Paddock CD, Genrich GL, Lamirande EW, Kapadia SU, Rose JK, Zaki SR,
  Subbarao K (2007) Utility of the aged BALB/c mouse model to demonstrate prevention and
  control strategies for severe acute respiratory syndrome coronavirus (SARS-CoV). Vaccine
  25(12):2173–2179
Watanabe T, Watanabe S, Kim JH, Hatta M, Kawaoka Y (2007) A novel approach to the development
  of effective H5N1 influenza A virus vaccines: the use of M2 cytoplasmic tail mutants. J Virol
  82(5):2486–2492
Weaver SC, Ferro C, Barrera R, Boshell J, Navarro JC (2004) Venezuelan equine encephalitis.
  Annu Rev Entomol 49:141–174

Webster DE, Thomas MC, Pickering R, Whyte A, Dry IB, Gorry PR, Wesselingh SL (2005) Is there a role for plant-made vaccines in the prevention of HIV/AIDS. Immunol Cell Biol 83(3):239–247

Weiss S (2003) Transfer of eukaryotic expression plasmids to mammalian hosts by attenuated Salmonella spp. Int J Med Microbiol 293(1):95–106

Wesley RD, Tang M, Lager KM (2004) Protection of weaned pigs by vaccination with human adenovirus 5 recombinant viruses expressing the hemagglutinin and the nucleoprotein of H3N2 swine influenza virus. Vaccine 22(25–26):3427–3434

White LJ, Parsons MM, Whitmore AC, Williams BM, de Silva A, Johnston RE (2007) An immunogenic and protective alphavirus replicon particle-based dengue vaccine overcomes maternal antibody interference in weanling mice. J Virol 81(19):10329–10339

Youngner SJ, Whitaker-Dowling P, Chambers TM, Rushlow KE, Sebring R (2001) Derivation and characterization of a live attenuated equine influenza vaccine virus. Am J Vet Res 62(8): 1290–1294

Yusibov V, Modelska A, Steplewski K, Agadjanyan M, Weiner D, Hooper DC, Koprowski H (1997) Antigens produced in plants by infection with chimeric plant viruses immunize against rabies virus and HIV-1. Proc Natl Acad Sci USA 94(11):5784–5788

Zhang X, Cassis-Ghavami F, Eller M, Currier J, Slike BM, Chen X, Tartaglia J, Marovich M, Spearman P (2007) Direct comparison of antigen production and induction of apoptosis by canarypox virus- and modified vaccinia virus Ankara-human immunodeficiency virus vaccine vectors. J Virol 81(13):7022–7033

Zuniga A, Wang Z, Liniger M, Hangartner L, Caballero M, Pavlovic J, Wild P, Viret JF, Glueck R, Billeter MA, Naim HY (2007) Attenuated measles virus as a vaccine vector. Vaccine 25(16):2974–2983

# Influenza Virus-Like Particles as Pandemic Vaccines

**S.M. Kang, P. Pushko, R.A. Bright, G. Smith, and R.W. Compans**

## Contents

**Abstract** There is an urgent need to develop novel approaches for vaccination against emerging pathogenic avian influenza viruses as a priority for pandemic preparedness. Influenza virus-like particles (VLPs) have been suggested and developed as a new generation of non-egg-based cell culture-derived vaccine candidates

S.M. Kang (✉) and R.W. Compans (✉)
Department of Microbiology & Immunology and Emory Vaccine Center, School of Medicine, Emory University, 1510 Clifton Rd, Atlanta, GA 30322, USA
e-mail: skang2@emory.edu
e-mail: compans@microbio.emory.edu

P. Pushko, R.A. Bright, and G. Smith
Novavax, Inc., Vaccine Technologies, 1 Taft Court, Rockville, MD, USA

R.W. Compans and W.A. Orenstein (eds.), *Vaccines for Pandemic Influenza*,
Current Topics in Microbiology and Immunology 333,
DOI 10.1007/978-3-540-92165-3_14, © Springer-Verlag Berlin Heidelberg 2009

against influenza infection. Influenza VLPs are formed by a self-assembly process incorporating structural proteins into budding particles composed of the hemagglutinin (HA), neuraminidase (NA) and M1 proteins, and may include additional influenza proteins such as M2. Animals vaccinated with VLPs were protected from morbidity and mortality resulting from lehal influenza infections. The protective mechanism of influenza VLP vaccines was similar to that of the currently licensed influenza vaccines inducing neutralizing antibodies and hemagglutination inhibition activities. Current studies demonstrate that influenza VLP approaches can be a promising alternative approach to developing a vaccine for pandemic influenza viruses. The first human clinical trial of a recombinant pandemic-like H5N1 influenza VLP vaccine was initiated in July 2007 (Bright et al., unpublished).

# 1    Introduction

Currently licensed inactivated influenza vaccines are composed of formalin-treated whole virus or detergent-split viral components. Vaccine strains are selected based on epidemiologic and antigenic considerations of circulating human strains and their anticipated prevalence during the coming year. To obtain high-yield vaccine seed viruses, the chosen strains are adapted to grow in embryonated eggs, or reassortant viruses are generated containing glycoprotein (HA, NA) genes of current strains and genes for internal proteins of influenza A/Puerto Rico/8/34 (H1N1) virus which confer high growth capacity in eggs (Robertson et al. 1992). A live attenuated trivalent influenza virus vaccine has also recently been licensed for intranasal administration to people 2–49 years of age. These live virus strains are cold adapted, temperature sensitive, and are attenuated so as not to produce influenza-like illness by limiting their replication to only the upper respiratory tract in humans. Reassortant strains developed by serial passage at sequentially lower temperatures acquire attenuated phenotypes as a result of multiple mutations in gene segments that encode viral internal proteins (Murphy and Coelingh 2002). However, there are still concerns related to the reversion of attenuated vaccine strains or incomplete attenuation, and the uncertainties of their pathogenic characteristics, particularly when combined with highly pathogenic avian influenza viruses with pandemic potential.

Nonreplicating virus-like particles (VLPs) resemble infectious virus particles in structure and morphology, and contain immunologically relevant viral structural proteins. VLPs have been produced from both nonenveloped and enveloped viruses. Among nonenveloped VLPs, human papillomavirus (HPV) VLPs are the most thoroughly studied, and the expression of the major capsid protein L1 resulted in the production of VLPs (Kirnbauer et al. 1992; Sasagawa et al. 1995). VLPs from viruses with lipid envelopes represent more difficult challenges. Given that envelopes are derived from the host cells, the choice of expression system can be a relevant issue. The formation of VLPs has been demonstrated using both mammalian and insect cells for enveloped viruses such as retroviruses (HIV, SIV) and hepatitis

C virus (Baumert et al. 1998; Gheysen et al. 1989; Yamshchikov et al. 1995). In addition, recent studies demonstrated that VLPs are capable of inducing protective immunity against viral infections including influenza viruses (Bright et al. 2007; Galarza et al. 2005; Pushko et al. 2005, 2007; Quan et al. 2007). Considering the urgent need to develop novel approaches to vaccination against highly pathogenic avian influenza viruses, VLPs can be a promising approach for generating vaccine candidates, particularly against newly emerging influenza viruses with pandemic potential. Here we review the current progress in the development of influenza VLP vaccines.

# 2  VLPs as Influenza Vaccine Candidates

## 2.1  Rationale for VLPs as a Pandemic Influenza Vaccine

Although there are licensed influenza vaccines, including inactivated whole, split, and subunit or live attenuated virus formats, limitations do exist particularly for potential pandemic influenza viruses. The current egg-based system for influenza vaccine manufacture has drawbacks that include recent problems in vaccine supply in response to the influenza season, local or systemic allergic reactions to egg-derived vaccine components, and a short duration of immunity. Also, there are known problems with growing highly pathogenic avian influenza viruses in embryonated eggs because they can kill the embryos, which can hamper virus production, and there are associated human safety concerns in working with live pathogenic viruses. In addition, diseases that affect chicken flocks due to an avian influenza virus outbreak could easily disrupt the supply line of eggs available for vaccine manufacturing. These factors, as well as the requirement for biosafety level 3 or higher containment facilities for safe handling of pathogenic avian influenza viruses, support the urgent need to develop a new influenza vaccine modality. Importantly, thus far the only FDA-approved recombinant protein viral vaccines are based on VLPs. The yeast-derived recombinant hepatitis B vaccine became the first human vaccine manufactured using recombinant DNA technology, and has been used for over a decade (Assad and Francis 1999). The development of human papillomavirus (HPV) VLPs has also resulted in successful clinical trials for preventing HPV infection (Ljubojevic 2006; Markowitz et al. 2007; Stanley 2006). Therefore, influenza VLPs can be a promising alternative vaccine, particularly for potential pandemic influenza viruses (Bright et al. 2008; Matassov et al. 2007; Pushko et al. 2005, 2007). Effective vaccines for pandemic influenza could prevent massive losses of human lives in the case of an influenza pandemic. It has been estimated that the 1918 pandemic outbreak of Spanish influenza virus killed up to 50 million people worldwide (de Jong and Hien 2006; Taubenberger et al. 2001). In recent years, 1918 Spanish influenza virus has been reconstructed to study the pathogen and to find effective ways to respond to such an extreme pathogenic potential (Tumpey et al. 2005).

## 2.2    Characteristics of VLPs as a Promising Vaccine Candidate for Influenza Virus

One attractive property of VLPs as a promising vaccine candidate is their safety. The noninfectious nature of VLPs (Fig. 1) and their lack of viral genomic material make them safe as an attractive candidate vaccine that can be useful for repeated adminis-trations and for use in all populations, including high-risk groups. In addition, VLPs as particulate antigens are an attractive target for antigen-presenting cells such as dendritic cells to capture the antigen for presentation to both T and B lymphocytes (Buonaguro et al. 2006; Da Silva et al. 2001; Lenz et al. 2003; Moron et al. 2002, 2003; Sailaja et al. 2007). As shown by electron microscopy (Figs. 2, 3b, and 4), influenza VLPs resemble intact virions in structure and morphology, and can contain immunologically relevant structural proteins (Fig. 1). Influenza VLPs are assembled on the cell surfaces via the budding process. The viral glycoproteins on VLPs are presented in a native conformation and are unmodified by fixatives.

Human isolates derived from infections with avian influenza viruses including the H5N1 viruses were highly pathogenic for chickens and lethal for chick embryos (De Benedictis et al. 2007; Onishchenko et al. 2006; Shortridge et al. 1998; Suarez et al. 1998; Subbarao et al. 1998). Although these viruses were shown to replicate in fertilized chicken eggs under conditions of relatively high temperature and by shortening the incubation time, the yield of virus was low (Takada et al. 1999). The pathogenic nature of H5 and H7 avian influenza viruses is linked to the presence of additional basic residues at the site of cleavage in the HA glycoprotein, a step required for HA activation (Buranathai et al. 2007; Horimoto and Kawaoka 1997; Kawaoka et al. 1987; Perdue et al. 1997). The presence of basic amino acids adjacent

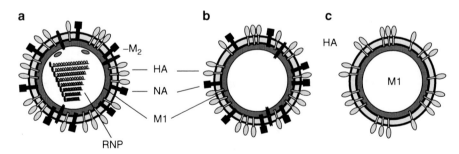

**Fig. 1** Schematic diagrams of an influenza A virus particle and VLPs. **a** A diagram of the influenza A virion. Three types of integral membrane proteins, hemagglutinin (HA), neuraminidase (NA), and M2 ion channel protein are inserted through the lipid bilayer of the viral membrane. The virion matrix protein M1 is thought to underlie the lipid bilayer. Within the envelope are eight segments of single-stranded genomic RNA associated with nucleocapsid and polymerase proteins (ribonu-cleoproteins, RNP). **b** A diagram of a VLP structure showing the major structural proteins (HA, NA, M1) and the lipid bilayer. VLPs resemble virus particles but are devoid of genetic materials. **c** A diagram of influenza VLPs containing M1 and HA only. Influenza VLPs containing M1 and HA alone were shown to induce protective immunity (Galarza et al. 2005; Quan et al. 2007)

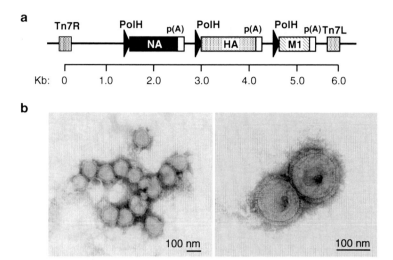

**Fig. 2 a** rBV transfer vector for coexpression of H9N2 influenza proteins and production of H9N2 influenza VLPs. Indicated are the polyhedrin promoter (PolH) and influenza genes. *HA*, hemagglutinin; *NA*, neuraminidase; *M1*, matrix protein. Positions of HA, NA, and M1 proteins on the surfaces of influenza VLPs are also indicated. **b** Negative staining electron microscopy of H9N2 influenza native VLPs. Influenza VLPs were generated from influenza A/Hong Kong/1073/99 (H9N2) HA, NA, and M1 proteins. *Bars* represent 100 nm. VLPs were negatively stained with 2% sodium phosphotungstate, pH 6.5. Micrographic courtesy of Dr. Ling Ye, Emory University

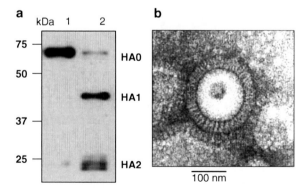

**Fig. 3 a** Western blot of VLPs before (*lane 1*) and after (*lane 2*) treatment with trypsin. HA and M1 were coexpressed in insect cells using a recombinant baculovirus expression system (Quan et al. 2007), and culture supernatants were harvested to purify influenza VLPs. Influenza VLPs produced in insect cells contain HA dominantly in the precursor form. After treatment with trypsin, the HA precursor is cleaved into HA1 and HA2. **b** Electron microscopic examination of influenza VLPs containing A/PR8 HA and M1, which resemble influenza virions in structure. Micrograph courtesy of Dr. Jac-Min Song, Emory University

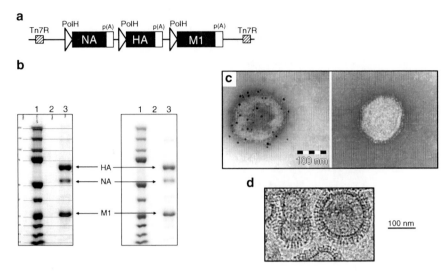

**Fig. 4 a** rBV transfer vector for coexpression of H5N1 influenza proteins and production of H5N1 influenza VLPs. Indicated are the PolH promoter and influenza HA, NA, and M1 genes. **b** Purified H5N1 VLPs, (*lane 3*), by western blotting (*left panel*) and Coomassie staining (*right panel*). Positions of HA, NA, and M1 proteins are indicated. **c** Electron microscopy using staining with gold-labeled H5N1-specific antiserum (*left panel*). On the *right panel*, negatively-stained H5N1 influenza VLPs are shown. **d** Cryoelectron microscopy of H5N1 influenza VLPs. Influenza VLPs were generated using influenza A/Indonesia/5/05 (H5N1) HA, NA, and M1 proteins. The HA protein was engineered to lack the multibasic cleavage site. *Bar* represents 100 nm. Courtesy of Dr. Terje Dokland (Dokland and Ng 2006)

to the cleavage site increases the tissue range of highly pathogenic avian viruses, resulting in replication of the virus in multiple organs and severe, usually fatal, systemic disease in chickens (Senne et al. 1996). To overcome the high pathogenicity of the virus, vaccine strains derived from avian influenza viruses with pandemic potential have been constructed that are devoid of this polybasic amino acid sequence of HA prior to generating vaccine stocks (Matsuoka et al. 2003; Subbarao et al. 2003; Webby et al. 2004). The effects of this mutation on immunogenicity remain to be determined. For influenza VLP production, this kind of genetic modification may not be necessary, and there would be much fewer safety concerns during VLP manufacture and production even with the highly pathogenic wild-type HA glycoprotein.

The self-assembled macrostructure of VLPs can present conformational epitopes of surface proteins to the immune system comparable to those of live virions. Chemical inactivation of influenza vaccine viruses may affect the immunogenicity of the vaccine, resulting in reduced vaccine efficacy. Therefore, the presence of unmodified HA on VLPs and the avoidance of the need for chemical inactivation are desirable features compared to inactivated whole-virus or split vaccines. In addition, a unique feature is that VLPs have high versatility in their ability to be manipulated to incorporate immunostimulatory and/or targeting molecules to enhance or broaden the immunogenicity of VLPs, as demonstrated (Sailaja et al. 2007; Skountzou et al. 2007).

# 3 Production of Influenza VLPs

## 3.1 Formation of VLPs

Influenza viruses are assembled at the plasma membranes of infected cells and released by the budding of newly assembled virions into the outside environment; intact virions are not observed intracellularly (Ali et al. 2000; Gomez-Puertas et al. 2000; Latham and Galarza 2001; Nayak et al. 2004; Roberts et al. 1998). For influenza virus budding to occur, viral structural components including the matrix protein (M1) and the viral ribonucleoprotein complex as well as the three transmembrane proteins of viral envelope [hemagglutinin (HA), neuraminidase (NA), and M2] should be produced and transported to the plasma membrane. During the final step in the growth cycle, these viral proteins interact with each other to initiate the budding processes, leading to morphogenesis of virus particles and the release of virions containing derived host lipids.

The influenza virus M1 protein has been suggested to play an important role in driving virus assembly and budding processes (Ali et al. 2000; Gomez-Puertas et al. 2000; Latham and Galarza 2001). Supporting the important role of M1 in assembly, mature virus particles lacking either HA or NA were found to be formed and released from infected cells (Liu et al. 1995; Pattnaik et al. 1986). M1 is proposed to interact with the cytoplasmic tail of transmembrane viral proteins and plasma membrane as well as the viral nucleocapsid (Ali et al. 2000). However, the major viral components required to control influenza VLP formation are still uncertain, and recent studies have yielded conflicting results. Interactions between M1 and the cytoplasmic tails of HA, NA, and M2 were reported to be necessary for efficient virus assembly, morphology, and budding, which suggests a role of these proteins independent of M1 (Jin et al. 1997; Zhang et al. 2000). These studies demonstrated that virus budding was impaired when the cytoplasmic tails of the viral glycoproteins were mutated. A recent study by Chen et al. (2007) reported that influenza virus HA and NA, but not M1, were required for assembly and budding of DNA vector-derived influenza VLPs, and that low levels of VLPs were found in the culture supernatants when HeLa cells, but not 293T cells, were transfected with DNA expressing M1 alone. Other investigators reported that M1 was sufficient for VLP formation when M1 alone was expressed from recombinant DNA via either vaccinia virus plus T7 RNA polymerase or baculovirus expression systems (Gomez-Puertas et al. 2000; Latham and Galarza 2001). These different results may result from differences in expression systems (DNA tranfection vs. recombinant viruses; mammalian vs. insect cells), and further studies are needed to better understand the contribution of each influenza component to the budding process of VLPs.

## 3.2 Expression Systems

The recombinant baculovirus (rBV) expression system yields high expression levels of recombinant proteins and allows subsequent large-scale manufacturing of a vaccine. A variety of recombinant proteins and VLPs have been produced by rBV expression, and VLP antigens were highly immunogenic, inducing both neutralizing antibodies and cellular immune responses. Examples include simian and human immunodeficiency virus (SIV/HIV) (Buonaguro et al. 2007; Deml et al. 1997a; Guo et al. 2003; Kang and Compans 2003; Wang et al. 2007; Yao et al. 2000), human papillomavirus (Harro et al. 2001), rotavirus (O'Neal et al. 1998), hepatitis type C virus (Murata et al. 2003; Qiao et al. 2003; Triyatni et al. 2002), and Ebola virus (Ye et al. 2006). It is also considered likely that there will be fewer safety concerns about VLPs produced in insect cells as compared with the use of alternative expression systems in mammalian cells. Baculoviruses are found in green vegetables and are not able to replicate in mammalian cells, and they are thought to present no threat to vaccinated individuals. In contrast, mammalian expression systems use cancer cell-derived cell lines. There are also additional concerns about the utilization of replicating recombinant viruses such as vaccinia or adenoviruses as an expression vehicle because of pre-existing immunity to the vector or unknown potential effects of vector gene expression in humans.

As shown in Fig. 3, the HA proteins of influenza A subtypes H1, H5, and H7 expressed in insect cells retained hemagglutination activity and generated the HA1 and HA2 subunits upon treatment with trypsin, indicating that baculovirus-expressed HA proteins were properly folded after expression (Crawford et al. 1999; Quan et al. 2007). Also, soluble NA produced in the rBV system formed tetramers and maintained an enzymatically active conformation (Fiers et al. 2001). These functional studies showed that rBV-derived influenza proteins were likely to maintain conformational integrity similar to that seen in native influenza virions, and that the rBV expression system could provide an attractive approach for manufacturing an influenza vaccine.

Influenza VLP production was also demonstrated by using other expression systems; recombinant vaccinia viruses (Ali et al. 2000), DNA plasmid transfection with T7 RNA polymerase-expressing vaccinia virus (Gomez-Puertas et al. 2000), and recombinant DNA expression vectors (Chen et al. 2007; Szecsi et al. 2006). For the production of VLPs containing HA in mammalian cells, coexpression of NA or exogenously added NA was required for the effective release of VLPs into culture media (Ali et al. 2000; Chen et al. 2007; Gomez-Puertas et al. 2000), whereas VLPs containing HA can be produced in insect cells in the absence of NA expression (Galarza et al. 2005; Guo et al. 2003; Latham and Galarza 2001; Quan et al. 2007). In contrast to mammalian cells, insect cells do not add sialic acids to N-glycans during the posttranslational modification (Lanford et al. 1989), which explains why VLPs containing HA can be released from the insect cell surfaces without requiring cleavage of sialic acid by the neuraminidase. However, it will be important to study the potential role of neuraminidase in the assembly, production yields, and antigenic properties of influenza VLPs in insect cell expression systems.

## 3.3   Expression Vectors for Insect Cells

The host Sf9 insect cell line used to manufacture recombinant influenza VLP vaccines was originally derived from the ovaries of the fall armyworm *Spodoptera frugiperda* and cloned in the early 1980s (Smith et al. 1983). The Sf9 cells are usually maintained in a serum-free, animal-product-free media; can be identified by karyotype and isoenzyme analysis; are free of contaminating microorganisms, adventitious agents, retroviruses, or C-type particles; and have been shown to be nontumorigenic.

In experiments that involved Sf9 cells and rBVs, two strategies have been used by different groups for the expression of influenza VLPs. In one strategy, influenza proteins were coexpressed from a single rBV expressing multiple influenza components (Bright et al. 2007; Latham and Galarza 2001; Pushko et al. 2005, 2007). The other strategy involved coinfection of the Sf9 cells with the two rBVs, one expressing the HA protein, whereas the other rBV expressed the M1 protein (Galarza et al. 2005; Quan et al. 2007). The highly active baculovirus polyhedrin promoter is generally used for the expression of influenza genes (Bright et al. 2007, 2008; Pushko et al. 2005). Alternatively, a pc/pS1 hybrid capsid-polyhedrin promoter has been used (Quan et al. 2007), or a combination of various promoters. For example, polyhedrin promoters have been used for the expression of HA and M1 proteins, whereas baculovirus P10 promoters have been involved in the expression of NA and M2 genes all cloned into a single baculovirus vector (Latham and Galarza 2001).

Recombinant influenza VLPs have been generated in Sf9 cells with baculovirus vectors expressing either four structural influenza genes HA, NA, M1, and M2 (Latham and Galarza 2001) or three HA, NA, and M1 genes cloned into a single baculovirus construct (Bright et al. 2007, 2008; Pushko et al. 2005, 2007). For example, A/Hong Kong/1073/99 (H9N2), an avian influenza virus believed to have pandemic potential if spread among humans, HA, NA, and M1 genes were sequenced and cloned into rBV (Pushko et al. 2005), with each gene within its own expression cassette that included a polyhedrin promoter and transcription termination sequences (Fig. 2a). Recombinant HA was expressed as HA0 in Sf9 cells, with no significant proteolytic cleavage into HA1 and HA2. Influenza VLPs were purified from culture media by sucrose gradient centrifugation and the presence of HA, NA, and M1 in VLPs was confirmed by SDS-PAGE, western blot, hemagglutination, and NA enzymatic activity assays. Electron microscopic analysis of negatively-stained samples showed H9N2 VLPs with a diameter of approximately 80–120 nm and surface spikes characteristic of influenza HA and NA proteins on influenza virions (Fig. 2b).

Recombinant VLPs have also been generated in Sf9 cells infected with rBV that expressed the HA, NA, and M1 of various clades of the H5N1 subtype of avian influenza with pandemic potential (Fig. 4a) (Bright et al. 2008). HA proteins were modified to remove the polybasic cleavage site at the HA1–HA2 junction that was characteristic of highly pathogenic avian influenza viruses. Modified HA proteins were capable of efficient assembly into VLPs in Sf9 insect cells along with native NA and M1 proteins. H5N1 VLPs purified from Sf9 cells and rBV contaminants were characterized by negative staining and immune electron microscopy, as shown in Fig. 4. Cryoelectron microscopy of purified H5N1 VLPs containing influenza HA, NA, and M1 proteins is shown on Fig. 4d. Further, influenza VLPs have also been successfully generated in Sf9 cells from HA and M1 proteins only (Galarza et al. 2005; Quan et al. 2007) or from HA and NA only in 293 cells (Chen et al. 2007).

Identifying the optimal protein composition of influenza VLPs to ensure efficient protective characteristics and high production yields of VLP-based influenza vaccines remains an important objective. Including additional influenza proteins such as M2 and/or NP may prove beneficial for inducing broader immune responses. However, coexpression of additional proteins in insect cells may reduce the overall levels of expression of VLPs due to promoter dilution effect or because of excessive metabolic burden (Roldao et al. 2007). Overexpression of M2 has also been reported to dramatically decrease the yields of VLPs (Gomez-Puertas et al. 1999), and these authors hypothesized that overexpression of M2, an ion channel protein, inhibited intracellular transport and drastically reduced the accumulation of coexpressed HA and hence reduced VLP yields.

## 3.4   Large-Scale Production and Purification of VLPs

One of many challenges facing the development of an influenza vaccine for pandemics is the ability to manufacture millions of doses in a relatively short period to meet a sudden demand for vaccine. To address this need, a manufacturing process is needed that would permit a rapid surge in capacity. In addition, supply distribution would become a formidable challenge during a time of international crisis; therefore, a suitable manufacturing process would make use of portable and disposable processes and equipment that can be rapidly deployed locally and globally in the event of an influenza pandemic.

The production of a recombinant H5N1 influenza VLP vaccine candidate studied in human clinical trials was performed under cGMP conditions (Smith and Robinson, unpublished). VLPs were produced in Sf9 insect cells infected with rBV and secreted in the serum-free culture medium, as described elsewhere (Bright et al. 2007, 2008). VLPs were separated from Sf9 cells, baculoviruses, and host cell contaminants using a scalable process that included tangential flow filtration, sucrose gradient centrifugation and chromatography procedures, and any residual live baculovirus was then inactivated by beta-propiolactone. Recombinant influenza VLP vaccines have been produced for both pandemic-like and seasonal strains of influenza using this process, and a phase I/IIa clinical trial of an A/Indonesia/5/05 (H5N1) influenza VLP vaccine was initiated in July, 2007 by Novavax, Inc. (Rockville, MD, USA).

# 4 Immune Responses Induced by Influenza VLPs

## 4.1 Humoral Immune Responses Elicited by Influenza VLPs

In preclinical studies, influenza VLPs were found to induce immune responses specific to influenza HA. Intranasal or intramuscular immunizations of mice with influenza VLPs containing HA (A/Udorn H3N2, A/PR8 H1N1, A/Fujian H3N2, A/Indonesia/5/05 H5N1) induced high titers of antibodies specific to the vaccine strains (Bright et al. 2007, 2008; Galarza et al. 2005; Quan et al. 2007). Immunization of mice with HA-negative control VLPs containing M1 alone did not induce influenza virus-specific antibodies and did not protect against challenge infection (Quan et al. 2007). In addition, VLPs were heat-treated and used for immunization to determine the requirements for VLP integrity and HA activity to induce immune responses. Heat treatment of VLPs resulted in loss of hemagglutination activity, and no significant levels of antibodies specific to PR8 virus were detected in groups of mice immunized with heat-treated influenza VLPs, despite the induction of antibodies capable of binding to heat-treated VLPs (Quan et al. 2007). Therefore, the integrity of influenza VLPs seems to be a critical factor in inducing functional antibodies that are protective.

IgG1, IgG2a, and IgG2b antibodies were found to be the major isotypes present at similarly high levels, indicating that T helper type 1 and 2 (Th1, Th2) immune responses were induced by intranasal immunization with influenza VLPs. Notably, twofold higher titers of antibodies were observed at five months compared to those at two weeks after immunization (Quan et al. 2007). In other studies, mice immunized with VLPs (A/Fujian, H3N2) showed IgG2a and IgG2b as dominant serum antibody isotypes with IgG1 as the third major antibody (Bright et al. 2007). In contrast to VLP vaccination, mice immunized with inactivated whole virus had a dominant IgG2a isotype and IgG1 as the second most dominant, but little IgG2b antibodies, and rHA primarily elicited an IgG1 response (Bright et al. 2007). These studies indicate that influenza VLPs are immunogenic and capable of inducing long-lived antibody responses characterized by both Th1- and Th2-type immune responses.

It is important to induce mucosal immune responses, since the respiratory mucosal surfaces (nose, trachea, and lung) are the natural route of entry and the primary replication site of influenza virus. High levels of mucosal IgG and IgA antibodies specific to influenza virus A/PR/8/34 were observed in all mucosal samples obtained from mice intranasally immunized with influenza HA (H1N1) VLPs (Quan et al. 2007). Significant increases in virus specific IgG antibody levels were also found in mucosal samples from mice immunized with PR8 HA VLPs following challenge infection with either A/PR8 or A/WSN as compared to those before challenge. Significant increases were also found in lung IgA antibodies in the A/PR8 HA VLP immunized group after PR8 or WSN challenge. Overall, these results indicate that mucosal immunization with influenza VLPs can induce good memory immune responses.

Intramuscular immunization of mice with influenza VLPs (A/Fujian, H3N2) induced broader serum immune responses than inactivated influenza whole virions or recombinant HA (rHA), as determined by hemagglutination inhibition assay using a panel of H3N2 human isolates (Bright et al. 2007). When equal amounts of HA in soluble rHA form and in VLPs were compared, VLPs induced 10- to 15-fold higher serum titers than rHA. More than a 20-fold HA antigen sparing effect was observed with VLPs by comparative immunogenicity studies of influenza VLPs and soluble rHA in terms of induction of functional antibodies that inhibited hemagglutination (Bright et al. 2007). Also, VLP immunization showed twofold higher hemagglutination inhibition titers than inactivated whole virus. This study (Bright et al. 2007) suggests that influenza VLPs are superior antigens to rHA, and even to whole inactivated influenza virus.

## 4.2   Cellular Immune Responses Induced by Influenza VLPs

VLPs are nonreplicating exogenous antigens, and thus they have been thought to be presented by antigen-presenting cells (APCs) using the major histocompatibility (MHC) class II presentation pathway. However, there is evidence that VLPs may cross over to the endogenous pathway to gain access to MHC class I, inducing CD8+ cytotoxic T cell activation (Deml et al. 1997b; Moron et al. 2003). Recombinant parvovirus VLPs are very efficiently captured by dendritic cells (DCs) and then localized in late endosomes of DCs via macropinocytosis (Moron et al. 2003). Processing of VLPs requires vacuolar acidification and proteasome activity, and transporter associated with antigen presentation (TAP) translocation, as well as neosynthesis of MHC class I molecules. Therefore, DCs can cross-present VLP antigens to CD8+ T cells to activate cytotoxic T cells. In addition, influenza VLPs and inactivated influenza virus were found to interact with DCs and monocyte/macrophage immune cells in vitro (Kang and Compans, unpublished data). Although nonreplicating antigens are weak in activating T cell immune responses, VLPs can activate both CD4+ and CD8+ T cells.

The magnitude of elicitation of virus-specific CD4+ and CD8+ T cell responses was studied while characterizing influenza VLP-induced immunity (Quan et al. 2007). Splenocytes from mice immunized with VLPs were stimulated with HA-specific MHC I- or MHC II-restricted peptides to quantify HA-specific CD4+ and CD8+ cells secreting Th1-type (IFN-$\gamma$, IL-2) and Th2 type (IL-4, IL-5) cytokines. Significant levels of IFN-$\gamma$ and IL-2 in response to MHC I or MHC II peptide stimulation were detected in mice immunized with VLPs, but not in naïve mice. CD4+ cells were found to secrete higher levels of the cytokines IL-4 and IL-5 than CD8+ cells. Upon virus infection, mice immunized with VLPs rapidly induced significantly higher levels of lymphocytes secreting IFN-$\gamma$, and IL-2 secreting CD4+ and CD8+ cells as compared to those observed prior to challenge. In contrast, naïve mice that received the same dose of A/PR8 infection did not induce cytokine-producing lymphocytes specific to HA peptides. These results suggest that influenza VLPs induce both Th1- and Th2-type cellular immune responses, which can expand rapidly in response to influenza virus infection.

## 4.3 Influenza VLPs Induce Protective Immunity

The primary goal of vaccination is to provide immunized hosts with protection against morbidity and mortality from unpredictably lethal infections. The mouse model is a well-accepted and characterized small animal model to evaluate influenza vaccine efficacy, though other animals such as ferrets and chickens are also used. Several subtypes of influenza VLPs have been shown to induce protective immunity in mice. It was reported that immunization with influenza VLPs containing the HA from H3N2 (A/Udorn) induced higher titers of antibodies specific to HA than those induced by intranasal inoculation with a sublethal dose of the challenge virus and conferred immunized mice with protection against lethal challenge with influenza A/HK/68 (H3N2, 5 $LD_{50}$) (Galarza et al. 2005). Similarly, influenza VLPs containing HA and neuraminidase (NA) of influenza A/HK/1073/99 (H9N2) elicited serum antibodies specific for the virus, which also contributed to a significant reduction of challenge virus replication in the animal (Pushko et al. 2005). Intranasal immunization with influenza VLPs containing an A/PR8 HA induced 100% protection against lethal challenge with the homologous strain A/PR8 or the heterologous stain A/WSN (Quan et al. 2007). Importantly, CD4+ or CD8+ T cell-deficient mice immunized intramuscularly with influenza A/PR8 HA VLPs were also protected against lethal infection with influenza A/PR8 virus (Kang and Compans, unpublished data). These studies indicate that influenza VLPs are highly immunogenic, inducing protective immunity under normal or T cell-deficient conditions.

## 4.4 Characterization of Protective Immune Correlates to Influenza VLPs

Neutralizing and hemagglutination inhibition activities against influenza virus are an indicator of the induction of functional antibodies that most likely confer protective immunity from viral infection. Preimmune and immune sera from HA-negative M1 VLPs showed no neutralizing activity, whereas immune sera collected at four weeks after the second immunization with VLPs containing A/PR8 HA showed neutralizing titers of over 3,200 (50% neutralization activity titer) against the homologous strain A/PR8 (Quan et al. 2008). Similarly, immune sera of A/Aichi HA (H3N2) VLPs displayed high neutralizing titers of over 3,200 against the homologous A/Aichi strain. A/PR8 HA VLP immune sera also exhibited significant levels of neutralizing titers of over 400 against the heterologous strain A/WSN, although lower than those against the homologous A/PR8 strain. Naïve mice that received A/PR8 HA VLP-immune sera collected five months postimmunization were protected against lethal challenge infections with A/PR8 or A/WSN (Quan et al. 2007). In contrast, A/Aichi H3 HA VLP-immune sera showed only negligible neutralizing titers against A/WSN. These results suggest that immunization with influenza VLPs, similar to inactivated influenza virus, induces protective neutralizing antibodies against influenza viruses which are mostly subtype specific with a certain degree of cross-reactivity with heterologous strains within a subtype.

In an important dose-sparing experiment, mice or ferrets were vaccinated intramuscularly with influenza VLPs containing HA and NA of A/Fujian (H3N2), based on HA concentration (15 μg–24 ng) in VLPs, and the immune responses were compared to responses elicited in animals vaccinated with recombinant HA (rHA) or inactivated whole-influenza virions (Bright et al. 2007). All vaccinated animals showed high titers of anti-HA antibodies regardless of the vaccine immunogen, and mice vaccinated with doses of VLPs containing 3 μg and 600 ng of HA also had antibodies against NA which have been shown in numerous studies to have an ability to contribute to enhancing the efficacy of influenza vaccines that contain an NA component. There was a correlation between the antigenic distance, related to the year of virus isolation, and the ability to prevent hemagglutination. Interestingly, VLPs elicited antibodies that recognized a broader panel of antigenically distinct H3N2 viral isolates compared to rHA or inactivated influenza virus in a hemagglutination-inhibition (HAI) assay (Bright et al. 2007).

To investigate whether vaccinated mice are cross-protected against a lethal challenge, and whether VLP-induced immune responses can lower viral load in vivo, VLP-vaccinated mice were challenged with homologous (A/PR8 or A/Aichi) or heterologous (WSN) strains (Quan et al. 2008). All mice immunized with A/PR8 HA VLPs survived a lethal virus challenge with A/PR8 as well as A/WSN without showing clinical signs of illness. In contrast, the naïve group showed a significant and progressive loss in body weight and shivering after challenge infection, indicating that these mice suffered severe illness due to A/PR8 or A/WSN viral infection. By days 7–9 postchallenge with A/PR8 or A/WSN strains, all mice in the naïve group had lost over 30% of their body weight and died. Similarly, A/Aichi HA VLP-immunized mice were protected against the homologous strain A/Aichi challenge infection but not the heterologous strains A/PR8 or A/WSN, which is consistent with the neutralizing titers in VLP immune sera.

# 5 Influenza VLPs as Pandemic Vaccines

There is increasing concern about a potential influenza pandemic, as highly virulent avian influenza viruses continue to circulate and spread worldwide with a high risk of crossing species-specific barriers. Development of effective vaccines for avian influenza is a priority in preparations for an influenza pandemic. In this regard, VLPs comprised of structural proteins of influenza A/HK/1073/99 (H9N2) can induce influenza-specific antibodies and inhibit replication of the influenza virus after challenge (Pushko et al. 2005, 2007). This insect cell-derived H9N2 VLP vaccine was immunogenic in three different animal models, including mice, rats, and ferrets, and protected these animals from challenge infections (Pushko et al. 2007). Inclusion of an adjuvant improved both virus neutralizing antibody and hemagglutination inhibition titers (Pushko et al. 2007). In a recent study, the protective efficacy provided by H5N1 VLPs was investigated in comparison with recombinant HA in the mouse model (Bright et al. 2008). The H5N1 VLP vaccine induced immune responses with binding and hemagglutination inhibition activities that were significantly higher than those obtained with the same dose of soluble HA. H5N1 VLP induced cross-clade protective immunity against lethal challenge, while mice receiving the recombinant HA vaccine showed a degree of morbidity and mortality under the same condition. Therefore, VLPs can be safe and effective vaccines for avian influenza viruses with pandemic potential, which will be superior to the soluble HA vaccine.

Avian influenza VLPs have also been assembled on replication-defective core particles derived from murine leukemia retrovirus using DNA expression plasmid transfection (Szecsi et al. 2006). These chimeric VLPs were engineered to contain the HA, NA, and M2 envelope proteins from highly pathogenic H5N1 or H7N1 avian influenza viruses and were shown to induce high titers of neutralizing antibodies in mice. The 1918 pandemic was the most devastating and widespread of all influenza pandemics known to date. There is still the possibility of the 1918 type virus re-emerging with high virulence and human transmissibility. Matassov et al. (2007) generated influenza VLPs containing the surface glycoproteins HA and NA of the 1918 influenza virus, and tested their immunogenicity in mice. Antibody titers specific to the 1918 HA protein in mice immunized with VLP vaccines were higher than in mice vaccinated with an inactivated swine virus (H1N1) control (Matassov et al. 2007). Vaccine efficacy was evaluated by challenging immunized mice with the antigenically related influenza virus A/swine/Iowa/15/30 (H1N1). Significantly lower viral titers were found in the nose and lungs of VLP-immunized mice than the placebo group as well as the inactivated virus group on days 4 and 6 postchallenge (Matassov et al. 2007). These results suggest that it is feasible to make a safe and immunogenic vaccine to protect even against the extremely virulent 1918 virus using insect cell-derived influenza VLPs. These studies demonstrate that VLPs can be an effective and safe vaccine against potential pandemic influenza viruses.

# 6 Strategies to Enhance the Immunogenicity of Influenza VLPs

VLP technology has a unique potential for decorating the surfaces of VLPs with antigens as well as targeting or immunostimulatory molecules. Although influenza VLPs themselves are highly immunogenic, VLPs can be engineered and developed to incorporate immunostimulatory molecules and/or dendritic targeting molecules. To enhance the immunogenicity of VLPs, GM-CSF (granulocyte macrophage colony stimulating factor), an adjuvant molecule, or DC growth factor flt3 ligand were modified so that these immunostimulatory molecules could be expressed in a membrane-anchored form and incorporated into budding VLPs (Skountzou et al. 2007). The engineered GM-CSF with a membrane-anchoring domain via GPI (glycosylphosphatidylinositol) was found to be expressed on the cell surface and incorporated into budding simian immunodeficiency virus (SIV) VLPs. Also, Flt3-ligand fused to the transmembrane domain of HIV Env was incorporated into HIV VLPs (Sailaja et al. 2007). These chimeric VLPs were shown to be more effective at activating immune cells, resulting in enhanced immunogenicity. In addition, GPI-anchored GM-CSF was effectively incorporated into HA/M1-containing influenza VLPs and significantly increased the immunogenicity of influenza VLPs compared to that of influenza VLPs without GM-CSF (Kang and Compans, unpublished data). Therefore, incorporating immunostimulatory molecules is a promising approach for further enhancing the immunogenicity and antigen-sparing effects of influenza VLPs.

# 7  Current Status of Product Development

Pandemic influenza presents significant challenges for vaccine development and supply. First, the window of time between identification of the causative strain and the need to begin vaccination is short. With traditional egg-based and mammalian cell-based vaccine production technologies, the time between strain identification and vaccine production will be approximately 5–6 months. Recombinant, cell culture-based technologies, such as influenza VLPs, offer the potential for vaccine production within 2–3 months, a time saving that may prevent many thousands, if not millions, of deaths.

A second challenge of pandemic influenza outbreaks is antigenic drift of influenza strains. Influenza virus genes coding for surface proteins continuously mutate, resulting in amino acid changes in immunologically important epitopes of the HA surface protein. Antigenic drift occurs throughout the waves of a pandemic and throughout annual influenza epidemics. Thus, vaccines that can provide a broader array of protection from drifted heterovariant strains of influenza are needed to address a rapidly evolving pandemic virus. Recombinant VLP vaccines have demonstrated the ability to protect against homologous and heterologous H5N1 viruses in animal models without an adjuvant (Bright et al. 2008). Immunogenicity and lethal challenge studies in ferrets have demonstrated that influenza VLP vaccines induce cross-protection and cross-reactivity against drifted strains of H5 and H3 influenza strains (Mahmood et al. 2008).

A third challenge of pandemic influenza outbreaks is the quantity of vaccine required to address the surge in demand for an effective vaccine. In theory, every human around the globe should be provided with a vaccine for such a potentially devastating disease. Most existing influenza vaccine manufacturers have adapted or modified their seasonal influenza vaccine production processes to create pandemic-like vaccine candidates. This process often includes constructing reassortant influenza viruses and/or multiple adaptation passages in eggs, which can result in additional mutations within the HA and NA that can diminish vaccine efficacy (Lugovtsev et al. 2005; Widjaja et al. 2006). In spite of significant efforts, a large gap remains between anticipated global demand and supply of vaccine in the event of a pandemic. Supply limitations are further complicated by the high doses reported to be required for pandemic-like influenza vaccines to induce seroprotective levels of immunity. The pandemic-like vaccine licensed in the United States induced only modest seroprotection rates, even at very high doses and with two injections of 90 μg HA (Bresson et al. 2006). The recombinant VLP approach offers a technology in which vaccines can be made using: (1) non-egg-adapted influenza genes, (2) an immunologically promising, high-yielding insect cell culture substrate, and (3) a simplified, portable manufacturing process that utilizes disposable equipment. Rapid scale-up and regional production may be possible given the anticipated simplicity of the manufacturing process.

# 8   Conclusion

Recent studies on VLPs have demonstrated that they can be a promising vaccine platform for both seasonal influenza and pandemic influenza viruses. Nonreplicating, noninfectious VLPs possess desirable safety properties and are highly immunogenic due to their particulate nature. Influenza VLPs have been shown to have the ability to induce neutralizing and hemagglutination inhibition activity against strains closely related to those included in the vaccine, which is similar to the protective effect of currently licensed influenza vaccine. Another protective advantage of influenza VLP vaccines is that they may provide a broader range of protection against antigenic variants of the virus. In addition, VLPs are likely to induce cellular immune responses that are known to play a contributing role in broadening protection, especially in high-risk groups such as the elderly population. Importantly, VLPs can also be engineered to improve their potential quality by incorporating immunostimulatory molecules as well as highly conserved influenza components such as the M2 ion channel protein or the nucleocapsid protein. Although the protective efficacy of VLP vaccines in human trials remains to be tested, Novavax recently initiated human clinical trials with a pandemic VLP vaccine (http://www.novavax.com).

**Acknowledgments** This work was supported by NIH Grants AI068003, AI074579. The authors express thanks to Drs. T. Dokland and L. Ye for the electron micrographic pictures, Mr. D. Yoo for the schematic diagram of influenza VLPs, and E.J. Collins for the preparation of the manuscript.

# References

Ali A, Avalos RT, Ponimaskin E, Nayak DP (2000) Influenza virus assembly: effect of influenza virus glycoproteins on the membrane association of M1 protein. J Virol 74(18):8709–8719
Assad S, Francis A (1999) Over a decade of experience with a yeast recombinant hepatitis B vaccine. Vaccine 18(1–2):57–67
Baumert TF, Ito S, Wong DT, Liang TJ (1998) Hepatitis C virus structural proteins assemble into virus-like particles in insect cells. J Virol 72(5):3827–3836

Bresson JL, Perronne C, Launay O, Gerdil C, Saville M, Wood J, Hoschler K, Zambon MC (2006) Safety and immunogenicity of an inactivated split-virion influenza A/Vietnam/1194/2004 (H5N1) vaccine: phase I randomised trial. Lancet 367(9523):1657–1664

Bright RA, Carter DM, Daniluk S, Toapanta FR, Ahmad A, Gavrilov V, Massare M, Pushko P, Mytle N, Rowe T, Smith G, Ross TM (2007) Influenza virus-like particles elicit broader immune responses than whole virion inactivated influenza virus or recombinant hemagglutinin. Vaccine 25(19):3871–3878

Bright RA, Carter DM, Crevar CJ, Toapanta FR, Steckbeck JD, Cole KS, Kumar NM, Pushko P, Smith G, Tumpey TM, Ross TM (2008) Cross-clade protective immune responses to influenza viruses with H5N1 HA and NA elicited by an influenza virus-like particle. PLoS One 3:e1501

Buonaguro L, Tornesello ML, Tagliamonte M, Gallo RC, Wang LX, Kamin-Lewis R, Abdelwahab S, Lewis GK, Buonaguro FM (2006) Baculovirus-derived human immunodeficiency virus type 1 virus-like particles activate dendritic cells and induce ex vivo T-cell responses. J Virol 80(18):9134–9143

Buonaguro L, Devito C, Tornesello ML, Schroder U, Wahren B, Hinkula J, Buonaguro FM (2007) DNA-VLP prime-boost intra-nasal immunization induces cellular and humoral anti-HIV-1 systemic and mucosal immunity with cross-clade neutralizing activity. Vaccine 25(32):5968–5977

Buranathai C, Amonsin A, Chaisigh A, Theamboonlers A, Pariyothorn N, Poovorawan Y (2007) Surveillance activities and molecular analysis of H5N1 highly pathogenic avian influenza viruses from Thailand, 2004–2005. Avian Dis 51(suppl 1):194–200

Chen BJ, Leser GP, Morita E, Lamb RA (2007) Influenza virus hemagglutinin and neuraminidase, but not the matrix protein, are required for assembly and budding of plasmid-derived virus-like particles. J Virol 81(13):7111–7123

Crawford J, Wilkinson B, Vosnesensky A, Smith G, Garcia M, Stone H, Perdue ML (1999) Baculovirus-derived hemagglutinin vaccines protect against lethal influenza infections by avian H5 and H7 subtypes. Vaccine 17(18):2265–2274

Da Silva DM, Velders MP, Nieland JD, Schiller JT, Nickoloff BJ, Kast WM (2001) Physical interaction of human papillomavirus virus-like particles with immune cells. Int Immunol 13(5):633–641

De Benedictis P, Joannis TM, Lombin LH, Shittu I, Beato MS, Rebonato V, Cattoli G, Capua I (2007) Field and laboratory findings of the first incursion of the Asian H5N1 highly pathogenic avian influenza virus in Africa. Avian Pathol 36(2):115–117

de Jong MD, Hien TT (2006) Avian influenza A (H5N1) J Clin Virol 35(1):2–13

Deml L, Kratochwil G, Osterrieder N, Knuchel R, Wolf H, Wagner R (1997a) Increased incorporation of chimeric human immunodeficiency virus type 1 gp120 proteins into Pr55gag virus-like particles by an Epstein–Barr virus gp220/350-derived transmembrane domain. Virology 235(1):10–25

Deml L, Schirmbeck R, Reimann J, Wolf H, Wagner R (1997b) Recombinant human immunodeficiency Pr55gag virus-like particles presenting chimeric envelope glycoproteins induce cytotoxic T-cells and neutralizing antibodies. Virology 235(1):26–39

Dokland T, Ng ML (2006) Electron microscopy of biology samples. In: Dokland T, Hutmacher DW, Ng ML, Schantz JP (eds) Techniques in microscopy for biomedical application. World Scientific Press, Singapore

Fiers W, Neirynck S, Deroo T, Saelens X, Jou WM (2001) Soluble recombinant influenza vaccines. Philos Trans R Soc Lond B Biol Sci 356(1416):1961–1963

Galarza JM, Latham T, Cupo A (2005) Virus-like particle (VLP) vaccine conferred complete protection against a lethal influenza virus challenge. Viral Immunol 18(1):244–251

Gheysen D, Jacobs E, de Foresta F, Thiriart C, Francotte M, Thines D, De Wilde M (1989) Assembly and release of HIV-1 precursor Pr55gag virus-like particles from recombinant baculovirus-infected insect cells. Cell 59(1):103–112

Gomez-Puertas P, Mena I, Castillo M, Vivo A, Perez-Pastrana E, Portela A (1999) Efficient formation of influenza virus-like particles: dependence on the expression levels of viral proteins. J Gen Virol 80(Pt 7):1635–1645

Gomez-Puertas P, Albo C, Perez-Pastrana E, Vivo A, Portela A (2000) Influenza virus matrix protein is the major driving force in virus budding. J Virol 74(24):11538–11547

Guo L, Lu X, Kang SM, Chen C, Compans RW, Yao Q (2003) Enhancement of mucosal immune responses by chimeric influenza HA/SHIV virus-like particles. Virology 313(2):502–513

Harro CD, Pang YY, Roden RB, Hildesheim A, Wang Z, Reynolds MJ, Mast TC, Robinson R, Murphy BR, Karron RA, Dillner J, Schiller JT, Lowy DR (2001) Safety and immunogenicity trial in adult volunteers of a human papillomavirus 16 L1 virus-like particle vaccine. J Natl Cancer Inst 93(4):284–292

Horimoto T, Kawaoka Y (1997) Biologic effects of introducing additional basic amino acid residues into the hemagglutinin cleavage site of a virulent avian influenza virus. Virus Res 50(1):35–40

Jin H, Leser GP, Zhang J, Lamb RA (1997) Influenza virus hemagglutinin and neuraminidase cytoplasmic tails control particle shape. EMBO J 16(6):1236–1247

Kang SM, Compans RW (2003) Enhancement of mucosal immunization with virus-like particles of simian immunodeficiency virus. J Virol 77(6):3615–3623

Kawaoka Y, Nestorowicz A, Alexander DJ, Webster RG (1987) Molecular analyses of the hemagglutinin genes of H5 influenza viruses: origin of a virulent turkey strain. Virology 158(1):218–227

Kirnbauer R, Booy F, Cheng N, Lowy DR, Schiller JT (1992) Papillomavirus L1 major capsid protein self-assembles into virus-like particles that are highly immunogenic. Proc Natl Acad Sci USA 89(24):12180–12184

Lanford RE, Luckow V, Kennedy RC, Dreesman GR, Notvall L, Summers MD (1989) Expression and characterization of hepatitis B virus surface antigen polypeptides in insect cells with a baculovirus expression system. J Virol 63(4):1549–1557

Latham T, Galarza JM (2001) Formation of wild-type and chimeric influenza virus-like particles following simultaneous expression of only four structural proteins. J Virol 75(13):6154–6165

Lenz P, Thompson CD, Day PM, Bacot SM, Lowy DR, Schiller JT (2003) Interaction of papillomavirus virus-like particles with human myeloid antigen-presenting cells. Clin Immunol 106(3):231–237

Liu C, Eichelberger MC, Compans RW, Air GM (1995) Influenza type A virus neuraminidase does not play a role in viral entry, replication, assembly, or budding. J Virol 69(2):1099–1106

Ljubojevic S (2006) The human papillomavirus vaccines. Acta Dermatovenerol Croat 14(3):208

Lugovtsev VY, Vodeiko GM, Levandowski RA (2005) Mutational pattern of influenza B viruses adapted to high growth replication in embryonated eggs. Virus Res 109(2):149–157

Mahmood K, Bright RA, Mytle N, Carter DM, Crevar CJ, Achenbach JE, Heaton PM, Tumpey TM, Ross TM (2008) H5N1 VLP vaccine induced protection in ferrets against lethal challenge with highly pathogenic H5N1 influenza viruses. Vaccine 26(42):5393–5399

Markowitz LE, Dunne EF, Saraiya M, Lawson HW, Chesson H, Unger ER (2007) Quadrivalent human papillomavirus vaccine: Recommendations of the Advisory Committee on Immunization Practices (ACIP). MMWR Recomm Rep 56(RR-2):1–24

Matassov D, Cupo A, Galarza JM (2007) A novel intranasal virus-like particle (VLP) vaccine designed to protect against the pandemic 1918 influenza A virus (H1N1). Viral Immunol 20(3):441–452

Matsuoka Y, Chen H, Cox N, Subbarao K, Beck J, Swayne D (2003) Safety evaluation in chickens of candidate human vaccines against potential pandemic strains of influenza. Avian Dis 47(suppl 3):926–930

Moron G, Rueda P, Casal I, Leclerc C (2002) CD8alpha- CD11b+ dendritic cells present exogenous virus-like particles to CD8+ T cells and subsequently express CD8alpha and CD205 molecules. J Exp Med 195(10):1233–1245

Moron VG, Rueda P, Sedlik C, Leclerc C (2003) In vivo, dendritic cells can cross-present virus-like particles using an endosome-to-cytosol pathway. J Immunol 171(5):2242–2250

Murata K, Lechmann M, Qiao M, Gunji T, Alter HJ, Liang TJ (2003) Immunization with hepatitis C virus-like particles protects mice from recombinant hepatitis C virus-vaccinia infection. Proc Natl Acad Sci USA 100(11):6753–6758

Murphy BR, Coelingh K (2002) Principles underlying the development and use of live attenuated cold-adapted influenza A and B virus vaccines. Viral Immunol 15(2):295–323

Nayak DP, Hui EK, Barman S (2004) Assembly and budding of influenza virus. Virus Res 106(2):147–165

O'Neal CM, Clements JD, Estes MK, Conner ME (1998) Rotavirus 2/6 virus-like particles administered intranasally with cholera toxin, *Escherichia coli* heat-labile toxin (LT), LT-R192G induce protection from rotavirus challenge. J Virol 72(4):3390–3393

Onishchenko GG, Shestopalov AM, Ternovoi VA, Evseenko VA, Durymanov AG, Rassadkin Iu N, Zaikovskaia AV, Zolotykh SI, Iurlov AK, Mikheev VN, Netesov SV, Drozdov IG (2006) Study of highly pathogenic H5N1 influenza virus isolated from sick and dead birds in Western Siberia. Zh Mikrobiol Epidemiol Immunobiol 5:47–54

Pattnaik AK, Brown DJ, Nayak DP (1986) Formation of influenza virus particles lacking hemagglutinin on the viral envelope. J Virol 60(3):994–1001

Perdue ML, Garcia M, Senne D, Fraire M (1997) Virulence-associated sequence duplication at the hemagglutinin cleavage site of avian influenza viruses. Virus Res 49(2):173

Pushko P, Tumpey TM, Bu F, Knell J, Robinson R, Smith G (2005) Influenza virus-like particles comprised of the HA, NA, M1 proteins of H9N2 influenza virus induce protective immune responses in BALB/c mice. Vaccine 23(50):5751–5759

Pushko P, Tumpey TM, Van Hoeven N, Belser JA, Robinson R, Nathan M, Smith G, Wright DC, Bright RA (2007) Evaluation of influenza virus-like particles and Novasome adjuvant as candidate vaccine for avian influenza. Vaccine 25(21):4283–4290

Qiao CL, Yu KZ, Jiang YP, Jia YQ, Tian GB, Liu M, Deng GH, Wang XR, Meng QW, Tang XY (2003) Protection of chickens against highly lethal H5N1 and H7N1 avian influenza viruses with a recombinant fowlpox virus co-expressing H5 haemagglutinin and N1 neuraminidase genes. Avian Pathol 32(1):25–32

Quan FS, Huang C, Compans RW, Kang SM (2007) Virus-like particle vaccine induces protective immunity against homologous and heterologous strains of influenza virus. J Virol 81(7):3514–3524

Quan FS, Steinhauer D, Huang C, Ross TM, Compans RW, Kang SM (2008) A bivalent influenza VLP vaccine confers complete inhibition of virus replication in lungs. Vaccine 26(26):3352–3361

Roberts PC, Lamb RA, Compans RW (1998) The M1 and M2 proteins of influenza A virus are important determinants in filamentous particle formation. Virology 240(1):127–137

Robertson JS, Nicolson C, Newman R, Major D, Dunleavy U, Wood JM (1992) High growth reassortant influenza vaccine viruses: new approaches to their control. Biologicals 20(3):213–220

Roldao A, Vieira HL, Charpilienne A, Poncet D, Roy P, Carrondo MJ, Alves PM, Oliveira R (2007) Modeling rotavirus-like particles production in a baculovirus expression vector system: Infection kinetics, baculovirus DNA replication, mRNA synthesis and protein production. J Biotechnol 128(4):875–894

Sailaja G, Skountzou I, Quan FS, Compans RW, Kang SM (2007) Human immunodeficiency virus-like particles activate multiple types of immune cells. Virology 362(2):331–241

Sasagawa T, Pushko P, Steers G, Gschmeissner SE, Hajibagheri MA, Finch J, Crawford L, Tommasino M (1995) Synthesis and assembly of virus-like particles of human papillomaviruses type 6 and type 16 in fission yeast *Schizosaccharomyces pombe*. Virology 206(1):126–135

Senne DA, Panigrahy B, Kawaoka Y, Pearson JE, Suss J, Lipkind M, Kida H, Webster RG (1996) Survey of the hemagglutinin (HA) cleavage site sequence of H5 and H7 avian influenza viruses: amino acid sequence at the HA cleavage site as a marker of pathogenicity potential. Avian Dis 40(2):425–437

Shortridge KF, Zhou NN, Guan Y, Gao P, Ito T, Kawaoka Y, Kodihalli S, Krauss S, Markwell D, Murti KG, Norwood M, Senne D, Sims L, Takada A, Webster RG (1998) Characterization of avian H5N1 influenza viruses from poultry in Hong Kong. Virology 252(2):331–342

Skountzou I, Quan FS, Gangadhara S, Ye L, Vzorov A, Selvaraj P, Jacob J, Compans RW, Kang SM (2007) Incorporation of glycosylphosphatidylinositol-anchored granulocyte-macrophage colony-stimulating factor or CD40 ligand enhances immunogenicity of chimeric simian immunodeficiency virus-like particles. J Virol 81(3):1083–1094

Smith GE, Fraser MJ, Summers MD (1983) Molecular engineering of the *Autographa californica* nuclear polyhedrosis virus genome: deletion mutations within the eutralizi gene. J Virol 46(2):584–593

Stanley MA (2006) Human papillomavirus vaccines. Rev Med Virol 16(3):139–149

Suarez DL, Perdue ML, Cox N, Rowe T, Bender C, Huang J, Swayne DE (1998) Comparisons of highly virulent H5N1 influenza A viruses isolated from humans and chickens from Hong Kong. J Virol 72(8):6678–6688

Subbarao K, Klimov A, Katz J, Regnery H, Lim W, Hall H, Perdue M, Swayne D, Bender C, Huang J, Hemphill M, Rowe T, Shaw M, Xu X, Fukuda K, Cox N (1998) Characterization of an avian influenza A (H5N1) virus isolated from a child with a fatal respiratory illness. Science 279(5349):393–396

Subbarao K, Chen H, Swayne D, Mingay L, Fodor E, Brownlee G, Xu X, Lu X, Katz J, Cox N, Matsuoka Y (2003) Evaluation of a genetically modified reassortant H5N1 influenza A virus vaccine candidate generated by plasmid-based reverse genetics. Virology 305(1):192–200

Szecsi J, Boson B, Johnsson P, Dupeyrot-Lacas P, Matrosovich M, Klenk HD, Klatzmann D, Volchkov V, Cosset FL (2006) Induction of eutralizing antibodies by virus-like particles harbouring surface proteins from highly pathogenic H5N1 and H7N1 influenza viruses. Virol J 3:70

Takada A, Kuboki N, Okazaki K, Ninomiya A, Tanaka H, Ozaki H, Itamura S, Nishimura H, Enami M, Tashiro M, Shortridge KF, Kida H (1999) Avirulent Avian influenza virus as a vaccine strain against a potential human pandemic. J Virol 73(10):8303–8307

Taubenberger JK, Reid AH, Janczewski TA, Fanning TG (2001) Integrating historical, clinical and molecular genetic data in order to explain the origin and virulence of the 1918 Spanish influenza virus. Philos Trans R Soc Lond B Biol Sci 356(1416):1829–1839

Triyatni M, Saunier B, Maruvada P, Davis AR, Ulianich L, Heller T, Patel A, Kohn LD, Liang TJ (2002) Interaction of hepatitis C virus-like particles and cells: a model system for studying viral binding and entry. J Virol 76(18):9335–9344

Tumpey TM, Garcia-Sastre A, Taubenberger JK, Palese P, Swayne DE, Pantin-Jackwood MJ, Schultz-Cherry S, Solorzano A, Van Rooijen N, Katz JM, Basler CF (2005) Pathogenicity of influenza viruses with genes from the 1918 pandemic virus: functional roles of alveolar macrophages and neutrophils in limiting virus replication and mortality in mice. J Virol 79(23):14933–14944

Wang BZ, Liu W, Kang SM, Alam M, Huang C, Ye L, Sun Y, Li Y, Kothe DL, Pushko P, Dokland T, Haynes BF, Smith G, Hahn BH, Compans RW (2007) Incorporation of high levels of chimeric human immunodeficiency virus envelope glycoproteins into virus-like particles. J Virol 81(20):10869–10878

Webby RJ, Perez DR, Coleman JS, Guan Y, Knight JH, Govorkova EA, McClain-Moss LR, Peiris JS, Rehg JE, Tuomanen EI, Webster RG (2004) Responsiveness to a pandemic alert: use of reverse genetics for rapid development of influenza vaccines. Lancet 363(9415):1099–1103

Widjaja L, Ilyushina N, Webster RG, Webby RJ (2006) Molecular changes associated with adaptation of human influenza A virus in embryonated chicken eggs. Virology 350(1):137–145

Yamshchikov GV, Ritter GD, Vey M, Compans RW (1995) Assembly of SIV virus-like particles containing envelope proteins using a baculovirus expression system. Virology 214(1):50–58

Yao Q, Kuhlmann FM, Eller R, Compans RW, Chen C (2000) Production and characterization of simian–human immunodeficiency virus-like particles. AIDS Res Hum Retroviruses 16(3):227–236

Ye L, Lin J, Sun Y, Bennouna S, Lo M, Wu Q, Bu Z, Pulendran B, Compans RW, Yang C (2006) Ebola virus-like particles produced in insect cells exhibit dendritic cell stimulating activity and induce neutralizing antibodies. Virology 351(2):260–270

Zhang J, Leser GP, Pekosz A, Lamb RA (2000) The cytoplasmic tails of the influenza virus spike glycoproteins are required for normal genome packaging. Virology 269(2):325–234

# Pandemic Influenza Vaccines

## Lauren J. DiMenna and Hildegund C.J. Ertl

### Contents

**Abstract** Since their compositions remain uncertain, universal pandemic vaccines are yet to be created. They would aim to protect globally against pandemic influenza viruses that have not yet evolved. Thus they differ from seasonal vaccines to influenza virus, which are updated annually in spring to incorporate the latest circulating viruses, and are then produced and delivered before the peak influenza season starts in late fall and winter. The efficacy of seasonal vaccines is linked to their ability to induce virus-neutralizing antibodies, which provide subtype-specific protection

L.J. DiMenna and H.C.J. Ertl (✉)

The Wistar Institute, 3601 Spruce St, Philadelphia, PA, 19104, USA

e-mail: ertl@wistar.upenn.edu

R.W. Compans and W.A. Orenstein (eds.), *Vaccines for Pandemic Influenza*,
Current Topics in Microbiology and Immunology 333,
DOI 10.1007/978-3-540-92165-3_15, © Springer-Verlag Berlin Heidelberg 2009

against influenza A viruses. If pandemic vaccines were designed to resemble current vaccines in terms of composition and mode of action, they would have to be developed, tested, and mass-produced after the onset of a pandemic, once the causative virus had been identified. The logistic problems of generating a pandemic vaccine from scratch, conducting preclinical testing, and producing billions of doses within a few months for global distribution are enormous and may well be insurmountable. Alternatively, the scientific community could step up efforts to generate a universal vaccine against influenza A viruses that provides broadly cross-reactive protection through the induction of antibodies or T cells to conserved regions of the virus.

# 1    Introduction

Influenza viruses belong to the family of *Orthomyxoviridae*, which includes negative single-stranded RNA viruses with segmented genomes. Among the three genera of influenza viruses (A, B, and C), influenza A and C viruses infect humans as well as other species, while influenza B virus mainly infects humans. The most common and serious infections of humans are caused by influenza A virus. Influenza A viruses are further divided into subtypes based on their hemagglutinin (HA) and neuraminidase (NA) genes, which encode the two viral surface proteins.

Influenza A virus typically infects epithelial cells that line the respiratory tract, but may also replicate in other tissues in different hosts, including conjunctiva, intestine, brain, liver, kidney, and gut. In general, influenza A virus infections are self-limiting in healthy human adults, and mainly cause life-threatening disease in the very young and in the elderly. Notwithstanding, this depends on the circulating type. Aquatic birds serve as the main reservoir of influenza A viruses and carry all of the known subtypes (H1-16, N1-9) without necessarily developing disease upon infection. The virus can adapt to other species such as poultry, pigs, horses, or humans. In humans, thus far the H1, H2, or H3, and N1 or N2 influenza viruses have established transmittable infections.

Influenza viruses mutate rapidly, and these mutations affect mainly (but not exclusively) the genes encoding the surface proteins. Point mutations that cause gradual changes are referred to as antigenic drift, and allow the virus to evade protective neutralizing antibody responses induced by previous infections. Most annual epidemics are caused by antigenic drift variants. Rearrangements of the HA- or NA-encoding gene segments between viral types circulating in humans and those endemic in animals result in more dramatic changes, also called antigenic shifts, and the pandemics of 1957 with H2N2 and 1968 with H3N2 were caused by such new types of influenza virus. According to the World Health Organization (WHO), a pandemic is the emergence of a serious new disease caused by an agent that spreads easily among humans. WHO recommends three measures to lessen the impact of the next influenza virus pandemic: (1) increased surveillance to allow for the earliest possible warning that a human pandemic has started; (2) early intervention to stall global spread and prevent further adaptations; and (3) development of an effective pandemic vaccine.

Available vaccines against influenza virus are seasonal vaccines that are updated annually to incorporate the latest circulating viruses. Seasonal vaccines are composed of three different influenza viruses, which are typically two subtypes of influenza A virus and one strain of influenza B virus. The vaccine composition is generally agreed upon in spring to allow for manufacturing and distribution before onset of the influenza season in late fall to winter. Seasonal vaccines are currently derived from egg-grown viruses that are either inactivated and then given systemically or attenuated by cold adaptation and given directly to the airways.

Pandemic vaccines, the focus of this chapter, are at this stage virtual vaccines of an unknown composition. They aim to protect against a newly evolved pandemic influenza virus. A pandemic vaccine may thus have to be manufactured at the onset of a pandemic, or alternatively one would need to devise a vaccine that induces broadly cross-reactive protection, unlike the current vaccines.

## 2 Influenza A Viruses

Influenza A viruses are enveloped spherical viruses which contain eight segments of single negative-stranded RNA. Segments 1, 2, and 3 encode the transcriptase complex composed of basic polymerases (PB)2 (segment 1) and PB1 (segment 2) and acid polymerase (PA, segment 3). Segment 4 encodes the hemagglutinin (HA), which has receptor-binding activity, promotes cell fusion, and is the major target for neutralizing antibodies. Segment 5 encodes the nucleoprotein (NP) which complexes the viral RNA to form the nucleocapsid. NP is a major target for cross-reactive $CD8^+$ T cells in mice and humans (Falk et al. 1991). Segment 6 encodes the viral neuraminidase (NA), a cell surface protein with enzymatic activity, which also provides a target for neutralizing antibodies. Drugs such as zanamivir and oseltamivir, which block the enzymatic cleavage of sialic acid residues by NA, are available and can be used to treat or prevent infections (Garman and Laver 2004). Segment 7 encodes matrix (M) protein 1 and 2. M2 has ion channel activity, which is blocked by the antiviral drug amantadine (Ison and Hayden 2001). M protein is also a target for cross-reactive $CD8^+$ T cells in humans, while the M2 ectodomain is a target for nonneutralizing but nevertheless protective antibodies (Zhang et al. 2006). Segment 8 encodes nonstructural proteins (NS) 1 and 2.

## 3 Previous Influenza A Virus Pandemics

The twentieth century experienced three major influenza virus pandemics (Table 1) and several small abortive pandemics, as well as pandemic threats and numerous outbreaks in animals also called epizootics or panzootics.

**Table 1** Recent influenza pandemics

| Pandemic | Subtype | Place of origin | Age group most affected (years) | Death toll |
|---|---|---|---|---|
| Spanish Flu 1918–1920 | H1N1 | USA | 20–40 | 50–100 million |
| Asian Flu 1957–1958 | H2N2 | China | 65+ | 1–4 million |
| Hong Kong Flu 1968–1969 | H3N2 | Hong Kong | 65+ | 500,000 |

## 3.1 Pandemics

### 3.1.1 Spanish Flu

The first pandemic of the twentieth century started in 1918 in the USA and then spread to Africa and Europe, first to France and then Spain, and subsequently to every part of the globe. This pandemic was caused by an H1N1 virus and is paradoxically and unfairly referred to as the Spanish Flu. The pandemic that started in March of 1918 and lasted until June of 1920 killed half a million Americans and somewhere between 50 and 100 million humans worldwide (Johnson and Mueller 2002). This virus infected nearly 50% of the population and killed 2.5% of all of those that became infected. It is estimated that 25 million people died during the first 25 weeks of the pandemic. Death rates were high in humans between the ages of 20–40, an age group which generally recovers easily from influenza A virus infection. During the initial stages of the pandemic, the early symptoms of infection, which included hemorrhages and lung edema followed by death within 24–48 h, were commonly misdiagnosed. The severity of the symptoms is assumed to have been caused by an excessive release of cytokines in response to the virus (Kash et al. 2004), which was most severe in healthy adults with sturdy immune systems. The 1918 H1N1 virus was recently isolated from victims preserved in permafrost, and upon sequencing the virus was rederived through genetic engineering (Tumpey et al. 2005). This allowed for an extensive characterization of the virus using modern tools of science. The 1918 H1N1 virus has several distinct features that may explain its unique virulence. Most types of influenza A virus require trypsin-like enzymes for cleavage of the viral HA, which in turn restricts their cellular tropism. The NA of the H1N1 virus of 1918 can directly or indirectly cleave HA, thus rendering this virus independent of trypsin-like enzymes (Steinhauer 1999). Increased virulence was further enabled by NS proteins, which allow the virus to disable the interferon (IFN) pathway (Seo et al. 2004), a crucial component of both innate and adaptive immunity. Human-to-human transmission, a prerequisite for a human pandemic, appears to have involved a switch in preferential binding of the HA protein from $\alpha$-2,3 sialic acid found in the avian enteric tract to $\alpha$-2,6 sialic acid present in the human respiratory tract (Tumpey et al. 2007). This altered receptor binding activity can be achieved experimentally through a single amino acid exchange at position 190 of the HA of the 1918 H1N1 subtype. Additional changes in the viral PB and PB2 proteins, which contain four amino acids that are conserved in human viruses and that differ from those prevalent in avians, are likely to have affected transmission between humans (Russell and Webster 2005).

### 3.1.2  Asian Flu

The 1957 pandemic, also referred to as the Asian Flu, originated from a recombination between a circulating human virus and a virus endemic in ducks. The virus was first isolated early in 1956 in Guizhou, China, and by February of 1957 had spread to Singapore, and to the USA by June of that year. This virus, an H2N2 virus, caused an estimated 1–4 million deaths worldwide (Dunn 1958). Death rates were highest in the elderly.

### 3.1.3  Hong Kong Flu

The 1968 pandemic, also called the Hong Kong Flu, was caused by an antigenic shift of an H2N2 virus to an H3N2 virus. This pandemic was comparatively mild, causing an estimated 500,000 human deaths (Cockburn et al. 1969; Kilbourne 2006). Again mortality was high in those above 65 years of age.

## 3.2  Pseudopandemics and Abortive Pandemics

In 1946 an H1N1 virus that was first seen in Japan and Korea spread to military bases in the USA (Lessler et al. 2007). Further spread was not observed. In 1977, an H1N1 virus spread rapidly from China and caused epidemic disease in children and young adults (<23 years) worldwide. Older humans were not affected, presumably due to protection from previous exposure to H1N1 viruses.

## 3.3  Pandemic Threats

### 3.3.1  Swine Flu

In the winter of 1976, a novel swine influenza virus subtype was detected in military recruits at Fort Dix, New Jersey. A total of 13 soldiers became symptomatically infected and one died. There was only limited spread to humans living outside the military base. Fearing a major pandemic, a vaccine was rapidly generated and administered to 40 million humans. A few months after mass vaccination had started, reports of Guillain–Barré syndrome in vaccine recipients started to accumulate, and by early 1977 (when vaccination was stopped) more than 500 cases of GBS had been reported, 25 of which were fatal (Langmuir et al. 1984).

### 3.3.2 Bird Flu

A highly pathogenic form of avian H5N1 virus was first detected in Asian poultry in 1997 (Centers for Disease Control and Prevention 1997). During this year, a total of 18 human cases were reported from Hong Kong, of which six were fatal. The virus rapidly caused pneumonia and multiple organ failure in infected individuals, which were mainly young adults. Culling of infected flocks of poultry initially appeared to have stopped further spread, but then in 2003 additional human cases with a similar H5N1 virus were recorded in Vietnam (Tran et al. 2004). As of January of 2008, 349 human cases of H5N1 virus infection with 216 deaths have been reported from Asia, Eurasia, and North Africa.[1] Most cases occurred in Indonesia, Vietnam, and Egypt. Highly pathogenic H5N1 virus was also isolated from poultry and wild birds in Europe, including the United Kingdom. Thus far, the virus has been transmitted by human contact with infected birds, and only a few isolated cases were suggestive of direct human-to-human transmission. Further mutations of H5N1 virus, either in the form of adaptive point mutations (i.e., antigenic shift) or through reassortment in humans concomitantly infected with a different influenza A virus, could eventually allow for sustained and efficient human-to-human transmission. Control measures have focused on culling of infected flocks of domestic birds and restriction of poultry trade between countries. Some countries implemented vaccination programs for poultry (Steel et al. 2008; Cristalli and Capua 2007). Other countries rejected the idea of bird vaccination due to fears that this may mask infections and allow for further mutations that may promote human transmissibility.

Similar to the 1918 H1N1 virus, pathogenic H5N1 virus activates HA through a trypsin-independent mechanism (Hulse et al. 2004). Pathogenic H5N1 virus has a multibasic cleavage site that can be digested by furin and furin-like proteases, which are more ubiquitously present in human tissues than the trypsin-like enzymes that cleave HA of current human influenza viruses. The NS1 protein of pathogenic subtypes of H5N1 virus renders the virus resistant to the activity of IFNs and tumor necrosis factor (TNF)-$\alpha$ (Seo et al. 2002). The H5N1 virus has changed since its first isolation in 1997. Such changes include resistance to the antiviral drug amantadine due to a M2 mutation first reported in 2004 from Thailand (Cheung et al. 2006). The virus has become more lethal for humans and mice, and has gained robustness against destruction in the environment. The virus has increased its host range and has been shown to cause disease in felines such as tigers (Keawcharoen et al. 2004), which are otherwise resistant to influenza A virus infections.

In 1999, an H9N2 virus, which also originated from poultry, caused illnesses in two children in Hong Kong. Both children survived and there was no serological evidence that the virus spread to their contacts.

---

[1] See http://www.who.int/csr/disease/avian_influenza/country/cases_table_2008_01_21/en/index.html.

## 3.4 Epizootics and Panzootics

A number of epizootics and panzootics have been caused by a wide variety of influenza viruses. In poultry, numerous outbreaks with highly pathogenic influenza viruses have been reported from all over the globe within the last 50 years. These outbreaks were caused by a variety of subtypes, such as H5N1, H7N2, H1N7, H7N3, H13N6, H5N9, H11N6, H3N8, H9N2, H5N2, H4N8, H10N7, H2N2, H8N4, H14N5, H6N5, H12N5, and others. H5N1 virus, which is currently endemic in Asia, Africa, and Europe, has within the last eight years caused the deaths of millions of birds, many of which were culled to prevent further spread and to protect humans.

Influenza virus outbreaks have been observed in other species. For example, from 1979 to 1980, several hundred harbor seals died along the coast of New England due to infection with a H7N7 virus (Geraci et al. 1982). As of 1997, H3N2 circulates in pigs (Gramer et al. 2007). Horses have been infected with H7N7 and H3N8 viruses (Amonsin et al. 2007; Oxburgh and Hagström 1999). The latter can also infect and kill canines. H5N1 has caused the deaths of felines, including tigers and domestic cats (Cristalli and Capua 2007; Steel et al. 2008).

Several of these viruses have infected humans without achieving the capacity for human-to-human transmission. In 2003, 89 people were infected with H7N7 influenza virus from poultry in the Netherlands (Koopmans et al. 2004).

In 2002–2003, two residents of US mid-Atlantic states showed serologic evidence of infection with H7N2 (Senne et al. 2006). In 2004, two poultry farm workers in British Colombia became infected with H7N3 virus (Tweed et al. 2004). In 2004, Egypt reported human infections with H10N7. Any subtype of the influenza virus thus has the potential to infect humans and to evolve into a pandemic virus, which has to be taken into account when designing pandemic vaccines.

## 4 Risk Factors for Severe Influenza Virus Infections

More than 90% of deaths during seasonal influenza virus outbreaks occur in the elderly ($\geq$65 years of age). Immunosenescence during aging leads to impaired immune responses, which increases the susceptibility of the aged to infectious agents. The elderly are affected by primary immunological changes, which are part of the natural aging process, and secondary immunological changes caused by underlying diseases and unhealthy life styles (Malaguarnera et al. 2001). Primary changes of the immune system in healthy elderly involve mainly T cells, though changes in natural killer (NK) cells and NK T cell function with age have been noted (Ginaldi et al. 1999c; Solana and Mariani 2000). T cells show clonal senescence, their potential for expansion is decreased, and their ability to produce certain cytokines and to respond to cytokines decreases. The proportion of T cells with a memory cell phenotype increases while numbers of naïve T cells decrease. Stimulation with new antigens appears to result in shortened immunological memory (Ginaldi et al. 1999b). The T cell repertoire loses diversity (Effros et al. 2003) due to chronic antigenic stimulation, leading to continued clonal expansion of some T cells, which undermines the homeostatic balance of the immune system. Primary B cell responses in the elderly are commonly low and short-lived, resulting in antibodies with low affinity (Ginaldi et al. 1999a). Formation of germinal centers is decreased, antigen transport is impaired, and follicular dendritic cells show atrophy and their capacity to form antigen depots is reduced (Zheng et al. 1997; Aydar et al. 2004). Autoantibodies are more common and the B cell repertoire becomes more restricted. Many of these changes reflect secondary effects due to an age-related decline of helper functions from $CD4^+$ T cells, which show reduced expression of critical costimulatory receptors that are essential for activation of B cells, germinal center formation and rearrangement, and hypermutation of immunoglobulin genes. Underlying chronic diseases dramatically increase the risk of serious complications of an influenza virus infection. Patients with one or two chronic diseases have 40- or 150-fold (respectively) greater risk for developing pneumonia upon influenza virus infection (Janssens and Krause 2004; Stott et al. 2001). Underlying chronic heart, lung, or liver diseases increase the risk of serious influenza virus infection in all age groups, not just the elderly.

Vaccines perform poorly in the elderly, commonly resulting in inadequate and short-lived titers of protective antibody responses (Biro 1978; Saurwein-Teissl et al. 2002). Current influenza virus vaccines provide 70–90% protection against a closely related virus in those <65 years of age, but only 30–40% protection in humans above the age of 65.

Young children, pregnant women, and immunosuppressed individuals also have an increased risk for influenza A virus-associated morbidity. Another risk factor is superinfection of the airways with bacterial pathogens, which can enhance virulence of the influenza virus through bacterial proteases (Callan et al. 1997). On the other hand, influenza virus can increase bacterial infection by destroying respiratory epithelium and increasing bacterial receptor (McCullers 2006). Other risk factors include living in institutionalized settings such as prisons or nursing homes, or working in healthcare, where the risk of exposure and the risk of further spread are increased.

## 5 Immune Responses to Influenza A Viruses

Vaccines aim to induce memory immune responses that, upon encountering the virus, are rapidly reactivated or recruited to either completely prevent an infection by causing so-called sterilizing immunity, or to rapidly control viral spread. It is thus important to understand which type of immune response provides reliable protection in order to specifically design immunogens that elicit this type of a response. Influenza virus pandemics unfortunately have an element of surprise on their side by their very nature, and it may be unrealistic to expect that at the onset of a pandemic, which can potentially spread around the globe within less than six months, sufficient doses of a reliable vaccine or efficacious antiviral drugs will be available to protect the entire human population. Other preventions, such as activation of protective innate immune responses in those at immediate risk for infection, may add to the repertoire we can call upon to combat the next influenza virus pandemic.

## 5.1 Innate Immunity

Innate immunity can provide resistance to influenza virus infection, as has been demonstrated in animals treated with immunomodulators such as baculovirus, lentidan, double-stranded RNA, or modified heat-labile toxin of *Escherichia coli* prior to infection (Abe et al. 2003; Irinoda et al. 1992; Saravolac et al. 2001; Williams et al. 2004). Clinical trials in children who were vaccinated with an attenuated influenza A virus vaccine after the onset of an influenza A virus out-break also suggested that protection was at least in part mediated by an innate immune response to the vaccine (Piedra et al. 2007).

Influenza A virus infection leads to the rapid increase of proinflammatory cytokines in nasal and pulmonary secretions (Jao et al. 1970; Gentile et al. 1998). The virus causes the activation and maturation of dendritic cells and stimulates plasmacytoid dendritic cells to secrete large amounts of type I IFNs (López et al. 2004; Cella et al. 2000). Influenza virus activates macrophages to secrete IL-1, 6 and 12 and TNF-$\alpha$ (Mak et al. 1982; Pirhonen et al. 1999). IL-12 in turn induces IFN-$\gamma$ production by NK cells. The early cytokine response to influenza virus can be pronounced and can result in significant pathology (Van Reeth et al. 2002).

Nevertheless, early cytokines such as interferons also provide resistance to influenza A viruses (Beilharz et al. 2007; Fattal-German and Bizzini 1992). NS1 of H5N1 renders the virus resistant to the antiviral activity of IFNs and TNF-$\alpha$ (Sekellick et al. 2000). Reassortant influenza A viruses carrying the NS1 of H5N1 induce increased levels of cytokines in mice and decreased levels of IL-10 (Lipatov et al. 2005a). Both macrophages and NK cells can kill infected cells and are crucial to early infection control (Zychlinsky et al. 1990; Tsuru et al. 1987), as are natural IgM and the early components of the classical pathway of complement, which together can neutralize influenza virus (Jayasekera et al. 2007).

## 5.2 Primary Adaptive Immune Responses to Influenza Virus Infection

Inhalation infection with influenza A virus triggers a mucosal immune response in the upper respiratory tract that is initiated within nasal-associated lymphoid tissue (NALT) in mice and within Waldeyer's ring (tonsils) in primates. In the lower respiratory tract, responses are induced in bronchus-associated lymphoid tissues. Responses can also be detected in distant lymphoid tissues such as spleen or blood.

Infection causes a local secretory IgA response as well as IgM and IgG antibodies directed mainly against the viral HA. Antibody-secreting cells can be detected in mice in the respiratory mucosa and in lung tissue within five days after infection. Dimeric IgA (dIgA) antibodies which are transcytosed across epithelial cells upon binding to their receptors can bind to de novo synthesized viral antigens and block viral assembly, thus contributing to viral clearance (Tamura and Kurata 2004).

Influenza virus-specific CD8[+] and CD4[+] T cells are induced upon intranasal application of influenza A virus (Roti et al. 2008; Swain et al. 2004). Viral clearance following a primary infection is mediated in part by CD8[+] T cells and in part by antibodies, which in turn require the activity of CD4[+] T helper cells for their induction. Lack of CD4[+] T cells does not affect induction of a primary CD8[+] T cell response to influenza A virus (Yap and Ada 1978; Mozdzanowska et al. 2005), although absence of CD4[+] T cells in general reduces the magnitude of the memory CD8[+] T cell pool and the CD8[+] T cell recall response. Neither IFN-$\gamma$ nor IFN-$\alpha/\beta$ appear to be essential for viral clearance (Price et al. 2000), although loss of both IFN pathways has been reported to exacerbate disease. Perforin is essential for viral clearance, and mice lacking perforin show delayed viral clearance and increased mortality to influenza A virus infection (Topham et al. 1997). Increased mortality was also observed in IL-1 receptor knockout mice (Szretter et al. 2007); these mice developed normal CD8[+] T cell responses and viral titers were only modestly above those of normal mice. IL-1 receptor knockout mice showed a defect in recruitment of inflammatory cells to the site of infection, most notably neutrophils and CD4[+] T cells.

## 5.3 Secondary Adaptive Immune Responses and Their Role in Protecting Against Infection

A secondary infection with influenza A virus can be prevented by local sIgA and can be blunted by rapid activation of memory B cells. Neutralizing IgA antibodies are thought to primarily prevent infection of the upper respiratory tract, while serum IgG plays a role in protecting against viral pneumonia (Tamura and Kurata 2004). Protective neutralizing antibody responses induced by infection or vaccination are subtype specific and do not provide protection against heterotypic challenge. Their ability to provide resistance to an antigenic drift subtype depends on the degree of antigenic variation between the viruses (Kaye et al. 1969).

It must pointed out, however, that although the role of neutralizing antibodies in providing resistance to influenza virus is not debated, it remains far from clear-cut. Some mouse studies showed that adaptive transfer of neutralizing secretory IgA protected the animals, while transfer of neutralizing antibodies of the IgG isotype was inefficient (Renegar and Small 1991). Other mouse studies showed that protection by H5-specific IgG1 monoclonal antibodies can be achieved against H5N1 infections (Hanson et al. 2006). Yet others reported protection by IgG antibodies that bound HA but failed to neutralize the virus (McLain and Dimmock 1989). One monoclonal neutralizing antibody was described that cross-reacted between H1 and

H2 and consequently protected animals upon passive transfer against infection with either virus (Okuno et al. 1994). In other virus infections, such as those with rabies virus, where neutralizing antibodies are known to play a dominant role in protection against infection and disease, protective titers of neutralizing antibodies have been defined. For rabies virus, a titer of or above 0.5 international units protects against challenge; this knowledge has greatly facilitated vaccination efforts. In contrast, it is still not known what titer of influenza A virus-neutralizing antibodies reliably provides protection against disease. In general, it is assumed that titers above 1:40 are protective, although numerous clinical trials have demonstrated that humans with lower titers were protected while others with higher titers developed symptomatic infections.

Protection against heterotypic challenge (i.e., challenge with a different subtype of influenza virus than that used for immunization) can be mediated by a number of mechanisms. As already mentioned above, some neutralizing antibodies can cross-neutralize several subtypes of influenza A virus. Nonneutralizing antibodies to the ectodomain of matrix protein (M2e) can protect against heterotypic challenge in animal models (Mozdzanowska et al. 2003). The 23 amino acid (aa) long M2e is conserved in its nine N-terminal amino acids and shows relative minor variability in the remaining sequences. This is likely to reflect a lack of selective pressure, as natural infections or traditional vaccines induce only low antibody responses to M2e (Feng et al. 2006). The currently circulating avian H5N1 and H7N2 subtypes show sequence variability with previous human isolates that affect M2e antibody-binding sites. For example, they show changes in amino acids at positions 10–16 of M2e (H5N1: PIRNEWG to PTRNGWG, or PTRNEWE) (Liu et al. 2005).

CD8[+] T cells induced by repeated infections appear to contribute little to natural resistance to influenza virus infection in humans. This may be linked to suboptimal stimulation of this T cell subset upon natural infection, as human volunteers with exceptionally high levels of circulating influenza A virus-specific CD8[+] T cells showed reduced viral shedding upon an experimental infection compared to those with low levels of pre-existing influenza A virus-specific CD8[+] T cells (Epstein 2006; Murasko et al. 2002).

In mice, a number of studies showed that CD8[+] T cells protect, while other showed that they fail to protect. Early studies from the group of G. Ada showed that adoptive transfer of influenza virus immune cells provided protection against challenge with a heterotypic subtype of the virus (Yap and Ada 1978). These studies were confirmed by R. Dutton and colleagues, who studied the efficacy of passively transferred, in vitro activated CD8[+] T cells isolated from mice transgenic (tg) for a T cell receptor (TcR) to the influenza A virus HA (Cerwenka et al. 1999). Transfer of naïve TcR-tg CD8[+] T cells failed to provide resistance to challenge. Protection against a lethal infection could be provided by the transfer of rested memory-like or effector TcR-tg CD8[+] T cells, although the latter effected more rapid viral clearance, which may indicate that the rested CD8[+] T cells needed to expand before they assumed effector functions. Protection was only mediated by CD8[+] T cells that were able to home to the infected respiratory tissues. Poxvirus vectors expressing the influenza A virus NP, which induce a CD8[+] T cell response (Andrew et al. 1986), were shown to induce some protection against heterotypic challenge (Endo et al. 1991; Altstein et al. 2006). Further studies showed that although vaccinia virus vectors expressing the influenza virus NP induced only limited protection in mice, adoptive transfer of T cells isolated from NP-immune mice and expanded in vitro were highly effective (Mbawuike et al. 2007). Yet another group reported that a vaccinia virus vector which expressed a sequence of NP that induced a sturdy CD8+ T cell response in mice, including in their lungs, completely failed to induce protective immunity as assessed by peak viral loads, morbidity or mortality (Lawson et al. 1994).

Heterotypic T cell-mediated protection was also reported after immunization of mice with an adjuvanted influenza virus vaccine (Sambhara et al. 1998) or with DNA vaccines expressing internal proteins of influenza virus (Saha et al. 2006; Fu et al. 1997). Another group reported that protection upon intranasal immunization with an adjuvanted nucleoprotein vaccine was mediated by T helper cells of the Th1 type rather than by CD8+ T cells (Tamura et al. 1996). Yet another group reported protection with an adenovirus vector expressing nucleoprotein (Roy et al. 2007). In our hands, subunit vaccines expressing the nucleoprotein induced strong CD8+ T cell responses that could readily be detected in spleen, blood, or even lungs of vaccinated mice. Nevertheless, vaccinated mice were not reliably protected against disease or death following challenge with influenza A virus (unpublished). Overall T cell protection studies largely agree that adoptive transfer of in vitro expanded CD8+ T cells provides protection against influenza virus. Results on the protective nature of in situ activated influenza virus-specific CD8+ T cells range from solid protection to complete absence of protection, even under circumstances where high numbers of influenza virus-specific CD8+ T cells were present in the airways at the time of challenge. The lack of consistency of protection through CD8+ T cells may reflect genetic differences in the mouse strain used for the experiments, differences in the dose or type of challenge virus, differences in the interval between vaccination and challenge, and/or differences in the functionality of CD8+ T cells induced by various approaches.

The take-home message for developing an influenza vaccine that is useful for preventing or ameliorating a pandemic therefore remains ambiguous. Neutralizing antibodies protect against HA provided there is sufficient homology between the vaccine and the infecting virus. Antibodies against M2e protect against a wider array of subtypes, as M2e is more conserved; nevertheless, M2e shows some variability, and protection through M2e-specific antibodies is not as robust as protection provided by neutralizing antibodies. The rules that govern CD8+ T cell-mediated protection against influenza virus remain ill-defined.

# 6 Influenza Virus Vaccines

Influenza virus was first isolated in 1933 (Smith et al. 1933), and effective vaccines were developed and tested by 1943–1944 and became available by 1945 (Francis et al. 1945a,b). Vaccines were thus not available during the Spanish Flu pandemic, but rapidly became available during the 1957 Asian Flu pandemic (Gundlefinger et al. 1958), when they were mainly used in military personnel. Although the 1968 Hong Kong Flu subtype was identified rapidly, vaccine production was delayed, and a vaccine was not available during the outbreak.

Until recently, all available influenza vaccines were trivalent inactivated (killed) virus vaccines. Initially whole-virus vaccines were used, which were then replaced by 2001 by the less reactogenic split-virus vaccines. In June of 2003, a live attenuated, cold-adapted, temperature-sensitive, trivalent influenza virus vaccine was licensed in the United States for use in humans between 2 and 49 years of age. Multiple clinical trials have been performed in adults (Demicheli et al. 2004), children (Smith et al. 2006), and the elderly (Jefferson et al. 2005) to assess the efficacy of influenza vaccines. Studies on live vaccines are still limited, but to date they suggest that such vaccines may be more effective than inactivated vaccines in some cohorts (Treanor et al. 1999).

One manuscript published an analysis of trials involving a total of 59,566 adults (Demicheli et al. 2004) which showed that the live attenuated vaccines reduced the number of cases of serologically confirmed influenza by 48% while the inactivated vaccines had a vaccine efficacy of 70%. The yearly recommended vaccines had low effectiveness against clinical influenza cases or time off work, the later a nonspecific outcome that included illness caused by influenza as well as other agents. The authors concluded that universal immunization of healthy adults is not supported by their results.

Fifty-one studies involving 263,987 children were included in an analysis of influenza virus vaccine efficacy in children (Smith et al. 2006). The attenuated vaccines showed an efficacy of 79% in children older than two years, while inactivated vaccines had a lower efficacy of 59%. In children under two, the efficacy of inactivated vaccine was similar to placebo. In another study, results from 19 randomized clinical studies covering a total of 247,517 children were analyzed and reported to show an overall vaccination efficacy of 36% against clinical disease, 67% against laboratory-confirmed cases, and 51% against acute otitis media. Between-study variability was related to the children's age and study quality. For example, when studies from the USSR were excluded from the analysis, the overall efficacy of the vaccine in preventing clinical cases increased from 36% to 61% (Manzoli et al. 2007).

Indirect evidence for the effectiveness of annual influenza virus vaccination of children can be gained from Japan, where as of 1957 school children were vaccinated annually. Vaccination became mandatory in the 1970s and was discontinued in 1994. During the time of mandatory vaccination, mortality among the elderly declined markedly, presumably due to reduced exposure to their infected grandchildren.

Sixty-four studies were analyzed to determine the efficacy of influenza vaccination in the elderly (Jefferson et al. 2005; Rivetti et al. 2006). In homes for elderly individuals, the effectiveness of vaccines against disease caused by influenza virus could not be demonstrated. When the vaccines were closely matched to the circulating virus subtype, they prevented pneumonia, hospital admission, and deaths. In elderly individuals living in the community, vaccines were not significantly effective against clinical influenza or pneumonia that were not laboratory confirmed. The authors concluded that vaccination was useful in long-term care facilities but not necessarily in community settings. Another large analysis of community-living elderly came to the opposite conclusion. This analysis showed that vaccination was associated with a 27% reduction in the risk of hospitalization for pneumonia and a 48% reduction in the risk of death (Nichol et al. 2007). Other smaller studies showed that immunization of frail elderly did not reduce the rate of hospital admissions due to acute respiratory illnesses (Jordan et al. 2006), and that vaccination failed to reduce the overall mortality of the elderly (Rizzo et al. 2007).

In summary, although annual influenza virus vaccination is highly recommended, especially for high-risk populations, results of clinical trials designed to prove their efficacy remain controversial and thus far do not fully support the notion that vaccination affords reliable protection against influenza virus infection and its sequelae.

# 7 Pandemic Influenza Virus Vaccines

WHO has summarized a number of global pandemic phases that have been adopted in federal and regional response plans and serve to define the type of responses required. Details on these phases and suggested courses of action can be obtained online (see also Table 2).[2] These phases are as follows:

A. Interpandemic period: *Phase I:* No new influenza virus subtypes have been detected in humans but they may be present in animals. Risk of human infection is considered low. *Phase II:* An influenza virus subtype circulating in animals poses a high risk to humans.
B. Pandemic alert period: *Phase III:* Animal-to-human infection(s) with a new subtype, *Phase IV:* Small and localized cluster(s) with limited human-to-human transmission; *Phase V:* Larger but still localized cluster(s) of human-to-human transmission.
C. Pandemic period: *Phase VI:* Increased and sustained human-to-human transmission.
D. Postpandemic period.

As of early 2008, the USA is currently in an interpandemic period, while parts of Asia, Africa, and Eurasia have entered Phase III(/IV) of a pandemic alert period; pathogenic avian H5N1 virus has repeatedly infected humans without causing proven human-to-human transmission yet. Small clusters of human infections that may reflect human-to-human transmission have been observed.

In anticipation of an influenza virus pandemic that would kill up to an estimated 1.9 million Americans and require the hospitalization of an estimated 10 million Americans, in November of 2005 the Department of Health and Human Services issued a pandemic influenza plan,[3] and State Governments developed blueprints for

---

[2] See http://www.who.int/csr/disease/avian_influenza/phase/en/index.html

[3] See http://www.hhs.gov/pandemicflu/plan/overview.html#es.

**Table 2** WHO pandemic classifications

| Interpandemic period | Phase I | Novel influenza subtypes present in animals. Low risk of human infection |
|---|---|---|
| | Phase II | Humans at high risk of animal subtype |
| Pandemic alert period | Phase III | Animal-to-human transmission of a novel influenza subtype |
| | Phase IV | Small clusters of human-to-human transmission |
| | Phase V | Larger contained clusters with human-to-human transmission |
| Pandemic period | Phase VI | Human-to-human transmission of virus is sustained and spreading |
| Postpandemic period | | Threat of human-to-human transmission has subsided |

local pandemic response plans. Funding was provided to increase infrastructure, enhance vaccine production capability, and to augment basic knowledge on influenza virus pathogenesis and host responses.

How much experience do we have with pandemic influenza virus vaccines? As mentioned above, vaccines for the Spanish and Hong Kong Flu pandemic were not available at that time, and the vaccine that was available during the Asian Flu pandemic was mainly used in military personnel (Dull et al. 1960). In summary, our experience with the global use of a vaccine for pandemic influenza virus is nonexistent. One could envision four scenarios for the role of a vaccine in the next influenza virus pandemic: (a) an ideal outcome in which the world could be vaccinated with a universal vaccine that would never allow another pandemic to strike, (b) an optimistic outlook in which sufficient doses of a vaccine are produced in advance in order to rapidly immunize those at the epicenter of the pandemic and those at high risk, before additional vaccine for global immunization could be produced and distributed, (c) a pragmatic attitude that prepares as effectively as possible for the next pandemic without necessarily expecting that a vaccine will be on hand at the start of the pandemic, and (d) a worst-case scenario, in which the next pandemic influenza virus will outsmart us.

## 7.1 Ideal Outcome: Universal Vaccine for Influenza Virus

In an ideal scenario, scientists would develop a universal vaccine for influenza virus, industry would rapidly get involved in conducting large-scale trials needed for licensure, and then, with the aid of governments and philanthropic agencies, initiate a worldwide vaccination program before the next pandemic subtype of influenza virus evolves. Ideally, the vaccine would be adjuvanted to induce robust, long-lasting immunity not only in healthy adults but also in high-risk populations such as the elderly, infants, or those suffering from chronic diseases. It is hoped such a vaccine would prevent the development of any future influenza virus pandemics (Table 3).

**Table 3** Influenza vaccines

| Vaccine | Type vaccine | Antigen | Correlate of protection |
| --- | --- | --- | --- |
| Pandemic | Inactivated influenza virus | All viral proteins | Neutralizing antibodies |
| | Attenuated influenza virus | All viral proteins | Neutralizing antibodies, T cells (?) |
| | Subunit (viral vectors, DNA vaccines) | HA | Neutralizing antibodies |
| Prepandemic | Subunit (viral vectors, DNA vaccines, fusion proteins, peptides) | M2e | Nonneutralizing antibodies |
| | Subunit (viral vectors, DNA vaccines) | NP, M | T cells |

In experimental animals, some vaccines affect protection against heterotypic challenge with influenza virus, such as vaccines based on M2e (Mozdzanowska et al. 2003; Liu et al. 2004; Slepushkin et al. 1995; Fan et al. 2004; Frace et al. 1999; Neirynck et al. 1999; De Filette et al. 2006; EurekAlert 2007). Protection through M2e-expressing vaccines is mediated by humoral immunity and can be achieved by passive transfer of monoclonal M2e-specific antibodies prior to virus challenge (Mozdzanowska et al. 2003). One vaccine developed by W. Gerhard and colleagues was based on an M2e peptide linked to universal T helper cell epitopes. Others developed M2e vaccines based on papilloma virus-like particles (Ginaldi et al. 1999a), or fusion proteins linking M2e to hepatitis B virus core protein (De Filette et al. 2006). All of these subunit vaccines elicited antibodies to M2e in animals that protected against subsequent challenge with different types of influenza A virus, and the M2e-hepatitis B virus core fusion protein vaccine has now entered a phase I trial (EurekAlert 2007). The immunogenicity of an M2e vaccine could be increased by adjuvants such as Toll-like receptor 5 ligands (Huleatt et al. 2008). In one study, passively immunized animals were challenged with influenza A viruses that were identical or that differed in their M2e sequence; animals were protected against viruses that expressed the same M2e sequence but not against subtypes with M2e variants, (Fan et al. 2004). Several M2e sequences corresponding to the H1N1, H5N1, and H9N2 influenza subtypes were formulated using a liposome-based vaccine technology and evaluated as potential immunogens for the development of a "universal" influenza vaccine. Mice immunized with the polyvalent liposomal M2e survived challenges with different subtypes of influenza virus, and antiserum from immunized mice provided passive protection to naïve mice (Ernst et al. 2006). One study on a DNA vaccine expressing M2e fused to the nucleoprotein of influenza A virus reported increased mortality in vaccinated pigs, indicating that a poorly immunogenic vaccine (and DNA vaccines are commonly poorly immunogenic, especially in larger species) may exacerbate influenza virus-associated pathology (Heinen et al. 2002). In most studies, vaccines expressed one sequence of M2e. Notwithstanding, although M2e is far less variable than HA, it is not completely conserved, and mutants such as those present in recent H5N1 variants have been observed, suggesting that a universal M2e-based vaccine for influenza A virus should incorporate several common variants of M2e, including those that are present in the currently circulating pathogenic H5N1 viruses (Aydar et al. 2004). M2e vaccines induce some protective immunity, although this protection wanes against high challenge doses of virulent virus. M2e vaccines thus need to be optimized further, either through the use of novel adjuvants, or by their incorporation into more immunogenic vaccine carriers. Once this is achieved, M2e vaccines may well become part of an ideal universal vaccine for influenza virus, alleviating the need for a pandemic vaccine.

Under some circumstances, CD8$^+$ T cells directed against conserved sequences of influenza A virus provide protection against heterotypic challenge; however, under other circumstances, they fail to protect. Influenza virus antigens such as NP and M proteins (which carry conserved epitopes of influenza A virus) and vaccine carriers that induce robust CD8$^+$ T cell responses to such epitopes are readily available—the missing link remains a solid knowledge of what distinguishes a protective CD8$^+$ T cell from one that is ineffective or, even worse, exacerbates disease. Once this knowledge is gained, a universal influenza vaccine based on antigens that aim to induce T cell responses could be developed and deployed, either alone or in combination with an M2e-expressing vaccine.

Currently there is no universal vaccine for influenza virus in the industrial pipeline, and WHO estimates that it will take at least another 5–10 years before such vaccines become available.[4]

## 7.2  Optimistic Outlook: Prepandemic Vaccines

The highly pathogenic H5N1 virus that is endemic in wild birds in Asia, Africa, and Europe, and has spread to poultry and from there to humans, is currently viewed as a major candidate to evolve into the next pandemic subtype, through mutations that allow for efficient human-to-human transmission. Several entities have started to develop vaccines based on current subtypes of avian influenza virus under (a) the assumption that H5N1 would evolve into a pandemic virus, and (b) the optimistic conjecture that the pandemic virus would have sufficient homology with currently circulating viruses to allow for cross-protective immunity (Table 3).

The Asian highly pathogenic avian H5N1 virus has divided into two antigenic clades. Clade 1 includes human and bird isolates from Vietnam, Thailand, and Cambodia and bird isolates from Laos and Malaysia. Clade 2 viruses include bird isolates from China, Indonesia, Japan, South Korea, the Middle East, Europe and Africa, and were primarily responsible for human H5N1 infections during 2005–2006. Clade 2 is further subdivided into six subclades with a distinct geographic distribution. Over time, the pool of H5N1 viruses that could potentially evolve into a pandemic form is diversifying rapidly, making it very difficult to decide on a specific virus as the basis for a vaccine.

---

[4] See http://www.who.int/immunization/newsroom/PI_QAs/en/index.html.

Initial vaccines were developed for protection against the H5N1 subtype that was isolated from humans in Hong Kong in February 2003, but this virus has changed substantially, so these vaccines are now most likely no longer useful (Suguitan et al. 2006). In April 2004, WHO made a H5N1 prototype seed virus available to manufacturers. In August 2006, WHO changed the prototype and now offers three new prototype viruses. Future changes of the reference virus to accommodate additional mutations are expected.

Developing vaccines to H5N1 based on traditional approaches was a challenging task. The highly virulent H5N1 viruses rapidly kill embryonated chicken eggs, which are used to propagate the influenza A viruses for the annual vaccines. A number of manufacturers thus started to develop cell culture systems based for example on Vero or MDCK cells to propagate H5N1 influenza virus. Cell-grown influenza virus vaccines were tested in humans and showed immunogenicity and safety profiles that were comparable to those of egg-grown vaccines (Halperin et al. 2002). Others used reverse genetics to develop reassortant viruses in which gene segments encoding HA and NA were derived from highly pathogenic H5N1 virus, and all other genes were derived from the H1N1 virus A/PR/8/34, which was isolated in Puerto Rico in 1934 and is commonly used in animal studies (Lipatov et al. 2005b; Subbarao et al. 2003). The HA gene was further modified to replace the stretch of six basic amino acids at the cleavage site that can be digested by furin (Shi et al. 2007), and the resulting virus is avirulent in chickens and can be grown readily in eggs.

Most vaccines for highly pathogenic H5N1 tested to date were based on inactivated or attenuated virus used with or without adjuvant (Matsuoka et al. 2003; Subbarao et al. 2003; Lipatov et al. 2005b; Stephenson et al. 2005). These vaccines achieved protection in mice, ferrets, or birds against pathogenic subtypes of influenza A virus expressing the same or a closely related HA through the induction of neutralizing antibodies. In a human clinical trial with an inactivated H5N1 influenza virus vaccine attenuated through reverse genetics and changes of the HA cleavage site to allow propagation in eggs, protective titers of neutralizing Abs could be induced in volunteers after two doses of the vaccine (Treanor et al. 2006). Unfortunately, the dose that was needed to induce immune responses was six times that used for current influenza A virus vaccines. In a subsequent larger trial, the vaccine was adjuvanted with aluminum hydroxide, which did not improve the vaccine's immunogenicity (Bresson et al. 2006). Others reported the opposite results (Leroux-Roels et al. 2007). Additional clinical trials were conducted with inactivated whole-virus vaccine, which caused seroconversion in ~80% of vaccines that received the highest vaccine dose (10 mg), again indicating that the HA of H5N1 viruses is not a potent inducer of neutralizing antibody responses (Lin et al. 2006). The immunogenicity of H5N1 vaccine could be increased by adding MF59 adjuvant (Nicholson et al. 2001). Clinical trials have also been initiated with an attenuated H5N1 vaccine.[5]

---

[5] See http://www.nih.gov/news/pr/sep2006/niaid.//.htm

A number of groups have developed subunit vaccines for H5N1 virus. A DNA vaccine encoding H5 provided partial protection against challenge with H5N1 virus (Bright et al. 2003), while DNA vaccines encoding NP or M were comparatively ineffective (Epstein et al. 2002). DNA vaccine priming followed by a booster immunization with a replication-defective vector of adenovirus of the human serotype 5 (AdHu5), both expressing NP, augmented specific T cell responses and provided superior protection against challenge (Epstein et al. 2005). Two groups explored AdHu5 vectors expressing H5; they were shown to induce B and T cell responses against HA which protected against challenge with a pathogenic H5N1 virus (Gao et al. 2006; Hoelscher et al. 2006). Nevertheless, it should be pointed out that seroprevalence rates of neutralizing antibodies to AdHu5 are high in humans, especially those living in Asia or Africa, and that such antibodies strongly dampen antibody responses to the transgene product expressed by an AdHu5 vector. Fowl pox vectors (Qiao et al. 2006) and alpha virus replicons (Schultz-Cherry et al. 2000) expressing H5 were also shown to induce protective immunity against H5N1 influenza viruses.

In 2003, a H7N7 virus caused an outbreak in poultry in the Netherlands during which 88 humans became infected and mainly developed conjunctivitis, while one died of complications due to pneumonia. The virus isolated from the fatal case showed a mutation in the polymerase gene that was similar to that of highly pathogenic H5N1 (Munster et al. 2007). A reassortant vaccine expressing H7 and N7 on the A/PR8 background was developed, and an inactivated adjuvanted form of this vaccine induced neutralizing antibodies and protection in mice after two doses (de Wit et al. 2005).

A low-pathogenic subtype of H7N2 has been circulating in poultry in the northeastern USA since 1994, while highly pathogenic avian influenza has sporadically appeared, such as in an outbreak in Chile (H7N3) in 2002, an outbreak in the United States (H5N2) in 2004, and an outbreak in Canada (H7N3) in 2004 (Senne 2007). These viruses readily become pathogenic through some mutations (Lee et al. 2006) and thus pose a pandemic threat. A reassortant vaccine has been generated against H7N2 and was shown to induce protective immunity in mice and ferrets (Pappas et al. 2007).

Thus far, only one H5N1 influenza vaccine has been licensed by the United States Food and Drug Administration (FDA), while a number of other candidate vaccines against H5N1 avian influenza are in clinical trials and should be licensed in the near future. Whether or not these vaccines will be protective against the next pandemic virus is unknown. Initiating widespread vaccination before the actual pandemic starts would thus raise ethical questions—any vaccine, even one that is well tolerated, carries risks for the recipients. Without any clear indication that H5N1 is turning into a pandemic virus, the risk of vaccination would surpass the benefit to the vaccinated individual. This was demonstrated during the swine flu vaccine debacle of 1976, when vaccination against a virus that never spread caused a serious, crippling disease in hundreds of recipients. In addition to causing harm to these unfortunate individuals, this incident continues to provide ammunition to the vocal community of vaccine opponents that seem to have forgotten the haunting images of humans disfigured by poxviruses or wards full of polio virus-infected children on iron lungs. In contrast, a universal vaccine could be given before a pandemic, as such a vaccine would prevent seasonal influenza, thus providing a tangible benefit to its recipients.

## 7.3   Pragmatic Attitude

*Scio me nihil scire* (I know that I don't know) is a famous saying attributed to the Greek philosopher Socrates by Plato. If we take a Socratic view of the form and shape of the next pandemic influenza virus, mass production of a vaccine that induces protection through subtype-specific antibodies before a pandemic virus has actually evolved makes no sense. Making sure that an infrastructure is in place to rapidly and efficiently respond to a pandemic is, on the other hand, a prudent approach, and global agencies such as WHO in concert with governments are preparing for the next pandemic. Constant monitoring of evolving subtypes, new human infections and potential human-to-human spread in order to detect a pandemic at the earliest possible time is a vital task, and this been established. The sharing of virus isolates to identify potential vaccine candidates is important and requires international collaboration. Indonesia, which has the highest incidence of H5N1-related human deaths, initially refused to provide H5N1 samples to WHO in order to focus attention on their concern that while developing countries provide new viral isolates, any resulting vaccines produced by commercial companies would likely be used primarily in developed countries. By March 2007, Indonesia, which was the only country that took this stance, reversed its policy.

To date, global vaccine production capacity is insufficient, as seasonal influenza vaccines are only used by a small portion of the global population. A number of vaccine manufacturers have started to increase their production capacity, and it is expected that the current capacity will double by 2009. The long-term goal is to increase production capacity to three billion doses per year. It is also expected that manufacturing will commence in less developed countries. Idiosyncrasies of the actual pandemic vaccine, such as the required dose and the potential need for repeat injections, will determine whether this capacity will suffice. This is being addressed by attempts to increase the immunogenicity of influenza virus vaccines through novel adjuvants. Poorly growing vaccine subtypes could also offset the speed of production, and this is being tackled by developing cell-culture-based systems and through the use of reverse genetics to achieve rapid attenuation of influenza viruses. Recent studies indicated that antibodies to currently circulating viruses show some cross-reactivity with H5N1.[6] These studies must be confirmed. If indeed annual vaccinations with certain types of vaccine offer some degree of heterotypic protection

---

[6] Sandbulte MR, et al. 2007 Cross-reactive neuraminidase antibodies afford partial protection against H5N1 in mice and are presection unexposed humans PLoS Med

a broadening of seasonal influenza vaccine coverage would certainly be warranted, especially in countries that have advanced to a "pandemic alert period."

Progression through all of the steps of vaccine development, from preclinical trials in rodents through to clinical trials in humans and then licensing, generally takes 5–10 years. The FDA, which regulates vaccine licensure in the USA, has formulated guidelines for industry for the accelerated licensure of pandemic influenza vaccines based on the induction of neutralizing antibodies to hemagglutinin in order to ensure that regulatory aspects do not hinder the rapid deployment of a pandemic vaccine. Vaccines that do not contain viral hemagglutinin are not covered by these guidelines.

Nevertheless, even if the production of a vaccine starts on the day that a pandemic virus has been identified, it will still take 4–6 months until the very first dose of vaccine is available. Other control measures are therefore needed to limit damage until the vaccines become available for everyone. In the USA, the federal government and state governments have formulated pandemic preparedness plans to be followed in the event of a pandemic. These plans not only list the responsibilities of government entities and individuals, but also address the use of limited pharmacological agents and other types of control measures. Similar to vaccines, the availability of antiviral drugs is expected to be limited. Antiviral drugs may slow the pandemic if used in a timely manner at the epicenter of the pandemic. They may also be extraordinarily useful for protecting persons that provide essential healthcare and for maintaining vital infrastructure.

Assuming a delay of at least 6–12 months before sufficient doses of vaccine are available for global mass vaccination, vaccines will have to be rationed at the beginning of the pandemic. Governments will issue lists of high-priority personnel that are to be vaccinated first. Although these lists vary from state to state in the USA, they typically include hospital and health department staff, emergency medical service personnel and household members, law enforcement personnel, fire fighters, medical laboratory workers, emergency management personnel, long-term care facility staff, utility workers (gas, electric, water, waste management, etc.), communications personnel, fuel and food suppliers, public transportation and air travel personnel, corrections workers, morticians/coroners/medical examiners, pharmacists, Red Cross field workers, US postal service staff, persons involved with vaccine production and delivery, etc. It has been suggested that once more vaccine becomes available, healthy working adults should be vaccinated before high-risk populations such as children or the elderly.

It is still to be decided who will ultimately purchase the pandemic vaccine—federal or state governments, who could clearly facilitate orderly distribution, or the private sector. If the private sector carries the cost, insurance companies will have to take a stance on cost coverage, and plans will have to be developed for the uninsured.

Once a vaccine becomes available, other problems will arise. Some of these issues are being addressed by governments, and the examples below apply to the USA. A pandemic vaccine would not be expected to undergo the vigorous safety testing typical of other vaccines. Manufacturers that develop pandemic vaccines can request indemnification from the Secretary of Health and Human Services for "an activity that involves unusually hazardous risks and for which insurance is not available or sufficient to cover those risks."

A vaccine will have to be distributed rapidly and in an orderly manner. The US government ruled that it may mobilize the PHS Commissioned Corps to distribute vaccines to federal agencies with direct patient care responsibilities, or to states, tribes, and other localities through the National Disaster Medical System and through agreements between the federal government, states, and localities.

Liability protection must be put in place. The US government has ruled that federal employee administrators are covered by the federal government and could make claims through the Federal Tort Claims Act. State employees may be covered for malpractice or tort claims coverage under state law. Federal contractor and private sector employees distributing the vaccine would be expected to carry malpractice insurance or they could be covered by the Volunteer Protection Act, State Good Samaritan Act, or State Emergency Compact provisions. If a person is injured following administration of a pandemic vaccine or antiviral medication in connection with his/her employment, compensation may be available under a state's worker's compensation program. For federal employees, compensation may be available under the Federal Employees' Compensation Act.

Assuming that vaccines will not be available at the onset of the next pandemic, other nonpharmaceutical measures are being discussed to (1) limit international spread, (2) reduce spread within national and local populations, and (3) reduce an individual person's risk for infection.

Influenza viruses are typically shed 24–48 h before the onset of disease, and virus titers peak within the first three days after onset of symptoms and then decline by day 7–8. It is possible but not yet proven that the virus spreads by shedding a small amount of virions before the first symptoms occur, which could markedly reduce the effectiveness of most quarantine measures. Infection occurs predominantly via droplets formed by coughing or sneezing individuals. Infection by aerosolized virus is less common. The virus can also be transmitted via infected hands or surfaces. The wearing of masks and the employment of appropriate sanitizing measures to clean hands and infected surfaces are thus useful actions for protecting an individual (Jefferson et al. 2008).

It is thought that temporary protection of populations may in part be achieved by implementing quarantine measures. In the pandemic of 1918, some island countries enacted maritime quarantines (Markel et al. 2007). Australia and Madagascar were able to delay the start of the pandemic by several months, while Samoa and New Caledonia remained completely free of the pandemic. Quarantine measures were attempted on land, but they were unsuccessful. Quarantine was tried again in 1957, but it largely failed. Quarantine was very successful in stopping the SARS epidemic of 2003 (Hsieh et al. 2007). The SARS virus has a longer incubation time than influenza virus, and peak virus titers are not reached until several days after the onset of symptoms. Isolating cases thus proved an effective way to prevent the further spread of SARS, but it is unlikely to work against influenza virus. WHO recommends exit screening for international travelers that leave countries that are affected by an influenza virus pandemic. It is unclear if and to what extent this may delay the spread of the virus. However, any delay would be useful in allowing extra time for vaccine production.

The US government has issued an Executive Order adding potentially pandemic influenza viruses to the list of quarantinable diseases, which empowers the Centers for Disease Control and Prevention (CDC) to detain, medically examine, or conditionally release individuals that are reasonably believed to be carrying a communicable disease. The intent of this order is to enable the United States to respond efficiently and effectively in the case of an outbreak by pandemic influenza viruses. This order gives legal authority to the Department of Health and Human Services to isolate a passenger arriving on board an international vessel that show evidence of infection with a novel influenza virus, even if that passenger refuses to cooperate. The federal government (such as the CDC) generally defers to the state and local health authorities in the use of their own quarantine powers. State-implemented interventions would likely include the isolation and treatment of infected individuals, voluntary home quarantine for members of the households of infected individuals, the closure of schools and universities, and the encouragement of social distancing through the cancellation of large public gatherings.

## 7.4   Worst Case Scenario

WHO estimates that in the worst case scenario more than 70 million people could die as a consequence of the next influenza virus pandemic. Other estimates are higher: 180–360 million deaths. To put this number into perspective, 360 million is approximately the total population of South America, or half of the population of Europe. Death tolls will primarily depend on the virulence of the next pandemic—current clades of highly pathogenic H5N1 virus kill >60% of those infected. It would also depend on the ease with which the virus is transmitted between humans, and last but not least on the effectiveness of control measures. Late detection of a pandemic, which could evolve in a war-ravaged country that lacks surveillance, would shorten the time interval available for the development of a vaccine. Vaccine production could be further delayed by problems with the seed virus, such as toxicity toward eggs or cell substrates, poor immunogenicity of the vaccine, or unacceptable reactogenicity. The modern world, with its high degree of social and economic interdependency, has not yet experienced a major pandemic that disrupts crucial aspects of local and global infrastructure. Lack of available workers and restricted movement could threaten essential services, and the failure of any one system could trigger others to fail too, causing cascading breakdowns.

Highly pathogenic H5N1 influenza viruses have been circulating for more than ten years and many have come to believe that they are unlikely to jump species. Although this may well be the case, the consequences of relaxing our efforts to prepare for the next pandemic could be horrific.

## 8  Summary

In 1969, when smallpox virus was on the brink of extinction and polio was disappearing from developed countries, the US Surgeon General, William Stewart, told Congress that it was time to "close the books on infectious diseases." Unfortunately, Congress listened and shifted federal funding from microbiology/virology to cancer and cardiovascular diseases. Ever-increasing liability costs made the industry more and more reluctant to stay involved in vaccine production, and the number of companies that produce vaccines has now become so limited that annual vaccine shortages are common (Markel et al. 2007). Infectious agents have continued to take a major toll on human lives. Since 1969 more than 30 new microbes have emerged, such as HIV-1, which has claimed over 22 million lives thus far. Even old microbes such as influenza virus continue to take a toll on human lives; for example, in the USA alone, seasonal influenza causes over 200,000 hospitalizations and over 30,000 death each year. The world will experience a new influenza virus pandemic. No-one can predict when it will happen, where it will start, what virus will cause it, and what the global and local impacts will be. The only aspect of the next pandemic we can predict with certainty is that it will happen.

Complacency, not only in the USA but worldwide, has weakened the infrastructure for efficiently combating newly emerging infectious agents, and this infrastructure needs to be rebuilt. Communications with the public must be improved globally in a manner that informs accurately without alarming unduly. It is clear that this has not yet been achieved, as exemplified by a recent H5N1 outbreak in West Bengal in India,[7] where children were reported to have unprotected contact with birds that died due to infection with H5N1.

Our continued lack of knowledge about the very basic question of the immunobiology of influenza viruses and the efficacy of vaccines in different cohorts is mind-boggling. We still do not fully understand correlates of protection against influenza virus, and debates on the role of $CD8^+$ T cells in providing protection are continuing. This knowledge needs to be generated, especially in order to enable the development of a universal influenza virus vaccine, the "holy grail" that could prevent future influenza virus pandemics.

## References

Abe T, Takahashi H, Hamazaki H, Miyano-Kurosaki N, Matsuura Y, Takaku H (2003) Baculovirus induces an innate immune response and confers protection from lethal influenza virus infection in mice. J Immunol 171:1133–1139

Altstein AD, Gitelman AK, Smirnov YA, Piskareva LM, Zakharova LG, Pashvykina GV, Shmarov MM, Zhirnov OP, Varich NP, Ilyinskii PO, Shneider AM (2006) Immunization with influenza

---

[7] www.recombinomics.com/News/01160801/H5N1.Birbhun_Suspect.z.htm

A NP-expressing vaccinia virus recombinant protects mice against experimental infection with human and avian influenza viruses. Arch Virol 151:921–931

Amonsin A, Songserm T, Chutinimitkul S, Jam-On R, Sae-Heng N, Pariyothorn N, Payungporn S, Theamboonlers A, Poovorawan Y (2007) Genetic analysis of influenza A virus (H5N1) derived from domestic cat and dog in Thailand. Arch Virol 152:1925–1933

Andrew ME, Coupar BE, Ada GL, Boyle DB (1986) Cell-mediated immune responses to influenza virus antigens expressed by vaccinia virus recombinants. Microb Pathog 1:443–452

Aydar Y, Balogh P, Tew JG, Szakal AK (2004) Follicular dendritic cells in aging, a "bottle-neck" in the humoral immune response. Ageing Res Rev 3:15–29

Beilharz MW, Cummins JM, Bennett AL (2007) Protection from lethal influenza virus challenge by oral type 1 interferon. Biochem Biophys Res Commun 355:740–744

Biro J (1978) Age-related change of the absolute number of IgG-, and IgM-bearing B-lymphocytes and T-lymphocytes in human peripheral blood following influenza vaccination. Aktuelle Gerontol 8:81–83

Bresson JL, Perronne C, Launay O, Gerdil C, Saville M, Wood J et al (2006) Safety and immunogenicity of an inactivated split-virion influenza A/Vietnam/1194/2004 (H5N1) vaccine: phase 1 randomised trial. Lancet 367:1657–1664

Bright RA, Ross TM, Subbarao K, Robinson HL, Katz JM (2003) Impact of glycosylation on the immunogenicity of a DNA-based influenza H5 HA vaccine. Virology 308:270–278

Callan RJ, Hartmann FA, West SE, Hinshaw VS (1997) Cleavage of influenza A virus H1 hemagglutinin by swine respiratory bacterial proteases. J Virol 71:7579–7585

Cella M, Facchetti F, Lanzavecchia A, Colonna M (2000) Plasmacytoid dendritic cells activated by influenza virus and CD40L drive a potent TH1 polarization. Nat Immunol 1:305–310

Centers for Disease Control and Prevention (1997) Isolation of avian influenza A(H5N1) viruses from humans—Hong Kong, May–December 1997. MMWR Morb Mortal Wkly Rep 46:1204–1207

Cerwenka A, Morgan TM, Dutton RW (1999) Naive, effector, and memory CD8 T cells in protection against pulmonary influenza virus infection: homing properties rather than initial frequencies are crucial. J Immunol 163:5535–5543

Cheung CL, Rayner JM, Smith GJ, Wang P, Naipospos TS, Zhang J, Yuen KY, Webster RG, Peiris JS, Guan Y, Chen H (2006) Distribution of amantadine-resistant H5N1 avian influenza variants in Asia. J Infect Dis 193:1626–1629

Cockburn WC, Delon PJ, Ferreira W (1969) Origin and progress of the 1968–69 Hong Kong influenza epidemic. Bull World Health Organ 41:345–348

Cristalli A, Capua I (2007) Practical problems in controlling H5N1 high pathogenicity avian influenza at village level in Vietnam and introduction of biosecurity measures. Avian Dis 51:461–462

De Filette M, Ramne A, Birkett A, Lycke N, Lowenadler B, Min Jou W, Saelens X, Fiers W (2006) The universal influenza vaccine M2e-HBc administered intranasally in combination with the adjuvant CTA1-DD provides complete protection. Vaccine 24:544–551

de Wit E, Munster VJ, Spronken MI, Bestebroer TM, Baas C, Beyer WE, Rimmelzwaan GF, Osterhaus AD, Fouchier RA (2005) Protection of mice against lethal infection with highly pathogenic H7N7 influenza A virus by using a recombinant low-pathogenicity vaccine strain. J Virol 79:12401–12407

Demicheli V, Rivetti D, Deeks J, Jefferson T (2004) Vaccines for preventing influenza in healthy adults. Cochrane Database Syst Rev CD001269

Dunn FL (1958) Pandemic influenza in 1957; review of international spread of new Asian strain. J Am Med Assoc 166:1140–1148

Dull HB, Jensen J, Rakich JH, Cohen A, Henderson DA, Pirkle CI (1960) Monovalent Asian influenza vaccine: evaluation of its use during two waves of epidemic Asian influenza in partly immunized penitentiary population. JAMA 172:1223–1229

Effros RB, Cai Z, Linton PJ (2003) CD8 T cells and aging. Crit Rev Immunol 23:45–64

Endo A, Itamura S, Iinuma H, Funahashi S, Shida H, Koide F, Nerome K, Oya A (1991) Homotypic and heterotypic protection against influenza virus infection in mice by recombinant vaccinia virus expressing the haemagglutinin or nucleoprotein of influenza virus. J Gen Virol 72:699–703

Epstein SL (2006) Prior H1N1 influenza infection and susceptibility of Cleveland Family Study participants during the H2N2 pandemic of 1957: an experiment of nature. J Infect Dis 193:49–53

Epstein SL, Tumpey TM, Misplon JA, Lo CY, Cooper LA, Subbarao K, Renshaw M, Sambhara S, Katz JM (2002) DNA vaccine expressing conserved influenza virus proteins protective against H5N1 challenge infection in mice. Emerg Infect Dis 8:796–801

Epstein SL, Kong WP, Misplon JA, Lo CY, Tumpey TM, Xu L, Nabel GJ (2005) Protection against multiple influenza A subtypes by vaccination with highly conserved nucleoprotein. Vaccine 23:5404–5410

Ernst WA, Kim HJ, Tumpey TM, Jansen AD, Tai W, Cramer DV et al. (2006) Protection against H1, H5, H6 and H9 influenza A infection with liposomal matrix 2 epitope vaccines. Vaccine 24:5158–5168

EurekAlert (2007) Universal flu vaccine being tested on humans. http://www.eurekalert.org/pub_releases/2007-07/vfii-nuf071707.php

Falk K, Rötzschke O, Deres K, Metzger J, Jung G, Rammensee HG (1991) Identification of naturally processed viral nonapeptides allows their quantification in infected cells and suggests an allele-specific T cell epitope forecast. J Exp Med 174:425–434

Fan J, Liang X, Horton MS, Perry HC, Citron MP, Heidecker GJ, Fu TM, Joyce J, Przysiecki CT, Keller PM, Garsky VM, Ionescu R, Rippeon Y, Shi L, Chastain MA, Condra JH, Davies ME, Liao J, Emini EA, Shiver JW (2004) Preclinical study of influenza virus A M2 peptide conjugate vaccines in mice, ferrets, and rhesus monkeys. Vaccine 22:2993–3003

Fattal-German M, Bizzini B (1992) Assessment of the anti-viral effect of a short-term oral treatment of mice with live *Saccharomyces cerevisiae* cells. Dev Biol Stand 77:115–120

Feng J, Zhang M, Mozdzanowska K, Zharikova D, Hoff H, Wunner W, Couch RB, Gerhard W (2006) Influenza A virus infection engenders a poor antibody response against the ectodomain of matrix protein 2. Virol J 3:102

Frace AM, Klimov AI, Rowe T, Black RA, Katz JM (1999) Modified M2 proteins produce heterotypic immunity against influenza A virus. Vaccine 17:2237–2244

Francis T, Salk JE, Pearson HE, Brown PN (1945a) Protective effect of vaccination against induced influenza A. J Clin Invest 24:536–546

Francis T, Salk JE, Pearson HE, Brown PN (1945b) Protective effect of vaccination against induced influenza B. J Clin Invest 24:547–553

Fu TM, Friedman A, Ulmer JB, Liu MA, Donnelly JJ (1997) Protective cellular immunity: cytotoxic T-lymphocyte responses against dominant and recessive epitopes of influenza virus nucleoprotein induced by DNA immunization. J Virol 71:2715–2721

Gao W, Soloff AC, Lu X, Montecalvo A, Nguyen DC, Matsuoka Y, Robbins PD, Swayne DE, Donis RO, Katz JM, Barratt-Boyes SM, Gambotto A (2006) Protection of mice and poultry from lethal H5N1 avian influenza virus through adenovirus-based immunization. J Virol 80:1959–1964

Gentile D, Doyle W, Whiteside T, Fireman P, Hayden FG, Skoner D (1998) Increased interleukin-6 levels in nasal lavage samples following experimental influenza A virus infection. Clin Diagn Lab Immunol 5:604–608

Geraci JR, St Aubin DJ, Barker IK, Webster RG, Hinshaw VS, Bean WJ, Ruhnke HL, Prescott JH, Early G, Baker AS, Madoff S, Schooley RT (1982) Mass mortality of harbor seals: pneumonia associated with influenza A virus. Science 215:1129–1131

Ginaldi L, De Martinis M, D'Ostilio A, Marini L, Loreto MF, Corsi MP, Quaglino D (1999a) The immune system in the elderly: I. Specific humoral immunity. Immunol Res 20:101–108

Ginaldi L, De Martinis M, D'Ostilio A, Marini L, Loreto MF, Martorelli V, Quaglino D (1999b) The immune system in the elderly: II. Specific cellular immunity. Immunol Res 20:109–115

Ginaldi L, De Martinis M, D'Ostilio A, Marini L, Loreto MF, Quaglino D (1999c) The immune system in the elderly: III. Innate immunity. Immunol Res 20:117–126

Gramer MR, Lee JH, Choi YK, Goyal SM, Joo HS (2007) Serologic and genetic characterization of North American H3N2 swine influenza A viruses. Can J Vet Res 71:201–206

Gundlefinger BF, Stille WT, Bell JA (1958) Effectiveness of influenza vaccines during an epidemic of Asian influenza. New Engl J Med 259:1005–1009

Halperin SA, Smith B, Mabrouk T, Germain M, Trepanier P, Hassell T et al. (2002) Safety and immunogenicity of a trivalent, inactivated, mammalian cell culture-derived influenza vaccine in healthy adults, seniors, and children. Vaccine 20:1240–1247

Hanson BJ, Boon AC, Lim AP, Webb A, Ooi EE, Webby RJ (2006) Passive immunoprophylaxis and therapy with humanized monoclonal antibody specific for influenza A H5 hemagglutinin in mice. Respir Res 7:126

Heinen PP, Rijsewijk FA, de Boer-Luijtze EA, Bianchi AT (2002) Vaccination of pigs with a DNA construct expressing an influenza virus M2-nucleoprotein fusion protein exacerbates disease after challenge with influenza A virus. J Gen Virol 83:1851–1859

Hoelscher MA, Garg S, Bangari DS, Belser JA, Lu X, Stephenson I, Bright RA, Katz JM, Mittal SK, Sambhara S (2006) Development of adenoviral-vector-based pandemic influenza vaccine against antigenically distinct human H5N1 strains in mice. Lancet 367:475–481

Hsieh YH, King CC, Chen CW, Ho MS, Hsu SB, Wu YC (2007) Impact of quarantine on the 2003 SARS outbreak: a retrospective modeling study. J Theor Biol 244:729–736

Huleatt JW, Nakaar V, Desai P, Huang Y, Hewitt D, Jacobs A, Tang J, McDonald W, Song L, Evans RK, Umlauf S, Tussey L, Powell TJ (2008) Potent immunogenicity and efficacy of a universal influenza vaccine candidate comprising a recombinant fusion protein linking influenza M2e to the TLR5 ligand flagellin. Vaccine 26:201–214

Hulse DJ, Webster RG, Russell RJ, Perez DR (2004) Molecular determinants within the surface proteins involved in the pathogenicity of H5N1 influenza viruses in chickens. J Virol 78:9954–9964

Irinoda K, Masihi KN, Chihara G, Kaneko Y, Katori T (1992) Stimulation of microbicidal host defence mechanisms against aerosol influenza virus infection by lentinan. Int J Immunopharmacol 14:971–977

Ison MG, Hayden FG (2001) Therapeutic options for the management of influenza. Curr Opin Pharmacol 1:482–490

Janssens JP, Krause KH (2004) Pneumonia in the very old. Lancet Infect Dis 4:112–124

Jao RL, Wheelock EF, Jackson GG (1970) Production of interferon in volunteers infected with Asian influenza. J Infect Dis 121:419–426

Jayasekera JP, Moseman EA, Carroll MC (2007) Natural antibody and complement mediate neutralization of influenza virus in the absence of prior immunity. J Virol 81:3487–3494

Jefferson T, Rivetti D, Rivetti A, Rudin M, Di Pietrantonj C, Demicheli V (2005) Efficacy and effectiveness of influenza vaccines in elderly people: a systematic review. Lancet 366:1165–1174

Jefferson T, Foxlee R, Del Mar C, Dooley L, Ferroni E, Hewak B, Prabhala A, Nair S, Rivetti A (2008) Physical interventions to interrupt or reduce the spread of respiratory viruses: systematic review. BMJ 336:77–80

Johnson NP, Mueller J (2002) Updating the accounts: global mortality of the 1918–1920 "Spanish" influenza pandemic. Bull Hist Med 76:105–115

Jordan RE, Hawker JI, Ayres JG, Tunnicliffe W, Adab P, Olowokure B, Kai J, McManus RJ, Salter R, Cheng KK (2006) Influenza-related mortality in the Italian elderly: no decline associated with increasing vaccination coverage. Vaccine 24:6468–6475

Kash JC, Basler CF, García-Sastre A, Carter V, Billharz R, Swayne DE, Przygodzki RM, Taubenberger JK, Katze MG, Tumpey TM (2004) Global host immune response: pathogenesis and transcriptional profiling of type A influenza viruses expressing the hemagglutinin and neuraminidase genes from the 1918 pandemic virus. J Virol 78:9499–9511

Kaye HS, Dowdle WR, McQueen JL (1969) Studies on inactivated influenza vaccines. I. The effect of dosage on antibody response and protection against homotypic and heterotypic influenza virus challenge in mice. Am J Epidemiol 90:162–169

Keawcharoen J, Oraveerakul K, Kuiken T, Fouchier RA, Amonsin A, Payungporn S, Noppornpanth S, Wattanodorn S, Theamboonlers A, Tantilertcharoen R, Pattanarangsan R, Arya N, Ratanakorn P, Osterhaus DM, Poovorawan Y (2004) Avian influenza H5N1 in tigers and leopards. Emerg Infect Dis 10:2189–2191

Kilbourne ED (2006) Influenza pandemics of the 20th century. Emerg Infect Dis 12:9–14

Koopmans M, Wilbrink B, Conyn M, Natrop G, van der Nat H, Vennema H, Meijer A, van Steenbergen J, Fouchier R, Osterhaus A, Bosman A (2004) Transmission of H7N7 avian influenza A virus to human beings during a large outbreak in commercial poultry farms in the Netherlands. Lancet 363:587–593

Langmuir AD, Bregman DJ, Kurland LT, Nathanson N, Victor M (1984) An epidemiologic and clinical evaluation of Guillain–Barré syndrome reported in association with the administration of swine influenza vaccines. Am J Epidemiol 119:841–879

Lawson CM, Bennink JR, Restifo NP, Yewdell JW, Murphy BR (1994) Primary pulmonary cytotoxic T lymphocytes induced by immunization with a vaccinia virus recombinant expressing influenza A virus nucleoprotein peptide do not protect mice against challenge. J Virol 68:3505–3511

Lee CW, Lee YJ, Senne DA, Suarez DL (2006) Pathogenic potential of North American H7N2 avian influenza virus: a mutagenesis study using reverse genetics. Virology 353:388–395

Leroux-Roels I, Borkowski A, Vanwolleghem T, Dramé M, Clement F, Hons E, Devaster JM, Leroux-Roels G (2007) Antigen sparing and cross-reactive immunity with an adjuvanted rH5N1 prototype pandemic influenza vaccine: a randomised controlled trial. Lancet 370:580–589

Lessler J, Cummings DA, Fishman S, Vora A, Burke DS (2007) Transmissibility of swine flu at Fort Dix, 1976. J R Soc Interface 4:755–762

Lin J, Zhang J, Dong X, Fang H, Chen J, Su N, Gao Q, Zhang Z, Liu Y, Wang Z, Yang M, Sun R, Li C, Lin S, Ji M, Liu Y, Wang X, Wood J, Feng Z, Wang Y, Yin W (2006) Safety and immunogenicity of an inactivated adjuvanted whole-virion influenza A (H5N1) vaccine: a phase I randomised controlled trial. Lancet 368:991–997

Lipatov AS, Andreansky S, Webby RJ, Hulse DJ, Rehg JE, Krauss S, Perez DR, Doherty PC, Webster RG, Sangster MY (2005a) Pathogenesis of Hong Kong H5N1 influenza virus NS gene reassortants in mice: the role of cytokines and B and T cell responses. J Gen Virol 86:1121–1130

Lipatov AS, Webby RJ, Govorkova EA, Krauss S, Webster RG (2005b) Efficacy of H5 influenza vaccines produced by reverse genetics in a lethal mouse model. J Infect Dis 191:1216–1220

Liu W, Zou P, Chen YH (2004) Monoclonal antibodies recognizing EVETPIRN epitope of influenza A virus M2 protein could protect mice from lethal influenza A virus challenge. Immunol Lett 93:131–136

Liu W, Zou P, Ding J, Lu Y, Chen YH (2005) Sequence comparison between the extracellular domain of M2 protein human and avian influenza A virus provides new information for bivalent influenza vaccine design. Microbes Infect 7:171–177

López CB, Moltedo B, Alexopoulou L, Bonifaz L, Flavell RA, Moran TM (2004) TLR-independent induction of dendritic cell maturation and adaptive immunity by negative-strand RNA viruses. J Immunol 173:6882–6889

Mak NK, Leung KN, Ada GL (1982) The generation of 'cytotoxic' macrophages in mice during infection with influenza A or Sendai virus. Scand J Immunol 15:553–561

Malaguarnera L, Ferlito L, Imbesi RM, Gulizia GS, Di Mauro S, Maugeri D, Malaguarnera M, Messina A (2001) Immunosenescence: a review. Arch Gerontol Geriatr 32:1–14

Manzoli L, Schioppa F, Boccia A, Villari P (2007) The efficacy of influenza vaccine for healthy children: a meta-analysis evaluating potential sources of variation in efficacy estimates including study quality. Pediatr Infect Dis J 26:97–106

Markel H, Lipman HB, Navarro JA, Sloan A, Michalsen JR, Stern AM, Cetron MS (2007) Nonpharmaceutical interventions implemented by US cities during the 1918–1919 influenza pandemic. JAMA 298:644–654

Matsuoka Y, Chen H, Cox N, Subbarao K, Beck J, Swayne D (2003) Safety evaluation in chickens of candidate human vaccines against potential pandemic strains of influenza. Avian Dis 47:926–930

Mbawuike IN, Zhang Y, Couch RB (2007) Control of mucosal virus infection by influenza nucleoprotein-specific CD8+ cytotoxic T lymphocytes. Respir Res 8:44

McCullers JA (2006) Insights into the interaction between influenza virus and pneumococcus. Clin Microbiol Rev 19:571–582

McLain L, Dimmock NJ (1989) Protection of mice from lethal influenza by adoptive transfer of non-neutralizing haemagglutination-inhibiting IgG obtained from the lungs of infected animals treated with defective interfering virus. J Gen Virol 70:2615–2624

Mozdzanowska K, Feng J, Eid M, Kragol G, Cudic M, Otvos L Jr, Gerhard W (2003) Induction of influenza type A virus-specific resistance by immunization of mice with a synthetic multiple antigenic peptide vaccine that contains ectodomains of matrix protein 2. Vaccine 21:2616–2626

Mozdzanowska K, Furchner M, Zharikova D, Feng J, Gerhard W (2005) Roles of CD4+ T cell-independent and -dependent antibody responses in the control of influenza virus infection: evidence for noncognate CD4+ T cell activities that enhance the therapeutic activity of antiviral antibodies. J Virol 79:5943–5951

Munster VJ, de Wit E, van Riel D, Beyer WE, Rimmelzwaan GF, Osterhaus AD, Kuiken T, Fouchier RA (2007) The molecular basis of the pathogenicity of the Dutch highly pathogenic human influenza A H7N7 viruses. J Infect Dis 196:258–265

Murasko DM, Bernstein ED, Gardner EM, Gross P, Munk G, Dran S, Abrutyn E (2002) Role of humoral and cell-mediated immunity in protection from influenza disease after immunization of healthy elderly. Exp Gerontol 37:427–439

Neirynck S, Deroo T, Saelens X, Vanlandschoot P, Jou WM, Fiers W (1999) A universal influenza A vaccine based on the extracellular domain of the M2 protein. Nat Med 5:1157–1163

Nichol KL, Nordin JD, Nelson DB, Mullooly JP, Hak E (2007) Effectiveness of influenza vaccine in the community-dwelling elderly. N Engl J Med 357:1373–1381

Nicholson KG, Colegate AE, Podda A, Stephenson I, Wood J, Ypma E, Zambon MC (2001) Safety and antigenicity of non-adjuvanted and MF59-adjuvanted influenza A/Duck/Singapore/97 (H5N3) vaccine: a randomised trial of two potential vaccines against H5N1 influenza. Lancet 357:1937–1943

Okuno Y, Matsumoto K, Isegawa Y, Ueda S (1994) Protection against the mouse-adapted A/FM/1/47 strain of influenza A virus in mice by a monoclonal antibody with cross-neutralizing activity among H1 and H2 strains. J Virol 68:517–520

Oxburgh L, Hagström A (1999) A PCR based method for the identification of equine influenza virus from clinical samples. Vet Microbiol 67:161–174

Pappas C, Matsuoka Y, Swayne DE, Donis RO (2007) Development and evaluation of an Influenza virus subtype H7N2 vaccine candidate for pandemic preparedness. Clin Vaccine Immunol 14:1425–1432

Piedra PA, Gaglani MJ, Kozinetz CA, Herschler GB, Fewlass C, Harvey D, Zimmerman N, Glezen WP (2007) Trivalent live attenuated intranasal influenza vaccine administered during the 2003–2004 influenza type A (H3N2) outbreak provided immediate, direct, and indirect protection in children. Pediatrics 120:e553–e564

Pirhonen J, Sareneva T, Kurimoto M, Julkunen I, Matikainen S (1999) Virus infection activates IL-1 beta and IL-18 production in human macrophages by a caspase-1-dependent pathway. J Immunol 162:7322–7329

Price GE, Gaszewska-Mastarlarz A, Moskophidis D (2000) The role of alpha/beta and gamma interferons in development of immunity to influenza A virus in mice. J Virol 74:3996–4003

Qiao C, Yu K, Jiang Y, Li C, Tian G, Wang X, Chen H (2006) Development of a recombinant fowlpox virus vector-based vaccine of H5N1 subtype avian influenza. Dev Biol 124:127–132

Renegar KB, Small PA Jr (1991) Passive transfer of local immunity to influenza virus infection by IgA antibody. J Immunol 146:1972–1978

Rivetti D, Jefferson T, Thomas R, Rudin M, Rivetti A, Di Pietrantonj C, Demicheli V (2006) Vaccines for preventing influenza in the elderly. Cochrane Database Syst Rev 3:CD004876

Rizzo C, Viboud C, Montomoli E, Simonsen L, Miller MA (2007) A case-control study of elderly patients with acute respiratory illness: effect of influenza vaccination on admission to hospital in winter 2003–2004. Vaccine 25:7909–7913

Roti M, Yang J, Berger D, Huston L, James EA, Kwok WW (2008) Healthy human subjects have CD4+ T cells directed against H5N1 influenza virus. J Immunol 180:1758–1768

Roy S, Kobinger GP, Lin J, Figueredo J, Calcedo R, Kobasa D, Wilson JM (2007) Partial protection against H5N1 influenza in mice with a single dose of a chimpanzee adenovirus vector expressing nucleoprotein. Vaccine 25:6845–6851

Russell CJ, Webster RG (2005) The genesis of a pandemic influenza virus. Cell 123:368–371

Saha S, Yoshida S, Ohba K, Matsui K, Matsuda T, Takeshita F, Umeda K, Tamura Y, Okuda K, Klinman D, Xin KQ, Okuda K (2006) A fused gene of nucleoprotein (NP) and herpes simplex virus genes (VP22) induces highly protective immunity against different subtypes of influenza virus. Virology 354:48–57

Sambhara S, Woods S, Arpino R, Kurichh A, Tamane A, Underdown B, Klein M, Lövgren Bengtsson K, Morein B, Burt D (1998) Heterotypic protection against influenza by immunostimulating complexes is associated with the induction of cross-reactive cytotoxic T lymphocytes. J Infect Dis 177:1266–1274

Saravolac EG, Sabuda D, Crist C, Blasetti K, Schnell G, Yang H, Kende M, Levy HB, Wong JP (2001) Immunoprophylactic strategies against respiratory influenza virus infection. Vaccine 19:2227–2232

Saurwein-Teissl M, Lung TL, Marx F, Gschosser C, Asch E, Blasko I, Parson W, Bock G, Schonitzer D, Trannoy E, Grubeck-Loebenstein B (2002) Lack of antibody production following immunization in old age: association with CD8(+)CD28(−) T cell clonal expansions and an imbalance in the production of Th1 and Th2 cytokines. J Immunol 168:5893–5899

Schultz-Cherry S, Dybing JK, Davis NL, Williamson C, Suarez DL, Johnston R, Perdue ML (2000) Influenza virus (A/HK/156/97) hemagglutinin expressed by an alphavirus replicon system protects chickens against lethal infection with Hong Kong-origin H5N1 viruses. Virology 278:55–59

Sekellick MJ, Carra SA, Bowman A, Hopkins DA, Marcus PI (2000) Transient resistance of influenza virus to interferon action attributed to random multiple packaging and activity of NS genes. J Interferon Cytokine Res 20:963–970

Senne DA (2007) Avian influenza in North and South America, 2002–2005. Avian Dis 51:167–173

Senne DA, Suarez DL, Stallnecht DE, Pedersen JC, Panigrahy B (2006) Ecology and epidemiology of avian influenza in North and South America. Dev Biol 124:37–44

Seo SH, Hoffmann E, Webster RG (2002) Lethal H5N1 influenza viruses escape host anti-viral cytokine responses. Nat Med 8:950–954

Seo SH, Hoffmann E, Webster RG (2004) The NS1 gene of H5N1 influenza viruses circumvents the host anti-viral cytokine responses. Virus Res 103:107–113

Shi H, Liu XF, Zhang X, Chen S, Sun L, Lu J (2007) Generation of an attenuated H5N1 avian influenza virus vaccine with all eight genes from avian viruses. Vaccine 25:7379–7384

Slepushkin VA, Katz JM, Black RA, Gamble WC, Rota PA, Cox NJ (1995) Protection of mice against influenza A virus challenge by vaccination with baculovirus-expressed M2 protein. Vaccine 13:1399–1402

Smith W, Andrewes CH, Laidlaw PP (1933) A virus obtained from influenza patients. Lancet 2:66–68

Smith S, Demicheli V, Di Pietrantonj C, Harnden A, Jefferson T, Matheson N, Rivetti A (2006) Vaccines for preventing influenza in healthy children. Cochrane Database Syst Rev CD004879

Solana R, Mariani E (2000) NK and NK/T cells in human senescence. Vaccine 18:1613–1620

Steel J, Burmakina SV, Thomas C, Spackman E, García-Sastre A, Swayne DE, Palese P (2008) A combination in-ovo vaccine for avian influenza virus and Newcastle disease virus. Vaccine 26:522–531

Steinhauer DA (1999) Role of hemagglutinin cleavage for the pathogenicity of influenza virus. Virology 258:1–20

Stephenson I, Bugarini R, Nicholson KG, Podda A, Wood JM, Zambon MC, Katz JM (2005) Crossreactivity to highly pathogenic avian influenza H5N1 viruses after vaccination with nonadjuvanted and MF59-adjuvanted influenza A/Duck/Singapore/97 (H5N3) vaccine: a potential priming strategy. J Infect Dis 191:1210–1215

Stott DJ, Carman WF, Elder AG (2001) Influenza in old age. Age Ageing 30:361–363

Subbarao K, Chen H, Swayne D, Mingay L, Fodor E, Brownlee G, Xu X, Lu X, Katz J, Cox N, Matsuoka Y (2003) Evaluation of a genetically modified reassortant H5N1 influenza A virus vaccine candidate generated by plasmid-based reverse genetics. Virology 305:192–200

Suguitan AL Jr, McAuliffe J, Mills KL, Jin H, Duke G, Lu B, Luke CJ, Murphy B, Swayne DE, Kemble G, Subbarao K (2006) Live, attenuated influenza A H5N1 candidate vaccines provide broad cross-protection in mice and ferrets. PLoS Med 3:e360

Swain SL, Dutton RW, Woodland DL (2004) T cell responses to influenza virus infection: effector and memory cells. Viral Immunol 17:197–209

Szretter KJ, Gangappa S, Lu X, Smith C, Shieh WJ, Zaki SR, Sambhara S, Tumpey TM, Katz JM (2007) Role of host cytokine responses in the pathogenesis of avian H5N1 influenza viruses in mice. J Virol 81:2736–2744

Tamura S, Kurata T (2004) Defense mechanisms against influenza virus infection in the respiratory tract mucosa. Jpn J Infect Dis 57:236–247

Tamura S, Miyata K, Matsuo K, Asanuma H, Takahashi H, Nakajima K, Suzuki Y, Aizawa C, Kurata T (1996) Acceleration of influenza virus clearance by Th1 cells in the nasal site of mice immunized intranasally with adjuvant-combined recombinant nucleoprotein. J Immunol 156:3892–3900

Topham DJ, Tripp RA, Doherty PC (1997) CD8+ T cells clear influenza virus by perforin or Fas-dependent processes. J Immunol 159:5197–5200

Tran TH, Nguyen TL, Nguyen TD, Luong TS, Pham PM, Nguyen VC, Pham TS, Vo CD, Le TQ, Ngo TT, Dao BK, Le PP, Nguyen TT, Hoang TL, Cao VT, Le TG, Nguyen DT, Le HN, Nguyen KT, Le HS, Le VT, Christiane D, Tran TT, Menno de J, Schultsz C, Cheng P, Lim W, Horby P, Farrar J; World Health Organization International Avian Influenza Investigative Team (2004) Avian influenza A (H5N1) in 10 patients in Vietnam. N Engl J Med 350:1179–1188

Treanor J, Kotloff K, Betts R, Belshe R, Newman F, Iacuzio D, Wittes J, Bryant M (1999) Evaluation of trivalent, live, cold-adapted (CAIV-T) and inactivated (TIV) influenza vaccines in prevention of virus infection and illness following challenge of adults with wild-type influenza A (H1N1), A (H3N2), and B viruses. Vaccine 18:899–906

Treanor JJ, Campbell JD, Zangwill KM, Rowe T, Wolff M (2006) Safety and immunogenicity of a recombinant of an inactivated subvirion influenza A (H5N1) vaccine. N Engl J Med 354:1343–1351

Tsuru S, Fujisawa H, Taniguchi M, Zinnaka Y, Nomoto K (1987) Mechanism of protection during the early phase of a generalized viral infection. II. Contribution of polymorphonuclear leukocytes to protection against intravenous infection with influenza virus. J Gen Virol 68:419–424

Tumpey TM, Basler CF, Aguilar PV, Zeng H, Solórzano A, Swayne DE, Cox NJ, Katz JM, Taubenberger JK, Palese P, García-Sastre A (2005) Characterization of the reconstructed 1918 Spanish influenza pandemic virus. Science 310:77–80

Tumpey TM, Maines TR, Van Hoeven N, Glaser L, Solórzano A, Pappas C, Cox NJ, Swayne DE, Palese P, Katz JM, García-Sastre A (2007) A two-amino acid change in the hemagglutinin of the 1918 influenza virus abolishes transmission. Science 315:655–659

Tweed SA, Skowronski DM, David ST, Larder A, Petric M, Lees W, Li Y, Katz J, Krajden M, Tellier R, Halpert C, Hirst M, Astell C, Lawrence D, Mak A (2004) Human illness from avian influenza H7N3, British Columbia. Emerg Infect Dis 10:2196–2199

Van Reeth K, Van Gucht S, Pensaert M (2002) In vivo studies on cytokine involvement during acute viral respiratory disease of swine: troublesome but rewarding. Vet Immunol Immunopathol 87:161–168

Williams AE, Edwards L, Humphreys IR, Snelgrove R, Rae A, Rappuoli R, Hussell T (2004) Innate imprinting by the modified heat-labile toxin of Escherichia coli (LTK63) provides generic protection against lung infectious disease. J Immunol 173:7435–7443

Yap KL, Ada GL (1978) The recovery of mice from influenza virus infection: adoptive transfer of immunity with immune T lymphocytes. Scand J Immunol 7:389–397

Zhang M, Zharikova D, Mozdzanowska K, Otvos L, Gerhard W (2006) Fine specificity and sequence of antibodies directed against the ectodomain of matrix protein 2 of influenza A virus. Mol Immunol 43:2195–2206

Zheng B, Han S, Takahashi Y, Kelsoe G (1997) Immunosenescence and germinal center reaction. Immunol Rev 160:63–77

Zychlinsky A, Karim M, Nonacs R, Young JD (1990) A homogeneous population of lymphokine-activated killer (LAK) cells is incapable of killing virus-, bacteria-, or parasite-infected macrophages. Cell Immunol 125:261–267

# Adjuvants for Pandemic Influenza Vaccines

**Robert L. Atmar and Wendy A. Keitel**

**Contents**

**Abstract** The use of adjuvants is being explored as a means of improving vaccine immunogenicity. This is particularly important for the development of vaccines against potential pandemic influenza virus strains. Adjuvants act by prolonging the exposure time of antigen to the immune system, enhancing the delivery of antigen to antigen-presenting cells, or providing immunostimulatory signals that potentiate the immune response. Aluminum salts are the only licensed adjuvant in the United States, but the combination of these salts with inactivated influenza A/H5N1 antigens has had little effect on seroresponses. Several oil-in-water adjuvants, including MF59 and AS03, have significantly enhanced immune responses in healthy adult vaccine recipients to inactivated influenza A/H5N1. Additional studies are needed in vulnerable populations (younger and elderly persons, pregnant women, and immunocompromised patients) to confirm the safety and enhanced immunogenicity

R.L. Atmar (✉) and W.A. Keitel
Departments of Medicine and Molecular Virology and Microbiology,
Baylor College of Medicine 280, One Baylor Plaza, Houston, TX 77030, USA
e-mail: ratmar@bcm.edu

R.W. Compans and W.A. Orenstein (eds.), *Vaccines for Pandemic Influenza*,                 323
Current Topics in Microbiology and Immunology 333,
DOI 10.1007/978-3-540-92165-3_16, © Springer-Verlag Berlin Heidelberg 2009

of these promising formulations. A number of other adjuvants are under investigation to evaluate their ability to improve the immunogenicity of inactivated vaccines targeting influenza A/H5N1.

# 1  Introduction

Adjuvants are substances that enhance the ability of an antigen (immunogen) to induce an immune response. The term is derived from the Latin word *adjuvare*, and it was first used by Ramon in the mid-1920s to describe the improved antitoxin responses to diphtheria and tetanus toxoids when these antigens were administered to horses in combination with tapioca compared to when the antigens were administered alone (Ramon 1926; Vogel and Hem 2008). Since then, a number of substances have been found to have adjuvant properties. The utility and safety of these adjuvants has largely been determined empirically. However, as our understanding of the immune system has grown, strategies are being used to develop adjuvants that target specific arms of the immune response.

The ideal adjuvant should increase a vaccine's immunogenicity without adversely affecting the safety of the immunogen. Some adjuvants have failed because they were associated with unacceptable toxicities, even though they led to significantly improved immune responses. Examples of adverse events following administration of adjuvanted inactivated influenza virus vaccines include the formation of sterile abscesses and cysts at the injection site associated with mineral oil (Aguilar and Rodriguez 2007), systemic febrile reactions associated with MF59 combined with muramyl tripeptide conjugated to phosphatidylethanolamine (MTP-PE) (Keitel et al. 1993), and the excess occurrence of Bell's palsy associated with the use of intranasally delivered vaccine adjuvanted with the *Escherichia coli* heat-labile enterotoxin (Mutsch et al. 2004).

Serum IgG antibody to the influenza virus hemagglutinin (HA) plays a major role in immunity to influenza virus infection (Couch et al. 1984). Resistance to infection correlates directly with both serum hemagglutination inhibition (HAI) and neutralization (Nt) antibody levels, and these assays are the primary approaches that are used to assess the immunogenicity of vaccine candidates. Although serum HAI antibody levels of 32 or 40 are commonly stated to be "protective," this is a level at which there is only an ~50% reduction in the risk of infection by circulating interpandemic strains. Nevertheless, this level of serum antibody is used as a target in vaccine studies, and the percentage of subjects achieving a titer of 40 or greater is called the seroprotection rate (European Committee for Proprietary Medicinal Products 2003). It is not known whether similar levels of protection will be afforded by an HAI antibody titer of 40 against a pandemic strain, such as A/H5N1 or A/H9N2. Other measures of vaccine immunogenicity include the frequency of fourfold or greater increases in antibody level (seroconversion rate), the geometric mean titer (GMT) achieved after vaccination, and the ratio of GMTs after and before each vaccination. Both the European Union Committee for Medicinal Products for

**Table 1** Classification of adjuvants evaluated with influenza vaccine antigens

|  |  | Significant adjuvant activity | |
|---|---|---|---|
| Type | Examples | Animal studies | Human studies |
| Mineral salts | Aluminum hydroxide, Aluminum phosphate, Alum | Yes | Inconsistent |
| Emulsions/surfactants | MF59 | Yes | Yes |
|  | AS03 | Yes | Yes |
|  | QS21 (saponin) | Yes | No |
| Particulates | Liposomes / virosomes | Yes | No |
|  | Immunostimulatory complexes (ISCOMs) | Yes | No |
|  | Biodegradable polymers (PLG) | Yes | No |
| Microbial-derived | Exotoxin (cholera toxin [CT]; *E. coli* heat labile enterotoxin) | Yes | Yes |
|  | Endotoxin (monophosphoryl lipd A [MPL]) | Yes | Not tested |
|  | Flagellin | Yes | Under evaluation |
|  | Bacterial DNA (CpG oligonucleotides) | Yes | Yes |
|  | Muramyl tripeptide | Yes | Yes |
| Other PRR agonists | Poly I:C | Yes | Not tested |
| Cytokines | IL-2 | Yes | Yes |
|  | GM-CSF | Yes | No |
|  | Type 1 interferons | Yes | No |

Human Use (CHMP) (European Committee for Proprietary Medicinal Products 2003) and the US Food and Drug Administration (FDA) (2007) have used these parameters to establish performance criteria for the licensure of pandemic influenza vaccine candidates (enumerated in Keitel and Atmar 2009).

Protection from influenza virus infection also correlates with local (nasal) IgA levels (Clements et al. 1983; Belshe et al. 2000), leading to topical immunization strategies to enhance mucosal IgA responses. Cell-mediated immunity (CMI) may also provide some protection from disease (Murasko et al. 2002), and steps to improve CMI responses are a third approach for protecting against influenza virus infections. Adjuvants are being used to enhance immune responses in each of these compartments (Table 1).

# 2  General Mechanisms of Action of Adjuvants

Both innate and adaptive immunity are involved in protection against invasive pathogens. The innate immune response is not pathogen-specific and can be initiated immediately following exposure to a pathogen. The adaptive immune response is generated more slowly (over days to weeks) and is involved in pathogen-specific antigen recognition and in the establishment of immunologic memory following infection or vaccination. These two arms of the immune system were once thought to be independent of each other, but it is now apparent that proinflammatory and costimulatory molecules induced by the innate immune response can modulate the adaptive immune response (O'Hagan and Valiante 2003). Pathogen-specific molecular patterns (PAMPs), such as lipopolysaccharide from Gram-negative bacteria, unmethylated bacterial DNA (CpG), and bacterial peptidoglycans (e.g., muramyl dipeptide), are recognized by pattern recognition receptors (PRRs) such as toll-like receptors (TLRs). Binding of a PRR by its ligand triggers a signaling cascade that ultimately results in the production of specific cytokines and chemokines and leads to distinct antipathogen responses (Akira et al. 2006). The resulting cytokine/chemokine milieu influences the development of the host's adaptive immune response.

Antigen-presenting cells (APCs; e.g., dendritic cells, macrophages) initially take up and process pathogen-specific antigens and present these antigens to T and B cells. The interaction of PAMPs with PRRs on the APCs leads to their activation, resulting in the expression of costimulatory molecules (such as CD40 and CD80/86) and the production of proinflammatory cytokines. These in turn result in the activation of antigen-specific T cells. B cells also express PRRs and can function as APCs. Activated antigen-specific CD8+ T cells differentiate into effector cells that are able to kill cells expressing the antigen. CD4+ T cells differentiate along one of two major pathways: Th1 cells secrete interferon-gamma and IL-2 and provide cytokine signals that lead to the proliferation of CD8+ cells and help maximize killing by macrophages; Th2 cells secrete IL-4, IL-5, and IL-10 and promote B cell proliferation and production of antibodies (including IgA and IgE).

Based upon our understanding of the host immune response, adjuvants can be classified into two main groups: delivery systems and immune potentiators (O'Hagan and Valiante 2003) (Table 2). Delivery systems help target vaccine antigens to APCs in a variety of ways. A depot effect is achieved by adjuvants (e.g., mineral salts, emulsions) that slowly release antigen from the inoculation site, increasing the opportunity for APCs to interact with the antigen. In addition to a depot effect (i.e., slow release of antigen), adjuvants may protect the antigen from destruction (e.g., liposomes). On the other hand, particulate antigens are taken up more easily by APCs such that these antigens are more readily processed and presented to immune effector cells (Aguilar and Rodriguez 2007).

**Table 2** Adjuvant mechanisms of action

| Mechanism of action | Example(s) |
|---|---|
| Delivery system | |
|   Depot effect, delayed clearance | Aluminum salts, emulsions, saponins |
|   Enhance antigen presentation | Particulates, emulsions, aluminum salts, liposomes |
|   Protect antigen from destruction | Liposomes |
| Immunopotentiation | |
|   Bind PRRs | Microbial derived |
|   Activate Nalp3 inflammasome | Aluminum salts |
|   Induce APC maturation | Emulsions, ISCOMs |
|   Cytokine | Cytokine, GM-CSF |

**Table 3** Pattern recognition receptors (PRRs) targeted by different adjuvants (Takeda and Akira 2003; Aguilar and Rodriguez 2007; Guy 2007)

| PRR | Cellular location of PRR | Natural ligand | Adjuvant |
|---|---|---|---|
| TLR1/TLR2 heterodimer | Cell surface | Bacterial triacylated lipoproteins | *E. coli* heat-labile enterotoxin (B subunit) |
| TLR2/TLR6 heterodimer | Cell surface | Lipoteichoic acids, bacterial diacylated lipoproteins, fungal zymosan | Macrophage-activating lipopeptide-2 |
| TLR3 | Endosome/ lysosome | Double-stranded RNA | Poly(I:C) |
| TLR4 | Cell surface | Gram-negative bacterial lipopolysaccharide | Monophosphoryl lipid A (MPL) |
| TLR5 | Cell surface | Flagellin | Flagellin fusion proteins |
| TLR7, TLR8 | Endosome/ lysosome | Single-stranded RNA | Imiquimod, resiquimod |
| TLR9 | Endosome/ lysosome | Bacterial (unmethylated) CpG DNA | CpG oligonucleotides |
| NOD1 | Cytoplasm | Bacterial peptidoglycan | Diaminopimelic acid (DAP) |
| NOD2 | Cytoplasm | Bacterial peptidoglycan | Muramyl dipeptide (MDP) |

Adjuvants that act as immune potentiators activate APCs to promote the secretion of proinflammatory mediators that provide an environment conducive to an improved immune response. The most common means of APC activation is through binding of the PRRs. PRRs include TLRs that are associated with cellular membranes and nucleotide-binding oligomerization domains (NODs) that are found in the cytoplasm. TLRs are expressed in different cellular compartments, with TLR1, TLR2, and TLR4 being found on the surface of the cell and TLR3, TLR7, TLR8, and TLR9 being present in intracellular (e.g., endosomal) compartments (Takeda and Akira 2005). A variety of microbial-derived products that bind to different PRRs have been evaluated for their adjuvant activity (Table 3).

Other approaches that lead to immunopotentiation include the direct administration of cytokines with vaccine antigens (especially as a mucosal adjuvant), and the use of substances that have TLR-independent immunostimulatory properties that increase the local secretion of cytokines and chemokines and lead to the activation and maturation of APCs (Aguilar and Rodriguez 2007; O'Hagan ). Genes that express specific cytokines are also effective adjuvants when given as part of DNA vaccines.

# 3 Mineral Salts

Aluminum salts are the most common adjuvant used in vaccines for humans, and they are the only adjuvant currently licensed in the United States. There are several compounds that are used and these are commonly known as aluminum hydroxide, aluminum phosphate and alum. The first two terms are misnomers in that they do not reflect the chemical structure of the adjuvant; chemically, the aluminum hydroxide adjuvant is aluminum oxyhydroxide while the aluminum phosphate adjuvant is aluminum hydroxyphosphate, with the relative amounts of hydroxyl and phosphate groups varying depending on the degree of phosphate substitution (Hem and HogenEsch 2007). Alum refers to aluminum adjuvant for which aluminum potassium disulfate is used as the source of aluminum cations. Calcium phosphate is another mineral salt that has been used as an adjuvant (Vogel and Hem 2008).

There are several potential mechanisms by which aluminum salts can potentiate the immune response: (1) aluminum salts function as an antigen depot, slowly releasing antigen over time; (2) aluminum salts induce local inflammation, attracting APCs to the site of inoculation; and (3) antigen adsorbed to aluminum salts appears to the immune system as particulate antigen rather than soluble antigen, improving uptake by APCs. (Vogel and Hem 2008). The most important factor in the adjuvanticity of aluminum salts appears to be their activation of the Nalp3 inflammasome, which leads to the production of the pro-inflammatory cytokines, IL-1 and IL-18. The adjuvant properties of these salts is lost when the inflammasome cannot be activated (Eisenbarth et al. 2008). Aluminum-containing adjuvants push the host immune response towards a Th2 bias.

The adjuvanticity of aluminum-based salts in combination with influenza antigens has been evaluated over several decades. Many early studies failed to demonstrate improved immune responses in primed individuals receiving adjuvanted vaccine (Davenport et al. 1968; Gerth and Mok-Hsu 1981; Potter 1982), but other studies showed modest enhancement in antibody responses, especially in unprimed individuals (Hennessy and Davenport 1974; Nicholson et al. 1979; Pressler et al. 1982). More recently, Hehme et al. (2002) evaluated low-dose (1.9–7.5 mcg) whole-virus A/H2N2 and A/H9N2 vaccines adjuvanted with aluminum phosphate and compared them to a 15-mcg dose of nonadjuvanted subvirion vaccine. They found that similar HAI antibody responses were achieved in all groups, but the study design does not allow an assessment of the adjuvanticity of the aluminum phosphate.

Initial studies of candidate A/H5N1 vaccines demonstrated that high dosage levels were required to achieve antibody responses in the majority of study subjects (Treanor et al. 2006). Because earlier studies in people either provided conflicting evidence about the adjuvanticity of aluminum salts when combined with influenza antigens or did not have appropriate controls to address this question, studies were designed to evaluate whether aluminum salts enhance the immunogenicity of inactivated influenza A/H5N1 vaccines (Table 4).

Several studies of aluminum-adjuvanted subvirion A/H5N1 vaccines compared immune responses in healthy adult populations. Bresson et al. (2006) found that persons receiving the highest dosage (30 mcg) of adjuvanted vaccine had the greatest immune response as measured by seroconversion and geometric mean titer, but there was no apparent effect of adjuvant at lower dosage levels. Keitel et al. (2008a) observed more frequent responses in the group receiving 7.5 mcg of adjuvanted vaccine, but the nonadjuvanted 7.5-mcg dosage group had an unexpectedly low (0%) seroresponse frequency. There were no differences among other groups. Neither

**Table 4** Human clinical trials of candidate adjuvanted pandemic influenza vaccines

| Adjuvant | Antigen | Adjuvanted dosages tested (mcg) | Study population (N) | Vaccine regimen | Results | Reference |
|---|---|---|---|---|---|---|
| *Aluminum salts* | | | | | | |
| Subvirion | | | | | | |
| Aluminum hydroxide (600 mcg) prepared at bedside | A/Vietnam/1194/04 (H5N1) [NIBRG-14] | 7.5, 15, 30 | 18–40 year-old healthy adults (300) | 2 doses, 21 days apart | Highest adjuvanted dose elicited greatest antibody response, but no significant differences demonstrated between adjuvanted and nonadjuvanted vaccine | Bresson et al. (2006) |
| Aluminum hydroxide (600 mcg) | A/Vietnam/1203/04 (H5N1) | 3.75, 7.5, 15, 45 | 18–49 year-old healthy adults (600) | 2 doses, 28 days apart | Immune response frequency and magnitude increased with higher dosages. Only 7.5-mcg dosage had a significantly greater immune response vs. comparable nonadjuvanted group | Keitel et al. (2008a) |
| Aluminum hydroxide (600 mcg) | A/Vietnam/1203/04 (H5N1) | 3.75, 7.5, 15, 45 | ≥65 year-old healthy adults (600) | 2 doses, 28 days apart | Immune response frequency and magnitude increased with higher dosages. No adjuvant effect of aluminum hydroxide was observed | Brady et al. (2007) |
| Aluminum hydroxide (350–880 mcg) | A/Vietnam/1203/04 (H5N1) | 7.5, 15, 30 | 18–64 years old healthy adults (330) | 2 doses, 28 days apart | No adjuvant effect of aluminum hydroxide | Bernstein et al. (2008) |

(continued)

**Table 4** (continued)

| Adjuvant | Antigen | Adjuvanted dosages tested (mcg) | Study population (N) | Vaccine regimen | Results | Reference |
|---|---|---|---|---|---|---|
| *Whole virus* | | | | | | |
| Aluminum phosphate (500 mcg) | A/Singapore/1/57 (H2N2) | 1.9, 3.8, 7.5 | 18–30 years old healthy adults (198) | 2 doses, 21 days apart | Serum HAI antibody responses similar to those achieved using a 15-mcg subvirion vaccine as comparator | Hehme et al. (2002) |
| Aluminum phosphate (500 mcg) | A/Hong Kong/1073/99 (H9N2) | 1.9, 3.8, 7.5 | 18–60 year-old healthy adults (194) | 2 doses, 21 days apart | Serum HAI antibody responses similar to those achieved using a 15-mcg subvirion vaccine as comparator | Hehme et al. (2002) |
| Aluminum hydroxide (250 mcg) | A/Vietnam/1194/04 (H5N1) [NIBRG-14] | 1.25, 2.5, 5, 10 | 18–60 year-old healthy adults (120) | 2 doses, 28 days apart | Dose response, with highest dosage (10 mcg) meeting European regulatory requirements for licensure of interpandemic influenza vaccine; no nonadjuvanted comparator | Lin et al. (2006) |
| Aluminum phosphate (310 mcg) | A/Vietnam/1194/04 (H5N1) [NIBRG-14] | 6 | >18 year-old adults (146) | 1 dose | Single dosage tested; no nonadjuvanted comparator. 67% had fourfold or greater seroresponse at 21 days. | Vajo et al. (2007) |
| Aluminum hydroxide (350 mcg) | A/Vietnam/1203/04 (H5N1) grown in Vero cells | 7.5, 15 | 18–40 year-old healthy adults (210) | 2 doses, 28 days apart | Immune responses lower in adjuvanted dosage groups | Keitel et al. (2008b) |
| *Oil-in-water adjuvants* | | | | | | |
| *Subvirion* | | | | | | |
| MF59 (9.75 mg squalene) | A/duck/Singapore/97 (H5N3) | 7.5, 15, 30 | 18–40 year-old healthy adults (65) | 2 doses, 21 days apart | Significantly better immune responses in adjuvanted compared to nonadjuvanted group | Nicholson et al. (2001) |

| Adjuvant | Virus strain | Dosage (mcg) | Population | Schedule | Findings | Reference |
|---|---|---|---|---|---|---|
| MF59 (9.75 mg squalene) | A/duck/Singapore/97 (H5N3) | 7.5, 15, 30 | 18–45 year-old healthy adults (26) | 1 dose (boost) | Significantly better immune responses in adjuvanted compared to nonadjuvanted group | Stephenson et al. (2003) |
| MF59 (9.75 mg squalene) | A/chicken/Hong Kong/G9/97 (H9N2) | 3.75, 7.5, 15, 30 | 18–34 year-old healthy adults (96) | 2 doses, 28 days apart | All adjuvanted groups had comparable and high geometric mean antibody titers that were significantly higher than nonadjuvanted groups | Atmar et al. (2006) |
| MF59 (4.9–9.75 mg squalene) | A/Vietnam/1203/04 (H5N1) | 7.5, 15 | 18–64 years old healthy adults (394) | 2 doses, 28 days apart | Significant adjuvant effect at both dosage levels tested compared to no adjuvant or aluminum hydroxide. Immune response to adjuvanted 15-mcg dosage significantly higher than nonadjuvanted 45-mcg dosage | Bernstein et al. (2008) |
| AS03 | A/Vietnam/1194/04 (H5N1) [NIBRG-14] | 3.8, 7.5, 15, 30 | 18–60 year-old healthy adults (400) | 2 doses, 21 days apart | Significant adjuvant effect at all dosage levels. Immune response following single dose of adjuvanted vaccine significantly better than two doses of nonadjuvanted vaccine | LeRoux-Roels et al. (2007) |
| AdjA | A/Vietnam/1194/04 (H5N1) [NIBRG-14] | 1.9, 3.8, 7.5, 15 | 18–40 year-old adults (265) | 2 doses, 21 days apart | Significant adjuvant effect of all doses compared to nonadjuvanted 15-mcg dose. 3.8-mcg and higher adjuvanted dosages all with similar antibody responses | Levie et al. (2008) |

Bernstein et al. (2008) nor Brady et al. (2007) found evidence of a significant adjuvant effect at a range of dosage levels in healthy (18–64 years) and older (≥65 years) adults, respectively. From these studies, it does not appear that there is biologically meaningful adjuvant effect of aluminum-containing salts in subvirion influenza A/H5N1 vaccines.

The picture with aluminum-adjuvanted whole-virus vaccines is less clear because most of the studies of these vaccines reported to date have not included a comparable nonadjuvanted control group. All of the dosage levels (1.25–10 mcg) evaluated by Lin et al. (2006) and the single dosage level evaluated by Vajo et al. (2007) had no nonadjuvanted comparator. Thus, although potentially acceptable seroresponse frequencies (seroconversion rates) were observed in these trials, it is not possible to evaluate the contribution of the aluminum salt to the vaccine's immunogenicity. Keitel et al. (2008b) found significantly poorer immune responses in groups receiving adjuvanted vaccine compared to nonadjuvanted groups, but another study with the same vaccine did not show any differences between adjuvanted and nonadjuvanted groups (Ehrlich et al. 2008).

# 4 Emulsions/Surfactants

Emulsions are another common group of substances used in vaccines as adjuvants. An emulsion is made by mixing two immiscible substances in the presence of one or more emulsifiers or surfactants that stabilize the mixture. Emulsions used in vaccines can be either water-in-oil or oil-in-water emulsions. The type of emulsion is determined by the relative amounts of each immiscible substance and the type of surfactant used (Guy 2007). Water-in-oil emulsions have water dispersed in an oil phase. Examples include complete Freund's adjuvant, incomplete Freund's adjuvant, adjuvant 65 and Montanide ISA51. In general, although the water-in-oil emulsions have adjuvant effects, they have been too toxic for use in humans due to the induction of local inflammatory reactions (granulomas, cyst formation, ulceration) at the injection site (Aguilar and Rodriguez 2007). However, oil-in-water emulsions are promising adjuvants that are currently being used in combination with influenza virus antigens.

The mechanisms by which emulsions have adjuvant activity are unknown. Proposed mechanisms include a depot effect leading to retention of antigen at the injection site, induction of APC maturation by triggering the local production of cytokines and chemokines, and improved presentation of antigen to APCs. The relative importance of these potential mechanisms may vary with different emulsions. For example, an animal study in which MF59 was injected in combination with a herpes simplex antigen (gD2) showed that more than 60% of the MF59 was cleared from the muscle within 4 h and the antigen was cleared from the injection site independent of the presence of the adjuvant (Schultze et al. 2008). These data suggest that a depot effect is not a primary mechanism of action for MF59.

## 4.1 Oil-in-Water Emulsions

MF59 is the most widely evaluated oil-in-water emulsion used as an adjuvant. It is a microfluidized emulsion consisting of the oil squalene (4.3% w/v) with the two surfactants Tween 80 (polyoxyethylene sorbitan monooleate; 0.85% w/v) and Span-85 (sorbitan trioleate; 0.5% w/v) in a sodium citrate buffer (10 nM). MF59 has been used as an adjuvant in combination with a range of different antigens, and it is licensed in combination with inactivated influenza vaccine as a seasonal (interpandemic) vaccine for the elderly in Europe (O'Hagan 2007). More than 27 million doses of MF59-adjuvanted influenza vaccine have been distributed, and postmarketing surveillance studies have failed to identify any safety concerns related to vaccine use (Schultze et al. 2008).

The adjuvant activity of MF59 has been evaluated in a number of studies with potential pandemic influenza vaccine strains (Table 4). Nicholson et al. (2001) first demonstrated that healthy young adults responded poorly to an inactivated influenza A/H5N3 vaccine, and addition of MF59 to the vaccine significantly improved antibody responses. Sixteen months later 26 of the original 65 subjects received a third dose of the same vaccine preparation they had been administered originally, and those that received vaccine adjuvanted with MF59 had significantly better antibody responses than those who received nonadjuvanted vaccine (Stephenson et al. 2003).

Atmar et al. (2006) evaluated a range of dosages (3.75–30 mcg) of inactivated influenza A/H9N2 vaccine with and without MF59 in young healthy adults. His group observed that all adjuvanted vaccine groups had comparable and high geometric mean antibody titers that were significantly higher than nonadjuvanted groups. The serum antibody responses achieved after a single dose of adjuvanted vaccine were greater than those observed after two doses of nonadjuvanted vaccine. Even the antibody responses achieved with the lowest dosage examined (3.75 mcg) exceeded those achieved with the highest nonadjuvanted dosage (30 mcg), and all of the adjuvanted dosages exceeded the immunogenicity guidelines proposed by the CHMP and FDA. Adjuvanted vaccine recipients had significantly more discomfort at the injection site than nonadjuvanted vaccine recipients, although most of the pain was classified as mild and the vaccine was considered to be well tolerated (Atmar et al. 2006).

The success of MF59 as an adjuvant with the influenza A/H5N3 and A/H9N2 antigens led Bernstein et al. (2008) to evaluate it in a multicenter evaluation of a candidate influenza A/H5N1 vaccine. Different dosages of nonadjuvanted (15, 30, or 45 mcg), alum-adjuvanted (7.5, 15, 30 mcg) or MF59-adjuvanted vaccine were given to healthy adults 18–64 years of age in the double-blind, placebo-controlled trial. Study participants received two doses of vaccine four weeks apart, and immunogenicity assessments were made four weeks after each dose. The alum-adjuvanted vaccines were preformulated while the MF59 was mixed with the influenza vaccine at the time of administration. The only vaccine group that achieved a serum HAI antibody titer of 40 or greater in at least half of the subjects after two doses of vaccine was the group that received 15 mcg of vaccine with MF59, and the responses (geometric mean titer, seroprotection frequency) in this group were significantly higher than those in the groups with the highest dosages of nonadjuvanted or alum-adjuvanted vaccine. The group that received 7.5 mcg of vaccine with MF59 also had a significantly better antibody response than the nonadjuvanted and alum-adjuvanted groups that received the same dosage, but the immune response was significantly lower than that observed for the 15 mcg plus MF59 dose. However, unlike the earlier studies with H5N3 and H9N2 vaccines, the amount of MF59 used in the 7.5-mcg dose was half of that used for the 15-mcg dose, suggesting that the amount of MF59 used in the vaccine affects its adjuvanticity (Bernstein et al. 2008).

AS03 is another emulsion that has adjuvant activity when combined with influenza virus antigens. It is a 10% oil-in-water (w/v) emulsion, with the oil phase containing 5% DL-alpha-tocopherol and squalene and the aqueous phase containing 2% Tween-80 (Leroux-Roels et al. 2007). Leroux-Roels and colleagues (2007) performed an observer-blinded evaluation of AS03-adjuvanted influenza A/H5N1 vaccine in healthy adults 18–60 years of age. They evaluated four different dosages of influenza antigen (3.8, 7.5, 15, and 30 mcg) while keeping the amount of adjuvant constant. Persons who received adjuvanted vaccine were more likely to note local injection site symptoms, such as pain, in the week following vaccination; most symptoms were classified as mild to moderate. Significantly improved antibody responses (as measured by seroconversion frequency, seroprotection frequency, GMTs, and mean GMT increases) were seen in all adjuvanted vaccine groups after each vaccine dose, and all adjuvanted vaccine groups exceeded the CHMP and FDA immune response criteria for seroconversion and seroprotection frequencies. In contrast, the highest nonadjuvanted dosage group only met one of the CHMP criteria (seroprotection rate > 40%), and none of the other nonadjuvanted vaccine groups met any of the CHMP criteria (Leroux-Roels et al. 2007).

Another oil-in-water adjuvant, AdjA, is also in clinical development. The adjuvant combined with 30-mcg doses of a subvirion influenza A/Vietnam/1194/04 provided partial protection following homologous challenge (Ruat et al. 2008). Levie et al. (2008) described a double-blind, two-dose study to evaluate several dosages (1.9–15 mcg) of adjuvanted vaccine in 18–40 year-old adults. All groups receiving adjuvanted vaccine had significantly higher serum HAI and neutralizing antibody responses compared to the group that received 7.5-mcg doses of nonadjuvanted vaccine. Groups receiving 3.9, 7.5, or 15 mcg of adjuvanted vaccine all had similar levels of antibody response that were greater than those observed with the 1.9-mcg dosage, but even the latter dosage group met CHMP immunogenicity criteria.

Another potential advantage of oil-in-water emulsion adjuvants is that they may be more likely to elicit antibody responses that are cross-protective against other strains. In the AS03 study described above, seroresponses to influenza A/H5N1 viruses belonging to other clades were also present in postimmunization sera (Leroux-Roels et al. 2007, 2008). In a ferret model, AS03-adjuvanted influenza A/H5N1 vaccine from a clade 1 virus (influenza A/Vietnam/1194/04) protected animals from death following challenge with a clade 2 strain (influenza A/Indonesia/5/05) (Baras et al. 2008). Cross-reactive HAI antibody responses were also observed with increasing frequency as the dosage of MF59-adjuvanted influenza A/H9N2 vaccine increased and in the study of the AdjA adjuvant with a clade 2 strain of influenza A/H5N1 (Atmar et al. 2006; Levie et al. 2008). These observations are important because an emerging pandemic strain may have undergone antigenic drift or belong to a different clade than the strain used to produce vaccine stock-piles, and the ability of a vaccine to induce cross-reactive antibody responses may allow it to afford protection against the pandemic strain that would not be afforded by a vaccine that does not generate such responses.

## 4.2   Saponins

Quil A is a natural product derived as an aqueous extract from the bark of the South American tree *Quillaja saponaria*. Although Quil A is too toxic to be used in humans, the saponin QS-21 is a purified, less toxic derivative of Quil A that has been evaluated as an adjuvant (Aguilar and Rodriguez 2007). QS-21 had good adjuvant activity in combination with inactivated influenza vaccine in mice, inducing increased antibody levels and protection compared to aqueous vaccine alone (Wyde et al. 2001). However, when QS21 was evaluated in primed adults in combination with a trivalent inactivated influenza vaccine, antibody and cell-mediated immune responses were similar to those seen in persons receiving aqueous, nonadjuvanted vaccine alone (Mbawuike et al. 2007). In addition, the majority of subjects who received the adjuvanted vaccine had moderate pain at the injection site. Semisynthetic saponin derivatives are also being evaluated as potential adjuvants (Marciani et al. 2003).

## 5   Particulates

Particulate antigen can elicit a greater immune response than soluble antigen, so strategies to develop particulate antigen delivery systems are being pursued (Singh et al. 2007). Improved antigen uptake by APCs is one potential explanation for the better immune responses, and the size, charge and presence of specific ligands that interact with receptors on APCs are factors that can be modified to affect the effi-ciency of antigen delivery. Particulate delivery systems can also act as depot sys-tems in which antigens are delivered to APCs over an extended period of time. The antigen can be attached to the surface of the particulate through adsorption or covalent linkage, or it can be encapsulated within the carrier. Some of the particulate antigen delivery systems that have been most extensively explored include liposomes and virosomes, immunostimulatory complexes and biodegradable polymers, and these systems are discussed in further detail below.

## 5.1   Liposomes, Virosomes, and Virus-Like Particles

Liposomes are membranes that consist of lipid bilayers that surround an aqueous compartment; liposomes may be unilamellar (consist of a single lipid bilayer) or multilamellar (consist of more than one lipid bilayer). The lipids contain polar groups that abut the aqueous interface and nonpolar groups (cholesterol or long chain fatty acids) that interact with other lipid nonpolar groups to form the bilayer. The adjuvanticity of a liposome is influenced by the properties of its constituent lipids, including surface charge and fatty acid length. Positively charged liposomes (cationic) interact and are taken up more readily by APCs and in most circumstances are better adjuvants than neutral or negatively charged (anionic) liposomes (Christensen et al. 2007). The adjuvanticity of a liposome preparation combined with influenza antigens has been evaluated in elderly people. Powers (1997) failed to find any significant adjuvant effect of a monovalent H1N1 vaccine that was presented in an oligolamellar liposome made of dimyristoylphosphatidyl choline and cholesterol in a 7:3 molar ratio. On the other hand, a liposomal preparation has showed promising results against an avian influenza virus strain. The addition of a nonphospholipid liposomal preparation (Novasome) to recombinant H9 hemagglutinin or recombinant H9 virus-like particles (VLPs) increased the immunogenicity and protection achieved with these vaccines in a mouse model compared to nonadjuvanted vaccine (Pushko et al. 2007). Liposomes are also being used as a delivery system for DNA vaccines.

Virosomes are liposomes that contain viral envelope proteins. Large quantities of virus are grown and the envelope proteins are purified from the rest of the virion by extraction in detergent. Virosomes form spontaneously when the detergent is removed. Thus, virosomes are derived from native virus particles and are distinguished from recombinant VLPs that are produced in host cells following the in vitro expression of viral proteins (Moser et al. 2007). Virosomes are being evaluated as a delivery system for noninfluenzavirus antigens, but two influenza virosomal vaccines (produced by Solvay and Berna Biotech) have also been licensed for use in Europe for the prevention of influenza. The immunogenicity of one of these (Invivac, Solvay) in an elderly population was similar to that of standard nonadjuvanted subunit vaccine and to MF59-adjuvanted influenza vaccine (de Bruijn et al. 2006).

Immunostimulatory molecules have been combined with liposomes, virosomes, and VLPs (discussed separately by Bright et al. 2009) to improve the immunogenicity of antigens present in these vaccines. Ben-Yehuda et al. (2003a,b) loaded multilamellar liposomes made of dimyristoyl phosphatidylcholine and dimyristoyl phosphatidylglycerol with either subunit influenza virus antigens or IL-2. Both young adults and elderly persons had significantly improved antibody responses to vaccines antigens when they were given a 1:1 mix of the IL-2 liposomes and the liposomes containing influenza antigens compared to the seroresponses achieved with standard inactivated subunit influenza vaccine.

Ernst et al. (2006) vaccinated mice with a liposomal preparation containing monophosphoryl A (MPL) and peptides consisting of a proprietary hydrophobic domain (HD) fused to the ectodomain of the influenza M2 proteins (M2eA) representing a range of different influenza A virus subtypes. The combination of MPL and liposome improved the ELISA antibody responses to M2eA antigen compared to administration of the fusion peptide with either MPL alone or liposome alone. Mice vaccinated with the fusion peptides in the MPL-liposomal preparation had improved protection from death or pulmonary virus shedding following challenge with

H1N1, H5N1, H6N1, and H9N2 viruses compared to mice that received the liposomal preparation alone (without the influenza fusion peptides). This preparation is being explored as a potential universal pandemic vaccine.

Other immunostimulatory molecules have been evaluated as adjuvants in combination with virosomes or VLPs delivered intranasally. Matassov et al. (2007) used a baculovirus expression system to generate VLPs containing the hemagglutinin and neuraminidase proteins of the 1918 pandemic strain and the M1 and M2 proteins of the A/Udorn/73 (H3N2) strain. Mice immunized with the VLPs and the adjuvant CpG (see Sect. 6) had less clinical disease and more rapid clearance of the challenge virus (A/swine/Iowa/15/30) than did mice that received unadjuvanted VLPs, as well as improved immune responses to influenza VLPs containing the hemagglutinin of the 1918 pandemic strain in a mouse model (Matassov et al. 2007). An intranasal virosomal influenza vaccine adjuvanted with heat-labile *E. coli* enterotoxin was licensed for human use in Europe. Distribution of the vaccine was suspended after the occurrence of an excess number of cases of Bell's palsy among vaccine recipients (Mutsch et al. 2004).

## 5.2   Immunostimulatory Complexes

Immunostimulatory complexes (ISCOMs) are another approach for the particulate presentation of antigens. The original ISCOM-adjuvanted vaccines presented viral membrane-associated proteins in a multimeric form (Morein et al. 1984). The envelope proteins were enclosed in a micelle consisting of Quil A extract, cholesterol, and phospholipids. The complexes were difficult to produce and were reactogenic. As with QS21, a purified fraction of the *Q. saponaria* bark extract was combined with cholesterol and phopholipids to produce the proprietary ISCOMATRIX using simpler production methods (Drane et al. 2007). In addition to improving antigen uptake by APCs, ISCOMATRIX activates and leads to the expression of maturation markers on these cells. The result is the generation of both strong humoral and cellular immune responses with ISCOMATRIX-adjuvanted vaccines.

A variety of different ISCOMs has been evaluated in animal models of influenza virus infection and have been shown to improve both antibody and CTL responses to the immunogen. In addition, heterosubtypic protection of mice immunized with H1N1 antigen has been observed following challenge with H5N1 and H9N2 strains, presumably through the induction of cross-reactive cellular immunity (Sambhara et al. 2001). Another ISCOM-adjuvanted inactivated influenza vaccine protected macaques against virus shedding in the nose, throat, and lung following homologous intratracheal challenge with influenza A/Netherlands/18/94 (H3N2), while no or only partial protection was observed with nonadjuvanted vaccine (Rimelzwaan et al. 1997). Studies with ISCOMATRIX-adjuvanted influenza vaccines delivered either intramuscularly or intranasally to humans are currently under way (Drane et al. 2007).

## 5.3  Biodegradable Polymers

Another approach to the delivery of antigen in a particulate form has been the use of biodegradable polymers. Polylactide-co-glycolide (PLG), poly(DL-lactic coglycolic acid) (PLGA), poly(DL-lactide) (PLA), and poly(ortho esters) (POE) are biodegradable materials that have been used in humans as sutures and as drug delivery systems, and these polymers have also been evaluated as candidate adjuvants (Xiang et al. 2006). Potential mechanisms by which the these agents can improve immune responses include presentation of antigen as a particulate, protection of antigen from degradation, and prolonged exposure of APCs to antigen through controlled release over time. A problem with this approach has been the difficulty in preventing degradation of the antigen during the encapsulation process. Another strategy has been to adsorb antigen to the surface of the polymer (Singh et al. 2007). Mice immunized with influenza viral antigens adsorbed to the surface of PLG microspheres or PLA lamellae had HAI antibody titers 3- and 14-fold higher than those achieved with aqueous antigen (Coombes et al. 1998). These polymers are also being evaluated in combination with other potential adjuvants (Wack et al. 2008).

## 6  Microbial-Derived Agents

A number of different microbial products that activate the innate immune system by binding to PRRs are being investigated as potential adjuvants, and many of these are being evaluated in combination with influenza virus antigens. Binding of the PRRs (Table 3) leads to the release of immunomodulatory cytokines and chemokines that influence the resulting immune response. Many microbial-derived agents that had previously been shown to have adjuvant activity were subsequently identified to be TLR agonists, including unmethylated CpG, poly I:C, lipopolysaccharide, and lipopeptides (Guy 2007). New synthetic TLR agonists are also being developed and evaluated.

Cholera toxin (CT) and *E. coli* heat-labile enterotoxins (HLTs) are among the most potent known mucosal adjuvants. The toxins each consist of an A subunit that is noncovalently linked to the pentameric B subunits. The toxins bind to gangliosides through the B subunits, and crosslinking of the gangliosides may be one of the mechanisms that leads to their immunopotentiating effects (Connell 2007). CT biases the immune response towards a Th2 response, with increased production of IgA. The toxicity of CT and HLT, including the occurrence of Bell's palsy associated with intranasal HLT-adjuvanted influenza vaccine mentioned earlier, has largely precluded their use as adjuvants in humans, although HLT is still being investigated as a topical, transdermal adjuvant (Frech et al. 2005). Mutation of the A subunit (which is responsible for the toxic effects of CT and HLTs) to inactivate its enzymatic activity does not abrogate the adjuvanticity of the agents, and mutant HLT is being evaluated (Stephenson et al. 2006). The B subunit by itself also has adjuvant activity, with the B subunit of HLT binding to TLR2 (Connell 2007).

Monophosphoryl A (MPL) is a derivative of the *Salmonella minnesota* R595 lipopolysaccharide that functions as a TLR4 agonist and has been used as an adjuvant. It has been combined with aluminum salts and given parenterally to induce both humoral and cellular immune responses for a number of vaccines (e.g., hepatitis B, papillomavirus, herpes simplex) (Garçon et al. 2007). It has also been evaluated in a mouse model as a potential mucosal adjuvant (Baldridge et al. 2000). Intranasal administration of MPL-adjuvanted influenza vaccine induced higher levels of virus-specific mucosal IgA and was associated with significantly higher survival following lethal challenge compared to nonadjuvanted vaccine.

Flagellin is another microbial product, and it functions as a TLR5 agonist. A fusion peptide consisting of four copies of a consensus sequence of the ectodomain of the influenza A M2 protein and flagellin derived from *Salmonella typhimurium* has been evaluated in a mouse model of influenza. Antibody responses to M2e were significantly higher in mice vaccinated with the fusion peptide compared to those vaccinated with M2e alone, and the M2e fusion peptide provided significant protection from death following a PR8 challenge (Huleatt et al. 2008). This vaccine has also been reported to be immunogenic in people; further studies are ongoing (NCT00603811).

Unmethylated CpG is a surrogate for bacterial DNA and functions as a TLR9 agonist. Parenteral administration with inactivated influenza virus vaccine significantly increases serum antibody responses in mice, and it also has adjuvant activity when delivered intranasally (Moldoveanu et al. 1998). The adjuvant activity of a CpG oligonucleotide was also evaluated in combination with a licensed trivalent inactivated influenza vaccine in a Phase I clinical trial (Cooper et al. 2004). The largest adjuvant effects were seen in persons receiving 1/10 the standard dosage of vaccine, but the study was too small to draw any firm conclusions about the adjuvanticity of the CpG. The combination of CpG with other adjuvants is another approach to enhance immune responses to influenza that is under evaluation (Matassov et al. 2007; Wack et al. 2008).

Muramyl dipeptide (MDP) is the active component of the mycobacterial cell wall and is important in the adjuvant effect associated with Freund's complete adjuvant. Muramyl tripeptide conjugated to phosphtidylethanolamine (MTP-PE) is a synthetic derivative of MDP and has been evaluated as a potential adjuvant. When it was combined with MF59 and inactivated influenza virus vaccine in a Phase I evaluation, all five persons who received the adjuvanted vaccine had moderate to severe reactions, including fever, chills and injection site tenderness, swelling and erythema (Keitel et al. 1993). Although recipients of the adjuvanted vaccine had significantly better serologic responses than those in persons who received nonadjuvanted vaccine, the formulation was considered to be too reactogenic for further evaluation.

# 7 Other Approaches

Intranasal administration of a seasonal trivalent inactivated influenza vaccine adjuvanted with poly(I):poly(C(12)U), a TLR3 agonist, led to increased survival and decreased virus shedding in mice following challenge with several different influenza A/H5N1 strains (Ichinohe et al. 2007). The investigators postulate that the protective effects were due to the induction of mucosal antibody based upon the lack of induction of cross-reactive serum antibody or proliferative T cell responses. Although the exact mechanism of protection is not clear, the effects were not due to nonspecific resistance from the adjuvant, as no decrease in virus shedding was observed in mice inoculated intranasally only with the adjuvant.

A variety of different cytokines have been added directly to influenza virus antigens to improve antibody responses. IL-2, GM-CSF, and type 1 interferons have all had significant immunostimulatory activity in animal models (Babai et al. 2001; Bracci et al. 2006). However, studies in humans have either failed to demonstrate an adjuvant effect or have not had adequate controls to allow an interpretation of the impact of the cytokine (Ben-Yehuda et al. 2003a; Somani et al. 2002). Potential limitations of the direct administration of cytokines with immunogens include their short half-lives and the adverse events they elicit.

## 8   Summary and Conclusions

Adjuvants are increasingly being explored as a means to improve immune responses to interpandemic and pandemic influenza viruses. The combination of oil-in-water emulsions with inactivated influenza A/H5 antigens has significantly improved the serologic responses to the hemagglutinin and has also led to the generation of antibodies that cross-react with strains from other H5 clades. Other potential advantages of adjuvants include the possibility of decreasing the amount of antigen needed in the vaccine (dose sparing) and the generation of better immune responses among groups that generally respond poorly to inactivated antigens (e.g., immuno-compromised, elderly). The safety of the adjuvant is the other major issue before it can be considered for widespread use. Currently, only aluminum salts are licensed as adjuvants in the United States, although other adjuvants (MF59, MPL with alum) are licensed in other parts of the world. Large numbers of people will need to be studied to document the safety of any candidate adjuvants being considered for use in pandemic influenza vaccines. Nevertheless, the results of recent clinical trials are reason for optimism that adjuvants may significantly enhance our ability to respond to the next influenza pandemic.

## References

Aguilar JC, Rodriguez EG. (2007) Vaccine adjuvants, revisited.  Vaccine 25:3752–3762
Akira S, Uematsu S, Takeuchi O. (2006) Pathogen recognition and innate immunity. Cell 124:783–801
Atmar RL, Keitel WA, Patel SM, Katz JM, She D, El Sahly H, Pompey J, Cate TR, Couch RB. (2006) Safety and immunogenicity of nonadjuvanted and MF59-adjuvanted influenza A/H9N2 vaccine preparations. Clin Infect Dis 43:1135–1142

Babai I, Barenholz Y, Zakay-Rones Z, Greenbaum E, Samira S, Hayon I, Rochman M, Kedar E. (2001) A novel liposomal influenza vaccine (INFLUSOME-VAC) containing hemagglutinin-neuraminidase and IL-2 or GM-CSF induces protective anti-neuraminidase antibodies cross-reacting with a wide spectrum of influenza A viral strains. Vaccine 20:505–515

Baldridge JR, Yorgensen Y, Ward JR, Ulrich JT. (2000) Monophosphoryl lipid A enhances mucosal and systemic immunity to vaccine antigens following intranasal administration. Vaccine 18:2416–2425

Baras B, Stittelaar KJ, Simon JH, Thoolen RJ, Mossman SP, Pistoor FH, van Amerongen G, Wettendorff MA, Hanon E, Osterhaus AD. (2008) Cross-protection against lethal H5N1 challenge in ferrets with an adjuvanted pandemic influenza vaccine. PLoS ONE 3:e1401

Belshe RB, Gruber WC, Mendelman PM, Mehta HB, Mahmood K, Reisinger K, Treanor J, Zangwill K, Hayden FG, Bernstein DI, Kotloff K, King J, Piedra PA, Block SL, Yan L, Wolff M. (2000) Correlates of immune protection induced by live, attenuated, cold-adapted, trivalent, intranasal influenza virus vaccine. J Infect Dis 181:1133–1137

Ben-Yehuda A, Joseph A, Barenholz Y, Zeira E, Even-Chen S, Louria-Hayon I, Babai I, Zakay-Rones Z, Greenbaum E, Galprin I, Glück R, Zurbriggen R, Kedar E. (2003a) Immunogenicity and safety of a novel IL-2-supplemented liposomal influenza vaccine (INFLUSOME-VAC) in nursing-home residents. Vaccine 21:3169–3178

Ben-Yehuda A, Joseph A, Zeira E, Even-Chen S, Louria-Hayon I, Babai I, Zakay-Rones Z, Greenbaum E, Barenholz Y, Kedar E. (2003b) Immunogenicity and safety of a novel liposomal influenza subunit vaccine (INFLUSOME-VAC) in young adults. J Med Virol 69:560–567

Bernstein DI, Edwards KM, Dekker CL, Belshe R, Talbot HK, Graham IL, Noah DL, He F, Hill H. (2008) Effects of adjuvants on the safety and immunogenicity of an avian influenza H5N1 vaccine in adults. J Infect Dis 197:667–675

Bracci L, Canini I, Venditti M, Spada M, Puzelli S, Donatelli I, Belardelli F, Proietti E. (2006) Type I IFN as a vaccine adjuvant for both systemic and mucosal vaccination against influenza virus. Vaccine 24:S56–S57

Brady RC, Treanor JJ, Atmar RL, Chen WH, Winokur P, Belshe R. (2007) A phase I-II, randomized, controlled, dose-ranging study of the safety, reactogenicity, and immunogenicity of intramuscular inactivated influenza A/H5N1 vaccine given alone or with aluminum hydroxide to healthy elderly adults. In: Options for the Control of Influenza VI, Toronto, Ontario, Canada, 17–23 June 2007 (abstract P739)

Bresson JL, Perronne C, Launay O, Gerdil C, Saville M, Wood J, Höschler K, Zambon MC. (2006) Safety and immunogenicity of an inactivated split-virion influenza A/Vietnam/1194/2004 (H5N1) vaccine: phase I randomised trial. Lancet 367:1657–1664

Bright R, Pushko P, Smith G, Compans R. (2009) Influenza virus-like particles as pandemic vaccines. In: Compans RW, Orenstein WA. (eds) Vaccines for pandemic influenza. Current topics in microbiology, vol 333. Springer, Heidelberg

Christensen D, Korsholm KS, Rosenkrands I, Lindenstrøm T, Andersen P, Agger EM. (2007) Cationic liposomes as vaccine adjuvants. Expert Rev Vaccines 6:785–796

Clements ML, O'Donnell S, Levine MM, Chanock RM, Murphy BR. (1983) Dose response of A/Alaska/6/77 (H3N2) cold-adapted reassortant vaccine virus in adult volunteers: role of local antibody in resistance to infection with vaccine virus. Infect Immun 40:1044–1051

Connell T. (2007) Cholera toxin, LT-I, LT-IIa and LT-IIb: the critical role of ganglioside binding in immunomodulation by type I and type II heat-labile enterotoxins. Expert Rev Vaccines 6:821–834

Coombes AG, Major D, Wood JM, Hockley DJ, Minor PD, Davis SS. (1998) Resorbable lamellar particles of polylactide as adjuvants for influenza virus vaccines. Biomaterials 19:1073–1081

Cooper CL, Davis HL, Morris ML, Efler SM, Krieg AM, Li Y, Laframboise C, Al Adhami MJ, Khaliq Y, Seguin I, Cameron DW. (2004) Safety and immunogenicity of CPG 7909 injection as an adjuvant to Fluarix influenza vaccine. Vaccine 22:3136–3143

Couch RB, Kasel JA, Six TH, Cate TR, Zahradnik JM. (1984) Immunological reactions and resistance to infection with influenza virus. In: Stuart-Harris C, Potter CW. (eds) The molecular virology and epidemiology of influenza. Academic, New York, pp 119–144

Davenport FM, Hennessy AV, Askin FB. (1968) Lack of adjuvant effect of AlPO4 on purified influenza virus hemagglutinins in man. J Immunol 100:1139–1140

de Bruijn IA, Nauta J, Gerez L, Palache AM. (2006) The virosomal influenza vaccine Invivac: immunogenicity and tolerability compared to an adjuvanted influenza vaccine (Fluad) in elderly subjects. Vaccine 24:6629–6631

Drane D, Gittleson C, Boyle J, Maraskovsky E. (2007) ISCOMATRIX adjuvant for prophylactic and therapeutic vaccines. Expert Rev Vaccines 6:761–772

Ehrlich HJ, Müller M, Oh HM, Tambyah PA, Joukhadar C, Montomoli E, Fisher D, Berezuk G, Fritsch S, Löw-Baselli A, Vartian N, Bobrovsky R, Pavlova BG, Pöllabauér EM, Kistner O, Barrett PN; Baxter H5N1 Pandemic Influenza Vaccine Clinical Study Team. (2008) A clinical trial of a whole-virus H5N1 vaccine derived from cell culture. N Engl J Med 358:2573–2584

Eisenbarth SC, Colegio OR, O'Connor W, Sutterwala FS, Flavell RA. (2008) Crucial role for the Nalp3 inflammasome in the immunostimulatory properties of aluminium adjuvants. Nature 453:1122–1126

Ernst WA, Kim HJ, Tumpey TM, Jansen ADA, Tai W, Cramer DV, Adler-Moore JP, Fujii G. (2006) Protection against H1, H5, H6 and H9 influenza A infection with liposomal matrix 2 epitope vaccines. Vaccine 24:5158–5168

European Committee for Proprietary Medicinal Products. (2003) Guideline on dossier structure and content for pandemic influenza vaccine marketing authorization application (CPMP/VEG/4717/03)

Food and Drug Administration. (2007) Guidance for industry: clinical data needed to support the licensure of pandemic influenza vaccines. http://www.fda.gov/CbER/gdlns/panfluvac.pdf

Frech SA, Kenney RT, Spyr CA, Lazar H, Viret JF, Herzog C, Glück R, Glenn GM. (2005) Improved immune responses to influenza vaccination in the elderly using an immunostimulant patch. Vaccine 23:946–950

Garçon N, Chomez P, Van Mechelen M. (2007) GlaxoSmithKline adjuvant systems in vaccines: concepts, achievements and perspectives. Expert Rev Vaccines 6:723–739

Gerth H-J, Mok-Hsu YCh. (1981) Reactogenicity and serological response to polyvalent aqueous and Al (OH)3 adsorbed Tween-ether split product influenza vaccine in young adults 1979. Infection 9:85–90

Guy B. (2007) The perfect mix: recent progress in adjuvant research. Nat Rev Microbiol 5:505–517

Hehme N, Engelmann H, Kunzel W, Neumeier E, Sanger R. (2002) Pandemic preparedness: lessons learnt from H2N2 and H9N2 candidate vaccines. Med Microbiol Immunol 191:203–208

Hem SL, HogenEsch H. (2007) Relationship between physical and chemical properties of aluminum-containing adjuvants and immunoprecipitation. Expert Rev Vaccines 6:685–698

Hennessy AV, Davenport FM. (1974) Studies on vaccination of infants against influenza with influenza hemagglutinin. Proc Soc Exp Biol Med 146:200–204

Huleatt JW, Nakaar V, Desai P, Huang Y, Hewitt D, Jacobs A, Tang J, McDonald W, Song L, Evans RK, Umlauf S, Tussey L, Powell TJ. (2008) Potent immunogenicity and efficacy of a universal influenza vaccine candidate comprising a recombinant fusion protein linking influenza M2e to the TLR5 ligand flagellin. Vaccine 26:201–214

Ichinohe T, Tamura S, Kawaguchi A, Ninomiya A, Imai M, Itamura S, Odagiri T, Tashiro M, Takahashi H, Sawa H, Mitchell WM, Strayer DR, Carter WA, Chiba J, Kurata T, Sata T, Hasegawa H. (2007) Cross-protection against H5N1 influenza virus infection is afforded by intranasal inoculation with seasonal trivalent inactivated influenza vaccine. J Infect Dis 196:1313–1320

Keitel WA, Atmar RL. (2009) Vaccines for pandemic influenza: summary of recent clinical trials. In: Compans RW, Orenstein WA. (eds) Vaccines for pandemic influenza. Current topics in microbiology, vol 333. Springer, Heidelberg

Keitel W, Couch R, Bond N, Adair S, van Nest G, Dekker C. (1993) Pilot evaluation of influenza virus vaccine (IVV) combined with adjuvant. Vaccine 11:909–913

Keitel WA, Campbell JD, Treanor JJ, Walter EB, Patel SM, He F, Noah DL, Hill H. (2008a) Safety and immunogenicity of an inactivated influenza A/H5N1 vaccine given with or without aluminum hydroxide to healthy adults: results of a phase I-II randomized clinical trial. J Infect Dis 198:1309-1316

Keitel W, Dekker C, Mink C, Campbell J, Edwards K, Patel S. (2008b) A phase I, randomized, double-blind, placebo-controlled dose ranging clinical trial of the safety, reactogenicity and immunogenicity of immunization with inactivated Vero cell culture-derived influenza A/H5N1 vaccine given with or without aluminum hydroxide to healthy young adults. In: 11th Annual Conference on Vaccine Research, Baltimore, MD, 5–7 May 2008 (abstract)

Leroux-Roels I, Bernhard R, Gérard P, Dramé M, Hanon E, Leroux-Roels G. (2008) Broad clade 2 cross-reactive immunity induced by an adjuvanted clade 1 rH5N1 pandemic influenza vaccine. PLoS ONE 3:e1401

Leroux-Roels I, Borkowski A, Vanwolleghem T, Dramé M, Clement F, Hons E, Devaster JM, Leroux-Roels G. (2007) Antigen sparing and cross-reactive immunity with an adjuvanted rH5N1 prototype pandemic influenza vaccine: a randomised controlled trial. Lancet 370:580–589

Levie K, Leroux-Roels I, Hoppenbrouwers K, Kervyn AD, Vandermeulen C, Forgus S, Leroux-Roels G, Pichon S, Kusters I. (2008) An adjuvanted, low-dose, pandemic influenza A (H5N1) vaccine candidate is safe, immunogenic, and induces cross-reactive immune responses in healthy adults. J Infect Dis 198:642–649

Lin J, Zhang J, Dong X, Fang H, Chen J, Su N, Gao Q, Zhang Z, Liu Y, Wang Z, Yang M, Sun R, Li C, Lin S, Ji M, Liu Y, Wang X, Wood J, Feng Z, Wang Y, Yin W. (2006) Safety and immunogenicity of an inactivated adjuvanted whole-virion influenza A (H5N1) vaccine: a phase I randomised controlled trial. Lancet 368:991–997

Marciani DJ, Reynolds RC, Pathak AK, Finley-Woodman K, May RD. (2003) Fractionation, structural studies, and immunological characterization of the semi-synthetic *Quillaja* saponins derivative GPI-0100. Vaccine 21:3961–3971

Matassov D, Cupo A, Galarza JM. (2007) A novel intranasal virus-like particle (VLP) vaccine designed to protect against the pandemic 1918 influenza A virus (H1N1). Viral Immunol 20:441–452

Mbawuike I, Zang Y, Couch RB. (2007) Humoral and cell-mediated immune responses of humans to inactivated influenza vaccine with or without QS21 adjuvant. Vaccine 25:3263–3269

Moldoveanu Z, Love-Homan L, Huang WQ, Krieg AM. (1998) CpG DNA, a novel immune enhancer for systemic and mucosal immunization with influenza virus. Vaccine 16:1216–1224

Morein B, Sundquist B, Höglund S, Dalsgaard K, Osterhaus A. (1984) Iscom, a novel structure for antigenic presentation of membrane proteins from enveloped viruses. Nature. 308:457–460

Moser C, Amacker M, Kammer AR, Rasi S, Westerfeld N, Zurbriggen R. (2007) Influenza virosomes as a combined vaccine carrier and adjuvant system for prophylactic and therapeutic immunizations. Expert Rev Vaccines 6:711–721

Murasko DM, Bernstein ED, Gardner EM, Gross P, Munk G, Dran S, Abrutyn E. (2002) Role of humoral and cell-mediated immunity in protection from influenza disease after immunization of healthy elderly. Exp Gerontol 37:427–439

Mutsch M, Zhou W, Rhodes P, Bopp M, Chen RT, Linder T, Spyr C, Steffen R. (2004) Use of the inactivated intranasal influenza vaccine and the risk of Bell's palsy in Switzerland. N Engl J Med 350:896–903

Nicholson KG, Tyrrell DA, Harrison P, Potter CW, Jennings R, Clark A. (1979) Clinical studies of monovalent inactivated whole virus and subunit A/USSR/77 (H1N1) vaccine: serological responses and clinical reactions. J Biol Stand 7:123–136

Nicholson KG, Colegate AE, Podda A, Stephenson I, Wood J, Ypma E, Zambon MC. (2001) Safety and antigenicity of non-adjuvanted and MF59-adjuvanted influenza A/Duck/Singapore/97 (H5N3) vaccine: a randomised trial of two potential vaccines against H5N1 influenza. Lancet 357:1937–1943

O'Hagan DT. (2007) MF59 is a safe and potent adjuvant that enhances protection against influenza virus infection. Expert Rev Vaccines 6:699–710

O'Hagan DT, Valiante NM. (2003) Recent advances in the discovery and delivery of vaccine adjuvants. Nat Rev Drug Discov 2:727–735

Potter CW. (1982) Inactivated influenza virus vaccine. In: Beare AS (ed) Basic and applied influenza research. CRC, Boca Raton, FL, pp 119–158

Powers DC. (1997) Summary of a clinical trial with liposome-adjuvanted influenza A virus vaccine in elderly adults. Mech Ageing Dev 93:179–188

Pressler K, Peukert M, Schenk D, Borgono M. (1982) Comparison of the antigenicity and tolerance of an influenza aluminium oxide adsorbate vaccine with an aqueous vaccine. Pharmatherapeutica 3:195–200

Pushko P, Tumpey TM, Van Hoeven N, Belser JA, Robinson R, Nathan M, Smith G, Wright DC, Bright RA. (2007) Evaluation of influenza virus-like particles and Novasome adjuvant as candidate vaccine for avian influenza. Vaccine 25:4283–4290

Ramon G. (1926) Procedes pour accroitre la production des antitoxins. Ann Inst Paseur (Paris) 29:31–40

Rimmelzwaan GF, Baars M, van Beek R, van Amerongen G, Lövgren-Bengtsson K, Claas EC, Osterhaus AD. (1997) Induction of protective immunity against influenza virus in a macaque model: comparison of conventional and iscom vaccines. J Gen Virol 78:757–765

Ruat C, Caillet C, Bidaut A, Simon J, Osterhaus AD. (2008) Vaccination of macaques with adjuvanted formalin-inactivated influenza A virus (H5N1) vaccines: protection against H5N1 challenge without disease enhancement. J Virol 82:2565–2569

Sambhara S, Kurichh A, Miranda R, Tumpey T, Rowe T, Renshaw M, Arpino R, Tamane A, Kandil A, James O, Underdown B, Klein M, Katz J, Burt D. (2001) Heterosubtypic immunity against human influenza A viruses, including recently emerged avian H5 and H9 viruses, induced by FLU-ISCOM vaccine in mice requires both cytotoxic T-lymphocyte and macrophage function. Cell Immunol 211:143–153

Schultze V, D'Agosto V, Wack A, Novicki W, Zorn J, Hennig R. (2008) Safety of MF59™ adjuvant. Vaccine 26:3209–3222

Singh M, Chakrapani A, O'Hagan D. (2007) Nanoparticles and microparticles as vaccine-delivery systems. Expert Rev Vaccines 6:797–808

Somani J, Lonial S, Rosenthal H, Resnick S, Kakhniashvili I, Waller EK. (2002) A randomized, placebo-controlled trial of subcutaneous administration of GM-CSF as a vaccine adjuvant: effect on cellular and humoral immune responses. Vaccine 21:221-230

Stephenson I, Nicholson KG, Colegate A, Podda A, Wood J, Ypma E, Zambon M. (2003) Boosting immunity to influenza H5N1 with MF59-adjuvanted H5N3 A/Duck/Singapore/97 vaccine in a primed human population. Vaccine 21:1687–1693

Stephenson I, Zambon MC, Rudin A, Colegate A, Podda A, Bugarini R, Del Giudice G, Minutello A, Bonnington S, Holmgren J, Mills KH, Nicholson KG. (2006) Phase I evaluation of intranasal trivalent inactivated influenza vaccine with nontoxigenic *Escherichia coli* enterotoxin and novel biovector as mucosal adjuvants, using adult volunteers. J Virol 80:4962–4970

Takeda K, Akira S. (2003) Toll receptors and pathogen resistance. Cell Microbiol 5:143-153

Takeda K, Akira S. (2005) Toll-like receptors in innate immunity. Int Immunol 17:1–14

Treanor JJ, Campbell James D, Zangwill KM, Rowe T, Wolff M. (2006) Safety and immunogenicity of an inactivated subvirion influenza A (H5N1) vaccine. N Engl J Med 54:1343–1351

Vajo Z, Kosa L, Visontay I, Jankovics M, Jankovics I. (2007) Inactivated whole virus influenza A (H5N1) vaccine. Emerg Infect Dis 13:807–808

Vogel FR, Hem SL. (2008) Immunologic adjuvants. In: Plotkin SA, Orenstein WA, Offit PA (eds) Vaccines, 5th edn. Saunders, Philadelphia, PA, pp 59–71

Wack A, Baudner BC, Hilbert AK, Manini I, Nuti S, Tavarini S, Scheffczik H, Ugozzoli M, Singh M, Kazzaz J, Montomoli E, Del Giudice G, Rappuoli R, O'Hagan DT. (2008) Combination adjuvants for the induction of potent, long-lasting antibody and T-cell responses to influenza vaccine in mice. Vaccine 26:552–561

Wyde PR, Guzman E, Gilbert BE, Couch RB. (2001) Immunogenicity and protection in mice given inactivated influenza vaccine, MPL, QS-21 or QS-7. In: Osterhaus ADME, Cox N, Hampson AW. (eds) Options for the control of influenza IV. Excerpta Medica, New York, pp 999–1005

# Part IV
# Novel Approaches for Vaccine Delivery

# Transcutaneous Immunization with Influenza Vaccines

**Ioanna Skountzou and Sang-Moo Kang**

## Contents

**Abstract**  Transcutaneous immunization (TCI) is a novel vaccination route involving the topical application of vaccine antigens on the skin. The skin is an attractive site for vaccination because it is rich in various antigen-capturing immune cells. The outer skin barrier can be overcome through the use of mild chemical and/or physical treatments, including ethanol–water hydration and stripping, which allows for large vaccine molecules or even particulate antigens to gain access to the skin's immune cells. The use of toxin adjuvants such as cholera or heat-labile toxins was demonstrated to enhance the immunogenicity of vaccine antigens, probably due to their stimulatory effects on immune cells. Oleic acid or retinoic acid, known as permeation enhancers or immune modulators, were found to increase immune responses to

I. Skountzou (✉) and S.-M. Kang
Department of Microbiology & Immunology and Emory Vaccine Center, School of Medicine, Emory University, 1510 Clifton Rd., Atlanta, GA 30322, USA
e-mail: iskount@emory.edu

R.W. Compans and W.A. Orenstein (eds.), *Vaccines for Pandemic Influenza,*
Current Topics in Microbiology and Immunology 333,
DOI 10.1007/978-3-540-92165-3_17, © Springer-Verlag Berlin Heidelberg 2009

inactivated whole-influenza viral vaccines. The further development of more effective delivery systems and nontoxic adjuvants is needed to enhance the efficacy of this approach to vaccination.

# 1   Introduction

The delivery of drugs and vaccines with needles and syringes is a common practice in preventive and therapeutic healthcare worldwide. Vaccination is one of the most successful and cost-effective public health interventions; well over 2 million deaths are currently averted through immunization each year. However, although the development of steel needles fine enough to pierce the skin by C.G. Pravaz and Dr. A. Wood as early as the mid-nineteenth century and their subsequent use in injections has saved millions of lives (McGrew and McGrew 1985), there is concern in the scientific community that some vaccines that are designed to be delivered subcutaneously or intramuscularly by injection may not be optimally effective. An increasing number of investigators involved in vaccine development and delivery have focused on novel approaches to vaccination based on our growing knowledge of the functioning of the immune system, of newly discovered elements such as Toll-like receptor (TLR) agonists which act as immune enhancers, as well as the discovery of virus-specific host cell signaling mechanisms upon virus entry. The complexity of interactions between the pathogen and the host raises questions such as: what is the best way of vaccinating against a pathogen based on its tissue tropism? How can we avoid side effects? How can we achieve optimal protective immune responses? And where in the body do we want to see them? In addition, despite the efforts of the World Health Organization and governments, millions of children, particularly in the developing world, do not receive all routinely recommended childhood vaccines. The two main reasons for this are a lack of accessibility to high-quality immunization services and the limited resources of government health authorities who need to purchase and deliver vaccines.

There are excellent review articles (Partidos and Muller 2005; Mitragotri 2005) and WHO reports (WHO 2004) that address problems with the use of syringes, including the improper use of disposable needles and syringes (such as the abuse of needles in healthcare, their unsafe reuse without any sterilization, poor collection and disposal that expose healthcare workers and the community to the risk of needlestick injuries, and work accidents that lead to needlestick injuries). Miller and Pisani (1999) reported that each year unsafe injections cause an estimated 1.6 million early deaths. The transmission of bloodborne pathogens such as HBV, HCV, and HIV leads to disease, disability, and death a number of years after the unsafe injection, placing a huge socioeconomic toll to society (Mitragotri 2005). Twenty percent to 80% of new HBV infections are attributed to unsafe injections (Kermode 2004), and although autodisable technology can prevent the reuse of syringes and needles, it is not yet available worldwide (Giudice and Campbell 2006). Children are at a higher risk than adults of developing a chronic condition when infected because less than 10% of them show any clinical symptoms; thus the disease remains undetected and therefore untreated (Kane and Lasher 2002).

In addition to the safety concerns, a large percentage of children (20%) and a surprising number of adults (2%) suffer from distress or fear of injections (Giudice and Campbell 2006), resulting in a reluctance to receive immunizations. Many complain about pain and discomfort and some degree of reactogenicity toward particular vaccines (i.e., tetanus toxoid), which is more profound in the adult population (Moylett and Hanson 2004). This needle phobia and the discomfort suffered by both children and adults led to the development of combination vaccines, but unfortunately some combinations show a diminished response to certain antigens when coadministered.

Although at present the cost of needles and syringes used for vaccination is quite low ($0.06), the additional costs due to medical care and lost productivity resulting from iatrogenic bloodborne pathogen transmission in developing countries can increase this to as much as $26.77 per injection (Ekwueme et al. 2002). Thus, the use of alternative vaccination methods may be a more viable approach in the sense that there will be no further economic burden on the population, and there may be a cost reduction that would result in an affordable larger-scale immunization program, especially in developing countries (Ekwueme et al. 2002; Roth et al. 2003; PATH 2007).

Lastly, the shelf lives of many vaccines currently in use depend on the recommended storage temperature, which most often ranges between 2°C and 8°C. Vaccines are sensitive to extreme temperatures, and their potency depends on the use of proper storage conditions (Atkinson et al. 2002). The *cold chain* refers to the materials, equipment, and procedures required to maintain vaccines within this temperature range from the time that they are manufactured until they are given to patients (Weir and Hatch 2004). Such maintenance of vaccines costs 200–300 million dollars annually worldwide (Das 2004). All of these issues further emphasize the need to replace the injections with formulations given by other routes (see WHO 2006).

Novel vaccines delivered in the form of patches (Glenn et al. 2003a,) or jet injections (Chen et al. 2001b,c) may have longer shelf lives without the need for a cold chain, which would spare funds that could instead be used to immunize more children. These methods also have the potential for easy distribution to remote areas of the world that cannot provide the storage conditions required for products that require a cold chain, leading to a further reduction in the costs of vaccine distribution.

# 2 Skin as an Attractive Organ for Immunization

During the past 50 years, scientific breakthroughs in vaccine design and development, in immunology and in biotechnology have spurred interest in alternative methods of delivering vaccine directly to the mucosa and skin without needles and syringes. These novel vaccination routes involve the delivery of antigens to mucosal surfaces by absorption (oral, such as the oral polio vaccine developed by Sabin in the 1960s, initially as a monovalent and later as a trivalent form; intranasal and aerosol vaccines), to the epidermis by transcutaneous immunization (TCI), and to the skin dermis by scarification (smallpox vaccine), by jet injectors or other needle-free devices. The skin has only been explored for its immunogenic potential within the last decade. The route of vaccination through the skin presents several advantages, such as the rapid distribution of vaccines in large-scale epidemics or pandemics and in the case of a bioterrorism-related emergency, as well as the potential to use volunteers outside the healthcare system, in particular in developing countries, where trained personnel for conventional vaccinations and immunization campaigns is limited.

## 2.1 Skin Structure

The skin is the primary interface between the body and the environment. Due to its complex structure, the skin can function as a barrier, conferring protection against injuries, UV and invasions from foreign bodies and pathogens, but it also maintains homeostasis of the body through appropriate perspiration. Streilein (1983) described the cutaneous inductive sites for immunity in the skin as SALT (skin-associated lymphoid tissue). The skin plays a major role in the immune defenses of the host and it is part of the first line of immune defense against foreign intruders. Skin consists of three principal layers: the epidermis, the dermis, and the subcutis (Fig. 1). The epidermis is primarily a mechanical barrier against external physical, chemical, and mechanical stimuli, and it also prevents water loss and dehydration. The epidermis is 50–100 μm thick and consists of keratinized stratified squamous epithelium, and keratinocytes are the principal cell type that leads to its functional qualities. The epidermis displays several *strata* (layers), reflecting visible stages of keratinocyte differentiation and maturation in the basal to superficial direction. The upper layer is the *stratum corneum*, which is water resistant and consists of 10–20 μm thick layers of keratinocytes, consisting mostly of dead cells (corneocytes) embedded in a lipophilic matrix. As the dead cells slough off, they are continuously replaced by new cells from the stratum germinativum (basale). The *stratum lucidum* only appears in certain parts of the body, such as the palms of the hands and the soles of the feet, and also consists mainly of dead cells. The *stratum granulosum* has 1–3 layers of squamous cells and contains filaggrin, a protein that bundles keratin and prevents loss of nutrients. The *stratum spinosum* is composed of cells joined by desmosomes. They synthesize cytokeratin, which in its filamentous form anchors to desmosomes and provides adjacent cells with additional adherence, thus enhancing structural support and preventing skin abrasion. The stratum spinosum is permeable to water and is the "home" of antigen-presenting cells, which are crucial for the initiation of innate immune responses. Finally, the *stratum basale* consists of one layer of columnar epithelial cells (live keratinocytes) lying on the basement membrane, which undergo rapid mitosis to replenish the dead cells

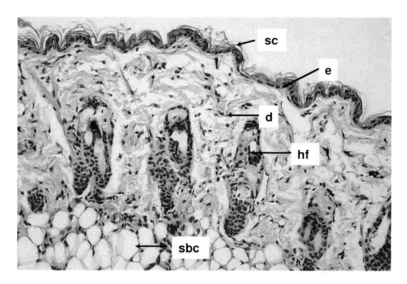

**Fig. 1** Layers of the skin. Eosin–hematoxylin stain of murine dorsal skin; transversal section. Skin samples were obtained from the midline of the anterior portion of the dorsal surface, fixed in 10% paraformaldehyde for 24 h, and further embedded in paraffin. Five-micron-thick transversal skin sections were stained with eosin and hematoxylin. The skin layers are (from outer to inner): the stratum corneum, the epidermis, the dermis, and the subcutis. The epidermis has 1–2 visible layers of viable keratinocytes, and the hair follicles are surrounded by the same cell population. The dermis is rich in fibroblasts, collagen and capillaries, and it is also home for innate immune cell populations such as macrophages and monocytes. The subcutis is rich in adipose cells. *sc*, stratum corneum; *e*, epidermis, *d*, dermis, *sb*, subcutis, *hf*, hair follicle

shed from the skin surface. The stratum basale also contains resident Langerhans cells and melanocytes.

The underlying dermis is a connective tissue layer that is more elastic than the epidermis because the cells are loosely interwoven. It harbors many sensory nerve endings that provide the senses of touch and heat, and contains hair follicles, sweat glands, sebaceous glands, apocrine glands, lymphatic vessels and blood vessels. The dermis also has an extracellular space with a jelly-like appearance, the interstitium. The interstitial cells are immune cells that can be recruited quickly to sites of antigen introduction into the skin. The subcutis or hypodermis is the subcutaneous adipose tissue.

The mouse model has been largely used for vaccine studies, over the past 20 years. It is therefore important to describe the major differences between the human and the mouse skin that may affect the interpretation of our data. The epithelium in fur-covered mouse skin is disproportionately composed of densely distributed hair follicles, whereas epithelium in human skin is much more interfollicular (Schon and Orfanos 1995). Mouse epidermis generally comprises only 2–3 keratinocyte layers, is only one quarter the thickness of human epidermis, and has a faster epidermal turnover (Berking et al. 2002). Human dermis is substantially thicker than mouse dermis and contains fewer hair follicles, and wounding is markedly different, as mouse skin regenerates effectively without significant scarring (Khavari 2006). Finally, mice have an entire cutaneous muscle layer, the panniculus carnosus, which is only present in a rudimentary form in humans as the platysma and other small muscles of the face and neck.

## 2.2 Immune Cell Populations in the Skin

Antigen-presenting cells (APCs), which reside in the skin in large numbers, include Langerhans cells (LC) and dermal dendritic cells (DDC). Langerhans cells are dendritic epidermal cells accounting for 2–8% of the total epidermal cell population and only 1% of the total skin cell population; they are in close proximity to the stratum corneum and represent a network of immune cells that underlies 25% of the skin's total surface area (Banchereau and Steinman 1998). They are immature APCs produced from bone marrow precursors that reach and populate the skin via the peripheral circulation (Romani et al. 1985). The LCs form a semicontiguous network in the epidermis, and are mostly located in the surroundings of hair follicles (Vogt et al. 2006). The underlying dermis layer is rich in mast cells, monocytes, and dendritic cells. Since there is no clear morphological distinction between DDCs and LCs, surface markers (cluster designation markers, CD, which label cell surface antigens) are used to differentiate them (Lenz et al. 1993). LCs express differential levels of CD11b, CD205$^{int/high}$ and more specifically CD207 (Langerin), whereas interstitial DCs and dermal DCs express CD11b$^{high}$ and CD205$^{low/int}$ and remain negative for CD207 (Itano et al. 2003; Valladeau et al. 2000). In humans, LCs strongly express CD1a. However, dermal DCs (DDCs) also express varying levels of CD1a, but at a lower level. In addition, LCs lack the monocyte–macrophage marker CD14 and the representative DDC marker FXIIIa (Larregina et al. 2001). LCs and DCs are also distinguished by their differential chemokine receptor expression, which has been studied on human and mouse LCs and DCs. Some of them are expressed in both species (Caux et al. 2000), and their upregulation is observed during the maturation and migration of LCs from tissues to draining lymph nodes (Sozzani et al. 2000).

Both epidermal LCs and dermal DCs are capable of stimulating primary immune responses in vivo and in vitro (Banchereau and Steinman 1998). The resting APCs express low levels of major histocompatibility molecules MHC I and II and costimulatory or adhesion molecules (Debenedictis et al. 2001). In the absence of antigen stimulation, LCs and dermal DCs are maintained locally (Merad et al. 2002), but upon encountering an antigen in the skin, they migrate via the afferent lymphatics to the draining lymph nodes (LN) and gut mucosa, where they present the processed antigenic peptides to antigen-specific T and B cells. This migration corresponds to a maturation process accompanied by the upregulation of molecules involved in antigen presentation, including MHC molecules, costimulatory molecules (B7-1, B7-2, CD40, and OX40L), and adhesion molecules (ICAM-1 and LFA-3)

(Banchereau and Steinman 1998; Norbury et al. 2002). Keratinocytes are also involved in innate and adaptive immune responses and secrete cytokines, antimicrobial peptides and chemokines (Uchi et al. 2000; Steinhoff et al. 2001) in response to various pathogens and their components. The cytokines play an important role because they affect the local microenvironment, thus maintaining epidermal home-ostasis (Williams and Kupper 1996). Although keratinocytes lack costimulatory molecules and they are not considered APCs, they constitutively express pattern-recognition receptors including TLRs on their surfaces, such as TLR1, TLR2, and TLR5, which induce keratinocyte activation by binding to various microbes and their components. These cells probably act as accessory cells for mitogen or superantigen-driven responses (Nickoloff et al. 1993) and also act as scavengers, by eliminating foreign bodies from the intercellular space of the epidermis (Wolff and Honigsmann 1971).

## 3 Approaches to Crossing the Skin Barrier for Vaccine Delivery

Because of the density, accessibility and antigen-presenting function of LCs, the skin epidermis is an ideal target for vaccine delivery. The epidermis offers an additional advantage: the scarcity of sensory nerve endings and therefore the absence of pain at the site of injection. However, the tight structure of the epidermis, mainly attributed to the desmosome-connected keratinocytes, makes it a very effective barrier against the entry of foreign intruders and hinders the entry of molecules larger than 500 kDa (Stingl and Steiner 1989; Bos and Meinardi 2000). Therefore, it is a challenge to develop effective methods to deliver vaccine antigens which circumvent the skin barrier.

There are two major categories of molecule application on the skin: (1) topical administration on the epidermis, and (2) transdermal application, which mainly delivers the molecules into the dermis.

In principle, there are three possible penetration pathways through the skin for topically applied substances: the intercellular penetration route, the intracellular penetration route (Combadiere and Mahe 2008), and the follicular penetration route (Vogt et al. 2005). Although the follicular penetration route has not been seriously considered for the process of vaccine delivery since hair is no more than 0.1% of the total skin surface area, more recent in vitro and in vivo studies suggest that follicular penetration is an important site of entry (Schaefer and Lademann 2001). Successful delivery of an antigen involves altering the skin by physical disruption and/or chemical permeation enhancers. Mild disruption of intact skin followed by the administration of molecules to the epidermis led to the activation of APCs and keratinocytes residing in that layer.

One of the first methods used to physically alter the skin was the removal of dead keratinocytes with alcohol-wetted cotton swabs. This form of very mild skin abrasion enhanced electrical conductivity when applied to lead attachment sites prior to electrocardiograms. Hydration of the stratum corneum, which can be accomplished by occluding and wetting the skin, also causes keratinocytes to swell and fluid to pool in the intercellular spaces. This allows antigens to pass through the skin more easily (Glenn et al. 2000). Other penetration-enhancing techniques involve physical disruption of the stratum corneum with tape stripping (Seo et al. 2000; Matyas et al. 2004) or emery paper (Belyakov et al. 2004). Experiments with murine models demonstrated that even shaving (Zhao et al. 2006) or the use of depilatory cream (Morel et al. 2004) causes mild abrasion of the skin and thus increases the permeability. Acetone (Kahlon et al. 2003) or other penetration enhancers can chemically alter the skin to increase its permeability. Cholera toxin works in this fashion (Glenn et al. 1998).

All of the aforementioned approaches are designed for the topical application of antigens. More recent efforts enhance penetration by modifying particle–skin interactions for transdermal delivery (Partidos 2003) and applying a driving force to deliver antigen or drugs into or through the dermis. These techniques include sonication (Tezel et al. 2005), electroporation/microporation (Zhao et al. 2006), microneedle arrays (Prausnitz 2004), jet injectors and particulate delivery systems such as liposomes, virosomes, transferosomes, nano- or microbeads and viral vectors. Some of these can enhance drug delivery through the skin up to fourfold. Charged substances, including RNA, DNA, and proteins, can enter through the epidermis with or without accompanying adjuvants in liposomes (Wood et al. 1992) or with iontophoresis. Transdermal delivery systems involve some degree of pain (Prausnitz 2004), and in some instances the rates and severity of local and systemic reactions as well as bleeding may be similar to delivery by needle and syringe.

TCI involves the application of vaccine antigen and often adjuvant to the more superficial layers of the skin, with subsequent penetration to immune cells that reside in the skin. Many TCI studies have used large immunogenic molecules such as cholera toxin (CT) (86 kDa) or *E. coli* heat-labile toxin (LT) (84 kDa) as antigens. Thus, although it was thought that the intact stratum corneum (SC) could not be penetrated by large molecules, these recent results indicate that crude patches and minor SC disruption or increased permeabilization can facilitate the entrance of very large antigens (Guerena-Burgueno et al. 2002; Skountzou et al. 2006; McKenzie et al. 2006).

# 4   Initiation of Epidermal Immune Responses by Vaccines

Breaching of the skin triggers a series of events that induce the secretion of several proinflammatory cytokines, such as IL-1α, IL-1β, TNF-α (tumor necrosis factor), and GM-CSF (granulocyte-macrophage colony-stimulating factor) by the keratinocytes (Kupper et al. 1996; Elias and Feingold 1992). In particular, IL-1β or TNF-α enhance the dissociation of LCs from the keratinocytes and the migration of LCs along with their concomitant maturation to the draining lymph nodes through the upregulation of alpha 6 integrins (Sect. 2.2) (Cumberbatch et al. 1997; Price et al. 1997), the downregulation of CCR1/CCR5 chemokine receptors and the subsequent homing of the LCs into the lymph nodes through the CCR7 receptor (Sozzani et al. 1998a,b). In addition, disruption of the skin initiates the induction of defensins. These are cationic antimicrobial peptides which are secreted by neutrophils and macrophages displaying a certain degree of adjuvanticity (Sozzani et al. 1998a,b), thus suggesting a regulatory role in the antigen-specific immune response in the epidermis (Biragyn et al. 2002). The same proinflammatory cytokines (IL-1β, TNF-α GM-CSF) which mobilize LCs from the epidermis along with IL-6 help repair damaged skin (Wood et al. 1992; Ruffini et al. 2002). On the other hand, IL-12 and IL-10 play a regulatory role in favoring either Th1 or Th2 immune responses, respectively. IL-12 induces Th1 responses by stimulating the secretion of IFN-γ (IL-12), and IL-10 suppresses them with the assistance of antigen-presenting cells (APCs) (Watford et al. 2003; Mosmann 1994). IL-10 and IL-4 secretion is triggered by skin disruption (Nickoloff and Naidu 1994; Kondo et al. 1998), and both cytokines contribute to the prevention of excessive inflammation of the skin. In addition, certain chemokines such as MIP-3, which is constitutively expressed in the skin, are upregulated after skin damage and recruit LCs (Dieu-Nosjean et al. 2000) to guarantee immunosurveillance. The mechanical disruption of the skin alone is therefore sufficient to trigger robust immune activation by T helper cells or T cytotoxic lymphocytes (CTL) (Seo et al. 2000).

## 4.1 Adjuvants

In TCI, the immune system's response to the antigens tested to date seems to be dependent on the presence of an adjuvant (Glenn et al. 1998; Scharton-Kersten et al. 2000; Chen et al. 2001a; Klimuk et al. 2004). Adjuvants are immunostimulating compounds that are used to augment the immune system's response to vaccine antigens (Klinman et al. 2004). Several studies have reported that potent adjuvants or their subunits administered as skin patches or by other methods elicit robust immune responses against antigens at the same time via skin delivery. These adjuvants were reported to enhance the systemic and mucosal antibody responses against the toxin or other coadministered antigens as well as antigen-specific T cell responses (Beignon et al. 2002; Baca-Estrada et al. 2000; Freytag and Clements 1999). The most commonly used adjuvants for TCI are the bacterial ADP-ribosylating exotoxins, including cholera toxin (CT), *Escherichia coli* heat-labile enterotoxin (LT), or mutants of those toxins (Scharton-Kersten et al. 2000). When such adjuvants are added to the site of antigen administration, they provide LCs with an activation signal to mature and become potent antigen-presenting cells (Glenn et al. 2003a,b). Coadministering CT or LT with vaccine antigens (including tetanus toxoid and influenza hemagglutinin) has been shown to elicit not only much higher antibody levels than antigens alone but also to favor CD4+ Th1 responses (Kahlon et al. 2003; Anjuere et al. 2003). New adjuvants such as CpG DNA oligonucleotides which bias CD4+ T cells toward Th1 immune responses can modulate the action of

CT and promote both Th1 and Th2 responses, whereas CT B (cholera toxin subunit B) promotes only Th1 (Klimuk et al. 2004; Dell et al. 2006). In the mid-1990s, a new class of low molecular weight synthetic antiviral compounds, imidazoquinolinamines (imiquimod/R-837), were reported to have strong immunostimulatory capacities (Johnston and Bystryn 2006). These are immune response modifiers in vitro and in vivo, and they demonstrate antiviral activity via endogenous cytokine production. Monocytes and keratinocytes are stimulated for cytokine secretion in the presence of these compounds via TLR7 receptors. Also, upon activation, LCs migrate to draining lymph nodes and further enhance T cell proliferation in response to imiquimod treatment. Imiquimod activates T and B cells and has been shown to prime cytotoxic T lymphocytes that play a major role in eliminating virus-infected cells. Further studies are needed to identify additional adjuvants that are effective in enhancing TCI.

## 4.2    Immunostimulant Patches

The impressive immune stimulatory effects of adjuvants in TCI led to the idea that skin-applied adjuvants may enhance immune responses to vaccines delivered by injection (Kahlon et al. 2003). These so-called immunostimulant patches may have safety benefits, since most adjuvants, including injected aluminum-based products, have significant side effects (Tierney et al. 2003). The combination of immunostimulant patches and jet injection could provide a needle-free means of vaccination that obviates the need for injection of adjuvant. The reactogenicity and safety of immunostimulant patches is currently under investigation. Although oral ingestion of LT causes diarrhea and intranasal administration caused facial nerve palsy (Mutsch et al. 2004), phase I trials using transcutaneous LT as both antigen and adjuvant showed that LT can be safely given and that it is generally well tolerated, although mild hypersensitivity skin reactions were observed (Glenn et al. 2007). Local rash is the most frequently observed side effect, which may be due to the particularly potent toxin.

## 4.3    Transcutaneous Immunization with Particulate Antigens

During the last decade, Glenn and coworkers have carried out pioneering studies demonstrating that TCI results in the induction of immune responses against coadministered antigens using potent adjuvant proteins such as CT, LT or its derivatives administered in the form of patches. Most of these studies determined immune responses against soluble antigens and/or strong immunostimulatory adjuvant proteins (Guerena-Burgueno et al. 2002; Godefroy et al. 2003; Glenn et al. 2007). It was generally considered that TCI with large particulate antigens would not be feasible due to the skin barrier. However El-Ghorr et al. (2000) reported that mice transcutaneously immunized with either CT plus UV-inactivated herpes simplex virus 1 (HSV-1) antigen or soluble HSV-1 antigen had elevated mucosal and serum antibody responses as well as cell-mediated immune responses. The application of all three vaccines to mouse skin resulted in the migration of LC from the epidermis. The whole inactivated HSV vaccine was considered more potent than the soluble HSV vaccine at generating a humoral immune response, whereas the efficacy of the two HSV vaccines was reversed when cell-mediated immunity was investigated. Both antigens delivered via TCI significantly reduced the size of herpetic HSV lesions, but protection was much higher in mice vaccinated with the CT plus HSV infected cell extract.

# 5 Application of Transcutaneous Immunization to Influenza Vaccines

In 2000, Chen et al. used a needle-free helium-powered PowderJect system to effectively deliver whole formalin inactivated A/Aichi/2/68 influenza virus or trivalent human influenza vaccine to the viable epidermis of Balb/c mice (Chen et al.). The same group later succeeded in coating influenza and hepatitis protein antigens onto the surfaces of gold particles, thus delivering the proteins in a particulate form to the skin of mice (Chen et al. 2001a,b,c). They reported that their approach was successful in inducing systemic and mucosal immune responses in the murine as well as in the nonhuman primate model (Chen et al. 2003).

Our group has demonstrated that the direct topical application of whole inactivated influenza virus is also capable of inducing mucosal and systemic immune responses by targeting the skin-resident LCs and DDCs (Skountzou et al. 2006). A transcutaneous patch that has preadsorbed inactivated influenza antigens is a very attractive vaccination modality for several reasons:

- *Rapid production and distribution of the vaccine.* Vaccines in the form of patches can be easily and safely distributed by pharmacies, clinics, and by mail, avoiding the creation of queues of patients in hospitals or doctors' offices, particularly during epidemics or pandemics
- *Safety and potential self-application.* Patients routinely use transcutaneous patches for other applications
- *Prolonged shelf-life and stability.* A transcutaneous patch can be manufactured similar to an ointment or cream, with stabilizing excipients to prolong the shelf life
- *Increased protective immunity.* By inducing mucosal responses, virus spread is restricted

There is a critical need to enhance the protective immunity induced by TCIs such that a low dose and preferably a single immunization can be used. Therefore, the use of adjuvants coadministered with the antigen is highly desirable in order to enhance the efficacy of immunization by targeting TLRs of skin-resident cells. Their activation triggers a cascade of cellular events that initiate the interplay between innate and adaptive immune responses. One common clinical scenario where adjuvants delivered via an immunostimulant patch could be beneficial is influenza virus vaccination in the elderly. In this population, vaccine does not fully protect recipients, whose low response rate may be attributed to the relatively compromised nature of the senescent immune system. A recent study examined immune responses in mice to an immunostimulant patch containing LT administered at the site of influenza virus vaccine injection. The patch was placed in order to target the lymph node nearest to the injection site. The patch led to enhanced antigen presentation, serum anti-influenza antibody responses, functional antibody responses, and T cell responses (Mkrtichyan et al. 2008).

We used TCI to deliver formalin-inactivated A/PR/8/34 (H1N1) influenza virus with or without coadministered adjuvant (CT) and immune modifiers (oleic acid and retinoic acid) directly onto the skin with simple pretreatment of the epidermis: trimming the dorsal surface hair of Balb/c mice and applying depilatory cream (Fig. 2). Our data demonstrated that, despite its large size (molecular weight ~250 million daltons; Compans et al. 1970), the virus was able to traverse the skin barrier. Topical application of influenza virus to the skin induced systemic humoral and mucosal immune responses as well as cytokine production, even without the use of any adjuvants, and conferred protective immunity. Coadministration with CT, a potent adjuvant for TCI, increased immune responses against the influenza virus antigen. The penetration enhancers or potential immunomodulators oleic acid (OA) and retinoic acid (RA) were investigated to determine whether pretreatment with these compounds would result in enhancement of immune responses. OA has been widely tested for the enhancement of transdermal penetration of mainly hydrophilic but also lipophilic molecules (Kim et al. 1996), and it is involved in the perturbation of the stratum corneum bilayer, generating pore or lacunae formation on the surfaces of epidermal corneocytes (Touitou et al. 2002). Biologically active metabolites of vitamin A (retinoids) have been shown to repair damaged skin (Kligman et al. 1993), modulate the immune responses (Mohty et al. 2003), and activate the keratinocytes located in the stratum corneum of the epidermis, initiating cutaneous inflammation (Varani et al. 2001b).

## 5.1   Humoral Immune Responses Induced by TCI

Whole formalin inactivated A/Aichi/2/68 influenza virus or trivalent subvirion human influenza vaccine induced elevated serum antibody responses in mice, as estimated with ELISA, and protection against homologous or heterologous challenge (Chen et al. 2000). A single dose of 5 μg of antigen delivered via epidermal immunization was able to induce significantly higher antibody responses than intramuscular or subcutaneous injection. Serum antibody responses to influenza

**Fig. 2a–c** Skin morphology after transcutaneous immunization of Balb/c mice with inactivated A/PR/8/34 influenza virus. **a** Transversal skin section stained with eosin and hematoxylin 24 h after topical application of inactivated influenza virus; the collagen of the dermis is intensely pink due to fibroblast activation. **b** Coadministration of inactivated influenza virus with trans retinoic acid, showing a similar skin picture 24 h postapplication. **c** At 72 h a diffuse infiltration of the dermis by neutrophils, lymphocytes and mast cells is accompanied by a thickening of the epidermis and the perifollicular space due to increased numbers of keratinocytes. The stratum corneum, which was removed prior to transcutaneous immunization, is completely restored after three days

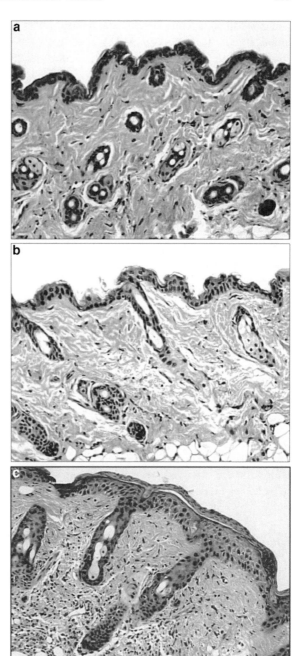

vaccine following epicutaneous immunization EPI were enhanced by codelivery of cholera toxin (CT), CpG DNA, or a combination of both. Higher humoral immune responses compared to intramuscular injection were observed in nonhuman primates in the presence of adjuvants such as QS-21 and CT (Chen et al. 2003).

In general, the antibody isotype profile reflects T cell immune responses, which are in turn influenced by the immunization procedure. Upon topical application of inactivated influenza virus to the skin epidermis, we observed primarily Th2-type responses, as determined from the IgG1/IgG2a isotype profile. Despite an increase in IgG2a responses upon the addition of CT, RA and OA, the dominant isotype was IgG1 (Table 1). The coadministration of antigen and certain adjuvants or immune modifiers to the epidermis amplifies humoral and cellular immune responses against the antigen and increases functional antibody titers, thus improving protection against live viral challenge (Skountzou et al. 2006; Yu et al. 2002; Glenn et al. 2003a,b). The enhancement of influenza-specific serum antibody responses and Th1 and Th2 cytokines observed in the presence of OA and RA suggests that these immune modulators might facilitate skin penetration of virus particles, allowing more frequent interactions with LCs or other APCs (macrophages, monocytes, DDCs), and the migration of antigen-loaded LCs to the draining lymph nodes (Touitou et al. 2002). Alternatively, since the epidermis is a site rich in keratinocytes, the latter can, upon encountering antigen and activation, in turn activate the LCs or other immune cell populations residing in the skin layers via cytokines or chemokines (Norval 2008).

There is mounting evidence that assays of functional antibodies against influenza virus are more reliable protective immunity indices than quantitative ELISA (Working Group 3 of the consultation for the development of a global action plan to increase the supply of influenza pandemic vaccines, WHO, Geneva, 2–3 May 2006). TCI of mice with inactivated A/PR/8/34 virus alone showed high hemagglutination inhibition (HAI) and neutralizing antibody titers; coadministration with CT significantly enhanced their levels, and the highest titers among the skin-immunized animals were observed when the antigen was combined with CT and oleic acid (Skountzou et al. 2006).

**Table 1** Humoral immune responses induced by transcutaneous immunization[a]

| Groups | PR8-specific IgG responses | Th2/Th1 ratio | HAI titers | Neutralization titers |
|---|---|---|---|---|
| Naive | 89 ± 25.4 | 0.8 | <10 | <5 |
| PR8 | 898.1 ± 872.5 | 66.6 | 218.7 | 135 |
| PR8/OA/RA | 2155.0 ± 733.4 | 80.4 | 160.0 | 100 |
| PR8/CT | 2211.3 ± 1029.3 | 8.9 | 266.7 | 160 |
| PR8/CT/RA | 4546.6 ± 1057.3 | 4.0 | 366.7 | 160 |
| PR8/CT/OA | 5714.2 ± 1522.6 | 9.8 | 486.7 | 160 |
| PR8 i.n. | 7570.0 ± 1465.3 | 0.5 | 403.7 | 270 |

[a] Serum humoral immune responses are induced by transcutaneous immunization (TCI) of Balb/c mice with inactivated A/PR/8/34 influenza virus. Mice were immunized with 100 μg of antigen alone or with cholera toxin (CT) or 50 μg retinoic acid (RA) or 1% v/v oleic acid (OA) in ethanol prior to antigen application. Details are provided in Skountzou et al. (2006)

## 5.2   Cellular Immune Responses Induced by TCI

Many current vaccines are known to induce robust humoral immune responses, but potent T cell immune responses against some infectious agents may also be needed. Targeting the skin lymphoid system is a smart strategy to induce potent T cell immune responses provided that an appropriate delivery method for antigens is chosen. El-Ghorr et al. (2000) demonstrated that the HSV-1 antigens extracted from infected Vero cells (CT + HSVag) and administered with TCI were a potent stimulator of cell-mediated immunity, giving rise to a strong delayed-type hypersensitivity response and lymphocyte proliferation in vitro.

Cytokine production can be used as an indicator of cellular immune responses. Since the majority of antigen-activated LCs migrate to the draining lymph node (LN), we studied the effect of TCI with inactivated influenza virus on the cytokines secreted from inguinal LN cell cultures isolated from immunized mice. We observed that TCI with A/PR/8/34 alone increased the production of mainly Th2 type cytokines such as IL-4, IL-5, and IL-6. The addition of CT amplified all types of cellular immune responses (Th1, Th2, proinflammatory), and in particular induced a tenfold increase in the level of IL-10 when compared to antigen delivery alone. Pretreatment with retinoic acid or oleic acid further elevated the proinflammatory and Th2-type cytokine levels and induced the secretion of substantial amounts of IL-10. Pretreatment with RA also increased the production of TNF-$\alpha$, which has been reported to be secreted by keratinocytes and to be implicated in the initiation of inflammation (Barker and Bowler 1991), whereas OA increased the number of IL-2 secreting cells when compared to the antigen/CT group.

TCI with influenza virus stimulated both spleen-derived CD4+ and CD8+ cells to secrete IL-4 and IFN-$\gamma$. CT enhanced the numbers of CD4+ cells secreting IL-4 in the presence of NP or HA Class II peptides and IFN-$\gamma$ secreting CD4+ cells in the presence of HA Class II peptides; OA exerted its immunomodulatory activity on IFN-$\gamma$-producing CD4+ and CD8+ cell populations, whereas RA enhanced only IFN-$\gamma$-producing CD4+ cells. The increased numbers of IFN-$\gamma$-secreting CD4+ cells may be responsible for limiting virus spread and for the efficient recoveries of infected mice. In the presence of CT, TCI activated more LN cells secreting IFN-$\gamma$ and IL-4 cytokines than intranasal immunization with inactivated influenza virus (Skountzou et al. 2006).

## 5.3   Induction of Mucosal Immune Responses
##         by Transcutaneous Immunization

It has been postulated that, upon taking up antigen, either LCs or dendritic cells which mainly reside in the dermis migrate not only to regional lymph nodes but also to mucosal lymphoid tissue, where they present processed or captured antigens to naïve T and B cells, resulting in their activation and differentiation to initiate adaptive immune responses (Enioutina et al. 2000; Yu et al. 2002). This hypothesis is supported by the observation that TCI with LT or CT induced migration of activated APCs from the skin to the proximal draining LN (Guebre-Xabier et al. 2003; El-Ghorr et al. 2000).

The induction of mucosal immune responses is very important for host defense against invading pathogens that exhibit a tropism for the mucosa. It was previously believed that only mucosal immunization at the sites of virus entry would elicit local immune responses that would block virus entry and replication. However epidermal powder immunization (EPI) elicited mucosal antibodies in addition to serum antibodies to an inactivated influenza virus vaccine (Chen et al. 2001a,b,c). Secretory immunoglobulin A (sIgA) antibodies were detected in the mucosal lavages of the small intestine, trachea, and vaginal tract (Chen et al. 2000). Other reports demonstrated similar results; TCI can induce IgG and IgA secretion from the mucosa at multiple sites, including the oral cavity, lung, gut, and the female reproductive tract (Belyakov et al. 2004; Berry et al. 2004). TCI also elicits systemic and mucosal antibody responses in humans (Varani et al. 2001a,b).

TCI with influenza virus alone or together with CT and OA or RA also induced virus-specific mucosal IgG and IgA antibodies. In saliva, mucosal antibody levels were low, but the addition of OA and RA induced significantly higher IgA responses compared to the PR8/CT group (Skountzou et al. 2006). This is a promising finding, because the replication of influenza virus in the upper respiratory tract can be greatly reduced by the local mucosal immune responses. The levels of vaginal IgG were higher than those of IgA, and in all vaccinated groups were similar to those obtained in an intranasal group, and the combined delivery of PR8 with CT and RA group exhibited the highest level of vaginal IgG. Recent studies suggest that certain mucosal surfaces have the same embryogenetic origin as the skin and therefore behave in a similar manner to it immunologically (Belyakov et al. 2004).

## 5.4   Protection of Mice from Challenge with Influenza Virus

Intranasal challenge of transcutaneously vaccinated animals with live influenza viruses demonstrated that TCI is capable of conferring long-lasting protection Virus can be rapidly cleared from the lungs of infected mice even when challenged three months after their last immunization (Skountzou et al. 2006; Chen et al. 2001a,b,c). The application of whole inactivated influenza virus alone without any adjuvant or immune modulator was sufficient to induce viral clearance and recovery. A major concern in vaccine development is the induction of strong memory responses that would help the host to fight invasion and replication of the foreign invaders. These anamnestic immune responses depend on the numbers and specificity of short or long-lived plasma cells that derive from the B-cell lymphocyte population. In the case of influenza vaccines, a robust memory response upon viral challenge depends on the influenza-specific memory B cells. Despite reports about the role of CT in TCI and in the augmentation of Th1 and Th2 responses, it seems that the presence of adjuvants does not affect memory B cell responses to influenza, since memory responses are antigen-specific. Although the coapplication of CT with the antigen does not speed recovery or virus clearance, pretreatment with immune modulators such as retinoic acid and oleic acid may provide better protection, because challenged animals showed only minimal signs of illness and more rapid clearance of the virus from their lungs compared to mice who did not receive CT.

## 6 Current Challenges and Future Directions of Transcutaneous Immunization

Needle-free immunization using the skin as a vaccination site merits increased attention for several reasons: it is safe because it minimizes the risks of pathogen transmission caused by self-injury needle accidents, it is painless and has fewer side effects than conventional immunization approaches, and it has great potential for rapid distribution to the public. In particular, TCI with a patch is easier to implement than other needle-free antigen delivery systems because it does not involve sophisticated technology, which may be expensive, limiting the use of such technology in developing countries. The patches can potentially be self-applied, which could be advantageous in remote areas with minimal or no organized health care systems. It is well accepted by patients and can quickly provide "herd" immunity. From a manufacturing point of view, TCI may enhance vaccine stability if the antigens are in a lyophilized form.

Several issues still need to be addressed before TCI can be transferred to the influenza vaccine pipeline. One very important question to answer is the optimal form of antigen to use for the influenza vaccine. The optimal vaccine formulation must be extensively investigated; along with the whole inactivated influenza virus, we should consider soluble hemagglutinin peptides and DNA vaccines. A long vaccine shelf life when the antigen is used in the form of a patch is critical. Vaccine is shipped and stored in remote places or developing countries, subjected to high temperatures and humidity. The doses that are needed for TCI to confer protection against homologous and heterologous influenza strains will have to be determined. In addition the role of adjuvants in potency will need to be investigated further. Mechanisms inducing protective mucosal and systemic immunity by TCI-delivered vaccines as well as the induction and assessment of early and late memory immune responses are critical for a successful vaccine. The prospect of delivering influenza vaccines in a transdermal patch, particularly during the "flu" season, is very appealing because it is a very practical vaccination method due to its speed of delivery and convenience. For all of these reasons, TCI against influenza virus is a very promising strategy for large-scale vaccination of populations, especially in epidemic life-threatening pandemics or outbreaks.

# References

Anjuere F, George-Chandy A, Audant F, Rousseau D, Holmgren J, Czerkinsky C (2003) Transcutaneous immunization with cholera toxin B subunit adjuvant suppresses IgE antibody responses via selective induction of Th1 immune responses. J Immunol 170:1586–1592

Atkinson WL, Pickering LK, Schwartz B, Weniger BG, Iskander JK, Watson JC (2002) General recommendations on immunization. Recommendations of the Advisory Committee on Immunization Practices (ACIP) and the American Academy of Family Physicians (AAFP). MMWR Recomm Rep 51:1–35

Baca-Estrada ME, Foldvari M, Ewen C, Badea I, Babiuk LA (2000) Effects of IL-12 on immune responses induced by transcutaneous immunization with antigens formulated in a novel lipid-based biphasic delivery system. Vaccine 18:1847–1854

Banchereau J, Steinman RM (1998) Dendritic cells and the control of immunity. Nature 392:245–252

Barker CJ, Bowler K (1991) Lipid composition of the membranes from cells of two rat tumors and its relationship to tumor thermosensitivity. Radiat Res 125:48–55

Beignon AS, Briand JP, Muller S, Partidos CD (2002) Immunization onto bare skin with synthetic peptides: immunomodulation with a CpG-containing oligodeoxynucleotide and effective priming of influenza virus-specific CD4+ T cells. Immunology 105:204–212

Belyakov IM, Hammond SA, Ahlers JD, Glenn GM, Berzofsky JA (2004) Transcutaneous immunization induces mucosal CTLs and protective immunity by migration of primed skin dendritic cells. J Clin Invest 113:998–1007

Berking C, Takemoto R, Binder RL, Hartman SM, Ruiter DJ, Gallagher PM, Lessin SR, Herlyn M (2002) Photocarcinogenesis in human adult skin grafts. Carcinogenesis 23:181–187

Berry LJ, Hickey DK, Skelding KA, Bao S, Rendina AM, Hansbro PM, Gockel CM, Beagley KW (2004) Transcutaneous immunization with combined cholera toxin and CpG adjuvant protects against *Chlamydia muridarum* genital tract infection. Infect Immun 72:1019–1028

Biragyn A, Ruffini PA, Leifer CA, Klyushnenkova E, Shakhov A, Chertov O, Shirakawa AK, Farber JM, Segal DM, Oppenheim JJ, Kwak LW (2002) Toll-like receptor 4-dependent activation of dendritic cells by beta-defensin 2. Science 298:1025–1029

Bos JD, Meinardi MM (2000) The 500 Dalton rule for the skin penetration of chemical compounds and drugs. Exp Dermatol 9:165–169

Caux C, Ait-Yahia S, Chemin K, de Bouteiller O, Dieu-Nosjean MC, Homey B, Massacrier C, Vanbervliet B, Zlotnik A, Vicari A (2000) Dendritic cell biology and regulation of dendritic cell trafficking by chemokines. Springer Semin Immunopathol 22:345–369

Chen D, Endres RL, Erickson CA, Weis KF, McGregor MW, Kawaoka Y, Payne LG (2000) Epidermal immunization by a needle-free powder delivery technology: immunogenicity of influenza vaccine and protection in mice. Nat Med 6:1187–1190

Chen D, Erickson CA, Endres RL, Periwal SB, Chu Q, Shu C, Maa YF, Payne LG (2001a) Adjuvantation of epidermal powder immunization. Vaccine 19:2908–2917

Chen D, Periwal SB, Larrivee K, Zuleger C, Erickson CA, Endres RL, Payne LG (2001b) Serum and mucosal immune responses to an inactivated influenza virus vaccine induced by epidermal powder immunization. J Virol 75:7956–7965

Chen D, Weis KF, Chu Q, Erickson C, Endres R, Lively CR, Osorio J, Payne LG (2001c) Epidermal powder immunization induces both cytotoxic T-lymphocyte and antibody responses to protein antigens of influenza and hepatitis B viruses. J Virol 75:11630–11640

Chen D, Endres R, Maa YF, Kensil CR, Whitaker-Dowling P, Trichel A, Youngner JS, Payne LG (2003) Epidermal powder immunization of mice and monkeys with an influenza vaccine. Vaccine 21:2830–2836

Combadiere B, Mahe B (2008) Particle-based vaccines for transcutaneous vaccination. Comp Immunol Microbiol Infect Dis 31(2–3):293–315

Compans RW, Klenk HD, Caliguiri LA, Choppin PW (1970) Influenza virus proteins. Analysis I of polypeptides of the virion and identification of spike glycoproteins. Virology 42:880–889

Cumberbatch M, Dearman RJ, Kimber I (1997) Langerhans cells require signals from both tumour necrosis factor-alpha and interleukin-1 beta for migration. Immunology 92:388–395

Das P (2004) Revolutionary vaccine technology breaks the cold chain. Lancet Infect Dis 4:719

Debenedictis C, Joubeh S, Zhang G, Barria M, Ghohestani RF (2001) Immune functions of the skin. Clin Dermatol 19:573–585

Dell K, Koesters R, Gissmann L (2006) Transcutaneous immunization in mice: induction of T-helper and cytotoxic T lymphocyte responses and protection against human papillomavirus-induced tumors. Int J Cancer 118:364–372

Dieu-Nosjean MC, Massacrier C, Homey B, Vanbervliet B, Pin JJ, Vicari A, Lebecque S, Dezutter-Dambuyant C, Schmitt D, Zlotnik A, Caux C (2000) Macrophage inflammatory protein 3alpha is expressed at inflamed epithelial surfaces and is the most potent chemokine known in attracting Langerhans cell precursors. J Exp Med 192:705–718

Ekwueme DU, Weniger BG, Chen RT (2002) Model-based estimates of risks of disease transmission and economic costs of seven injection devices in sub-Saharan Africa. Bull World Health Organ 80:859–870

El-Ghorr AA, Williams RM, Heap C, Norval M (2000) Transcutaneous immunisation with herpes simplex virus stimulates immunity in mice. FEMS Immunol Med Microbiol 29:255–261

Elias PM, Feingold KR (1992) Lipids and the epidermal water barrier: metabolism, regulation, and pathophysiology. Semin Dermatol 11:176–182

Enioutina EY, Visic D, Daynes RA (2000) The induction of systemic and mucosal immune responses to antigen-adjuvant compositions administered into the skin: alterations in the migratory properties of dendritic cells appears to be important for stimulating mucosal immunity. Vaccine 18:2753–2767

Freytag LC, Clements JD (1999) Bacterial toxins as mucosal adjuvants. Curr Top Microbiol Immunol 236:215–236

Giudice EL, Campbell JD (2006) Needle-free vaccine delivery. Adv Drug Deliv Rev 58:68–89

Glenn GM, Scharton-Kersten T, Vassell R, Mallett CP, Hale TL, Alving CR (1998) Transcutaneous immunization with cholera toxin protects mice against lethal mucosal toxin challenge. J Immunol 161:3211–3214

Glenn GM, Taylor DN, Li X, Frankel S, Montemarano A, Alving CR (2000) Transcutaneous immunization: a human vaccine delivery strategy using a patch. Nat Med 6:1403–1406

Glenn GM, Kenney RT, Ellingsworth LR, Frech SA, Hammond SA, Zoeteweij JP (2003a) Transcutaneous immunization and immunostimulant strategies: capitalizing on the immunocompetence of the skin. Expert Rev Vaccines 2:253–267

Glenn GM, Kenney RT, Hammond SA, Ellingsworth LR (2003b) Transcutaneous immunization and immunostimulant strategies. Immunol Allergy Clin North Am 23:787–813

Glenn GM, Villar CP, Flyer DC, Bourgeois AL, McKenzie R, Lavker RM, Frech SA (2007) Safety and immunogenicity of an enterotoxigenic Escherichia coli vaccine patch containing heat-labile toxin: use of skin pretreatment to disrupt the stratum corneum. Infect Immun 75:2163–2170

Godefroy S, Goestch L, Plotnicky H-Gilquin, Nguyen TN, Schmitt D, Staquet MJ, Corvaia N (2003) Immunization onto shaved skin with a bacterial enterotoxin adjuvant protects mice against respiratory syncytial virus (RSV). Vaccine 21:1665–1671

Guebre-Xabier M, Hammond SA, Epperson DE, Yu J, Ellingsworth L, Glenn GM (2003) Immunostimulant patch containing heat-labile enterotoxin from Escherichia coli enhances immune responses to injected influenza virus vaccine through activation of skin dendritic cells. J Virol 77:5218–5225

Guerena-Burgueno F, Hall ER, Taylor DN, Cassels FJ, Scott DA, Wolf MK, Roberts ZJ, Nesterova GV, Alving CR, Glenn GM (2002) Safety and immunogenicity of a prototype enterotoxigenic Escherichia coli vaccine administered transcutaneously. Infect Immun 70:1874–1880

Itano AA, McSorley SJ, Reinhardt RL, Ehst BD, Ingulli E, Rudensky AY, Jenkins MK (2003) Distinct dendritic cell populations sequentially present antigen to CD4 T cells and stimulate different aspects of cell-mediated immunity. Immunity 19:47–57

Johnston D, Bystryn JC (2006) Topical imiquimod is a potent adjuvant to a weakly-immunogenic protein prototype vaccine. Vaccine 24:1958–1965

Kahlon R, Hu Y, Orteu CH, Kifayet A, Trudeau JD, Tan R, Dutz JP (2003) Optimization of epicutaneous immunization for the induction of CTL. Vaccine 21:2890–2899

Kane M, Lasher H (2002) The case for childhood immunization (Occasional Paper 5). Children's Vaccine Program at PATH, Seattle (see http://www.path.org/vaccineresources/files/CVP_Occ_Paper5.pdf)

Kermode M (2004) Unsafe injections in low-income country health settings: need for injection safety promotion to prevent the spread of blood-borne viruses. Health Promot Int 19:95–103

Khavari PA (2006) Modelling cancer in human skin tissue. Nat Rev Cancer 6:270–280

Kim SS, Chen YM, O'Leary E, Witzgall R, Vidal M, Bonventre JV (1996) A novel member of the RING finger family, KRIP-1, associates with the KRAB-A transcriptional repressor domain of zinc finger proteins. Proc Natl Acad Sci USA 93:15299–15304

Kligman AM, Dogadkina D, Lavker RM (1993) Effects of topical tretinoin on non-sun-exposed protected skin of the elderly. J Am Acad Dermatol 29:25–33

Klimuk SK, Najar HM, Semple SC, Aslanian S, Dutz JP (2004) Epicutaneous application of CpG oligodeoxynucleotides with peptide or protein antigen promotes the generation of CTL. J Invest Dermatol 122:1042–1049

Klinman DM, Currie D, Gursel I, Verthelyi D (2004) Use of CpG oligodeoxynucleotides as immune adjuvants. Immunol Rev 199:201–216

Kondo H, Ichikawa Y, Imokawa G (1998) Percutaneous sensitization with allergens through barrier-disrupted skin elicits a Th2-dominant cytokine response. Eur J Immunol 28:769–779

Kupper T, Harasani X, Trakanvanich V, Pfitzer P (1996) Cytomorphologic findings in fine needle aspiration specimens of eosinophilic renal cell carcinoma. Gen Diagn Pathol 141:249–253

Larregina AT, Morelli AE, Spencer LA, Logar AJ, Watkins SC, Thomson AW, Falo LD Jr, (2001) Dermal-resident CD14+ cells differentiate into Langerhans cells. Nat Immunol 2:1151–1158

Lenz A, Heine M, Schuler G, Romani N (1993) Human and murine dermis contain dendritic cells. Isolation by means of a novel method and phenotypical and functional characterization. J Clin Invest 92:2587–2596

Matyas GR, Friedlander AM, Glenn GM, Little S, Yu J, Alving CR (2004) Needle-free skin patch vaccination method for anthrax. Infect Immun 72:1181–1183

McGrew R, McGrew M (1985) Encyclopedia of medical history. McGraw Hill, New York

McKenzie R, Bourgeois AL, Engstrom F, Hall E, Chang HS, Gomes JG, Kyle JL, Cassels F, Turner AK, Randall R, Darsley M, Lee C, Bedford P, Shimko J, Sack DA (2006) Comparative safety and immunogenicity of two attenuated enterotoxigenic *Escherichia coli* vaccine strains in healthy adults. Infect Immun 74:994–1000

Merad M, Manz MG, Karsunky H, Wagers A, Peters W, Charo I, Weissman IL, Cyster JG, Engleman EG (2002) Langerhans cells renew in the skin throughout life under steady-state conditions. Nat Immunol 3:1135–1141

Miller MA, Pisani E (1999) The cost of unsafe injections. Bull World Health Organ 77:808–811

Mitragotri S (2005) Immunization without needles. Nat Rev Immunol 5:905–916

Mohty M, Morbelli S, Isnardon D, Sainty D, Arnoulet C, Gaugler B, Olive D (2003) All-trans retinoic acid skews monocyte differentiation into interleukin-12-secreting dendritic-like cells. Br J Haematol 122(5):829–836

Mkrtichyan M, Ghochikyan A, Movsesyan N, Karapetyan A, Begoyan G, Yu J, Glenn GM, Ross TM, Agadjanyan MG, Cribbs DH (2008) Immunostimulant adjuvant patch enhances humoral and cellular immune responses to DNA immunization. DNA Cell Biol 27(1):19–24

Morel PA, Falkner D, Plowey J, Larregina AT, Falo LD (2004) DNA immunisation: altering the cellular localisation of expressed protein and the immunisation route allows manipulation of the immune response. Vaccine 22:447–456

Mosmann TR (1994) Properties and functions of interleukin-10. Adv Immunol 56:1–26

Moylett EH, Hanson IC (2004) Mechanistic actions of the risks and adverse events associated with vaccine administration. J Allergy Clin Immunol 114:1010–1020; quiz 1021

Mutsch M, Zhou W, Rhodes P, Bopp M, Chen RT, Linder T, Spyr C, Steffen R (2004) Use of the inactivated intranasal influenza vaccine and the risk of Bell's palsy in Switzerland. N Engl J Med 350:896–903

Nickoloff BJ, Naidu Y (1994) Perturbation of epidermal barrier function correlates with initiation of cytokine cascade in human skin. J Am Acad Dermatol 30:535–546

Nickoloff BJ, Mitra RS, Green J, Zheng XG, Shimizu Y, Thompson C, Turka LA (1993) Accessory cell function of keratinocytes for superantigens. Dependence on lymphocyte function-associated antigen-1/intercellular adhesion molecule-1 interaction. J Immunol 150:2148–2159

Norbury CC, Malide D, Gibbs JS, Bennink JR, Yewdell JW (2002) Visualizing priming of virus-specific CD8+ T cells by infected dendritic cells in vivo. Nat Immunol 3:265–271

Norval M (2008) The photobiology of human skin. In: Photobiology: the science of life and light, 2nd edn. Springer, New York, pp 553–576

Partidos CD (2003) Delivering vaccines into the skin without needles and syringes. Expert Rev Vaccines 2:753–761

Partidos CD, Muller S (2005) Decision-making at the surface of the intact or barrier disrupted skin: potential applications for vaccination or therapy. Cell Mol Life Sci 62:1418–1424

PATH (2007) Influenza vaccine strategies for broad global access: key findings and project methodology. PATH, Seattle (see http://www.path.org/vaccineresources/details.php?i=611)

Prausnitz MR (2004) Microneedles for transdermal drug delivery. Adv Drug Deliv Rev 56:581–587

Price AA, Cumberbatch M, Kimber I, Ager A (1997) Alpha 6 integrins are required for Langerhans cell migration from the epidermis. J Exp Med 186:1725–1735

Romani N, Tschachler E, Schuler G, Aberer W, Ceredig R, Elbe A, Wolff K, Fritsch PO, Stingl G (1985) Morphological and phenotypical characterization of bone marrow-derived dendritic Thy-1-positive epidermal cells of the mouse. J Invest Dermatol 85:91s–95s

Roth Y, Chapnik JS, Cole P (2003) Feasibility of aerosol vaccination in humans. Ann Otol Rhinol Laryngol 112:264–270

Ruffini PA, Biragyn A, Kwak LW (2002) Recent advances in multiple myeloma immunotherapy. Biomed Pharmacother 56:129–132

Schaefer H, Lademann J (2001) The role of follicular penetration. A differential view. Skin Pharmacol Appl Skin Physiol 14(suppl 1):23–27

Scharton-Kersten T, Yu J, Vassell R, O'Hagan D, Alving CR, Glenn GM (2000) Transcutaneous immunization with bacterial ADP-ribosylating exotoxins, subunits, and unrelated adjuvants. Infect Immun 68:5306–5313

Schon MP, Orfanos CE (1995) Transformation of human keratinocytes is characterized by quantitative and qualitative alterations of the T-16 antigen (Trop-2, MOv-16). Int J Cancer 60:88–92

Seo N, Tokura Y, Nishijima T, Hashizume H, Furukawa F, Takigawa M (2000) Percutaneous peptide immunization via corneum barrier-disrupted murine skin for experimental tumor immunoprophylaxis. Proc Natl Acad Sci U S A 97:371–376

Skountzou I, Quan FS, Jacob J, Compans RW, Kang SM (2006) Transcutaneous immunization with inactivated influenza virus induces protective immune responses. Vaccine 24:6110–6119

Sozzani S, Allavena P, D'Amico G, Luini W, Bianchi G, Kataura M, Imai T, Yoshie O, Bonecchi R, Mantovani A (1998a) Differential regulation of chemokine receptors during dendritic cell maturation: a model for their trafficking properties. J Immunol 161:1083–1086

Sozzani S, Bonecchi R, D'Amico G, Luini W, Bernasconi S, Allavena P, Mantovani A (1998b) Old and new chemokines. Pharmacological regulation of chemokine production and receptor expression: mini-review. J Chemother 10:142–145

Sozzani S, Allavena P, Vecchi A, Mantovani A (2000) Chemokines and dendritic cell traffic. J Clin Immunol 20:151–160

Steinhoff M, Brzoska T, Luger TA (2001) Keratinocytes in epidermal immune responses. Curr Opin Allergy Clin Immunol 1:469–476

Stingl G, Steiner G (1989) Immunological host defense of the skin. Curr Probl Dermatol 18:22–30

Streilein JW (1983) Skin-associated lymphoid tissues (SALT): origins and functions. J Invest Dermatol 80(suppl):12s–16s

Tezel A, Paliwal S, Shen Z, Mitragotri S (2005) Low-frequency ultrasound as a transcutaneous immunization adjuvant. Vaccine 23:3800–3807

Tierney R, Beignon AS, Rappuoli R, Muller S, Sesardic D, Partidos CD (2003) Transcutaneous immunization with tetanus toxoid and mutants of *Escherichia coli* heat-labile enterotoxin as adjuvants elicits strong protective antibody responses. J Infect Dis 188:753–758

Touitou E, Godin B, Karl Y, Bujanover S, Becker Y (2002) Oleic acid, a skin penetration enhancer, affects Langerhans cells and corneocytes. J Control Release 80:1–7

Uchi H, Terao H, Koga T, Furue M (2000) Cytokines and chemokines in the epidermis. J Dermatol Sci 24(suppl 1):S29–S38

Valladeau J, Ravel O, Dezutter-Dambuyant C, Moore K, Kleijmeer M, Liu Y, Duvert-Frances V, Vincent C, Schmitt D, Davoust J, Caux C, Lebecque S, Saeland S (2000) Langerin, a novel C-type lectin specific to Langerhans cells, is an endocytic receptor that induces the formation of Birbeck granules. Immunity 12:71–81

Varani J, Zeigler M, Dame MK, Kang S, Fisher GJ, Voorhees JJ, Stoll SW, Elder JT (2001b) Heparin-binding epidermal-growth-factor-like growth factor activation of keratinocyte ErbB receptors mediates epidermal hyperplasia, a prominent side-effect of retinoid therapy. J Invest Dermatol 117:1335–1341

Vogt A, Mandt N, Lademann J, Schaefer H, Blume-Peytavi U (2005) Follicular targeting—a promising tool in selective dermatotherapy. J Investig Dermatol Symp Proc 10:252–255

Vogt A, Combadiere B, Hadam S, Stieler KM, Lademann J, Schaefer H, Autran B, Sterry W, Blume-Peytavi U (2006) 40 nm, but not 750 or 1,500 nm, nanoparticles enter epidermal CD1a+ cells after transcutaneous application on human skin. J Invest Dermatol 126:1316–1322

Watford WT, Moriguchi M, Morinobu A, O'Shea JJ (2003) The biology of IL-12: coordinating innate and adaptive immune responses. Cytokine Growth Factor Rev 14:361–368

Weir E, Hatch K (2004) Preventing cold chain failure: vaccine storage and handling. CMAJ 171:1050

Williams IR, Kupper TS (1996) Immunity at the surface: homeostatic mechanisms of the skin immune system. Life Sci 58:1485–1507

Wolff K, Honigsmann H (1971) Permeability of the epidermis and the phagocytic activity of keratinocytes. Ultrastructural studies with thorotrast as a marker. J Ultrastruct Res 36:176–190

Wood LC, Jackson SM, Elias PM, Grunfeld C, Feingold KR (1992) Cutaneous barrier perturbation stimulates cytokine production in the epidermis of mice. J Clin Invest 90:482–487

WHO (2004) Safety of infections: global facts and figures (Factsheet 275). World Health Organization, Geneva (see http://www.path.org/vaccineresources/details.php?i=275)

WHO (2006) Injection safety (Factsheet 231). World Health Organization, Geneva (http://whqlibdoc. who.int/fact_sheet/2006/FS_231.pdf)

Yu J, Cassels F, Scharton-Kersten T, Hammond SA, Hartman A, Angov E, Corthesy B, Alving C, Glenn G (2002) Transcutaneous immunization using colonization factor and heat-labile enterotoxin induces correlates of protective immunity for enterotoxigenic *Escherichia coli*. Infect Immun 70:1056–1068

Zhao YL, Murthy SN, Manjili MH, Guan LJ, Sen A, Hui SW (2006) Induction of cytotoxic T-lymphocytes by electroporation-enhanced needle-free skin immunization. Vaccine 24:1282–1290

# Microneedle-Based Vaccines

**Mark R. Prausnitz, John A. Mikszta, Michel Cormier, and Alexander K. Andrianov**

## Contents

**Abstract**  The threat of pandemic influenza and other public health needs motivate the development of better vaccine delivery systems. To address this need, microneedles have been developed as micron-scale needles fabricated using low-cost

M.R. Prausnitz (✉)
School of Chemical and Biomolecular Engineering, Georgia Institute
of Technology, 311 Ferst Drive, Atlanta, GA 30332-0100, USA
e-mail: prausnitz@gatech.edu

J.A. Mikszta
BD Technologies, 21 Davis Drive, Research Triangle Park, NC, 27709, USA

M. Cormier
KMG Pharma LLC, 278 Andsbury Avenue, Mountain View, CA 94043, USA

A.K. Andrianov
Apogee Technology, 129 Morgan Drive, Norwood, MA 02062, USA

R.W. Compans and W.A. Orenstein (eds.), *Vaccines for Pandemic Influenza*,
Current Topics in Microbiology and Immunology 333,
DOI 10.1007/978-3-540-92165-3_18, © Springer-Verlag Berlin Heidelberg 2009

manufacturing methods that administer vaccine into the skin using a simple device that may be suitable for self-administration. Delivery using solid or hollow microneedles can be accomplished by (1) piercing the skin and then applying a vaccine formulation or patch onto the permeabilized skin, (2) coating or encapsulating vaccine onto or within microneedles for rapid, or delayed, dissolution and release in the skin, and (3) injection into the skin using a modified syringe or pump. Extensive clinical experience with smallpox, TB, and other vaccines has shown that vaccine delivery into the skin using conventional intradermal injection is generally safe and effective and often elicits the same immune responses at lower doses compared to intramuscular injection. Animal experiments using microneedles have shown similar benefits. Microneedles have been used to deliver whole, inactivated virus; trivalent split antigen vaccines; and DNA plasmids encoding the influenza hemagglutinin to rodents, and strong antibody responses were elicited. In addition, ChimeriVax™-JE against yellow fever was administered to nonhuman primates by microneedles and generated protective levels of neutralizing antibodies that were more than seven times greater than those obtained with subcutaneous delivery; DNA plasmids encoding hepatitis B surface antigen were administered to mice and antibody and T cell responses at least as strong as hypodermic injections were generated; recombinant protective antigen of *Bacillus anthracis* was administered to rabbits and provided complete protection from lethal aerosol anthrax spore challenge at a lower dose than intramuscular injection; and DNA plasmids encoding four vaccinia virus genes administered to mice in combination with electroporation generated neutralizing antibodies that apparently included both Th1 and Th2 responses. Dose sparing with microneedles was specifically studied in mice with the model vaccine ovalbumin. At low dose (1 µg), specific antibody titers from microneedles were one order of magnitude greater than subcutaneous injection and two orders of magnitude greater than intramuscular injection. At higher doses, antibody responses increased for all delivery methods. At the highest levels (20–80 µg), the route of administration had no significant effect on the immune response. Concerning safety, no infections or other serious adverse events have been observed in well over 1,000 microneedle insertions in human and animal subjects. Bleeding generally does not occur for short microneedles (<1 mm). Highly localized, mild, and transient erythema is often observed. Microneedle pain has been reported as nonexistent to mild, and always much less than a hypodermic needle control. Overall, these studies suggest that microneedles may provide a safe and effective method of delivering vaccines with the possible added attributes of requiring lower vaccine doses, permitting low-cost manufacturing, and enabling simple distribution and administration.

# 1 Introduction

The threat of a human influenza pandemic has greatly increased in recent years with the emergence of highly virulent avian influenza viruses (Fauci 2006). Influenza experts agree that another influenza pandemic is inevitable and may be imminent (Webby and Webster 2003). Vaccines are a critical component of pandemic influenza preparedness, yet the development and supply of such vaccines can be limited by a number of challenges, including inadequate production capabilities.

Preliminary findings have identified the H2, H5, H6, H7, and H9 subtypes of influenza A as those most likely transmitted to humans (Webby and Webster 2003). The current widespread circulation of H5N1 viruses among avian populations creates an unprecedented opportunity to prepare for the next pandemic threat. However, the major difficulty in the development of effective vaccines based on these viruses stems from the fact that, for unknown reasons, HA proteins of avian subtypes of influenza A viruses are not as immunogenic as those of human subtypes. Thus, to achieve the protective level of immunity, the amount of HA in pandemic vaccines needs to be increased and is likely to exceed the 15 μg present in the currently used interpandemic vaccines (Subbarao et al. 2006). Such an increase in the dose will put additional strain on the manufacturing capacity, potentially leading to a decrease in the availability of vaccine in the event of pandemic.

The need to develop technologies that would result in a reduction of the antigen dose required to elicit protective antibody titers is evident. Exploration of such dose-sparing technologies includes the development of effective adjuvants and the use of alternative routes of vaccine administration, such as intradermal. Intradermal injection of influenza vaccine could be a highly desirable antigen-sparing strategy. Conventional injection of vaccine bypasses the skin's immune system and delivers the antigen into the muscle or subcutaneous tissue, where there is no appreciable resident population of antigen-presenting cells. Alternatively, delivery of antigen to the skin, an anatomic space that contains a large number of epidermal Langerhans cells and dermal dendritic cells, has the potential for greater immunogenicity. Dendritic cells are thought to induce cell-mediated immune responses. However, they have also been shown to enhance antibody production by B cells, which is especially important for vaccines against influenza (La Montagne and Fauci 2004). Recent studies in young adults demonstrated that intradermal administration of one-fifth the standard intramuscular dose of influenza vaccine elicited immunogenicity that was similar to or better than that elicited by intramuscular injection (Belshe et al. 2004; Kenney et al. 2004).

Intradermal immunization, though, is facing technical challenges that must be addressed in order to effectively administer such vaccines. The approach requires either special training of personnel (La Montagne and Fauci 2004), which can be difficult to achieve in clinical environments, or the development of technologies that do not involve the use of conventional needles. To overcome the skin's stratum corneum barrier and increase skin permeability, various alternative approaches have been explored, which include both chemical and physical techniques.

The use of microneedles, submillimeter structures designed to pierce the skin and deliver vaccines or drugs in the epidermis or dermis compartments, is an especially attractive option for intradermal delivery (Prausnitz et al. 2005). Such microneedles are typically constructed as a combination of vaccine or drug formulation with a supporting material, such as metal or polymer, which provides the required mechanical strength. Solid-state vaccine formulations can take the form of coatings or they may even constitute the entire microneedle. These formulations are designed to dissolve or degrade once inserted in the skin. An alternative approach utilizes hollow microneedles through which the liquid formulation can be infused or injected into the skin. Regardless of the design, the geometry of microneedles is modulated to enable targeted delivery of the antigen to the skin layer rich in Langerhans cells, and to provide the basis for a significant antigen-sparing potential.

The advancement of microneedles for the delivery of vaccines against pandemic influenza can include other important benefits. First, the use of microneedles containing solid-state vaccine delivery formulations can result in significantly improved shelf lives of such delivery systems due to the inherently better stability of solid-state protein formulations compared to their conventional solution counterparts. As a result, more efficient distributions and stockpiling capabilities, much needed for vaccines against pandemic influenza, are anticipated. Second, microneedle arrays can potentially be self-administered and safely disposed of, which can be critical in the event of a shortage of medical personnel.

The current status of microneedle technology and its potential role in the development of effective vaccines against pandemic influenza are discussed in this chapter.

## 2  Microneedle Designs and Delivery Concepts

Four different microneedle designs have been developed for minimally invasive delivery of vaccines and other pharmaceutical compounds to the skin, as discussed in previous review articles (Birchall 2006; Cormier and Daddona 2003; Coulman et al. 2006a; McAllister et al. 2000, Prausnitz 2004, 2005; Prausnitz et al. 2003, 2005, 2008; Reed and Lye 2004; Sivamani et al. 2007). Most microneedle designs have been realized using fabrication tools adapted from the microelectronics industry or other established techniques that lend themselves to inexpensive mass production as a single-use, disposable device. Microneedle materials are generally metals and polymers that are already FDA-approved for implantation or parenteral delivery for other applications.

The first design developed involves arrays of hundreds of microneedles protruding a few hundred microns from a base substrate, which are used either to pierce or to scrape microscopic holes in the skin's outer layer of stratum corneum, which is just 10–20 μm thick but provides the skin's dominant barrier to percutaneous absorption (Fig. 1) (Bronaugh and Maibach 2005; Prausnitz and Langer 2008). By piercing the skin, transdermal permeability has been increased by as much as four orders of magnitude, and this approach has been shown to deliver compounds including proteins (e.g., bovine serum albumin), genetic material (e.g., oligonucleotides and plasmid DNA) and latex particles of viral dimensions in vitro and in vivo (Birchall et al. 2005; Chabri et al. 2004; Coulman et al. 2006b; Henry et al. 1998; Lin et al.

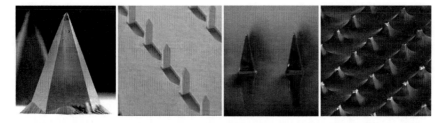

**Fig. 1** Solid microneedles used either to pierce or to scrape microscopic holes in the skin. *Shown from left to right:* platinum-coated silicon microneedle measuring 170 µm in height (image courtesy of James Birchall, Cardiff University); metal microneedles measuring 700 µm in height (image courtesy of Harvinder Gill, Georgia Institute of Technology); dissolving polymer microneedles measuring 650 µm in height (image courtesy of Sean Sullivan, Georgia Institute of Technology); blunt-tip polymer microneedles measuring 150 µm in height to scrape the skin (image courtesy of John Mikszta, BD Technologies)

2001; Martanto et al. 2004; McAllister et al. 2003; Park et al. 2005; Pearton et al. 2008; Teo et al. 2005; Verbaan et al. 2007; Wu et al. 2006, 2007). By scraping the skin, protein (e.g., the recombinant protective antigen of *Bacillus anthracis*), plasmid DNA (e.g., encoding hepatitis B surface antigen) and live attenuated virus (e.g., viral vector encoding Japanese encephalitis antigens) vaccines have been delivered to animal models, as discussed below (Dean et al. 2005; Mikszta et al. 2002, 2005). These methods make micron-scale holes in the skin, which are much larger than the size of the subunit as well as viral vaccines, but should nonetheless be small enough to avoid safety concerns, which is consistent with the lack of pain or complications observed in studies of human subjects and animals (Gardeniers et al. 2003; Gill et al. 2008; Kaushik et al. 2001; Mikszta et al. 2002).

Solid microneedles can also be prepared with dry vaccine coatings (Fig. 2). These coatings can be applied using gentle conditions at room temperature with aqueous solvents and excipients approved for parenteral delivery (Gill and Prausnitz 2007a,b). Although shown to be stable during storage for up to months, these coatings can dissolve from microneedles within the skin on a time scale of seconds. Using this approach, various model compounds, including proteins, DNA, and viruses, have been coated and delivered to the skin in vitro; a model vaccine, ovalbumin (OVA), has also been delivered to animals in vivo, as discussed below (Cormier et al. 2004, Gill and Prausnitz 2007a,b, 2008; Hooper et al. 2007; Matriano et al. 2002; Shirkhanzadeh 2005; Widera et al. 2006; Xie et al. 2005).

In contrast to coated microneedles, which apply a vaccine-encapsulated polymer coating onto a metal microneedle shaft, microneedles have also been prepared completely out of polymer with encapsulated vaccine (Fig. 3). By optimizing the design, these polymer microneedles can be made strong enough to insert into the skin. By using polymers that safely degrade or dissolve in the skin, microneedles can be inserted into the skin and left in place for a few minutes, after which the needles

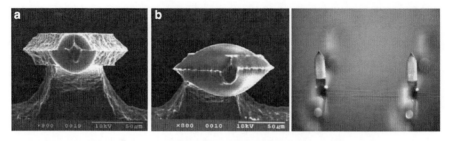

**Fig. 2a–b** Solid microneedles coated with model compounds. *Shown from left to right:* metal microneedles measuring 225 μm in height coated with approximately (**a**) 1.4 ng and (**b**) 19 ng of ovalbumin and viewed from above, looking from their tips down their shafts (images courtesy of Michel Cormier, Alza Corporation); metal microneedles measuring 700 μm in height, each coated with approximately 2 μg vitamin B (image courtesy of Harvinder Gill, Georgia Institute of Technology)

**Fig. 3** Dissolving or degrading polymer microneedles that encapsulate model compounds. *Shown from left to right:* dissolving polymer microneedles measuring 600 μm in height encapsulating sulforhodamine (image courtesy of Jeong-Woo Lee, Georgia Institute of Technology); biodegradable polymer microneedles measuring 600 μm in height encapsulating calcein (image courtesy of Jung-Hwan Park, Georgia Institute of Technology); array of biodegradable polymer microneedles held between two fingers (image courtesy of Gary Meek, Georgia Institute of Technology)

and their vaccine payload have dissolved in the skin and only the device backing remains to be discarded (i.e., without biohazardous sharps). Loadings of up to 2 μg per microneedle have been demonstrated, with dissolution efficiencies of more than 90% within the skin and little residue on the skin's surface (Gill and Prausnitz 2007a; Widera et al. 2006). This concept has been demonstrated in vitro and in vivo for the delivery of insulin, erythropoietin, and other model compounds (Ito et al. 2006a,b; Lee et al. 2008; Miyano et al. 2005; Park et al. 2006; Sullivan et al. 2008).

A final approach to vaccine delivery involves hollow microneedles (Fig. 4). In this case, one or more hollow needles are used to flow a liquid formulation into

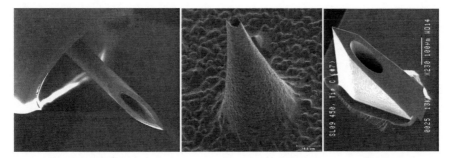

**Fig. 4** Hollow microneedles for injection into the skin. *Shown from left to right:* metal hypodermic needle protruding 1.5 mm from a specially designed hub for intradermal delivery (image courtesy of John Mikszta, BD Technologies); metal microneedle measuring 150 μm in height (image courtesy of Devin McAllister, Georgia Institute of Technology); silicon microneedle measuring 200 μm in height (image courtesy of Yotam Levin, NanoPass Technologies)

the skin. Submillimeter needles have been fabricated using microfabrication techniques, and 30–34 gauge (Ga) hypodermic needles measuring 1.0–1.5 mm in length have been produced by scaling down conventional needle manufacturing methods. A recent study examining skin thickness in humans has shown that a skin penetration depth of 1.5 mm is appropriate for intradermal delivery, irrespective of gender, age, ethnicity, or body mass index (Laurent et al. 2007). Hollow microneedles have been shown to infuse or inject a variety of compounds, including insulin, into the skin in vitro in animals and in humans, and have been used to deliver influenza and anthrax vaccines to animal models, as discussed below (Alarcon et al. 2007; Davis et al. 2005; Dean et al. 2005; Gardeniers et al. 2003; Martanto et al. 2006a,b; McAllister et al. 2003; Mikszta et al. 2005, 2006; Nordquist et al. 2007; Sivamani et al. 2005; Teo et al. 2005; Wang et al. 2006).

Selection of the best microneedle design depends on the specific application and on balancing the trade offs. Hollow microneedles offer the ability to deliver liquid vaccines without reformulation, but involve the added cost and size associated with coupling to an injection device (e.g., a syringe), the potential need for trained personnel for administration, and the need to either stabilize a liquid formulation during storage or the need to reconstitute the vaccine before injection. Coated microneedles can be prepared as a Band Aid-sized device that can probably be administered with little or no training using a solid formulation that may have good stability during storage, but they require specific vaccine formulation work and may require a device for reliable insertion into the skin. Encapsulated polymer microneedles have similar trade offs but the added advantage that needles that dissolve in the skin without producing medical sharps waste are employed. Finally, pretreating the skin with microneedles before applying, for example, a vaccine patch, while simple in principle, is associated with a potential risk of user error due to the two-step process.

# 3 Vaccination via the Skin

Microneedles are designed to facilitate intradermal delivery of vaccines, which is generally difficult to perform in a highly reproducible way in large populations. The notion of administering vaccines via the skin has existed for many centuries. The ancient Chinese practiced the art later known as "variolation," in which variola virus (smallpox virus) extracted from infected patients was scratched into the skin of healthy people (Ellner 1998). The intent was to deliberately induce smallpox infection, hopefully mild, which would protect against potentially more severe natural infection. While effective in many cases, the practice was associated with significant mortality. Later, the term "vaccination" was coined as a result of Edward Jenner's demonstration that scratching the related but less virulent vaccinia virus (cowpox virus) into the skin could effectively prevent smallpox virus infection (Ellner 1998). Even today, smallpox vaccination is accomplished by administering vaccinia virus to the skin using a bifurcated needle. What was not known by the ancient Chinese or by Edward Jenner and his followers is the fact that the skin is a very robust immune activating tissue due, in part, to the large concentration of potent antigen-presenting cells in the skin, notably the epidermal Langerhans cells and dermal dendritic cells (Huang 2007; Larregina and Falo 2005).

In addition to smallpox, a number of other vaccines have been administered to the skin using conventional needles and syringes according to the so-called "Mantoux technique" (Weniger and Papania 2008). This method is accomplished by inserting a standard 26 or 27 Ga needle into the skin at a very shallow angle with the bevel up. The needle is very carefully inserted just far enough into the skin to completely cover the bevel, and then 0.1 ml of fluid is injected into the skin, resulting in the formation of a raised wheal. This technique requires extensive training and is difficult to accomplish reproducibly. Furthermore, it is extremely difficult to precisely control the injection depth with this technique since the needle is inserted at an angle that is determined by the user; injections that are too deep deposit the vaccine into the subcutaneous (SC) tissue under the skin, while injections that are too shallow result in the leakage of part of the dose out of the skin during the injection or after the needle is removed.

To address the limited control and reproducibility of these existing methods of vaccine delivery to the skin, microneedle-based delivery has been developed to provide a means to more accurately and reproducibly access the skin in a less invasive fashion compared to what is possible using standard needles and the Mantoux technique. This feature could result in skin becoming the preferred site for the administration of a variety of vaccines.

## 3.1 Influenza Vaccine

Over the years, clinicians have used the Mantoux method for intradermal (ID) delivery of influenza vaccine and have shown this route to be effective

(Auewarakul et al. 2007; Belshe et al. 2007; Brown et al. 1977; Chiu et al. 2007; Halperin et al. 1979; Herbert et al. 1979; Kenney et al. 2004). Some studies have suggested that ID delivery of a low dose of influenza vaccine induces comparable levels of antibody to that obtained with standard intramuscular (IM) injection of up to five times more antigen (Belshe et al. 2004; Kenney et al. 2004). In a recent follow-up study, however, Belshe et al. (2007) did not observe such dose-sparing benefits associated with delivery by the Mantoux method. Studies to determine whether microneedles may provide more reproducible benefits for ID delivery are ongoing. Recent clinical results in elderly subjects support the notion that ID delivery using microneedles induces a superior immune response compared to IM (Lambert and Laurent 2008).

Preclinical studies have also supported the notion that microneedles can provide benefits over standard IM injection for influenza vaccine. A recent study examined the delivery of various types of influenza vaccines using a 1 mm long stainless steel hollow 34 Ga microneedle in rodents (Fig. 5) (Alarcon et al. 2007). Microneedle-based ID delivery was compared to IM injection using a conventional 27 Ga needle. Three types of influenza vaccines were examined: (1) whole inactivated virus, (2) trivalent split antigen vaccine, and (3) a DNA plasmid encoding the influenza hemagglutinin (HA). In addition, both high- and low-dose regimens were included for both routes of delivery and for each vaccine. The results demonstrated that microneedle-based ID delivery induces influenza-specific antibody responses that are at least as strong as those obtained by IM injection (Fig. 5) (Alarcon et al. 2007). Dose sparing was also evident; in many cases antibody responses induced by microneedle delivery remained elevated, while the corresponding responses elicited by IM injection dropped as the dose was reduced (Fig. 5) (Alarcon et al. 2007). Importantly, recent clinical trials have shown that microneedle-based ID delivery of influenza vaccine induces stronger humoral immune responses in the elderly as compared to IM injection (Lambert and Laurent 2008).

Additional studies carried out in the hairless guinea pig (HGP) have demonstrated that microneedle-coated trivalent influenza vaccine can induce primary anti-HA antibody responses to each strain comparable to their respective intramuscular injection controls (Maa et al. 2005). No significant difference, with respect to antibody responses, was seen among the various microneedle array designs used, which seems to at least partially confirm the results obtained with OVA. It also suggests that the length of the microneedles, in the range 225–600 μm, does not play an important role in the establishment of a solid immune response (Widera et al. 2006).

## 3.2 Other Vaccines

As noted above, smallpox vaccine has historically been administered to the skin by scarification using a bifurcated needle. Despite the general effectiveness of the approach, the use of live vaccinia virus makes the vaccine unsuitable for many populations (e.g., infants, pregnant women, immune-compromised/immune-suppressed individuals). In addition, vaccination often results in a severe skin reaction and a permanent

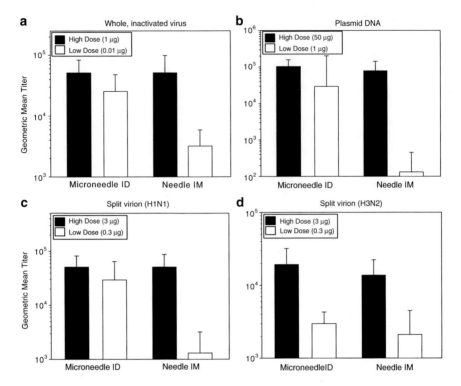

**Fig. 5a–d** Antibody response to influenza vaccines in rats. Data represent a subset of those originally reported in Alarcon et al. (2007). **a** Antibody response to whole inactivated influenza virus following immunization with either a high dose or low dose of vaccine. Data represent day 56 ELISA titers following immunization on day 0, day 21 and day 42. **b** Antibody response to influenza virus following immunization with either a high dose or low dose of plasmid DNA encoding influenza virus hemagglutinin. Data represent day 56 ELISA titers following immunization on day 0, day 21 and day 42. **c** Antibody response to the H1N1 strain of influenza virus following immunization with either a high dose or low dose of trivalent, split-virion vaccine. Data represent day 21 ELISA titers following a single immunization. **d** Antibody response to the H3N2 strain of influenza virus following immunization with either a high dose or low dose of trivalent, split-virion vaccine. Data represent day 21 ELISA titers following a single immunization

"pockmark" on the skin. Various investigators are exploring alternative vaccines and delivery methods in order to overcome these issues (Pickup 2007; Wiser et al. 2007). Hooper et al. recently described an approach using a solid microneedle array incorporated into a device for skin electroporation (Hooper et al. 2007). The microneedle array was coated with dried plasmid DNA encoding four vaccinia virus genes and applied to mouse skin, followed by electroporation. Mice generated a neutralizing antibody response that was at least as strong as by scarification and were protected from a lethal nasal challenge with vaccinia virus. Furthermore, both IgG1 and IgG2a antibodies were induced by this method of delivery, suggesting the induction of both Th1 and Th2 responses (Hooper et al. 2007).

The method of puncturing live virus into the skin by scarification has also been historically used with yellow fever viruses (Monath 2005). More recently, microneedles have been used to administer a live attenuated yellow fever virus vector in nonhuman primates (Dean et al. 2005). The vaccine, ChimeriVax™-JE, contains the yellow fever 17D strain genetically engineered to encode structural proteins of the related Japanese encephalitis (JE) virus (Monath et al. 2003). Successful delivery is associated with a low level of transient viremia resulting from uptake of the virus by host antigen-presenting cells. Delivery by both hollow and solid microneedles induced viremia that, in general, was of a higher frequency and duration than that obtained by SC injection using a standard needle (Table 1) (Dean et al. 2005). Notably, protective levels of neutralizing antibodies were detected in animals treated by all routes, with microneedle-based delivery inducing responses that were up to sevenfold greater than those yielded by SC injection using conventional needles (Dean et al. 2005). Interestingly, viremia resulting from delivery by scraping a solid microneedle array across the skin varied according to the method of delivery; one of three macaques became viremic when skin was pretreated with the array followed by topical application of the virus, while three of three macaques became viremic when the array was abraded through the droplet of vaccine on the surface of the skin. Additional studies are required in order to determine the mechanism for these differences as well as the clinical significance of the result.

Plasmid DNA has also been administered to mice using solid microneedle arrays according to the procedure whereby the array is rubbed across a droplet of vaccine on the surface of the skin. Mice treated with a plasmid encoding the hepatitis B surface antigen (HBsAg) generated antibody and T cell responses that were as strong or stronger than those obtained by injection using standard needles (Mikszta et al. 2002). In addition, both IgG1 and IgG2a subclasses of antibody were induced, suggesting a mixed and balanced helper T cell response from this method of delivery.

Protein antigens have also been administered using microneedles. In a rabbit study, various routes of delivery were compared for the recombinant protective antigen (rPA) of *Bacillus anthracis* (Mikszta et al. 2005). Rabbits immunized intradermally using hollow microneedles were completely protected from a lethal aerosol anthrax spore challenge, as were animals treated by the IM and intranasal (IN) routes. Although topical delivery to the skin following pretreatment with a

**Table 1** Viremia and neutralizing antibody response in nonhuman primates following immunization with ChimeriVax™-JE. Data represent a subset of those originally reported in Dean et al. (2005)

| Delivery | Viremia | | Nabs | |
|---|---|---|---|---|
| | No. of Responders | Duration (days) | No. of Responders | Titer |
| Needle | 1/3 | 3 | 3/3 | 1,067 |
| Hollow microneedle | 3/3 | 2–4 | 3/3 | 7,253 |
| Solid microneedle (preabrasion) | 1/3 | 7 | 1/3 | 3,467 |
| Solid microneedle (abrasion through vaccine) | 3/3 | 5–7 | 3/3 | 4,320 |

solid microneedle array induced antibodies at levels above those in animals treated topically without the device, only 33% protection was achieved. These results suggest that further improvements in the delivery method and/or the topical formulation will be required in order for this method to be comparable to injection for protein vaccines. In follow-up studies using the hollow microneedle, it was shown that 100% protection against lethal aerosol spore challenge could be achieved using as little as 10 μg of rPA, while IM injection of the same dose protected approximately 70% of rabbits (Mikszta et al. 2006). Dose sparing compared to IM injection was also evident during the early stages of the primary and secondary immune responses.

OVA has also been used as a model protein antigen for microneedle-based delivery. In this study, OVA was coated onto the surfaces of solid microneedles and allowed to dissolve from the microneedles within the skin of HGPs (Matriano et al. 2002; Widera et al. 2006). As discussed in greater detail in the next section, antibody responses generated by delivery using microneedles were at least as strong as intradermal, subcutaneous, and intramuscular delivery using conventional hypodermic needles.

More specifically, the immune response was found to be dose dependent, but mostly independent of depth of delivery (100–300 μm), density of microneedles (140–725 needles/cm$^2$), or area of application (2–4 cm$^2$) (Widera et al. 2006). It is surprising that the microneedle length did not have a dramatic effect on the immune outcome. The skin is highly stratified, with the highest abundance of dendritic cells, critical for antigen uptake and initiation of antigen-specific immune responses, located predominantly along the dermal–epidermal junction. On the other hand, it is well known that the dermis is also rich in other dendritic antigen-presenting cells (Nestle and Nickoloff 2007), and, in light of the results discussed here, it is likely that these cells also play a crucial role in the establishment of a solid immune response. With respect to the area of application and the density of microneedles, only minor differences were observed, which seems to indicate that here too the total dose is the most crucial parameter for establishing the antibody response (Widera et al. 2006).

# 4  Dose Sparing

A number of studies mentioned above indicate that microneedle-based delivery to the skin can generate the same immune response as delivery using higher vaccine doses via other routes. This dose-sparing ability was specifically studied using an antigen-coated microneedle array in the HGP model using the model antigen OVA (Matriano et al. 2002; Widera et al. 2006). The HGP is outbred, euthymic, and has been used in vaccine (Lowry et al. 1993; Ruble et al. 1994) and contact sensitization studies (Miyauchi and Horio 1992; Woodward et al. 1989). The following parameters were investigated: route of administration, dose of vaccine delivered, depth of vaccine delivery, density of microneedles on the array, and area of application.

The first set of studies (Fig. 6) used microneedles penetrating the skin to an average depth of about 100 μm, which corresponds to the thickness of the epidermis and the

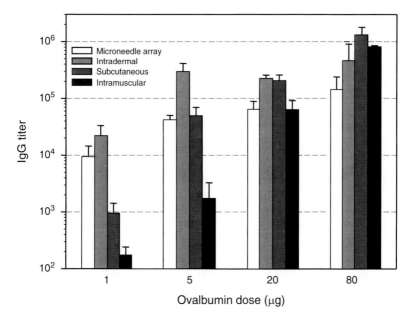

**Fig. 6** Anti-ovalbumin antibody response to ovalbumin immunization in hairless guinea pigs. Each animal received a primary immunization followed by a secondary immunization (booster) four weeks later with the same ovalbumin dose (From Matriano et al. 2002). The routes of administration were intracutaneous using coated microneedle arrays, ID, SC, and IM injection. The serum samples were collected a week after booster immunization and evaluated for the presence of anti-ovalbumin IgG antibodies by ELISA. The results are expressed as end-point antibody titers relative to unimmunized control sera

uppermost dermis layers. In these studies, which compared different routes of administration, it was found that when a low dose of antigen was delivered (1 μg), the immune response, as measured by specific antibody titers, was most efficient following microneedle and ID administration as compared to SC or IM administration. The immune response was more than one order of magnitude higher following microneedle-based administration vs. SC delivery, and about two orders of magnitude greater as compared to IM delivery.

The immune response increased with increasing dose (in the range of 1–80 μg administered antigen) for all routes of administration. This increase was most pronounced with the SC and IM routes of administration. As a result, at the highest doses studied (i.e., 20 and 80 μg), there were no significant differences between the different routes of administration, demonstrating that microneedles and ID delivery had a dose-sparing effect but did not further increase the immune response at high antigen doses. This may be the result of an upper limitation on the immune response with increasing dose, as has been observed with other antigens (Andersen et al. 1985; Diaz-Ortega et al. 1994). Indeed, it is well known that, from an antibody response standpoint, there is little benefit in increasing the dose of antigen above certain thresholds.

Dose-sparing potential was further demonstrated by incorporating the model adjuvant glucosaminylmuramyl dipeptide (GMDP) into the OVA coating, which produced a significant dose-sparing effect following coadministration of 1 µg antigen (Matriano et al. 2002). At this dose, the antibody titer generated with microneedles that codelivered OVA and GMDP approached the titer levels achieved with OVA doses of 20 µg in the absence of GMDP, which demonstrates a significant dose-sparing effect and indicates that the technology is amenable to the codelivery of antigen and adjuvant.

# 5  Safety

Microneedles are minimally invasive delivery systems that have been designed for safety. Safety concerns are chiefly related to the risk of infection. Other factors to consider, linked mostly to user acceptability and environmental issues, include pain, local bleeding, skin irritation, and residual vaccine left in the system and on the skin surface.

## 5.1  Infection

Based on experience with hypodermic needles, the greatest risk of infection comes from the use of contaminated needles (Atkinson et al. 2002). Thus, single-use microneedles that are sufficiently clean (e.g., sterilized) should pose little threat of infection. However, the skin is constantly in contact with environmental organisms and becomes readily colonized by certain microbial species (Roth and James 1989). Most skin microorganisms found in the most superficial layers of the epidermis and the upper parts of the hair follicles are nonpathogenic, but sometimes potentially pathogenic species such as *Staphylococcus aureus* and *Streptococcus pyogenes* can be found on the skin.

Any breach in the skin can provide an entry route for microorganisms that could cause local or even systemic infection. Microneedle arrays containing up to hundreds and possibly thousands of microneedles could therefore be problematic. However, the risk of infection is related to a large number of factors, including the size and number of breaches, the depth of the breaches, the number of microorganisms entering the skin and their nature, and the individual susceptibility of the patient. In clinical practice, it seems unlikely that the small, shallow injuries resulting from the use of microneedles would result in significant safety concerns. Indeed, the skin barrier is routinely breached during common experiences of minor abrasion, such as shaving, yet infection rarely occurs.

There are several preclinical and clinical studies that indicate that the risk of infection resulting from microneedle administration is minimal (Cormier and Daddona 2003; Cormier et al. 2004; Matriano et al. 2002; Widera et al. 2006).

In the animal studies discussed above, which were in some cases conducted with nonsterile systems, no signs of infection were reported. It is noteworthy that these animals were carefully monitored for typically many weeks after microneedle treatment under the careful supervision of veterinary staff with IACUC-mandated scrutiny. In addition, microneedle devices have been inserted into the skin well over 10,000 times in more than 7,000 human subjects through the collective experience of BD, Georgia Tech, Apogee and Zosano. In human studies conducted by the authors, there have been no microneedle-related infections and, indeed, no serious adverse events.

It is worth noting that these studies in animals and humans were conducted in healthy subjects and that, prior to application of the microneedle arrays, skin treatment sites were typically cleansed with alcohol wipes to reduce the skin's microbial bioburden (Adams et al. 2005; Barker and Ryan 1995). Therefore, it is unknown what the infection potential would be in widespread use or, for example, in immune-compromised patients. As a result of these uncertainties, and depending on the target population and clinical indication, it is possible that the regulatory agencies would require that any future commercial product would use a skin-cleansing pretreatment procedure as well as a low bioburden or sterile microneedle system.

As additional information, more than 125,000 microneedle devices that take the form of a cylindrical roller with solid microneedles protruding around the circumference have been sold around the world for cosmetic purposes (http://www.dermaroller.de). These devices are typically applied to the skin either by clinical personnel or by patients themselves, who may reuse them without sterilization or without even washing between uses. Although formal safety data on these microneedle rollers do not exist, the manufacturer reports no known adverse events so long as the microneedles are not reused (personal communication from Horst Liebl). Altogether, this extensive anecdotal evidence holds promise for the safety of microneedles.

## 5.2 Pain

Although pain is not a safety concern per se, it nonetheless affects patient acceptance and the perception of safety. Initial studies showed that the insertion of sharp-tipped microneedles or the scraping of blunt-tipped microneedles measuring 50–200 µm in length and assembled into arrays of up to 400 microneedles were generally regarded as painless procedures by human subjects (Kaushik et al. 2001; Mikszta et al. 2002).

More detailed studies showed that microneedle length correlates strongly with pain, such that an increase in microneedle length from 500 to 1,500 µm (i.e., a three-fold increase in length) resulted in a sevenfold increase in pain score from blinded human subjects (Gill et al. 2008). Increasing the number of microneedles from 5 to 50 (i.e., a tenfold increase) increased the pain score just 2.5-fold. Varying microneedle width, thickness, and tip angle did not have a significant effect on pain. In all cases, the pain score from microneedles was manyfold less than that from the 26 Ga hypodermic needle used as a positive control, and for the best microneedle design the pain score was just 1/20th that of the hypodermic needle. In another recent study, subject perception associated with the insertion of a 1.5 mm microneedle perpendicularly into the skin was shown to be significantly less than that observed when a standard needle was inserted into the skin at a shallow angle (Laurent et al. 2007).

## 5.3   Bleeding

The epidermis is devoid of vasculature, and the most superficial capillary bed is located in the upper dermis close to the dermal–epidermal junction (Barker and Ryan 1995). As a result, microneedles penetrating the skin deeper than about 100 μm could breach capillaries. Despite this expectation, most animal and human studies have not observed bleeding after microneedle treatment.

In a study examining the use of microneedles to extract interstitial fluid from the skin of hairless rats, microneedles were inserted into the skin hundreds of times to depths of 700–1,500 μm (Wang et al. 2005). In the absence of suction, no fluid was seen to exude from the skin. With the application of suction, interstitial fluid could be extracted, but this fluid generally appeared clear, without evidence of blood. Similar results were seen in replicate experiments in six human subjects.

In the vaccine delivery experiments discussed above, as well as other experiments involving the delivery of various therapeutic and model compounds in animal models including mice, hairless rats, and rhesus macaques, microneedle treatment has generally not been associated with bleeding (Alarcon et al. 2007; Davis et al. 2005; Dean et al. 2005; Gardeniers et al. 2003; Martanto et al. 2004, McAllister et al. 2003; Mikszta et al. 2002, 2006, 2005). However, in one study involving HGPs (Widera et al. 2006), 225 μm long microneedles were found to penetrate the skin to an average depth of about 150 μm and produced at most pinpoint bleeding, and 600 μm long microneedles were found to penetrate the skin to an average depth of about 300 μm, well beyond the location of the dermal capillary bed, and produced significant bleeding.

In human subjects, bleeding has also generally not been seen for microneedles ranging in size from 500 to 1,000 μm (Gill et al. 2008). However, 1.5 mm microneedles have been reported to sometimes leave a small droplet of blood on the skin after the insertion of solid microneedles (Gill et al. 2008) or injection using hollow microneedles (Barker and Ryan 1995).

## 5.4   Skin Irritation

No significant skin irritation (i.e., erythema and edema) has been noted in a number of studies involving the insertion of microneedles alone or for the delivery of biologics

(Cormier and Daddona 2003; Davis et al. 2005; Gardeniers et al. 2003; Lin et al. 2001; Martanto et al. 2004; Matriano et al. 2002; McAllister et al. 2003). However, experiments in human subjects have shown highly localized spots of redness that correspond exactly to the submillimeter sites of each microneedle insertion (Gill et al. 2008). This mild erythema was visible under close inspection, but is unlikely to be of cosmetic concern. In many subjects, it disappeared within minutes, and in most subjects within hours. A recent clinical evaluation of a 1.5 mm length microdelivery system indicated very minor local effects (e.g., redness, itching) that spontaneously resolved within 20–30 min without any long-term adverse events (Laurent et al. 2007).

Vaccine delivery using microneedles includes the added complexity of local immune response to the vaccine. For example, the average total skin score (combined erythema and edema, maximum combined score of 8) for all treatment groups after OVA immunization was just 0.3 at 24 h following primary immunization, but increased to 2.3 at 24 h following booster immunization (Widera et al. 2006). Although mild and transient, these scores indicated an immune-mediated skin response; possibly a delayed-type hypersensitivity response. Skin response following intradermal administration of vaccines has been routinely observed in the clinic (Belshe et al. 2004), and it is likely that they will be also observed following introduction of a vaccine into the skin using microneedles. These responses might cause minor discomfort at the site of application.

## 5.5   Microneedle and Skin Residuals

The implications of having residual vaccine left on the microneedles or at the surface of the skin depends on the specifics of the vaccine and its formulation. In any case, minimizing residual would be beneficial from a safety and environmental standpoint.

When microneedles are pierced into or scraped across the skin before or in the presence of a topical vaccine formulation, it is likely that a large fraction of the vaccine remains on the skin surface, although this has not been quantified. Following injection using a hollow microneedle, fluid hold-up in the syringe may be similar to that obtained following conventional hypodermic injection.

Following application of OVA-coated microneedles to HGP in vivo, measurements showed that 12% of the coated OVA remained on the microneedles, 11.5% of the total dose was found on the skin surface, and 48% was delivered into the skin (Widera et al. 2006). Although the antigen was administered into the uppermost layers of the skin, it was not extracted by extensive cleansing of the skin surface. A study using microneedles coated with vitamin B inserted into cadaver skin found that 91% of the dose was delivered into the skin, with 7% remaining on the microneedles and 2% on the skin surface (Gill and Prausnitz 2007a). This indicates that through proper engineering, contamination issues can be minimized to a considerable extent.

# 6   Logistical Issues

In addition to expected increases in vaccine efficacy and safety, the use of microneedles is expected to also provide logistical simplifications, which may be especially beneficial for vaccine administration during a pandemic. These simplifications include: (1) easier vaccine administration, including the possibility of self-administration; (2) more rapid vaccine distribution and easier stockpiling and disposal, and; (3) inexpensive manufacturing costs.

## 6.1   Vaccine Administration

Many microneedle designs envision a vaccine patch that would look and be applied similarly to a nicotine patch or a Band Aid. After peeling off a protective release liner to expose the microneedles, the patch is pressed to the skin and perhaps held in place by an adhesive incorporated onto the patch, if needed. Patch placement may be accomplished by hand or may require the use of a small device that presses the patch against the skin with a reproducible force to assure correct microneedle insertion into the skin (Davis et al. 2004; Yang and Zahn 2004). This tool could measure just a few centimeters in size and could be, for example, a disposable, plastic, spring-loaded device. After some seconds or minutes, the patch is removed and discarded. Studies carried out in human subjects have shown that microneedles can be inserted into the skin with little or no pain (Gill et al. 2008; Kaushik et al. 2001; Mikszta et al. 2002), which suggests that patient compliance can be increased.

Other designs involve hollow microneedles. Notably, a newly described hollow microneedle injection system was shown to be effective for intradermal delivery, even in the hands of untrained users (Laurent et al. 2007). Untrained users were at least as effective as extensively trained users in performing a correct intradermal injection. These results suggest that, with proper engineering, hollow microneedle delivery devices can achieve a high level of user-friendliness.

These simple delivery methods can shift the responsibility of vaccine administration to minimally trained personnel and possibly to patients themselves, leaving doctors and nurses in a supervisory role. This could have major advantages in a pandemic, where there is a critical bottleneck in assembling huge populations of patients at central locations and having sufficient medical personnel to administer vaccines, or possibly other therapies (Fauci 2006; Gostin 2006). Allowing minimally trained personnel to administer the vaccines would significantly ease this bottleneck, and allowing self-administration would increase throughput even more. Simple vaccine administration would also benefit seasonal influenza vaccination (e.g., patients pick up vaccine patches at the pharmacy for self-administration at home) and mass vaccination campaigns in the developing world (e.g., polio vaccination campaigns annually administer more than two billion doses over the course of a series of intensive "immunization days," which is possible only because of the simple delivery method enabled by oral administration of the polio vaccine (Centers for Disease Control and Prevention (CDC) 2007).

## 6.2   Vaccine Distribution

Because speed will be critical during a pandemic, the manufacture, stockpiling, distribution, and disposal of vaccine delivery systems need to be simplified and expedited. Using microneedles to deliver vaccine can expedite the response to a pandemic because intracutaneous delivery using microneedles is expected to require, less vaccine per dose, as discussed above. Given that production of the vaccine antigen will be a rate-limiting step when responding to an emerging threat, the need to manufacture less vaccine can significantly expedite the process of getting vaccine to the public (Fauci 2006; Gostin 2006; Ulmer et al. 2006).

Conventional vaccine delivery systems, which often involve a hypodermic needle, a syringe, a vial of lyophilized vaccine and a vial of diluent, can easily occupy tens of cubic centimeters including their packaging (Thomson 2007). In contrast, prefilled microneedle injection devices and, to an even greater extent, microneedle patches are expected to occupy much less space. Some of the thinnest patch-based systems are expected to be much less than one cubic centimeter in size and have a flat profile, which facilitates stacking. Moreover, with improvements in vaccine stabilization technologies, it may be possible to create microneedle vaccines that avoid the need for refrigeration. Altogether, this suggests that it should be possible to stockpile microneedle vaccines under much less costly conditions and possibly at a greater number of locations.

Distribution should also be facilitated by the small size and expected thermal stability of solid-state microneedle vaccines. The US Department of Health and Human Services has a goal of distributing drugs and vaccines via the postal service in a pandemic or other urgent scenario to every home in the United States within 12 h (Fauci 2006; Freking 2005). It would be easier to achieve this vision using a small, flat, light microneedle patch that can easily fit inside a standard envelope or small package.

Finally, microneedles may reduce the risks associated with the disposal of standard hypodermic needles, because inadvertent microneedle insertion is difficult and dissolving polymer microneedles leave no sharp waste behind.

## 6.3   Vaccine Patch Manufacturing

A variety of vaccine delivery systems are being studied (O'Hagan and Rappuoli 2004; Weniger and Papania 2008), but many are expected to be much more expensive than the conventional needle and syringe due to the cost of the delivery system itself and the added complexity of distributing and administering the vaccine. Microneedle devices are expected to have a cost similar to a conventional needle and syringe and to have significant cost savings associated with their distribution, due largely to the small size of the microneedle; reduced administration, due largely to the expected reduced need for trained personnel; and the smaller amounts of vaccine antigen required, due largely to the expectation that intracutaneous delivery can be dose sparing.

Different microneedle designs achieve their cost effectiveness through fabrication in different ways. Solid metal microneedles are typically fabricated by cutting metal sheets into the desired needle geometries using either chemical etchants or lasers (Gill and Prausnitz 2007a; Matriano et al. 2002). Vaccine coatings are then applied by dip-coating processes with safe, low-cost excipients. Polymer microneedles have been fabricated using molding methods in which the micromolds are typically prepared in a microelectronics cleanroom environment and then repeatedly reused to mold microneedles in a conventional manufacturing environment by adapting in situ polymerization, solvent casting and injection molding/embossing techniques (Lee et al. 2008; Miyano et al. 2005; Park et al. 2006; Sullivan et al. 2008). Hollow microneedles have been fabricated by direct etching in the cleanroom, reproduction using electroplating onto micromolds, or adaptation of conventional hypodermic needle manufacturing methods (Gardeniers et al. 2003; McAllister et al. 2003; Mikszta et al. 2005; Sivamani et al. 2005; Teo et al. 2005). All of these methods lend themselves to scale-up for mass production at low cost, where processing is generally the dominant cost, because the small size of microneedles means that material costs are much less.

# 7  Conclusions

Microneedles provide a number of advantages and opportunities to deliver vaccines more effectively, especially in the event of a pandemic influenza outbreak. Intracutaneous delivery has been shown in a number of studies to be not only effective but also to reduce the amount of antigen needed in a dose-sparing manner. Microneedles have yielded similar findings with a number of different antigens, including vaccines against seasonal influenza in Phase III clinical trials. Because of their small size and the inert materials used in their construction, microneedles have not raised safety concerns. Overall, it appears likely that microneedles can provide a safe and effective method to deliver influenza vaccine.

In addition, microneedles offer other advantages compared to conventional needle-and-syringe and other delivery methods that may also be safe and effective. For example, the patch-based format of many microneedle designs should facilitate simple vaccine administration and possibly self-administration by patients themselves. The small size of microneedle systems should also facilitate storage and rapid distribution to central locations or even to individual households by the postal service. Combined with the expected low manufacturing costs, these attributes suggest that microneedles are an attractive platform for vaccine delivery and can play an important role in the medical response to an influenza pandemic.

**Acknowledgments** This work was supported in part by the Georgia Research Alliance and National Institutes of Health grants 1U01AI074579 and 1R01EB006369. Mark Prausnitz is the Emerson-Lewis Faculty Fellow and is a member of the Center for Drug Design, Development, and Delivery and the Institute for Bioengineering and Biosciences at the Georgia Institute of Technology.

# References

Adams D, Quayum M, Worthington T, Lambert P, Elliott T (2005) Evaluation of a 2% chlorhexidine gluconate in 70% isopropyl alcohol skin disinfectant. J Hosp Infect 61:287–290

Alarcon JB, Hartley AW, Harvey NG, Mikszta JA (2007) Preclinical evaluation of microneedle technology for intradermal delivery of influenza vaccines. Clin Vaccine Immunol 14:375–381

Andersen KE, Boman A, Volund A, Wahlberg JE (1985) Induction of formaldehyde contact sensitivity: dose response relationship in the guinea pig maximization test. Acta Derm Venereol 65:472–478

Atkinson WL, Pickering LK, Schwartz B, Weniger BG, Iskander JK, Watson JC (2002) General recommendations on immunization. Recommendations of the Advisory Committee on Immunization Practices (ACIP) and the American Academy of Family Physicians (AAFP). MMWR Recomm Rep 51:1–35

Auewarakul P, Kositanont U, Sornsathapornkul P, Tothong P, Kanyok R, Thongcharoen P (2007) Antibody responses after dose-sparing intradermal influenza vaccination. Vaccine 25:659–663

Barker JH, Ryan TJ (1995) Skin microcirculation In: Barker JH, Anderson GL, Menger MD (eds) Clinically applied microcirculation research. CRC, Boca Raton, pp 315–338

Belshe RB, Newman FK, Cannon J, Duane C, Treanor J, Van Hoecke C, Howe BJ, Dubin G (2004) Serum antibody responses after intradermal vaccination against influenza. N Engl J Med 351:2286–2294

Belshe RB, Newman FK, Wilkins K, Graham IL, Babusis E, Ewell M, Frey SE (2007) Comparative immunogenicity of trivalent influenza vaccine administered by intradermal or intramuscular route in healthy adults. Vaccine 25:6755–6763

Birchall J, Coulman S, Pearton M, Allender C, Brain K, Anstey A, Gateley C, Wilke N, Morrissey A (2005) Cutaneous DNA delivery and gene expression in ex vivo human skin explants via wet-etch micro-fabricated micro-needles. J Drug Target 13:415–421

Birchall JC (2006) Microneedle array technology: the time is right but is the science ready? Expert Rev Med Devices 3:1–4

Bronaugh RL, Maibach HI (2005) Percutaneous absorption. Marcel Dekker, New York

Brown H, Kasel JA, Freeman DM, Moise LD, Grose NP, Couch RB (1977) The immunizing effect of influenza A/New Jersey/76 (Hsw1N1) virus vaccine administered intradermally and intramuscularly to adults. J Infect Dis 136(suppl):S466–S471

Centers for Disease Control and Prevention (CDC) (2007) Progress toward interruption of wild poliovirus transmission worldwide, January 2006–May 2007. MMWR Morb Mortal Wkly Rep 56:682–685

Chabri F, Bouris K, Jones T, Barrow D, Hann A, Allender C, Brain K, Birchall J (2004) Microfabricated silicon microneedles for nonviral cutaneous gene delivery. Br J Dermatol 150:869–877

Chiu SS, Peiris JS, Chan KH, Wong WH, Lau YL (2007) Immunogenicity and safety of intradermal influenza immunization at a reduced dose in healthy children. Pediatrics 119:1076–1082

Cormier M, Daddona PE (2003) Macroflux technology for transdermal delivery of therapeutic proteins and vaccines. In: Rathbone MJ, Hadgraft J, Roberts MS (eds) Modified-release drug delivery technology. Marcel Dekker, New York, pp 589–598

Cormier M, Johnson B, Ameri M, Nyam K, Libiran L, Zhang DD, Daddona P (2004) Transdermal delivery of desmopressin using a coated microneedle array patch system. J Control Release 97:503–511

Coulman S, Allender C, Birchall J (2006a) Microneedles and other physical methods for overcoming the stratum corneum barrier for cutaneous gene therapy. Crit Rev Ther Drug Carrier Syst 23:205–258

Coulman SA, Barrow D, Anstey A, Gateley C, Morrissey A, Willke N, Allender C, Brain K, Birchall JC (2006b) Minimally invasive delivery of macromolecules and plasmid DNA via microneedles. Curr Drug Deliv 3:65–75

Davis SP, Landis BJ, Adams ZH, Allen MG, Prausnitz MR (2004) Insertion of microneedles into skin: measurement and prediction of insertion force and needle fracture force. J Biomech 37:1155–1163

Davis SP, Martanto W, Allen MG, Prausnitz MR (2005) Transdermal insulin delivery to diabetic rats through microneedles. IEEE Trans Biomed Eng 52:909–915

Dean CH, Alarcon JB, Waterston AM, Draper K, Early R, Guirakhoo F, Monath TP, Mikszta JA (2005) Cutaneous delivery of a live, attenuated chimeric flavivirus vaccine against Japanese encephalitis (ChimeriVax-JE) in non-human primates. Hum Vaccin 1:106–111

Diaz-Ortega JL, Forsey T, Clements CJ, Milstien J (1994) The relationship between dose and response of standard measles vaccines. Biologicals 22:35–44

Ellner PD (1998) Smallpox: gone but not forgotten. Infection 26:263–269

Fauci AS (2006) Pandemic influenza threat and preparedness. Emerg Infect Dis 12:73–77

Freking K (2005) Health News: HHS may use mail to deliver emergency meds. http://www.health-news.org/breaking/2885/hhs-may-use-mail-to-deliver-emergency-meds.html. Accessed 30 Oct 2007

Gardeniers JGE, Luttge R, Berenschot JW, de Boer MJ, Yeshurun Y, Hefetz M, van 't Oever R, van den Berg A (2003) Silicon micromachined hollow microneedles for transdermal liquid transport. J MEMS 6:855–862

Gill HS, Prausnitz MR (2007a) Coated microneedles for transdermal delivery. J Control Release 117:227–237

Gill HS, Prausnitz MR (2007b) Coating formulations for microneedles. Pharm Res 24:1369–1380

Gill HS, Prausnitz MR (2008) Pocketed microneedles for drug delivery to the skin. J Phys Chem Solids 69:1537–1541

Gill HS, Denson DD, Burris B, Prausnitz MR (2008) Effect of microneedle design on pain in human subjects Clin J Pain 24:585–594

Gostin LO (2006) Medical countermeasures for pandemic influenza: ethics and the law. JAMA 295:554–556

Halperin W, Weiss WI, Altman R, Diamond MA, Black KJ, Iaci AW, Black HC, Goldfield M (1979) A comparison of the intradermal and subcutaneous routes of influenza vaccination with A/New Jersey/76 (swine flu) and A/Victoria/75: report of a study and review of the literature. Am J Public Health 69:1247–1251

Henry S, McAllister DV, Allen MG, Prausnitz MR (1998) Microfabricated microneedles: a novel approach to transdermal drug delivery. J Pharm Sci 87:922–925

Herbert FA, Larke RP, Markstad EL (1979) Comparison of responses to influenza A/New Jersey/76-A/Victoria/75 virus vaccine administered intradermally or subcutaneously to adults with chronic respiratory disease. J Infect Dis 140:234–238

Hooper JW, Golden JW, Ferro AM, King AD (2007) Smallpox DNA vaccine delivered by novel skin electroporation device protects mice against intranasal poxvirus challenge. Vaccine 25:1814–1823

Huang CM (2007) Topical vaccination: the skin as a unique portal to adaptive immune responses. Semin Immunopathol 29:71–80

Ito Y, Hagiwara E, Saeki A, Sugioka N, Takada K (2006a) Feasibility of microneedles for percutaneous absorption of insulin. Eur J Pharm Sci 29:82–88

Ito Y, Yoshimitsu J, Shiroyama K, Sugioka N, Takada K (2006b) Self-dissolving microneedles for the percutaneous absorption of EPO in mice. J Drug Target 14:255–261

Kaushik S, Hord AH, Denson DD, McAllister DV, Smitra S, Allen MG, Prausnitz MR (2001) Lack of pain associated with microfabricated microneedles. Anesth Analg 92:502–504

Kenney RT, Frech SA, Muenz LR, Villar CP, Glenn GM (2004) Dose sparing with intradermal injection of influenza vaccine. N Engl J Med 351:2295–2301

La Montagne JR, Fauci AS (2004) Intradermal influenza vaccination—can less be more? N Engl J Med 351:2330–2332

Lambert P-H, Laurent PE (2008) Intradermal vaccine delivery: will new delivery systems transform vaccine administration? Vaccine 26:3197–3208

Larregina AT, Falo LD Jr (2005) Changing paradigms in cutaneous immunology: adapting with dendritic cells. J Invest Dermatol 124:1–12

Laurent PE, Bonnet S, Alchas P, Regolini P, Mikszta JA, Pettis R, Harvey NG (2007) Evaluation of the clinical performance of a new intradermal vaccine administration technique and associated delivery system. Vaccine 25:8833–8842

Lee J-W, Park J-H, Prausnitz MR (2008) Dissolving microneedles for transdermal drug delivery. Biomaterials 29:2113–2124

Lin W, Cormier M, Samiee A, Griffin A, Johnson B, Teng C, Hardee GE, Daddona P (2001) Transdermal delivery of antisense oligonucleotides with microprojection patch (Macroflux) technology. Pharm Res 18:1789–1793

Lowry PW, Sabella C, Koropchak CM, Watson BN, Thackray HM, Abbruzzi GM, Arvin AM (1993) Investigation of the pathogenesis of varicella-zoster virus infection in guinea pigs by using polymerase chain reaction. J Infect Dis 167:78–83

Maa Y-F, Sellers S, Matriano J, Ramdas A (2005) Apparatus and method for transdermal delivery of influenza vaccine. US Pat Appl 20050220854

Martanto W, Davis S, Holiday N, Wang J, Gill H, Prausnitz M (2004) Transdermal delivery of insulin using microneedles in vivo. Pharm Res 21:947–952

Martanto W, Moore JS, Couse T, Prausnitz MR (2006a) Mechanism of fluid infusion during microneedle insertion and retraction. J Control Release 112:357–361

Martanto W, Moore JS, Kashlan O, Kamath R, Wang PM, O'Neal JM, Prausnitz MR (2006b) Microinfusion using hollow microneedles. Pharm Res 23:104–113

Matriano JA, Cormier M, Johnson J, Young WA, Buttery M, Nyam K, Daddona PE (2002) Macroflux microprojection array patch technology: a new and efficient approach for intracutaneous immunization. Pharm Res 19:63–70

McAllister DV, Allen MG, Prausnitz MR (2000) Microfabricated microneedles for gene and drug delivery. Annu Rev Biomed Eng 2:289–313

McAllister DV, Wang PM, Davis SP, Park J-H, Canatella PJ, Allen MG, Prausnitz MR (2003) Microfabricated needles for transdermal delivery of macromolecules and nanoparticles: fabrication methods and transport studies. Proc Natl Acad Sci USA 100:13755–13760

Mikszta JA, Alarcon JB, Brittingham JM, Sutter DE, Pettis RJ, Harvey NG (2002) Improved genetic immunization via micromechanical disruption of skin-barrier function and targeted epidermal delivery. Nat Med 8:415–419

Mikszta JA, Sullivan VJ, Dean C, Waterston AM, Alarcon JB, Dekker JP 3rd, Brittingham JM, Huang J, Hwang CR, Ferriter M, Jiang G, Mar K, Saikh KU, Stiles BG, Roy CJ, Ulrich RG, Harvey NG (2005) Protective immunization against inhalational anthrax: a comparison of minimally invasive delivery platforms. J Infect Dis 191:278–288

Mikszta JA, Dekker JP 3rd, Harvey NG, Dean CH, Brittingham JM, Huang J, Sullivan VJ, Dyas B, Roy CJ, Ulrich RG (2006) Microneedle-based intradermal delivery of the anthrax recombinant protective antigen vaccine. Infect Immun 74:6806–6810

Miyano T, Tobinaga Y, Kanno T, Matsuzaki Y, Takeda H, Wakui M, Hanada K (2005) Sugar micro needles as transdermic drug delivery system. Biomed Microdevices 7:185–188

Miyauchi H, Horio T (1992) A new animal model for contact dermatitis: the hairless guinea pig. J Dermatol 19:140–145

Monath TP (2005) Yellow fever vaccine. Expert Rev Vaccines 4:553–574

Monath TP, Guirakhoo F, Nichols R, Yoksan S, Schrader R, Murphy C, Blum P, Woodward S, McCarthy K, Mathis D, Johnson C, Bedford P (2003) Chimeric live, attenuated vaccine against Japanese encephalitis (ChimeriVax-JE): phase 2 clinical trials for safety and immunogenicity, effect of vaccine dose and schedule, and memory response to challenge with inactivated Japanese encephalitis antigen. J Infect Dis 188:1213–1230

Nestle FO, Nickoloff BJ (2007) Deepening our understanding of immune sentinels in the skin. J Clin Invest 117:2382–2385

Nordquist L, Roxhed N, Griss P, Stemme G (2007) Novel microneedle patches for active insulin delivery are efficient in maintaining glycaemic control: an initial comparison with subcutaneous administration. Pharm Res 24:1381–1388

O'Hagan DT, Rappuoli R (2004) Novel approaches to vaccine delivery. Pharm Res 21:1519–1530

Park J-H, Allen MG, Prausnitz MR (2005) Biodegradable polymer microneedles: fabrication, mechanics and transdermal drug delivery. J Control Release 104:51–66

Park JH, Allen MG, Prausnitz MR (2006) Polymer microneedles for controlled-release drug delivery. Pharm Res 23:1008–1019

Pearton M, Allender C, Brain K, Anstey A, Gateley C, Wilke N, Morrissey A, Birchall J (2008) Gene delivery to the epidermal cells of human skin explants using microfabricated microneedles and hydrogel formulations. Pharm Res 25(2):407–416

Pickup DJ (2007) Understanding orthopoxvirus interference with host immune responses to inform novel vaccine design. Expert Rev Vaccines 6:87–95

Prausnitz MR (2004) Microneedles for transdermal drug delivery. Adv Drug Deliv Rev 56:581–587

Prausnitz M (2005) Assessment of microneedles for transdermal drug delivery. In: Bronaugh R, Maibach H (eds) Percutaneous absorption. Marcel Dekker, New York, pp 497–507

Prausnitz M, Ackley D, Gyory J (2003) Microneedles for transdermal drug delivery. In: Rathbone M, Hadgraft J, Roberts M (eds) Modified release drug delivery systems. Marcel Dekker, New York, pp 513–522

Prausnitz MR, Langer R (2008) Transdermal drug delivery. Nat Biotech 26:1261–1268

Prausnitz M, Mikszta J, Raeder-Devens J (2005) Microneedles. In: Smith E, Maibach H (eds) Percutaneous penetration enhancers. CRC, Boca Raton, FL, pp 239–255

Prausnitz MR, Gill HS, Park J-H (2008) Microneedles for drug delivery. In: Rathbone MJ, Hadgraft J, Roberts MS, Lane ME (eds) Modified release drug delivery. Informa Healthcare, New York

Reed ML, Lye W-K (2004) Microsystems for drug and gene delivery. Proc IEEE 92:56–75

Roth RR, James WD (1989) Microbiology of the skin: resident flora, ecology, infection. J Am Acad Dermatol 20:367–390

Ruble DL, Elliott JJ, Waag DM, Jaax GP (1994) A refined guinea pig model for evaluating delayed-type hypersensitivity reactions caused by Q fever vaccines. Lab Anim Sci 44:608–612

Shirkhanzadeh M (2005) Microneedles coated with porous calcium phosphate ceramics: effective vehicles for transdermal delivery of solid trehalose. J Mater Sci Mater Med 16:37–45

Sivamani RK, Stoeber B, Wu GC, Zhai H, Liepmann D, Maibach H (2005) Clinical microneedle injection of methyl nicotinate: stratum corneum penetration. Skin Res Technol 11:152–156

Sivamani RK, Liepmann D, Maibach HI (2007) Microneedles and transdermal applications. Expert Opin Drug Deliv 4:19–25

Subbarao K, Murphy BR, Fauci AS (2006) Development of effective vaccines against pandemic influenza. Immunity 24:5–9

Sullivan SP, Murthy N, Prausnitz MR (2008) Minimally invasive protein delivery with rapidly dissolving polymer microneedles. Adv Mat 20:933–938

Teo MA, Shearwood C, Ng KC, Lu J, Moochhala S (2005) In vitro and in vivo characterization of MEMS microneedles. Biomed Microdevices 7:47–52

Thomson PDR (2007) Physicians' desk reference. Thomson PDR, Montvale

Ulmer JB, Valley U, Rappuoli R (2006) Vaccine manufacturing: challenges and solutions. Nat Biotechnol 24:1377–1383

Verbaan FJ, Bal SM, van den Berg DJ, Groenink WH, Verpoorten H, Luttge R, Bouwstra JA (2007) Assembled microneedle arrays enhance the transport of compounds varying over a large range of molecular weight across human dermatomed skin. J Control Release 117:238–245

Wang P, Cornwell M, Prausnitz M (2005) Minimally invasive extraction of dermal interstitial fluid for glucose monitoring using glass microneedles. Diabetes Technol Ther 7:131–141

Wang PM, Cornwell M, Hill J, Prausnitz MR (2006) Precise microinjection into skin using hollow microneedles. J Invest Dermatol 126:1080–1087

Webby RJ, Webster RG (2003) Are we ready for pandemic influenza? Science 302:1519–1522

Weniger BG, Papania M (2008) Alternative vaccine delivery methods. In: Plotkn S, Orenstein W, Offit P (eds) Vaccines. Elsevier, Philadelphia

Widera G, Johnson J, Kim L, Libiran L, Nyam K, Daddona PE, Cormier M (2006) Effect of delivery parameters on immunization to ovalbumin following intracutaneous administration by a coated microneedle array patch system. Vaccine 24:1653–1664

Wiser I, Balicer RD, Cohen D (2007) An update on smallpox vaccine candidates and their role in bioterrorism related vaccination strategies. Vaccine 25:976–984

Woodward DF, Nieves AL, Williams LS, Spada CS, Hawley SB, Duenes JL (1989) A new hairless strain of guinea pig: characterization of the cutaneous morphology and pharmacology. In: Maibach HI, Lowe NJ (eds) Models in dermatology. Karger, Basel, pp 71–78

Wu XM, Todo H, Sugibayashi K (2006) Effects of pretreatment of needle puncture and sandpaper abrasion on the in vitro skin permeation of fluorescein isothiocyanate (FITC)-dextran. Int J Pharm 316:102–108

Wu XM, Todo H, Sugibayashi K (2007) Enhancement of skin permeation of high molecular compounds by a combination of microneedle pretreatment and iontophoresis. J Control Release 118:189–195

Xie Y, Xu B, Gao Y (2005) Controlled transdermal delivery of model drug compounds by MEMS microneedle array. Nanomedicine 1:184–190

Yang M, Zahn JD (2004) Microneedle insertion force reduction using vibratory actuation. Biomed Microdevices 6:177–182

# Part V
# Vaccine Evaluation, Production and Distribution

# Animal Models for Evaluation
# of Influenza Vaccines

**Ralph A. Tripp and S. Mark Tompkins**

**Contents**

**Abstract** Influenza viruses are emerging and re-emerging viruses that cause worldwide epidemics and pandemics. Despite substantial knowledge of the mechanisms of infection and immunity, only modest progress has been made in human influenza vaccine development. The rational basis for influenza vaccine development originates in animal models that have helped us to understand influenza species barriers, virus–host interactions, factors that affect transmission, disease pathogenesis, and disease intervention strategies. As influenza evolution can surmount species barriers and disease intervention strategies that include vaccines, our need for appropriate animal models and potentially new host species will evolve to meet these adaptive challenges. This chapter discusses animal models for evaluating vaccines and discusses the challenges and strengths of these models.

R.A. Tripp(✉) and S.M. Tompkins
Center for Disease Intervention, Animal Health Research Center,
University of Georgia, 111 Carlton St., Athens, GA 30602, USA
e-mail: ratripp@uga.edu

R.W. Compans and W.A. Orenstein (eds.), *Vaccines for Pandemic Influenza*,
Current Topics in Microbiology and Immunology 333,
DOI 10.1007/978-3-540-92165-3_19, © Springer-Verlag Berlin Heidelberg 2009

## Abbreviations

| | |
|---|---|
| CAIV | Cold-adapted and attenuated reassortant influenza vaccine |
| CTL | Cytotoxic T cell |
| DNA | Deoxyribonucleotides |
| HI | Hemagglutination-inhibiting |
| LAIV | Live attenuated influenza virus |
| M1 | Matrix 1 protein |
| M2 | Matrix 2 protein |
| NHP | Nonhuman primate |
| NP | Nucleoprotein |
| OAS | Original antigenic sin |
| PR5 | Puerto Rico 5 |
| PR8-f | PR8 that had been passaged 91 times in ferrets |
| PR8-m | PR8 that had been passaged 332 times in mice |
| SA$\alpha$2,3Gal | Sialic acids with an $\alpha$2,3 linkage |
| SA$\alpha$2,6Gal | Sialic acids with an $\alpha$2,6 linkage |
| TIV | Trivalent inactivated vaccine |

## 1 Introduction

A variety of animal models have been critical to the foundation of human influenza vaccine development. Animal models are used to characterize the host and its immune response to infection, disease course, pathogenesis, and transmission of infectious diseases, and they also enable the development of diagnostics, therapeutics, and vaccines. Indeed, diseases lacking animal models are poorly understood in comparison to those with a good animal model. Animal models also enable preclinical testing of the safety and efficacy of investigational drugs and the safety and immunogenicity of investigational vaccines. Despite the number of scientific and medical barriers that animal models have helped to overcome, there are also political and social barriers that need to be addressed for vaccine development in particular, such as age bias, vaccine supply ignorance and fear of vaccines, an emerging anti-vaccine movement, issues with social reimbursement of vaccine costs, and inadequate systems and procedures for implementing vaccination. The following sections summarize the role of animal models and their contributions to human influenza vaccine development.

### 1.1 Isolation of Influenza Virus

Animal models have played an important role in our understanding of the spectrum of disease caused by influenza viruses. During the early twentieth century, viruses were generally identified and isolated by inoculation and passage in experimental animals (Eyler 2006). Likewise, the first influenza virus to be characterized (by Richard Shope in 1930; Shope 1931) was an H1N1 virus isolated from the lungs of diseased hogs, which was filtered and transferred to naïve swine, resulting in acute respiratory infection (Shope ,b). The first human influenza virus isolate, A/WS/33 (named after Wilson Smith who isolated the virus), was identified by infecting ferrets with filtered throat washings. The initial ferret infection showed that the disease could be transmitted by contact with infected animals or passaged by experimental infection with nasal washings from diseased ferrets (Smith et al. 1933). It was also shown that transmission of human influenza to ferrets was possible using sputum from patients collected during a 1934 epidemic in Puerto Rico (Francis 1934). This H1N1 influenza virus isolate, named Puerto Rico 5 (PR5), was passaged repeatedly in ferrets and was inadvertently transmitted back to a laboratory worker during the course of the animal studies (Francis 1934). Later, ferret passages of this virus were used to inoculate mice and caused variable disease; however, at the third mouse passage, the PR5 isolate was consistently lethal in mice (Francis 1934). The PR5 strain was lost, but PR8 (A/Puerto Rico/8/34) was subsequently derived (Francis 1937). By 1940, PR8 had been passaged 91 times in ferrets (PR8-f), and, after minimal passages in ferrets, 332 times in mice (PR8-m) (Horsfall et al. 1941). While the precise lineage may be uncertain, the PR8 strain of influenza (A/PR/8/34) remains a widely used laboratory strain. For the next 30 years, influenza virus was the most extensively studied viral pathogen of humans. The goal of this international effort was to develop a safe and efficacious vaccine. While some of this work was conducted in human trials, animal models were extensively used to maintain virus stocks, as well as in vaccine design, preliminary efficacy studies, and in the detection of antibodies against specific influenza viruses (Eyler 2006). By the early 1940s, World War II raised fears of a repeat of the Spanish influenza pandemic that was observed during World War I. These concerns drove the formation of the Commission on Influenza, which expanded the influenza vaccine program and focused ongoing research efforts.

## 2 Human Influenza Vaccines

### 2.1 The Early Years

The discovery of influenza A virus in 1933 (Smith 1933) and the development of an efficacious vaccine by the Commission on Influenza of the US Armed Forces Epidemiological Board during World War II marked the advent of intensive animal model studies in the development of influenza A vaccines (Francis 1953). However, once an early efficacious vaccine had been developed, limited attention was paid to additional influenza vaccine development until the 1946–1947 H1N1 influenza A epidemic in which there was lack of vaccine protection (Rasmussen et al. 1948). During the 1946–1947 H1N1 virus outbreak, it was noted that the antigenic specificity differed markedly from that of the viral antigens in the current vaccine based on findings using hemagglutination inhibition assays with ferret antisera (Hirst 1947a). Interestingly, during this scientific investigation it was noted that the viral antigenic specificity differed between individual ferret-derived antisera; thus, chickens were intraperitoneally injected with embryonated egg-passaged virus. The viruses did not proliferate in the chickens but gave potent antibody responses that were not biased in specificity compared to the different ferret antisera. These early studies of immunologic specificity among various influenza virus strains contributed to the breakthrough discovery that there was nonrandom progressive antigenic change in influenza A virus surface proteins isolated in successive years—a feature now termed antigenic drift (Hilleman et al. 1950). Emergence of influenza drift variants continues to be an issue with influenza vaccine efficacy, as evidenced by recent vaccine failures during the 2007–2008 influenza season (Branch 2008).

## 2.2 Vehicles for Scientific and Biomedical Discovery

The use of multiple animal species to model human disease was highlighted during World War II as the United States prepared to deal with the potential for biological warfare. The idea that vaccine countermeasures against viruses could be tested in valid animal models was intrinsic to the military research programs at that time and continues today. The use of animals as surrogates for humans in efficacy trials came under FDA scrutiny in the late 1950s because many therapeutics that were being introduced were not effective or had serious but undiscovered side effects (Anderson and Swearengen 2006). Today the use of animal models for vaccine efficacy studies are better understood, more tightly regulated, and offer a reasonable approach to developing safe and efficacious vaccines. There is a burgeoning need for animal models to evaluate influenza vaccine safety and efficacy, particularly as vaccine is increasingly used in young children, the immune suppressed, and the elderly—groups that have traditionally not responded well to the vaccine. In addition to the use of novel and sometimes complex influenza vaccine development strategies, as well as the push toward cell-based influenza vaccine development, it is important to have ways to study influenza vaccine safety and effectiveness prior to human studies and use. As vaccine development relies heavily on appropriate animal model studies, it is becoming clearer that the translation of animal model findings to the human condition is not straightforward and has limitations.

Our understanding of the immunogenic potential of human influenza vaccines has relied on results learned from animal models. To better understand some of the mechanisms that lead to vaccine inadequacy or failure, substantial research has focused on determining the relationship between laboratory and clinical measures of protection induced by modern influenza vaccines. These studies are often specific to the type of the influenza virus vaccine e.g., inactivated vs. live attenuated. For the inactivated product, indirect methods of potency quantitation have been used for evaluation. For example, early techniques to quantitate the immunogenic potential of influenza vaccines in experimental animals included antigen extinction methods, tests based upon the intranasal vaccinating dose required to inhibit replication of unadapted influenza viruses in the lungs of mice, and a two-step antigen extinction technique involving the intranasal instillation of pooled immune serum and virus mixtures into mice (Barry et al. 1974; Kilbourne 1976; Tannock et al. 1981). These and related methods are cumbersome, poorly reproducible, and rely excessively on the virulence of the mouse-adapted challenge virus. Current methods of evaluating the immunity induced by vaccination, particularly against a single strain, employ the analysis of antigenic differences first measured by means of red blood cell agglutination (Hirst 1943). This commonly used assay provides a qualitative view of antigenic differences, but it is considered inappropriate for quantitative analysis. Our increasing understanding of the immune response to vaccination or infection in animal models has provided important insights into other considerations that are used to assess vaccine potency and efficacy, including neutralizing antibody titers, mucosal IgA responses, original antigenic sin, and CD8 cytotoxic T cell responses important in heterotypic immunity.

# 3   Animal Models in Human Vaccine Development

## 3.1   The Ferret Model

The ferret was the first animal model used for influenza virus research and continues to have a major role in vaccine development. The concept of antigenic drift of the influenza virus was first charted in ferret studies, and early influenza vaccination studies in ferrets revealed important findings regarding vaccine efficacy. For example, the concept of original antigenic sin (OAS), defined as the tendency for antibodies produced in response to primary exposure to influenza antigens to suppress the creation of new and different antibodies to a new version of the influenza virus, was first observed in the ferret model (Webster 1966; Webster et al. 1976). The early finding of OAS highlighted the importance of developing vaccines with sufficient antigenic distance so as to broaden vaccine efficacy. This is particularly important today, as human influenza vaccine design for commercial translation to humans is done annually under considerable time constraints. The use of the ferret model in human vaccine development is based on three principal features: (1) influenza infection in ferrets emulates many features of the disease observed in humans; (2) human influenza A and B viruses infect ferrets without adaptation, and; (3) the physical features of ferrets, including their airways and sneeze response make them amenable for characterizing aspects of disease (Maher and DeStefano 2004). Ferrets and humans have similar clinical courses of disease (Leigh et al. 1995), and, similar to humans, the severity and time course of the disease can vary with virus strain, age and health of the animal. Infection with seasonal human influenza viruses is generally localized to the upper respiratory tract. Illness is usually acute, with clinical illness lasting up to a week in healthy individuals. During the peak of fever, which corresponds with peak virus shedding, both humans and ferrets transmit virus to each other. In both cases, transmission can occur by aerosol droplet and direct or indirect contact (fomites) (Bridges et al. 2003). However, the ferret model does have caveats, including cost, housing requirements, and availability of immunological and related reagents, which limits widespread use.

Although the ferret is a small animal model (a three-month old male weighs <1 kg), the species has a long trachea which helps to separate the upper and lower respiratory tracts, a feature similar to humans (Maher and DeStefano 2004). Importantly, influenza virus susceptibility and disease patterns seen in humans are generally recapitulated in ferrets. Influenza virus attaches via the N-acetylneuraminic acid (sialic acid; SA) linked to galactose sugars on surface glycoproteins. It is believed that influenza viruses that infect humans preferentially bind to sialic acids with an α2,6 linkage (SAα2,6Gal), while influenza viruses that infect avian species preferentially bind to sialic acids with an α2,3 linkage (SAα2,3Gal) (Palese and Shaw 2006). SAα2,6Gal receptors are found at a high density in the human respiratory tract (Baum and Paulson 1990; Matrosovich et al. 2004). The lower respiratory tract contains predominantly SAα2,6Gal, but there are also SAα2,3Gal linkages on bronchiolar cells and type II alveolar cells (Shinya et al. 2006). The ferret has a similar density and repertoire of sialic acid receptors (Leigh et al. 1989), and therefore has a similar influenza virus susceptibility (Leigh et al. 1995; Maines et al. 2006; Matrosovich et al. 2004; Piazza et al. 1991; Tumpey et al. 2007; van Riel et al. 2007).

The sialic acid expression and virus susceptibility profiles of ferrets and humans combined with their similar physical airway features translate to similar abilities to transmit influenza viruses. Ferrets are highly susceptible to human influenza virus infection and readily transmit the virus to naïve ferrets (Herlocher et al. 2001; Maher and DeStefano 2004; Maines et al. 2006; Tumpey et al. 2007) and humans (Francis 1934; Smith and Stuart-Harris 1936). For this reason, ferrets are an excellent model to study influenza virus transmission and disease intervention strategies; however, they are also a difficult model to work with. Influenza-naïve ferrets can be difficult to acquire, particularly during the influenza season, and naïve ferrets can readily become infected through environmental exposure if appropriate barrier conditions are not maintained during shipping and housing. Importantly, unlike some animal models of influenza infection, seropositive ferrets are generally susceptible to reinfection with variant viruses (Herlocher et al. 2001), although there is evidence of limited heterosubtypic immunity as well (Yetter et al. 1980).

## 3.2   The Immune Response in Ferrets

The immune response to influenza virus infection in ferrets is a double-edged sword—both a strength and a weakness—in the animal model. The ferret serum antibody response to influenza virus infection or vaccination is very similar to the response seen in humans; however, there are relatively few tools available for investigating parameters of the innate or cell-mediated immune response compared to the mouse model. The first isolation of human influenza virus in 1933 demonstrated that ferret immune serum would neutralize human influenza virus and that human immune serum would neutralize the virus during infection in ferrets (Smith et al. 1933). Years of influenza virus studies in the ferret model now predict that experimentally infected or vaccinated ferrets produce neutralizing or hemagglutination-inhibiting (HI) serum antibody responses with the same virus reactivity as would be generated in human antibody responses. For this reason, the cross-reactivity of ferret antisera to circulating human influenza virus strains is regularly used to identify strains to be included in annual formulations of the influenza virus vaccine (Jan and de Jong 2000). It is important to note that neutralizing serum antibody titers in ferrets do not correlate with prevention of upper respiratory tract infection; however, they do correlate with decreased severity of disease and prevention of lower respiratory tract infection and pneumonia. Mucosal antibody responses have also been shown to contribute to protection. The cellular immune response in ferrets has also been characterized, and similar cytotoxic T cell (CTL) responses have been noted to those of humans, indicating that CTLs play a major role in recovery from infection (Maher and DeStefano 2004). While extremely detailed studies of the immune response to influenza virus infection have been carried out in mice, these thorough studies have not been done in ferrets. This is due to a lack of immunologic reagents, including antibodies to cellular markers, cytokine reagents, and genomic tools. The absence of these tools, which are commonplace for murine studies, has limited the breadth of the ferret model. With the recent renewal of interest in influenza research and vaccine development, many of these reagents are now becoming available and will eventually eliminate this shortcoming in the ferret model.

Another related issue with the ferret model is the lack of inbred animals. Responses in ferrets are not uniform, which is both a strength and a weakness. Results may be more difficult to assess, due to variability; however, the conclusions may be more relevant to human studies for the very same reason. Several breeders are developing inbred and specific pathogen-free ferrets, which will overcome these potential hurdles, as previously noted.

Despite these issues, ferrets are currently the "gold standard" for influenza virus animal models. With concerns that H5N1 viruses might cause a pandemic, there has been a resurgence of interest in developing novel influenza vaccines, focused on H5 and a variety of platforms, including live attenuated, DNA, particle-based, inactivated, and adjuvanted vaccines. Each of these has been used in immunogenicity and challenge studies in ferrets (Subbarao and Luke 2007). These studies have presented a number of promising candidates, some of which are in clinical studies, and one of which is now licensed for use in the United States (FDA 2007). Moreover, studies comparing immunogenicity and protection in ferrets have uncovered an important issue concerning the classical correlates of protection and the actual level of protection from challenge with an H5N1 virus. Using the ferret model, it has been demonstrated that an inactivated whole-virion H5N1 vaccine could protect animals against infection with highly pathogenic H5N1 avian influenza despite inducing poor hemagglutination inhibition and virus neutralizing serum antibody titers (Lipatov et al. 2006). The disassociation of serum antibody responses from protection from challenge highlights the critical need for vaccine testing in animal models of disease.

## 3.3   The Murine Model

The first North American influenza isolate identified in 1934 was quickly moved from ferrets into mice and shown to cause disease in this model (Francis 1934). At the same time, researchers in Europe were demonstrating that mice were susceptible to both swine and human influenza viruses, and they showed that immune serum from immunized ferrets or horses could neutralize the infectivity of influenza virus prior to infection in mice (Andrewes et al. 1934). Since these seminal studies, mice have been widely used in all aspects of influenza virus research. The mouse model has several advantages over ferrets in that there are numerous inbred mouse strains that are commercially available, including mutant, congenic, transgenic, gene knockout, and combination mutant transgenic species. Also, the size and husbandry practices for mouse colonies make them affordable, mice have been extensively characterized, and there is an extensive array of reagents available for the study of immune responses (Novak et al. 1993). Together, these strengths allow researchers to execute in-depth studies using relatively large numbers of experimental subjects. The utility of the mouse model of influenza virus infection is reflected in the extraordinary immunologic discoveries made using this system. The study of influenza virus infection in mice has resulted in our fundamental understanding of MHC restriction, the innate immune response, immunodominance, humoral immunity, and immunologic memory.

The mouse model of influenza virus infection has notable weaknesses. First, most influenza viruses do not naturally cause disease in mice. There is no experimental evidence that human influenza viruses can be directly transmitted from humans to mice. The first successful influenza infections in mice occurred after only three passages in ferrets (Andrewes et al. 1934; Francis 1934). In later studies, human influenza A viruses were cultivated in embryonated chicken eggs prior to infection in mouse models. In these cases, the viruses replicated well but caused asymptomatic infections with little or no pathology, even when given at very high titers (Hirst 1947b; Novak et al. 1993). Murine infection with nonadapted influenza viruses has revealed that infection in mice is variable, but once established, replicating virus can be isolated from the lung, trachea, and nares for at least 5–6 days (Novak et al. 1993).

Repeated passage of human influenza viruses in mouse lungs can quickly adapt the virus to the mouse and result in virulent mouse-adapted viruses (Hirst 1947b; Novak et al. 1993; Smeenk and Brown 1994). Mouse-adapted viruses can cause severe pathology, morbidity and mortality, and lethal pneumonia caused by mouse-adapted influenza virus infection is similar to the pathology seen in human lower respiratory tract infections (Smeenk and Brown 1994). In some cases, limiting the inoculum and sedation of the mouse can limit the infection to the upper respiratory tract, resulting in apathogenic infection (Iida and Bang 1963; Novak et al. 1993). Whether infecting with wild-type or mouse-adapted influenza viruses, infected mice do not shed virus (Lowen et al. 2006). As mice can only be infected experimentally, the mouse model is not useful for transmission studies.

## 3.4 Vaccine Development in the Mouse Model

A substantial issue with using the mouse model for vaccine development is the relative ease in which vaccinated mice can be protected against challenge, as previously reviewed in studies of heterosubtypic immunity (Epstein 2003). In these studies, immune responses generated against conserved viral vaccine antigens, such as nucleoprotein (NP) or matrix (M1), were generally cell mediated (i.e., CTL specific for the NP or M1 proteins). However, related studies in humans have provided limited evidence that similar mechanisms of protection are efficacious (Epstein 2006; Steinhoff et al. 1993). While vaccine studies in murine models provide a wealth of information and an initial assessment of potential efficacy, there is concern that the findings will translate poorly to the clinic. Moreover, the rising concern regarding preventing transmission as a priority in vaccine development decreases the value of murine studies, as the mouse does not transmit influenza virus during infection.

## 3.5 Other Rodent Models

The guinea pig is a relatively new model for the study of influenza virus. Their use has been limited by the availability of the murine model; however, more recently the guinea pig has received attention as a potential model for influenza virus transmission (Lowen et al. 2006). Based upon an account of pneumonia in a laboratory guinea pig colony during the 1918 influenza epidemic, the susceptibility of the Hartley strain of guinea pigs to human influenza virus infection and their ability to transmit the virus to naïve animals was explored (Lowen et al. 2006). Wild-type, unadapted influenza virus was shown to replicate in both the upper and lower respiratory tracts of the Hartley strain guinea pigs, and to transmit to naïve animals via droplet. While high titers of virus were found in both the lungs and nasal secretions, the infection was completely asymptomatic. Interestingly, wild-type, unadapted influenza virus infection of strain 13 guinea pigs with the same virus resulted in clinical disease, although transmissibility was not addressed (Lowen et al. 2006). Similar to the ferret model, there are limited reagents available for guinea pigs. This, combined with the apparent absence of disease, reduces their value in vaccine studies; however, their size and the availability of specific pathogen-free inbred strains may make this model more appealing for prospective influenza virus transmission studies.

The cotton rat was first described as a model for influenza virus infection in 1987 (Eichelberger 2007). The cotton rat has a similar disease course to humans; however, there is no evidence of transmission. Influenza virus can be isolated from both the upper and lower respiratory tracts following intranasal infection (Ottolini et al. 2005). The cotton rat shows clinical signs of disease that include weight loss, and has pulmonary cellular infiltrates similar to humans with bronchopneumonia. A key strength of the cotton rat model is the ability to infect it with wild-type, unadapted influenza viruses (Eichelberger 2007). Moreover, while not as expansive as the mouse model, a variety of reagents are available for characterizing the immune response. These features make the cotton rat an appealing model for vaccine and immune response studies to influenza virus infection.

Syrian hamsters have also been used as a disease model for influenza virus infection. Like the cotton rat, the hamster is susceptible to infection with unadapted human influenza viruses. In contrast, the hamster supports higher titers of virus in the lung than in the upper respiratory tract (Heath et al. 1983). Other than these defining features, the Syrian hamster has limited application as an animal model for influenza virus. The other rodent species have equivalent or better features of disease and/or a broader utility because of the availability of reagents.

## 3.6   Nonhuman Primate Models

Serological studies have found that many native nonhuman primate species are seropositive for human influenza viruses (Clyde 1980), suggesting that they may be a natural host for infection and a potent model to study influenza virus. As such, a variety of nonhuman primate (NHP) species have been tested for their ability to support influenza virus infection and the disease associated with infection. Rhesus macaques are susceptible to human influenza virus infection. Interestingly, intranasal instillation of influenza virus has not been successful at establishing infection, but aerosol or intratracheal delivery causes infection, clinical symptoms (in some cases), and seroconversion (Berendt 1974). Seroconversion resulted in protection against repeated challenge. Variability in clinical symptoms was suggested to be related to strain virulence.

Squirrel monkeys have also been successfully used as models for influenza virus infection. A prominent example is provided by the studies done in the late 1970s in which squirrel monkeys were inoculated intratracheally with A/New Jersey/76, a swine virus isolated at Fort Dix that threatened to become pandemic. At the time no was information available on the transmissibility or pathogenicity of A/New Jersey/76. A decision was made to develop new vaccines, and NHP disease models were needed to test the immunogenicity of these proposed vaccines. Squirrel monkeys infected with A/New Jersey/76 were shown to shed virus and to develop clinical disease (Berendt and Hall 1977). Similar results were also shown in squirrel monkeys infected with A/Aichi/2/68 virus; symptoms and virus shedding were shown to be similar to what was seen in human infections (Murphy et al. 1980, 1982a,b, 1983). The similarities in disease between humans and squirrel monkeys have elevated the squirrel monkey model as a reasonable disease model to measure influenza virus virulence.

NHP models of influenza are generally less utilized than other models because of the lack of availability of animals, difficulty in handling, and the need for special facilities and veterinary care. However, there are advantages to NHP studies, including human reagent cross-reactivity, which can be used in Old World primates such as rhesus and cynomolgus macaques. Also, the size and similar physiology of many NHPs enable repeated sampling and monitoring of symptoms related to humans; the genetic relatedness to humans and outbred populations may enable more meaningful vaccine efficacy studies. These advantages, combined with the similarities to the disease observed in humans, make the NHP model of influenza virus infection a very powerful research tool.

## 3.7 Overview of Animal Models

In the 75 years since the first isolation of a human influenza virus, both ferrets and mice have continued to play a central role in our understanding of the host response to influenza virus infection, in developing correlates of protection against infection, and the development of vaccines and therapeutic drugs. Efforts towards the development of improved or even "universal" vaccines and (in the wake of drug resistance) new antiviral drugs continue. Mice and ferrets have an important role in these studies; however, there are other animal model options that can perhaps be used to better address the immunobiology of virus infection and the development of disease intervention strategies. These include other rodents (guinea pig, hedgehog, hamster, and cotton rat), birds, swine, nonhuman primates (rhesus macaque, cynomolgus macaques, squirrel monkeys, and others), and even humans. Even the observation by Frank MacFarlane Burnet that embryonated chicken eggs could support the growth of relatively pure, high-titer influenza virus stocks (Burnet 1940a,b), a critical step in influenza vaccine development (Eyler 2006), is arguably the development of a animal model. As studies continue and animal models develop, it is likely that the findings will lead to a better understanding of human influenza vaccine development, safety, and efficacy.

## 4 Human Vaccines: The End Game

The development of the first licensed killed influenza vaccine, led by the Commission on Influenza, relied on the cultivation and purification of the virus grown in the allantoic sac of embryonated hen's egg (Burnet 1941). This vaccine was prepared by purifying and concentrating the virus, then by absorption to and elution from red blood cells, and finally inactivation using formaldehyde (Hirst 1942). Subsequently, this crude but efficacious vaccine preparation was replaced by centrifuge-purified vaccine, which is still the basic format for much of today's influenza vaccine production (Stanley 1945). Killed influenza vaccines produced in eggs have proven to be safe, efficacious, and well tolerated, but caveats remain, such as the potential presence of residual egg proteins, the possibility that avian leukosis virus may be present in embryonated eggs used for vaccine production, and compromised production potential when highly pathogenic avian influenza virus is circulating, to name a few. To reduce some of the issues associated with killed vaccines, today's version consists of subvirion and purified surface antigen preparations made as a trivalent inactivated vaccine (TIV). Today's TIV contains one influenza A (H3N2) virus, one influenza A (H1N1) virus, and one influenza B virus, which may change from year to year based on global influenza surveillance and the emergence of new strains.

Subunit influenza vaccines are now used widely throughout the world and are the only inactivated vaccines used in the United States. These vaccines, given as a single dose, are adequate for boosting immunologic memory, but subunit vaccines such as split vaccines are often poorly immunogenic in persons who have not been primed through previous infection or vaccination (Hilleman 1977; Parkman et al. 1977; Wareing and Tannock 2002). The focusing of recent attention on the development of a universal subunit vaccine (i.e., a conserved M2 protein vaccine) is meant to prevent loss of vaccine effectiveness through antigenic drift and shift, because the M2 protein is highly antigenically conserved and it has been shown in mice that antibody directed against it prevents infection (Fan et al. 2004; Fiers et al. 2004; Neirynck et al. 1999; Slepushkin et al. 1995; Tompkins et al. 2007). Recombinant DNA plasmid vaccines, first demonstrated to vaccinate mice for humoral and cellular immunity to HA and NP, were shown to protect against lethal challenge with virulent PR8 virus (Donnelly et al. 1994; Montgomery et al. 1993; Ulmer et al. 1994). DNA vaccine approaches are still experimental. They are readily manipulated and manufactured, and vaccination results in antigens being expressed in the cell cytosol, where they are readily loaded by both class I and II histocompatibility antigens (Dean 2005; Laddy and Weiner 2006; Webster and Robinson 1997).

Live attenuated influenza virus (LAIV) vaccines have been used for many years in Russia with success (Aleksandrova et al. 1986; Desheva Iu et al. 2002; Kendal 1997a,b; Klimov et al. 1995; Rudenko et al. 1993; Zhilova et al. 1986). Intensive research in the United States led to the development of a cold-adapted and attenuated reassortant influenza vaccine (CAIV) into which any desired HA or NA can be inserted (Block 2004; Maassab et al. 1999). LAIV vaccines use a genetic reassortment method involving a combination of six genes from a master donor strain that code for internal viral proteins and two genes from contemporary wild virus strains that code for the desired HA and NA antigens (Ambrose et al. 2006; Belshe et al. 2004; Targonski and Poland 2004). The resulting vaccine viruses are attenuated, temperature sensitive, genetically stable and nontransmissible. They offer substantial advantages over TIV or subunit vaccines as they can are administered intranasally without the use of needles, induce a broad mucosal and cellular mediated immune response, and LAIV has demonstrated broader serum antibody responses than TIV, particularly against mismatched influenza A (Ambrose et al. 2006; Glezen 2006; Lynch and Walsh 2007; Nichol 2001; Piedra et al. 2005).

Although a variety of safe and effective human vaccines and vaccine platforms are now available, there is little doubt that vaccine strategies will evolve and that appropriate animal models will play an important role in these developments. Of the plethora of animal models to choose from, reagents, rationale, cost-effectiveness, and animal welfare issues will in part dictate the models chosen. Issues remain regarding the translation of findings from one animal model to another, and from animal models to humans, but much has been learned and many of the caveats recognized. Animal models will remain an integral part of human influenza vaccine development, safety, and efficacy studies, and can help to bridge the gaps in our understanding of the immunobiology of influenza virus infection.

**Acknowledgment** The authors would like to thank the Georgia Research Alliance for supporting their research efforts and enabling their animal modeling studies.

# References

Aleksandrova GI, Medvedeva TE, Polezhaev FI, Garmashova LM, Budilovskii GN (1986) Reactogenicity, genetic stability and effectiveness of a live recombinant influenza vaccine for children designed on the base of a cold-adapted attenuation donor A/Leningrad/134/47/57. Vopr Virusol 31:411–414

Ambrose CS, Walker RE, Connor EM (2006) Live attenuated influenza vaccine in children. Semin Pediatr Infect Dis 17:206–212

Anderson AO, Swearengen JR (2006) Scientific and ethical importance of animal models in biodefense research. CRC, Boca Raton

Andrewes CH, Laidlaw PP, Smith W (1934) The susceptibility of mice to the virus of human and swine influenza. Lancet 224:859–862

Barry DW, Staton E, Mayner RE (1974) Inactivated influenza vaccine efficacy: diminished antigenicity of split-product vaccines in mice. Infect Immun 10:1329–1336

Baum LG, Paulson JC (1990) Sialyloligosaccharides of the respiratory epithelium in the selection of human influenza virus receptor specificity. Acta Histochem Suppl 40:35–38

Belshe R, Lee MS, Walker RE, Stoddard J, Mendelman PM (2004) Safety, immunogenicity and efficacy of intranasal, live attenuated influenza vaccine. Expert Rev Vaccines 3:643–654

Berendt RF (1974) Simian model for the evaluation of immunity to influenza. Infect Immun 9:101–105

Berendt RF, Hall WC (1977) Reaction of squirrel monkeys to intratracheal inoculation with influenza/A/New Jersey/76. (swine) virus. Infect Immun 16:476–479

Block SL (2004) Role of influenza vaccine for healthy children in the US. Paediatr Drugs 6:199–209

Branch I. (ed). (2008) US flu activity report for week ending March 1. (Week 9), posted March 7. Centers for Disease Control and Prevention, Atlanta

Bridges C, Kuehnert M, Hall C (2003) Transmission of influenza: implications for control in health care settings. Clin Infect Dis 37:1094–1101

Burnet FM (1940a) Influenza virus infections of the chick embryo lung. Br J Exp Path 21:147–153

Burnet FM (1940b) Virus infections of the chick embryo by the amniotic route. 1. General character of the infections. Aust J Exp Biol Med Sci 18:353–360

Burnet FM (1941) Growth of influenza virus in the allantoic cavity of the chick embryo. Aust J Exp Biol Med Sci 19:291–295

Clyde WA Jr (1980) Experimental models for study of common respiratory viruses. Environ Health Perspect 35:107–112

Dean HJ (2005) Epidermal delivery of protein and DNA vaccines. Expert Opin Drug Deliv 2:227–236

Desheva Iu A, Danini GV, Grigor'eva EP, Donina SA, Kiseleva IV, Rekstin AR, Ermakova LA, Natsina VK, Nikolaeva VM, Lonskaia NI, El'shina GA, Zhavoronkov VG, Drinevskii VP, Erofeeva MK, Naikhin AN, Rudenko LG (2002) The investigation of the safety, genetic stability and immunogenicity of live influenza vaccine for adults in vaccination of 3–6 years old children. Vopr Virusol 47:21–24

Donnelly JJ, Ulmer JB, Liu MA (1994) Immunization with DNA. J Immunol Methods 176:145–152

Eichelberger MC (2007) The cotton rat as a model to study influenza pathogenesis and immunity. Viral Immunol 20:243–249

Epstein SL (2003) Control of influenza virus infection by immunity to conserved viral features. Expert Rev Anti Infect Ther 1:627–638

Epstein SL (2006) Prior H1N1 influenza infection and susceptibility of cleveland family study participants during the H2N2 pandemic of 1957: an experiment of nature. J Infect Dis 193:49–53

Eyler JM (2006) De Kruif's boast: vaccine trials and the construction of a virus. Bull Hist Med 80:409–438

Fan J, Liang X, Horton MS, Perry HC, Citron MP, Heidecker GJ, Fu TM, Joyce J, Przysiecki CT, Keller PM, Garsky VM, Ionescu R, Rippeon Y, Shi L, Chastain MA, Condra JH, Davies ME, Liao J, Emini EA, Shiver JW (2004) Preclinical study of influenza virus A M2 peptide conjugate vaccines in mice, ferrets, and rhesus monkeys. Vaccine 22:2993–3003

FDA (2007) Product approval information, influenza virus vaccine, H5N1. Office of Vaccines Research and Review, Center for Biologics Evaluation and Research, Rockville

Fiers W, De Filette M, Birkett A, Neirynck S, Min Jou W (2004) A "universal" human influenza A vaccine. Virus Res 103:173–176

Francis T Jr (1934) Transmission of influenza by a filterable virus. Science 80:457–459

Francis T (1937) Epidemiological studies in influenza. Am J Public Health Nations Health 27:211–225

Francis T (1953) Vaccination against influenza. Bull WHO 8:725–741

Glezen WP (2006) Herd protection against influenza. J Clin Virol 37:237–243

Heath AW, Addison C, Ali M, Teale D, Potter CW (1983) In vivo and in vitro hamster models in the assessment of virulence of recombinant influenza viruses. Antiviral Res 3:241–252

Herlocher ML, Elias S, Truscon R, Harrison S, Mindell D, Simon C, Monto A (2001) Ferrets as a transmission model for influenza: sequence changes in HA1 of type A (H3N2) virus. J Infect Dis 184:542–546

Hilleman MR (1977) Serologic responses to split and whole swine influenza virus vaccines in the light of the next influenza pandemic. J Infect Dis 136:S683–S685

Hilleman MR, Mason RP, Buescher EL (1950) Antigenic pattern of strains of influenza A and B. Proc Soc Exp Biol Med 75:829–835

Hirst GK (1942) Adsorption of influenza hemagglutinins and virus by red blood cells. J Exp Med 76:195–209

Hirst GK (1943) Studies of antigenic differences among strains of influenza A by means of red cell agglutination. J Exp Med 78:407–423

Hirst GK (1947a) Comparison of influenza virus strains from three epidemics. J Exp Med 86:367–381

Hirst GK (1947b) Studies on the mechanism of adaptation of influenza virus to mice. J Exp Med 86:357–366

Horsfall FL Jr, Lennette EH, Rickard ER (1941) A complex vaccine against influenza A virus: quantitative analysis of the antibody response produced by man. J Exp Med 73:335–355

Iida T, Bang FB (1963) Infection of the upper respiratory tract of mice with influenza A virus. Am J Epidemiol 77:169–176

Jan C, de Jong WE (2000) Mismatch between the 1997/1998 influenza vaccine and the major epidemic A(H3N2) virus strain as the cause of an inadequate vaccine-induced antibody response to this strain in the elderly. J Med Virol 61:94–99

Kendal AP (1997a) Cold-adapted live attenuated influenza vaccines developed in Russia: can they contribute to meeting the needs for influenza control in other countries. Eur J Epidemiol 13:591–609

Kendal AP (1997b) Cold-adapted live attenuated influenza vaccines developed in Russia: can they contribute to meeting the needs for influenza control in other countries? Eur J Epidemiol 13:591–609

Kilbourne ED (1976) Comparative efficacy of neuraminidase-specific and conventional influenza virus vaccines in induction of antibody to neuraminidase in humans. J Infect Dis 134:384–394

Klimov AI, Egorov AY, Gushchina MI, Medvedeva TE, Gamble WC, Rudenko LG, Alexandrova GI, Cox NJ (1995) Genetic stability of cold-adapted A/Leningrad/134/47/57. (H2N2) influenza virus: sequence analysis of live cold-adapted reassortant vaccine strains before and after replication in children. J Gen Virol 76(Pt 6):1521–1525

Laddy DJ, Weiner DB (2006) From plasmids to protection: a review of DNA vaccines against infectious diseases. Int Rev Immunol 25:99–123

Leigh MW, Cheng PW, Boat TF (1989) Developmental changes of ferret tracheal mucin composition and biosynthesis. Biochemistry 28:9440–9446

Leigh MW, Connor RJ, Kelm S, Baum LG, Paulson JC (1995) Receptor specificity of influenza virus influences severity of illness in ferrets. Vaccine 13:1468–1473

Lipatov A, Hoffmann E, Salomon R, Yen H-L, Webster R (2006) Cross-protectiveness and immunogenicity of influenza A/Duck/Singapore/3/97(H5) vaccines against infection with A/Vietnam/1203/04(H5N1) virus in ferrets. J Infect Dis 194:1040–1043

Lowen AC, Mubareka S, Tumpey TM, Garcia-Sastre A, Palese P (2006) The guinea pig as a transmission model for human influenza viruses. Proc Natl Acad Sci USA 103:9988–9992

Lynch JP 3rd, Walsh EE (2007) Influenza: evolving strategies in treatment and prevention. Semin Respir Crit Care Med 28:144–158

Maassab H, Herlocher ML, Bryant ML (1999) Live influenza vaccines. In: Plotkin OW, Mortimer SA. (ed) Vaccines. Saunders, Philadelphia, pp 909–927

Maher JA, DeStefano J (2004) The ferret: an animal model to study influenza virus. Lab Anim. (NY) 33:50–53

Maines TR, Chen LM, Matsuoka Y, Chen H, Rowe T, Ortin J, Falcon A, Nguyen TH, Mai le Q, Sedyaningsih ER, Harun S, Tumpey TM, Donis RO, Cox NJ, Subbarao K, Katz JM (2006) Lack of transmission of H5N1 avian-human reassortant influenza viruses in a ferret model. Proc Natl Acad Sci USA 103:12121–12126

Matrosovich MN, Matrosovich TY, Gray T, Roberts NA, Klenk H-D (2004) Human and avian influenza viruses target different cell types in cultures of human airway epithelium. Proc Natl Acad Sci USA 101:4620–4624

Montgomery DL, Shiver JW, Leander KR, Perry HC, Friedman A, Martinez D, Ulmer JB, Donnelly JJ, Liu MA (1993) Heterologous and homologous protection against influenza A by DNA vaccination: optimization of DNA vectors. DNA Cell Biol 12:777–783

Murphy BR, Sly DL, Hosier NT, London WT, Chanock RM (1980) Evaluation of three strains of influenza A virus in humans and in owl, cebus, and squirrel monkeys. Infect Immun 28:688–691

Murphy BR, Hinshaw VS, Sly DL, London WT, Hosier NT, Wood FT, Webster RG, Chanock RM (1982a) Virulence of avian influenza A viruses for squirrel monkeys. Infect Immun 37:1119–1126

Murphy BR, Sly DL, Tierney EL, Hosier NT, Massicot JG, London WT, Chanock RM, Webster RG, Hinshaw VS (1982b) Reassortant virus derived from avian and human influenza A viruses is attenuated and immunogenic in monkeys. Science 218:1330–1332

Murphy BR, Harper J, Sly DL, London WT, Miller NT, Webster RG (1983) Evaluation of the A/Seal/Mass/1/80 virus in squirrel monkeys. Infect Immun 42:424–426

Neirynck S, Deroo T, Saelens X, Vanlandschoot P, Jou WM, Fiers W (1999) A universal influenza A vaccine based on the extracellular domain of the M2 protein. Nat Med 5:1157–11563

Nichol KL (2001) Live attenuated influenza virus vaccines: new options for the prevention of influenza. Vaccine 19:4373–4377

Novak M, Moldoveanu Z, Schafer DP, Mestecky J, Compans RW (1993) Murine model for evaluation of protective immunity to influenza virus. Vaccine 11:55–60

Ottolini MG, Blanco JCG, Eichelberger MC, Porter DD, Pletneva L, Richardson JY, Prince GA (2005) The cotton rat provides a useful small-animal model for the study of influenza virus pathogenesis. J Gen Virol 86:2823–2830

Palese P, Shaw ML (2006) Orthomyxoviridae: the viruses and their replication. In: Knipe DM, Howley PM, Griffin DM, Lamb RA, Martin MA. (eds) Field's virology. Lippincott, Williams & Wilkins, Philadelphia, pp 1647–1689

Parkman PD, Hopps HE, Rastogi SC, Meyer H (1977) Summary of clinical trials of influenza virus vaccines in adults. J Infect Dis 136:S722–S730

Piazza FM, Carson JL, Hu SC, Leigh MW (1991) Attachment of influenza A virus to ferret tracheal epithelium at different maturational stages. Am J Respir Cell Mol Biol 4:82–87

Piedra PA, Gaglani MJ, Riggs M, Herschler G, Fewlass C, Watts M, Kozinetz C, Hessel C, Glezen WP (2005) Live attenuated influenza vaccine, trivalent, is safe in healthy children 18 months to

4 years, 5 to 9 years, and 10 to 18 years of age in a community-based, nonrandomized, open-label trial. Pediatrics 116:e397–e407

Rasmussen AF, Stokes JC, Smadel JE (1948) The army experience with influenza 1946–1947. Part II. Laboratory aspects. Am J Hyg 47:142–149

Rudenko LG, Slepushkin AN, Monto AS, Kendal AP, Grigorieva EP, Burtseva EP, Rekstin AR, Beljaev AL, Bragina VE, Cox N, et al. (1993) Efficacy of live attenuated and inactivated influenza vaccines in schoolchildren and their unvaccinated contacts in Novgorod, Russia. J Infect Dis 168:881–887

Shinya K, Ebina M, Yamada S, Ono M, Kasai N, Kawaoka Y (2006) Avian flu: influenza virus receptors in the human airway. Nature 440:435–436

Shope RE (1931a) Swine influenza: I. Experimental transmission and pathology. J Exp Med 54:349–359

Shope RE (1931b) Swine influenza: III. Filtration experiments and etiology. J Exp Med 54:373–385

Slepushkin VA, Katz JM, Black RA, Gamble WC, Rota PA, Cox NJ (1995) Protection of mice against influenza A virus challenge by vaccination with baculovirus-expressed M2 protein. Vaccine 13:1399–1402

Smeenk CA, Brown EG (1994) The influenza virus variant A/FM/1/47-MA possesses single amino acid replacements in the hemagglutinin, controlling virulence, and in the matrix protein, controlling virulence as well as growth. J Virol 68:530–534

Smith W, Stuart-Harris CH (1936) Influenza infection of man from the ferret. Lancet 228:121–123

Smith WA, Andrewes C, Laidlaw P (1933) A virus obtained from influenza patients. Lancet 2:66–68

Stanley WM (1945) The preparation and properties of influenza virus vaccine concentrated and purified by differential centrifugation. J Exp Med 81:193–211

Steinhoff MC, Fries LF, Karron RA, Clements ML, Murphy BR (1993) Effect of heterosubtypic immunity on infection with attenuated influenza A virus vaccines in young children. J Clin Microbiol 31:836–838

Subbarao K, Luke C (2007) H5N1 viruses and vaccines. PLoS Pathogens 3:e40

Tannock GA, Wark MC, Smith LE, Sutherland MM (1981) A clearance test in mice using non-adapted viruses to determine the immunogenicity of influenza strains. Arch Virol 70:91–101

Targonski PV, Poland GA (2004) Intranasal cold-adapted influenza virus vaccine combined with inactivated influenza virus vaccines: an extra boost for the elderly? Drugs Aging 21:349–359

Tompkins SM, Zhao ZS, Lo CY, Misplon JA, Liu T, Ye Z, Hogan RJ, Wu Z, Benton KA, Tumpey TM, Epstein SL (2007) Matrix protein 2 vaccination and protection against influenza viruses, including subtype H5N1. Emerg Infect Dis 13:426–435

Tumpey TM, Maines TR, Van Hoeven N, Glaser L, Solorzano A, Pappas C, Cox NJ, Swayne DE, Palese P, Katz JM, Garcia-Sastre A (2007) A two-amino acid change in the hemagglutinin of the 1918 influenza virus abolishes transmission. Science 315:655–659

Ulmer JB, Deck RR, DeWitt CM, Friedman A, Donnelly JJ, Liu MA (1994) Protective immunity by intramuscular injection of low doses of influenza virus DNA vaccines. Vaccine 12:1541–1544

van Riel D, Munster VJ, de Wit E, Rimmelzwaan GF, Fouchier RAM, Osterhaus ADME, Kuiken T (2007) Human and avian influenza viruses target different cells in the lower respiratory tract of humans and other mammals. Am J Pathol 171:1215–1223

Wareing MD, Tannock GA (2002) Influenza update: vaccine development and clinical trials. Curr Opin Pulm Med 8:209–213

Webster RG (1966) Original antigenic sin in ferrets: the response to sequential infections with influenza viruses. J Immunol 97:177–183

Webster RG, Robinson HL (1997) DNA vaccines: a review of developments. BioDrugs 8:273–292

Webster RG, Kasel JA, Couch RB, Laver WG (1976) Influenza virus subunit vaccines. II. Immunogenicity and original antigenic sin in humans. J Infect Dis 134:48–58

Yetter RA, Barber WH, Small PA, Jr. (1980) Heterotypic immunity to influenza in ferrets. Infect Immun 29:650–653

Zhilova GP, Ignat'eva GS, Orlov VA, Malikova EV, Maksakova VL (1986) Results of a study of the effectiveness of simultaneous immunization against influenza with live and inactivated vaccines. (1980–1983). Vopr Virusol 31:40–44

# Immunosenescence and Influenza Vaccine Efficacy

**Suryaprakash Sambhara and Janet E. McElhaney**

## Contents

**Abstract** A number of protective immune functions decline with age along with physiological and anatomical changes, contributing to the increased susceptibility of older adults to infectious diseases and suboptimal protective immune responses to vaccination. Influenza vaccination is the most cost-effective strategy to prevent complications from influenza viral infections; however, the immunogenicity and effectiveness of currently licensed vaccines in the United States is about 30–50% in preventing complications arising from influenza and preventing death from all causes during winter months in older adults. Hence, it is crucial to understand the molecular mechanisms that lead to immune dysfunction as a function of age so that appropriate strategies can be developed to enhance the disease resistance and immunogenicity of preventive vaccines, including influenza vaccines, for the older adult population.

S. Sambhara (✉)
Influenza Division, Centers for Disease Control and Prevention,
1600 Clifton Road, Atlanta, GA 30333, USA
e-mail: ssambhara@cdc.gov

J.E. McElhaney
Professor, Department of Medicine, University of British Columbia,
9B St. Paul's Hospital, 1081 Burrard Street, Vancouver, BC V6Y 1Y6, Canada

R.W. Compans and W.A. Orenstein (eds.), *Vaccines for Pandemic Influenza*,
Current Topics in Microbiology and Immunology 333,
DOI 10.1007/978-3-540-92165-3_20, © Springer-Verlag Berlin Heidelberg 2009

# 1  Introduction

A dramatic increase in the older adult population is occurring globally due to improved sanitation, preventive vaccination, development of effective antimicrobial drugs, and advances in medical sciences. This growth of the older adult population is having a major impact on healthcare, social services, and public health. In the year 2004, older adults accounted for 12.4% of the total population and required $531.5 billion in primary healthcare costs (Hartman et al. 2008). This represented almost 34% of all healthcare spending, as the cost of providing healthcare for an older adult aged 65 or above is 3–5 times greater than the cost for a younger adult. The older adult population in the USA is projected to almost double by 2030 2007(AoA). A decline in immune function leading to increased susceptibility to infectious diseases and poor adaptive immune response to vaccination is a key characteristic of aging (Miller 1996, 1997). For example, increased colonization of bacteria and yeast on the skin and mucosal surfaces, respiratory, and urogenital tracts, increased susceptibility to viral infections, and reactivation of latent viral and bacterial infections are all well documented (Gardner et al. 2006; Worley 2006; Ely et al. 2007; Htwe et al. 2007; Kovaiou et al. 2007; Simmons et al. 2007; van Duin and Shaw 2007). In general, infectious diseases such as severe acute respiratory syndrome (SARS), West Nile virus, respiratory syncytial virus (RSV), influenza, and pneumococcal infections tend to be more severe (with complications), often resulting in unfavorable outcomes among older adults when compared to those in healthy adults. In addition, the efficacy of preventive vaccines against bacterial and viral targets declines dramatically with the progression of age among older adults, clearly indicating that the dysregulated immune status referred to as "immunosenscence" is the consequence of altered physiological and anatomical functions (Ginaldi et al. 2001; Aw et al. 2007). Hence, in this review, we will address the status of innate and adaptive immune functions in aging, the current state of influenza vaccines and their efficacy in older adults, and strategies that need to be considered to protect them against influenza.

# 2  Immune Status in Aging

## 2.1  Innate Immunity

Innate immunity was considered "nonspecific" and received secondary importance when compared to antigen-specific adaptive immune functions until the late 1990s. However, with the discovery of Toll-like receptors (TLRs), innate immunity is now recognized to be crucial to the survival of species. Therefore, understanding innate immunity can offer newer insights into the development of novel immunomodulators and antimicrobials (Hoffmann et al. 1999; Medzhitov and Janeway 2000a,b; Imler and Hoffmann 2001). Since the discovery of TLRs, several other pathogen-sensing receptor families have been identified over the last decade (Bingle and Craven 2002; Kang et al. 2002; Lu et al. 2002; Holmskov et al. 2003; Yoneyama et al. 2004; Martinon and Tschopp 2005; Ting and Davis 2005; Brown 2006; Takaoka et al. 2007). These families evolved to overcome microbial strategies and their metabolic needs in order to eliminate them. TLRs are expressed either as soluble molecules on the cell membrane or in vesicular compartments, or in the cytosol, as shown in Fig. 1. These pathogen sensors recognize structural components of pathogens and activate signal transduction cascades, leading to gene transcription with several outcomes, such as activation of antibacterial and antiviral defenses, secretion of proinflammatory cytokines and chemokines, tissue repair in the event of damage, and activation of adaptive immune responses. In most cases, the precise structure or sequence of the pathogen signature that stimulate the innate immune receptors is not well defined.

The dynamic barrier against infectious diseases, the epithelial lining of skin, gastrointestinal, respiratory, and urogenital systems, prevents the colonization and entry of potential pathogens into the body's interior, which is sterile (Ganz 2002). These epithelial cells express several pattern recognition receptors and, upon recognition of the molecular signatures of pathogens, secrete antimicrobial substances that aid in the destruction of pathogenic microbes (Ganz et al. 1992; Schittek et al. 2001; Zanetti 2004). In addition to epithelial cell turnover that reduces the microbial load, mucosal secretions of respiratory, urogenital, and gastrointestinal tracts have antibacterial substances that also facilitate the elimination of colonization. Similarly, the antibacterial components of sweat and skin secretions reduce colonization of microbial load (Schittek et al. 2001; Zanetti 2004). Limited information is available on the status and functionality of skin, lung, and other mucosal epithelial layers among older adults. The epithelial cell turnover rate in skin slows down among the older adults, with a reduced secretion of sweat and sebum resulting in dryness and increased microbial colonization, especially with *Pseudomonas* and *Proteus* species

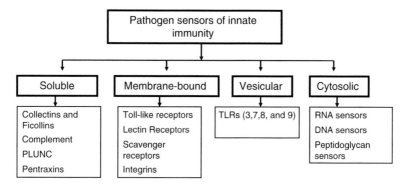

**Fig. 1** Pathogen sensors of the innate immune system. Several families of pathogen-sensing receptors that recognize conserved molecular signatures of pathogens are localized in various compartments within the cell as well as in body fluids. Engagement of these receptors leads to the activation of the innate immune system and the elimination of the pathogens

(Laube 2004). These altered physiological states may contribute to delayed skin wound healing (Thomas 2001; Reed et al. 2003; Sorensen et al. 2003; Gosain and DiPietro 2004; Laube 2004). Altered physiological and anatomical changes in the lungs also contribute to poor innate immunity, thereby increasing microbial colonization and the incidence of pneumonia (Meyer 2001). These changes include reduced elasticity and function of lung muscles, reduced mucociliary clearance rates, decreased oropharyngeal clearance of bacteria, decreased phagocytic activity of alveolar macrophages, and decreased mucosal secretions (Meyer et al. 1996; Meyer 2004, 2005). Similarly, there is increased localization of *Candida* species on the oral and urogenital mucosal surfaces with age among older adults (Shay and Ship 1995; Sobel 1997). In the case of influenza infection, no data are available on the status of innate immune responses at the epithelial barriers in aging. Hence, detailed studies addressing the statuses of pathogen-sensing mechanisms with age are required and will enable us to come up with the strategies to reduce microbial load at epithelial surfaces and to enhance disease resistance.

## 2.2   Pathogen Sensing and Antigen-Presenting Cells

The primary role of innate immunity is to prevent the entry of pathogens into the tissues; however, a number of factors such as dose of infecting pathogen and the immune and nutritional status of the individual determine if innate immunity is able to prevent colonization and infection. Once pathogens overcome the epithelial defenses and gain access into tissues, myeloid lineages of hemopoietic stem cells from bone marrow, namely tissue-resident macrophages and dendritic cells, recognize the pathogens. Innate immune receptors, either directly or through scavenger receptors or pathogens bound to soluble innate immune receptors, initiate phagocytosis and an inflammatory response. These interactions lead to the secretion of proinflammatory cytokines such as IL-6, TNF-α, and IL-8, which attract neutrophils and natural killer cells to the site of infection, thus creating an optimal priming environment to initiate an adaptive immune response. Dendritic cells (DCs) capture antigens from pathogens, mature, differentiate, and migrate to regional draining lymph nodes to stimulate antigen-specific T and B cells, the lymphoid lineages that originate from hemopoietic stem cells. Following antigen-specific clonal expansion of B and T cells, the invading pathogen is or the pathogen-infected cells are removed by specific antibody and T cells. Tissue-resident macrophages play a major role in pathogen sensing, elimination, and tissue repair. We have demonstrated previously that the expression and function of TLRs on peritoneal as well as splenic macrophages decline with age using a murine model or peripheral blood mononuclear cells (PBMCs) from humans (Renshaw et al. 2002; van Duin et al. 2007b; van Duin and Shaw 2007). These findings are consistent with the previous observations that macrophage function declines with age, although the molecular mechanisms were not clear (Plowden et al. 2004a,b; Sebastian et al. 2005). Not only does macrophage function decline with age, but so does their ability to process and present antigens, secrete proinflammatory cytokines and chemokines, provide costimulatory signals, and migrate to the site of infection, as documented in aged animal models (Plowden et al. 2004a,b). Although an age-related decline in the acute proinflammatory response of monocytes has been identified, other studies have demonstrated increased levels of proinflammatory cytokines in serum and in culture supernatants of in vitro stimulated monocyte cultures from healthy older adults compared to younger adults. These observations led Franceschi and colleagues to coin the term "inflammaging" indicating a low-grade chronic inflammatory state as a hallmark of aging (Franceschi et al. 2000; Franceschi 2007) and increased risk for adverse changes in health in older adults. This would predict high levels of proinflammatory cytokines in frail older adults, but just the opposite has been found; low levels of the cytokines have been associated with frailty. Differences in the observations may be accounted for based on the type (polyclonal vs. antigen- or ligand-specific) and duration (acute vs. chronic) of stimulus (van den Biggelaar et al. 2004). Although additional studies need to clarify the observed differences in the secretion of proin-flammatory cytokines between aged animal models, healthy older adults and frail older adults, it is clear that there are alterations in pro- and anti-inflammatory cytokine secretion and their balance with aging (Alberti et al. 2006). These altera-tions will affect both innate and adaptive immune functions (Fig. 2). In addition, the migration of antigen-bearing DCs is severely affected in aged animals, indicat-ing that the priming environment for adaptive immune responses is suboptimal (Linton et al. 2005). Although careful studies are yet to be performed, Langerhans cells in skin appear to decline in numbers with age, and their function also declines with age (Meyerson 1966; Laube 2004). In contrast, bone marrow-derived DCs generated with a cocktail of cytokines from aged animals or humans are found to be

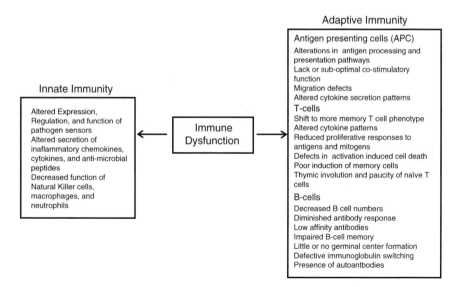

**Fig. 2** Immune dysfunction in older adults. The characteristics of the alterations observed in both the innate and adaptive immune compartments

as effective as those generated from their younger counterparts in recalling memory T cell responses. This would suggest that in vitro generation bypasses age-related defects that are seen with ex vivo DCs (Lung et al. 2000; Tesar et al. 2006). However, primary CD4 T-lymphocyte responses remain impaired in spite of normal DC function, suggesting increased antigen and costimulation thresholds of aged, naïve T-lymphocytes, consistent with our earlier published data (Haynes et al. 2000; Sambhara et al. 2001; van Duin et al. 2007a). Unpublished findings from our laboratory indicate a substantial delay in the mobilization of DCs and macrophages into lungs and regional draining lymph nodes following infection with influenza virus in aged compared to younger animals, suggesting an altered microenvironment. A decline in the expression of pathogen sensors (specifically TLRs) and the secretion of cytokines and chemokines was also observed that may influence the migration, activation, differentiation, and function of macrophages and dendritic cells. Indeed, a recently published study demonstrated poor induction of costimulatory molecules on monocytes of older adults following TLR stimulation, consistent with our observations in the murine model (Renshaw et al. 2002; van Duin and Shaw 2007). Using adjuvants that stimulate the innate immune system, providing costimulation, or supplementing with a cocktail of cytokines along with antigen at the time of immunization significantly improved adaptive immune responses by stimulating antigen-presenting cells (APCs) (Sambhara et al. 1998, 2001; Haynes et al. 2004). Hence, it is logical to formulate vaccines for older adults with adjuvants to induce an optimal priming environment for adaptive immune responses.

## 2.3    Adaptive Immunity

Hematopoietic stem cells (HSC) in the bone marrow give rise to both myeloid- and lymphoid-committed stem cells. While the myeloid lineage gives rise to monocytes, macrophages, and dendritic cells, lymphoid-committed stem cells give rise to T lymphocytes and B lymphocytes which go through "education and selection" in the thymus (in the case of T cells) and bone marrow (in the case of B cells), a process that removes potentially autoreactive clones. These cells are further educated in the periphery to be tolerant to self-antigens. Although there is some indirect evidence that HSC numbers decline with age, detailed studies are yet to be done to determine if HSC numbers, function or migration alter with age (Wang et al. 1995; Lamberts et al. 1997; de Haan and Van Zant 1999). Changes in the cellularity of bone marrow with aging have been clearly documented, and may alter the local cytokine milieu, thus affecting the proliferation, differentiation, and seeding of secondary lymphoid organs by lineage-specific stem cells (Liang et al. 2005).

### 2.3.1    Humoral Immunity in Aging

Antigen-specific adaptive immune responses against influenza virus infection or vaccination are mediated by B lymphocytes and T lymphocytes; both contribute to humoral and cellular immunity to influenza. T helper cells secrete cytokines for B lymphocyte differentiation and class switching. Following the recognition of antigens with their surface immunoglobulin receptors, B lymphocytes undergo differentiation to become plasma cells that secrete antibody. Antibodies against the major surface glycoprotein of influenza viruses, the hemagglutinin (HA), neutralize the virus by binding to conformational determinants on HA, and prevent infection. Antibodies directed against the second major surface glycoprotein, the neuraminidase (NA), can limit virus release from an infected cell and can therefore reduce virus replication. The functionality of anti-HA antibodies is usually determined by the hemagglutination-inhibition (HAI) test, and in some cases by virus-neutralization tests. A HAI titer of $\geq 1{:}40$ is correlated with a 50% protection rate in a population against influenza viral infections (Wood et al. 1997). Due to the high mutation rate of this RNA virus and the selection pressure of pre-existing antibody in humans that acts on circulating viruses, influenza viruses accumulate mutations in HA and NA genes, leading to antigenic drift, which requires that the strains of influenza contained in the vaccine must be updated every year to antigenically match the circulating strains. In general, it is known that humoral immune responses induced by influenza vaccination decline with age. However, humoral immune responses as measured by HAI titers in community-dwelling "healthy older adults" and centenarians are similar to those observed in younger adults, indicating that aging alone does not affect antibody responses against influenza vaccination. Other contributing factors to the decline in antibody responses include comorbid conditions such as chronic diseases and frailty, as well as poor nutrition, stress, and limited physical activity. Pre-existing humoral immunity due to annual vaccination of older adults does not appear to impact the antibody responses to subsequent vaccinations and does not explain the poor vaccine efficacy. One possibility is that the quality and duration, rather than the magnitude of the antibody response, may be affected; however, results from a recent study indicate that this may not be the case (de Bruijn et al. 1999; Gardner et al. 2001; Iorio et al. 2007). Hence, additional markers of the immune response may be needed to predict vaccine efficacy in the older adult population. Earlier studies from our laboratory have shown that serum antibody titers did not correlate with the susceptibility to influenza virus infection among older adults, suggesting that both antibodies and cellular immunity contribute to clinical protection against influenza illness. Although the antibody responses are strain-specific within a subtype, they do provide cross-protection against viruses of the same subtype via antibody-dependent cell-mediated cytotoxicity carried out by NK cells or macrophages.

### 2.3.2 Cellular Immunity in Aging

Unlike B lymphocytes, T lymphocytes recognize peptide fragments derived from the antigens that are presented with major histocompatibility complex molecules by professional antigen-presenting cells such as dendritic cells. T lymphocytes consist of CD4 T helper cells and CD8 cytotoxic T cells. While CD4 T lymphocytes recognize peptides that are processed from exogenous antigens (e.g., killed virus) presented with class II MHC molecules, CD8 T lymphocytes recognize peptide fragments derived from endogenous antigens (e.g., peptides derived from virus replicating inside the cell) that are presented with class I MHC molecules. Depending on the pattern of cytokines they secrete, CD4 T lymphocytes are further classified as T helper 1 (Th1), T helper 2 (Th2), T helper 3 (Th3), and T helper 17 (Th17) cells. T lymphocytes recognize peptide fragments derived from both surface glycoproteins and internal proteins. While surface glycoproteins (HA and NA) vary due to antigenic drift or shift, the internal proteins, namely the nucleoprotein, matrix protein and others, are fairly conserved within the subtype of influenza viruses. It has been shown that although T lymphocytes will not prevent infection, cytotoxic T lymphocytes kill virus-infected cells and aid in viral clearance, thus contributing to clinical protection against influenza illness (Yap et al. 1978). Hence, the activation of both CD4 and CD8 T lymphocytes will provide cross-protection against variant viruses within a subtype. Virus infection induces robust T lymphocyte responses, which persist for a very long time and provide cross-protection in mice. The magnitude and durability of T lymphocyte responses depend on the route of infection/immunization, whether or not the vaccine is formulated to induce or recall especially CD8 T lymphocyte responses. The current inactivated split-virus influenza vaccines provide only exogenous antigens for stimulation of T lymphocytes and thus are poor inducers of CD8 T lymphocyte responses. Activating or recalling CD8 T lymphocyte responses by formulating vaccines with adjuvants which will stimulate antigen-presenting cells creates an optimal priming environment and activates T lymphocytes to provide broader protection against serologically distinct viruses. CD4 T helper cells provide growth factors for B and CD8 T lymphocytes, thereby occupying a central role in the induction of humoral and cellular immune responses. Th1 and Th2 cells were defined based on the secretion of IFN-$\gamma$. While Th1 cells secrete IFN-$\gamma$ following stimulation by IL-12, Th2 cells stimulated by IL-4 secrete IL-4, IL-5, and IL-13. The decline of naïve T cells in the repertoire due to thymic involution and accumulation of dysfunctional memory T cells is well established, but the mechanism for these observations goes beyond that which can be explained by thymic involution alone. Interleukin 7 appears to play an important role in T cell survival in thymic recombination events, and in expanding positively selected thymocytes (Hare et al. 2000; Huang et al. 2001). An age-related reduction in production of IL-7 within the thymus may be responsible for the age-related decline in thymic output of naïve T cells (Andrew and Aspinall 2002; Ortman et al. 2002). In humans, accumulation of an anergic CD28− T cell population with age, especially among the CD8 T cell subset, has been documented (Boucher et al. 1998; Sansoni et al. 2008). The molecular mechanisms leading to the loss of CD28 are not known (Boucher et al. 1998; Sansoni et al. 2008). The CD28− T cells are anergic to stimulation with antigen or mitogen. In murine studies, it has been clearly shown that the clonal expansion and function of naïve CD4 or CD8 T cells is significantly reduced when compared to their younger counterparts (Plowden et al. 2004a,b; Jiang et al. 2007).

In addition to T helper and cytotoxic T cells, Th3 or Treg cells that are CD4+CD25+ Fox3+ have been shown to play an important role in regulating immune responses (Dejaco et al. 2006; Hill et al. 2007). A recently published report and our unpublished findings show a significant increase in the Treg population and function with age, which may be contributing to poor adaptive immune responses (Zhao et al. 2007). However, a direct demonstration of the role of Tregs in the decline in immune responsiveness with aging is lacking, although our preliminary results indicate that depleting the Treg subset prior to immunization or infection with A/PR/8/34 virus enhanced both humoral and cellular immune responses in aged mice when compared to the control aged mice. An increased number of Tregs with age may aid in controlling the initiation of autoimmune disorders, but may come at the cost of reducing effective immune responses against infectious agents. The evolutionary significance of this finding is not clear. The functionality of CD4+ Th17 cells is beginning to be elucidated in mice, and very limited information is available on their role in humans and the impact of aging on the function of this subset (Bi et al. 2007; Nakae et al. 2007; Chen and O'Shea 2008).

CTL activity has been shown to be important for recovery from influenza virus infection in the absence of seroprotective antibodies to the infecting virus strain (McElhaney et al. 2006). CD8+ T lymphocytes recognize peptide fragments derived from viral proteins that are bound to class I MHC molecules and lyse the influenza virus-infected cells. The lysis of target cells can be mediated by perforin or by granule-mediated or Fas-mediated mechanisms (Apasov et al. 1993). CD8 T cell cytolytic activity is normally measured by labeling the MHC-compatible target cells (which are either pulsed with relevant peptides or infected with virus) with $^{51}$Cr and determining the amount of $^{51}$Cr released into the medium 4–5 h after the addition of CD8 T cells (Martz et al. 1974). Another assay to assess CTL activity is the measurement of granzyme B activity in lysates of influenza virus-stimulated PBMC; low levels of granzyme B have been correlated with risk for influenza illness in older adults (McElhaney et al. 1996, 2006).

## 3  Influenza Vaccine Efficacy in the Older Adult Population

Annually, influenza epidemics cause three to five million cases of severe illness with about 250,000–500,000 deaths worldwide (World Health Organization 2008). In an average year in the United States, complications from influenza infections result in about 250,000 hospitalizations and 36,000 deaths, with the majority of the fatalities occurring among the elderly population (Thompson et al. 2004; Simonsen et al. 2005). Complications from influenza viral infections resulting in hospitalizations and death are greatest among older adults, people with chronic medical conditions or immunological disorders, and infants and young children (i.e., ≤ 2 years of age) whose immune systems are still maturing (Fiore et al. 2007). Vaccination is the primary strategy for reducing the morbidity and mortality associated with human influenza. An inactivated detergent-split trivalent influenza vaccine (TIV) containing two influenza A viruses (H1N1 and H3N2) and a type B virus as well as a live-attenuated nasal influenza vaccine containing all three components are marketed in the USA. While injectable vaccine is recommended for people at risk, including persons aged 50 years and older, live influenza vaccine is only recommended for persons 2–49 years of age (FDA 2007). Because older adults are a high-risk group for influenza-related deaths, the goal is to vaccinate 90% of this population (DHHS 2000). However, recent vaccination rates are stagnant and coverage still hovers around 65% (National Center for Health Statistics 2003). In healthy, younger adults, the vaccine may be 70–90% effective in preventing influenza-like illness if the vaccine antigen is antigenically closely matched with the circulating epidemic strain (Gross 2002). However, vaccine efficacy is substantially reduced to 30–50% in preventing complications from influenza infections among older adults (Nichol et al. 2007). The mortality benefits from influenza vaccination of older adults is a hotly debated topic (Simonsen et al. 2007). A meta-analysis of 18 cohorts of older adults in one HMO comprising data for ten seasons from 1990–1991 through 1999–2000 indicates that vaccination resulted in a 27% reduction in the risk for hospitalization due to influenza and a 48% reduction in the risk for death (Vu et al. 2002). However, the outcomes used for these studies included hospitalizations for pneumonia or influenza and death from any cause, which are not influenza-specific. Despite increased vaccination coverage of older adults since 1980, there was no decrease in influenza-related excess mortality rates among older adults in the USA (Thompson et al. 2003; Simonsen et al. 2005). Similarly, the results from studies of Netherlands and Italian groups suggest that vaccination did not result in a reduction in excess mortality due to influenza-like illnesses, although there was not enough statistical power to generalize those findings (Govaert et al. 1994; Rizzo et al. 2006). Ideally, a randomized placebo-controlled clinical trial with clearly defined clinical outcomes such as culture-positive influenza illnesses rather than influenza-like illness and pneumonia and all-cause mortality is required to evaluate the benefit of vaccination of older adults. However, policy decisions regarding the vaccination of all older adults make a placebo-controlled study ethically unacceptable to investigate the mortality benefits of influenza vaccination (Smith and Shay 2006). It has been shown previously that influenza vaccination is 49% and 32% effective in preventing hospitalizations from pneumonia or influenza and 55% and 64% effective in preventing death from any cause among older adults at low or intermediate risk, respectively. However, among older adults who are at high risk due to comorbid conditions, vaccination is 29% and 49% effective in preventing hospitalization and death, respectively. Furthermore, when efficacy and effectiveness of vaccination among older adults are stratified by age, a different picture emerges. The efficacy of vaccination in preventing illness and hospitalization decreases with advancing age and when associated with comorbid conditions, suggesting that old older adults do not mount optimal protective immune response to vaccination. Factors that impact vaccine efficacy are presented

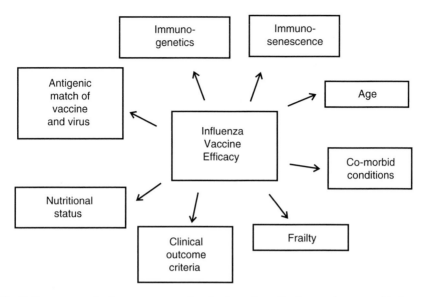

**Fig. 3** Factors contributing to poor or suboptimal vaccine effectiveness in seasonal influenza vaccination of older adults

in Fig. 3, and immunosenescence is discussed above along with other factors that influence the outcome.

## 4   Active and Passive Immunization Strategies for Older Adults

A number of strategies to induce protective immune responses against influenza are presented in Fig. 4. Although passive antibody for influenza has not been considered a potential approach for both preventive and therapeutic needs, this approach has its own merits, especially when older adults who are at high risk or frail older adults who exhibit severe immune dysfunction are the target group. In addition, if the infection is caused by drug-resistant strains of influenza or a pandemic strain, passive therapy with human polyclonal antibodies offers a potential therapeutic benefit (Traggiai et al. 2004; Lanzavecchia et al. 2007; Simmons et al. 2007). Currently, transgenic animals that carry human immunoglobulin genes make human polyclonal immunoglobulins when immunized with antigens from infectious disease agents are available and these animals can serve as a potential tool to generate influenza strain-specific human antibodies for passive transfer (Fishwild et al. 1996; Tomizuka et al. 2000; Kuroiwa et al. 2002; Buelow and van Schooten 2006).

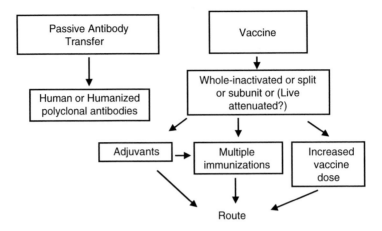

**Fig. 4** Passive and active immunization strategies for older adults against influenza

It is clear that influenza vaccine needs to be formulated differently for older adults to overcome the age-related decline in immune function and enhance the immunogenicity and protective levels of both humoral and cellular immune responses. Although MF59-adjuvanted TIV vaccine is marketed in Europe for older adults and has been shown to be safe and immunogenic, it is not yet approved in the USA (Podda and Del Giudice 2003). Newer adjuvant systems such as ASO3 have been shown to enhance the immunogenicity of H5N1 virus vaccines, and may provide a potential benefit to the older adult population if formulated with seasonal vaccines. (Treanor et al. 2006, 2007; Leroux-Roels et al. 2007; Sambhara and Poland 2007). Increasing the vaccine dose from 15 μg of HA of each vaccine component of a TIV vaccine is a potential option for enhancing the levels of protective antibodies. In a recent multisite, phase II, randomized, double-blind clinical study, older adults who received 60 μg of HA of each component were shown to generate higher levels of HAI and neutralizing antibodies when compared to those who received the standard vaccine dose of 15 μg of HA of each of the components. However, the vaccinees who received the higher dose of HA experienced more local and systemic reactions than those who received the standard vaccine dose (Keitel et al. 2006; Couch et al. 2007). This result is consistent with earlier studies and the concern that increasing the vaccine dose produces unacceptable local reactions, an effect that may be overcome with an adjuvanted vaccine to improve responses in older adults. Another possibility is to vaccinate older individuals more than once during the influenza season in order to boost antibody responses. In a small clinical study, revaccination of older adults twelve weeks later did not enhance HI titers, suggesting that such an approach may not be a viable alternative (Buxton et al. 2001). However, one of the caveats of this study was the lack of baseline titers of the vaccinees who received the second dose of the vaccine. Hence, additional studies are needed to evaluate if such an approach is a viable strategy to enhance the levels of protective antibodies in the older adult population.

# 5  Summary and Conclusions

Influenza is a vaccine-preventable disease, and the benefits of vaccination in preventing infection and complications arising from infection among adults are clearly documented. It is clear that the immune response declines with age due to alterations in innate and adaptive immune functions and that the vaccine does not provide adequate protection in this population. Hence, for older adults—a major target population for annual influenza vaccination, vaccine efficacy at preventing infection is low and the risk of serious complications from these infections is compounded by increasing age and comorbid conditions. Thus, efforts should be directed at formulating vaccines with adjuvants specifically for older adults to overcome immunosenescence, and passive immunization strategies with human polyclonal antibodies should be considered. In addition to conventional serological assays in the selection of vaccine candidates, other parameters namely, induction of cellular immune responses as well as activation of the innate immune system to facilitate an optimal microenvironment for the mobilization, activation, differentiation, maturation, and migration of antigen-presenting cells should be considered.

# References

Alberti S, Cevenini E et al (2006) Age-dependent modifications of Type 1 and Type 2 cytokines within virgin and memory CD4+ T cells in humans. Mech Ageing Dev 127(6):560–566

Andrew D, Aspinall R (2002) Age-associated thymic atrophy is linked to a decline in IL-7 production. Exp Gerontol 37(2–3):455–463

AoA. (2007) A profile of older Americans: 2007. Administration on Aging, US Department of Health and Human Services. http://www.aoa.gov/prof/Statistics/profile/2007/2007profile.pdf. Accessed 28 Feb 2008

Apasov S, Redegeld F et al. (1993) Cell-mediated cytotoxicity: contact and secreted factors. Curr Opin Immunol 5(3):404–410

Aw D, Silva AB et al (2007) Immunosenescence: emerging challenges for an ageing population. Immunology 120(4):435–446

Bi Y, Liu G et al (2007) Th17 cell induction and immune regulatory effects. J Cell Physiol 211(2):273–278

Bingle CD, Craven CJ (2002) PLUNC: a novel family of candidate host defence proteins expressed in the upper airways and nasopharynx. Hum Mol Genet 11(8):937–943

Boucher N, Dufeu-Duchesne T et al (1998) CD28 expression in T cell aging and human longevity. Exp Gerontol 33(3):267–282

Brown GD (2006) Dectin-1: a signalling non-TLR pattern-recognition receptor. Nat Rev Immunol 6(1):33–43

Buelow R, van Schooten W (2006) The future of antibody therapy. Ernst Schering Found Symp Proc 4:83–106

Buxton JA, Skowronski DM et al (2001) Influenza revaccination of elderly travelers: antibody response to single influenza vaccination and revaccination at 12 weeks. J Infect Dis 184(2):188–191

Chen Z, O'Shea JJ (2008) Regulation of IL-17 production in human lymphocytes. Cytokine 41(2):71–78

Couch RB, Winokur P et al (2007) Safety and immunogenicity of a high dosage trivalent influenza vaccine among elderly subjects. Vaccine 25(44):7656–7663

de Bruijn IA, Remarque EJ et al (1999) Quality and quantity of the humoral immune response in healthy elderly and young subjects after annually repeated influenza vaccination. J Infect Dis 179(1):31–36

de Haan G, Van Zant G (1999) Dynamic changes in mouse hematopoietic stem cell numbers during aging. Blood 93(10):3294–3301

Dejaco C, Duftner C et al (2006) Are regulatory T cells linked with aging? Exp Gerontol 41(4):339–345

DHHS (2000) Understanding and improving health and objectives for improving health. US Department of Health and Human Services, Washington DC. http://www.healthypeople.gov/Document/tableofcontents.htm#under. Accessed on 10 March 2008

Ely KH, Roberts AD et al (2007) Aging and CD8+ T cell immunity to respiratory virus infections. Exp Gerontol 42(5):427–431

FDA. (2007) FDA approves nasal influenza vaccine for use in younger children. http://www.fda.gov/bbs/topics/NEWS/2007/NEW01705.html. Accessed 12 Feb 2008

Fiore AE, Shay DK et al (2007) Prevention and control of influenza. Recommendations of the Advisory Committee on Immunization Practices (ACIP), 2007. MMWR Recomm Rep 56(RR-6):1–54

Fishwild DM, O'Donnell SL et al (1996) High-avidity human IgG kappa monoclonal antibodies from a novel strain of minilocus transgenic mice. Nat Biotechnol 14(7):845–851

Franceschi C (2007) Inflammaging as a major characteristic of old people: can it be prevented or cured? Nutr Rev 65(12 Pt 2):S173–S176

Franceschi C, Bonafe M et al (2000) Inflammaging. An evolutionary perspective on immunosenescence. Ann N Y Acad Sci 908:244–254

Ganz T (2002) Epithelia: not just physical barriers. Proc Natl Acad Sci USA 99(6):3357–3358

Ganz T, Oren A et al (1992) Defensins: microbicidal and cytotoxic peptides of mammalian host defense cells. Med Microbiol Immunol (Berl) 181(2):99–105

Gardner EM, Bernstein ED et al (2001) Characterization of antibody responses to annual influenza vaccination over four years in a healthy elderly population. Vaccine 19(32):4610–4617

Gardner EM, Gonzalez EW et al (2006) Age-related changes in the immune response to influenza vaccination in a racially diverse, healthy elderly population. Vaccine 24(10):1609–1614

Ginaldi L, Loreto MF et al (2001) Immunosenescence and infectious diseases. Microbes Infect 3(10):851–857

Gosain A, DiPietro LA (2004) Aging and wound healing. World J Surg 28(3):321–326

Govaert TM, Thijs CT et al (1994) The efficacy of influenza vaccination in elderly individuals. A randomized double-blind placebo-controlled trial. JAMA 272(21):1661–1665

Gross PA (2002) Review: inactivated vaccines provide the greatest protection against influenza in healthy persons. ACP J Club 136(3):103

Hare KJ, Jenkinson EJ et al (2000) An essential role for the IL-7 receptor during intrathymic expansion of the positively selected neonatal T cell repertoire. J Immunol 165(5):2410–2414

Hartman M, Catlin A et al (2008) US health spending by age, selected years through 2004. Health Affairs 27(1):w1–w12

Haynes L, Eaton SM et al (2000) The defects in effector generation associated with aging can be reversed by addition of IL-2 but not other related gamma(c)-receptor binding cytokines. Vaccine 18(16):1649–1653

Haynes L, Eaton SM et al (2004) Inflammatory cytokines overcome age-related defects in CD4 T cell responses in vivo. J Immunol 172(9):5194–5199

Hill JA, Benoist C et al (2007) Treg cells: guardians for life. Nat Immunol 8(2):124–125

Hoffmann JA, Kafatos FC et al (1999) Phylogenetic perspectives in innate immunity. Science 284(5418):1313–1318

Holmskov U, Thiel S et al (2003) Collectins and ficolins: Humoral lectins of the innate immune defense. Annu Rev Immunol 21(1):547–578

Htwe TH, Mushtaq A et al (2007) Infection in the elderly. Infect Dis Clin North Am 21(3):711–743

Huang J, Durum SK et al (2001) Cutting edge: histone acetylation and recombination at the TCR gamma locus follows IL-7 induction. J Immunol 167(11):6073–6077

Imler JL, Hoffmann JA (2001) Toll receptors in innate immunity. Trends Cell Biol 11(7):304–311

Iorio AM, Camilloni B et al (2007) Effects of repeated annual influenza vaccination on antibody responses against unchanged vaccine antigens in elderly frail institutionalized volunteers. Gerontology 53(6):411–418

Jiang J, Gross D et al (2007) Aging affects initiation and continuation of T cell proliferation. Mech Ageing Dev 128(4):332–339

Kang DC, Gopalkrishnan RV et al (2002) Mda-5: an interferon-inducible putative RNA helicase with double-stranded RNA-dependent ATPase activity and melanoma growth-suppressive properties. Proc Natl Acad Sci USA 99(2):637–642

Keitel WA, Atmar RL et al (2006) Safety of high doses of influenza vaccine and effect on antibody responses in elderly persons. Arch Intern Med 166(10):1121–1127

Kovaiou RD, Herndler-Brandstetter D et al (2007) Age-related changes in immunity: implications for vaccination in the elderly. Expert Rev Mol Med 9(3):1–17

Kuroiwa Y, Kasinathan P et al (2002) Cloned transchromosomic calves producing human immunoglobulin. Nat Biotechnol 20(9):889–894

Lamberts SW, van den Beld AW et al (1997) The endocrinology of aging. Science 278(5337):419–424

Lanzavecchia A, Corti D et al (2007) Human monoclonal antibodies by immortalization of memory B cells 88. Curr Opin Biotechnol. 18(6):523–528

Laube S (2004) Skin infections and ageing. Ageing Res Rev 3(1):69–89

Leroux-Roels I, Borkowski A et al (2007) Antigen sparing and cross-reactive immunity with an adjuvanted rH5N1 prototype pandemic influenza vaccine: a randomised controlled trial. Lancet 370(9587):580–589

Liang Y, Van G Zant et al (2005) Effects of aging on the homing and engraftment of murine hematopoietic stem and progenitor cells. Blood 106(4):1479–1487

Linton PJ, Li SP et al (2005) Intrinsic versus environmental influences on T cell responses in aging. Immunol Rev 205(1):207–219

Lu J, Teh C et al (2002) Collectins and ficolins: sugar pattern recognition molecules of the mammalian innate immune system. Biochim Biophys Acta 1572(2–3):387–400

Lung TL, Saurwein-Teissl M et al (2000) Unimpaired dendritic cells can be derived from monocytes in old age and can mobilize residual function in senescent T cells. Vaccine 18(16):1606–1612

Martinon F, Tschopp J (2005) NLRs join TLRs as innate sensors of pathogens. Trends Immunol 26(8):447–454

Martz E, Burakoff SJ et al (1974) Interruption of the sequential release of small and large molecules from tumor cells by low temperature during cytolysis mediated by immune T cells or complement. Proc Natl Acad Sci USA 71(1):177–181

McElhaney JE, Pinkoski MJ et al (1996) The cell-mediated cytotoxic response to influenza vaccination using an assay for granzyme B activity. J Immunol Methods 190(1):11–20

McElhaney JE, Xie D et al (2006) T cell responses are better correlates of vaccine protection in the elderly. J Immunol 176(10):6333–6339

Medzhitov R, Janeway C Jr (2000a) Innate immune recognition: mechanisms and pathways. Immunol Rev 173:89–97

Medzhitov R, Janeway C Jr (2000b) Innate immunity. N Engl J Med 343(5):338–344

Meyer KC (2001) The role of immunity in susceptibility to respiratory infection in the aging lung. Respir Physiol 128(1):23–31

Meyer KC (2004) Lung infections and aging. Ageing Res Rev 3(1):55–67

Meyer KC (2005) Aging. Proc Am Thorac Soc 2(5):433–439

Meyer KC, Ershler W et al (1996) Immune dysregulation in the aging human lung. Am J Respir Crit Care Med 153(3):1072–1079

Meyerson LB (1966) Aging and the skin. Ohio State Med J 62(5):453–456

Miller RA (1996) The aging immune system: primer and prospectus. Science 273(5271):70–74

Miller RA (1997) The aging immune system: subsets, signals, and survival. Aging (Milano) 9(4):23–24

Nakae S, Iwakura Y et al (2007) Phenotypic differences between Th1 and Th17 cells and negative regulation of Th1 cell differentiation by IL-17. J Leukoc Biol 81(5):1258–1268.

National Center for Health Statistics (2003) Early release of selected estimates based on data from the January–September National Health Interview Survey (NHIS). http://www.cdc.gov/nchs/about/major/nhis/released200303.htm#5. Accessed on 10, March 2008

Nichol KL, Nordin JD et al (2007) Effectiveness of influenza vaccine in the community-dwelling elderly. N Engl J Med 357(14):1373–1381

Ortman CL, Dittmar KA et al (2002) Molecular characterization of the mouse involuted thymus: aberrations in expression of transcription regulators in thymocyte and epithelial compartments. Int Immunol 14(7):813–822

Plowden J, Renshaw-Hoelscher M et al (2004a) Innate immunity in aging: impact on macrophage function. Aging Cell 3(4):161–167

Plowden J, Renshaw-Hoelscher M et al (2004b) Impaired antigen-induced CD8(+) T cell clonal expansion in aging is due to defects in antigen presenting cell function. Cell Immunol 229(2):86–92

Podda A, Del Giudice G (2003) MF59-adjuvanted vaccines: increased immunogenicity with an optimal safety profile. Expert Rev Vaccines 2(2):197–203

Reed MJ, Koike T et al (2003) Wound repair in aging. A review. Methods Mol Med 78:217–237

Renshaw M, Rockwell J et al (2002) Cutting edge: impaired Toll-like receptor expression and function in aging. J Immunol 169(9):4697–4701

Rizzo C, Viboud C et al (2006) Influenza-related mortality in the Italian elderly: no decline associated with increasing vaccination coverage. Vaccine 24(42–43):6468–6475

Sambhara S, Kurichh A et al (1998) Enhanced immune responses and resistance against infection in aged mice conferred by Flu-ISCOMs vaccine correlate with up-regulation of costimulatory molecule CD86. Vaccine 16(18):1698–1704

Sambhara S, Kurichh A et al (2001) Severe impairment of primary but not memory responses to influenza viral antigens in aged mice: costimulation in vivo partially reverses impaired primary immune responses. Cell Immunol 210(1):1–4

Sambhara S, Poland GA (2007) Breaking the immunogenicity barrier of bird flu vaccines. Lancet 370(9587):544–545

Sansoni P, Vescovini R et al (2008) The immune system in extreme longevity. Exp Gerontol 43(2):61–65

Schittek B, Hipfel R et al (2001) Dermcidin: a novel human antibiotic peptide secreted by sweat glands. Nat Immunol 2(12):1133–1137

Sebastian C, Espia M et al (2005) MacrophAging: a cellular and molecular review. Immunobiology 210(2–4):121–126

Shay K, Ship JA (1995) The importance of oral health in the older patient. J Am Geriatr Soc 43(12):1414–1422

Simmons CP, Bernasconi NL et al (2007) Prophylactic and therapeutic efficacy of human monoclonal antibodies against H5N1 influenza 362. PLoS Med 4(5):e178

Simonsen L, Reichert TA et al (2005) Impact of influenza vaccination on seasonal mortality in the US elderly population. Arch Intern Med 165(3):265–272

Simonsen L, Taylor RJ et al (2007) Mortality benefits of influenza vaccination in elderly people: an ongoing controversy. Lancet Infect Dis 7(10):658–666

Smith NM, Shay DK (2006) Influenza vaccination for elderly people and their care workers. Lancet 368(9549):1752–1753

Sobel JD (1997) Pathogenesis of urinary tract infection. Role of host defenses. Infect Dis Clin North Am 11(3):531–549

Sorensen OE, Cowland JB et al (2003) Wound healing and expression of antimicrobial peptides/polypeptides in human keratinocytes, a consequence of common growth factors. J Immunol 170(11):5583–5589

Takaoka A, Wang Z et al (2007) DAI (DLM-1/ZBP1) is a cytosolic DNA sensor and an activator of innate immune response. Nature 448:501–505

Tesar BM, Walker WE et al (2006) Murine myeloid dendritic cell-dependent toll-like receptor immunity is preserved with aging. Aging Cell 5(6):473–486

World Health Organization (2008) Influenza. http://www.who.int/mediacentre/factsheets/fs211/en/. Accessed 12 Feb 2008

Thomas DR (2001) Age-related changes in wound healing. Drugs Aging 18(8):607–620

Thompson WW, Shay DK et al (2003) Mortality associated with influenza and respiratory syncytial virus in the United States. JAMA 289(2):179–186

Thompson WW, Shay DK et al (2004) Influenza-associated hospitalizations in the United States. JAMA 292(11):1333–1340

Ting JP, Davis BK (2005) CATERPILLER: a novel gene family important in immunity, cell death, and diseases. Annu Rev Immunol 23:387–414

Tomizuka K, Shinohara T et al (2000) Double trans-chromosomic mice: maintenance of two individual human chromosome fragments containing Ig heavy and kappa loci and expression of fully human antibodies. Proc Natl Acad Sci U S A 97(2):722–727

Traggiai E, Becker S et al (2004) An efficient method to make human monoclonal antibodies from memory B cells: potent neutralization of SARS coronavirus. Nat Med 10(8):871–875

Treanor JJ, Schiff GM et al (2006) Dose-related safety and immunogenicity of a trivalent baculovirus-expressed influenza-virus hemagglutinin vaccine in elderly adults. J Infect Dis 193(9):1223–1228

Treanor JJ, Schiff GM et al (2007) Safety and immunogenicity of a baculovirus-expressed hemagglutinin influenza vaccine: a randomized controlled trial. JAMA 297(14):1577–1582

van den Biggelaar AH, Huizinga TW et al (2004) Impaired innate immunity predicts frailty in old age. The Leiden 85-plus study. Exp Gerontol 39(9):1407–1414

van Duin D, Shaw AC (2007) Toll-like receptors in older adults. J Am Geriatr Soc 55(9):1438–1444

van Duin D, Allore HG et al (2007a) Prevaccine determination of the expression of costimulatory B7 molecules in activated monocytes predicts influenza vaccine responses in young and older adults. J Infect Dis 195(11):1590–1597

van Duin D, Mohanty S et al (2007b) Age-associated defect in human TLR-1/2 function. J Immunol 178(2):970–975

Vu T, Farish S et al (2002) A meta-analysis of effectiveness of influenza vaccine in persons aged 65 years and over living in the community. Vaccine 20(13–14):1831–1836

Wang CQ, Udupa KB et al (1995) Effect of age on marrow macrophage number and function. Aging (Milano) 7(5):379–384

Wood J, Schild G et al (1997) Application of an improved single-radial-immunodiffusion technique for the assay of haemagglutinin antigen content of whole virus and subunit influenza vaccines. Dev Biol Stand 39:193–200

Worley CA (2006) Aging skin and wound healing. Dermatol Nurs 18(3):265–266

Yap KL, Ada GL et al (1978) Transfer of specific cytotoxic T lymphocytes protects mice inoculated with influenza virus. Nature 273:413–420

Yoneyama M, Kikuchi M et al (2004) Th.e RNA helicase RIG-I has an essential function in double-stranded RNA-induced innate antiviral responses. Nat Immunol 5(7):730–737

Zanetti M (2004) Cathelicidins, multifunctional peptides of the innate immunity. J Leukoc Biol 75(1):39–48

Zhao L, Sun L et al (2007) Changes of CD4+CD25+Foxp3+ regulatory T cells in aged Balb/c mice. J Leukoc Biol 81(6):1386–1394

# Vaccines for Pandemic Influenza: Summary of Recent Clinical Trials

Wendy A. Keitel and Robert L. Atmar

## Contents

**Abstract**   The emergence of influenza A/H5N1 viruses in Asia has raised concerns about their potential to cause pandemic disease. Because vaccination is the primary strategy for the prevention of influenza, efforts are in progress to develop safe and immunogenic vaccines against these viruses and other potential pandemic influenza strains. Results of initial studies indicated that subunit influenza A/H5N1virus vaccines were poorly immunogenic, and that high dosages were needed to induce seroresponses in the majority of subjects. Addition of aluminum-containing adjuvants resulted in variable effects on the immunogenicity of H5 vaccines, but in general, clinically meaningful effects have not been observed. Intradermal immunization was not associated with significant enhancement in one study. More recent studies indicate that oil-in-water adjuvants significantly enhance immune responses when compared with nonadjuvanted preparations containing the same dosage of H5 or H9 hemagglutinin. In addition, these formulations elicit higher levels of cross-reactive antibodies vs. different H5N1 clades. Several whole-virus vaccines have been

W.A. Keitel (✉)
Professor, Molecular Virology & Microbiology and Medicine, Baylor College of Medicine, 280 One Baylor Plaza, Houston, TX 77030, USA

R.L. Atmar
Professor, Medicine and Molecular Virology & Microbiology, Baylor College of Medicine, 280 One Baylor Plaza, Houston, TX 77030, USA

R.W. Compans and W.A. Orenstein (eds.), *Vaccines for Pandemic Influenza*,
Current Topics in Microbiology and Immunology 333,
DOI 10.1007/978-3-540-92165-3_21, © Springer-Verlag Berlin Heidelberg 2009

demonstrated to stimulate high frequencies of responses at relatively low dosages; however, direct comparisons with subunit vaccines have not been made. Finally, candidate live attenuated vaccines are under evaluation in clinical trials. The results of these and future trials will help to identify formulations and immunization regimens for various populations, and will better prepare us to address the threat of both pandemic and interpandemic influenza.

# 1  Introduction

Pandemics of influenza have occurred at unpredictable intervals over the past few several centuries (Cunha 2004; Glezen 1996). The last major pandemic, the influenza A/H3N2 pandemic of 1968, resulted from the spread of a reassortant virus that contained genes from avian and human viruses (Bean et al. 1992). Recent studies have provided evidence that pandemic viruses may transmit directly from avian reservoirs into human populations after acquiring mutations that permit efficient transmission and replication in human tissues (Taubenberger et al. 2005), although this remains uncertain (Antonovics et al. 2006; Gibbs and Gibbs 2006). Over the past ten years, human infections caused by three avian viruses bearing novel hemagglutinins (HAs; influenza A/H5N1, A/H9N2, and A/H7N7) have been identified, raising concerns about their pandemic potential (Writing Committee 2008; Guan et al. 2004; Koopmans et al. 2004; Lin et al. 2000). Outbreaks of influenza A/H2N3 in swine in 2006 also remind us of the pandemic potential of viruses bearing this HA (Ma et al. 2007). In particular, the unprecedented epizootic of influenza A/H5N1 has stimulated international efforts to develop (and in the case of H5N1, stockpile) safe and effective vaccines for pandemic influenza control. The goal of this review is to update previous reviews with data from recent clinical trials of potential pandemic vaccines, with a particular focus on H5N1 vaccine development.

# 2  Progress and Challenges in Influenza A/H5N1
   Vaccine Development

Much progress has been made in the ten years since the emergence of highly pathogenic influenza A/H5N1 viruses in poultry with associated human infections in 1997, as outlined in Table 1. Epidemiologic, technical, political, and economic factors as well as conceptual advances have facilitated the development process. Expanded worldwide surveillance for influenza has identified at least nine H5N1 clades based on phylogenetic analyses of the HA: viruses from clades 0, 1, 2.1, 2.2, 2.3, and 7 have caused human infections, and strains from several of these clades are being considered as potential vaccine candidates (Writing Committee 2008).

**Table 1**  Progress in influenza A/H5N1 vaccine development and persistent and emerging challenges; 1997–2008

| Progress | Persistent and emerging challenges |
| --- | --- |
| Epidemiologic | Immunization needs remain poorly defined |
| ∘ Expanded surveillance for emerging viruses | Relatively poor immunogenicity of the H5 |
| Technical | hemagglutinin (HA) |
| ∘ Reverse genetics for seed virus production | Lack of assay standardization |
| ∘ Assay and reagent development | Antigenic drift of influenza A/H5N1 viruses |
| ∘ Expanded manufacturing capacity | Political and economic barriers |
| Political and economic | ∘ Funding |
| ∘ Enhanced international cooperation | ∘ Intellectual property issues |
| ∘ Technology transfer to developing countries | ∘ Liability concerns |
| ∘ Liability relief | |
| Conceptual | |
| ∘ Alternative approaches: adjuvants; substrates; conserved targets; delivery devices; etc. | |

Removal of the polybasic amino acid cleavage site in the HA using reverse genetics techniques permits the growth of normally lethal H5N1 viruses in eggs, the traditional substrate for vaccine production (Neumann et al. 1999). With the leadership of the World Health Organization, the successful promotion of enhanced international cooperation and technology transfer to developing countries will help to prepare all nations to respond to a pandemic. Alternative approaches to vaccine development are being explored, such as inclusion of adjuvants, use of tissue culture-based substrates for vaccine production and expression of recombinant HA, evaluation of conserved targets (e.g., NP and M2e) as vaccine antigens, and assessment of novel delivery devices and alternative routes of vaccine administration.

Despite much progress, a number of challenges remain (Table 1). Although the level of serum hemagglutination inhibition (HAI) and neutralizing (Neut) antibody are correlated with protection from interpandemic influenza, the correlates of protection for pandemic strains remain uncertain; therefore, it is difficult to establish immunization needs. The H5 HA appears to be poorly immunogenic relative to other HAs, and nonadjuvanted subunit vaccines require high dosages of HA to stimulate responses in a majority of subjects. Furthermore, traditional serologic assays for the detection of immune responses to H5N1 viruses are insensitive, in part due to differences between avian and human influenza virus receptors (Shinya et al. 2006). Antibody assays performed in laboratories around the world also lack standardization, making comparisons between clinical trials difficult (Stephenson et al. 2007). Antigenic drift of H5N1 viruses poses challenges related to the development of prepandemic vaccines that may fail to match the ultimate pandemic virus (Chen et al. 2006). Finally, economic and political barriers continue to hamper the vaccine development process, including limited resources, liability concerns and intellectual property issues. Notably, many of the achievements and obstacles described above

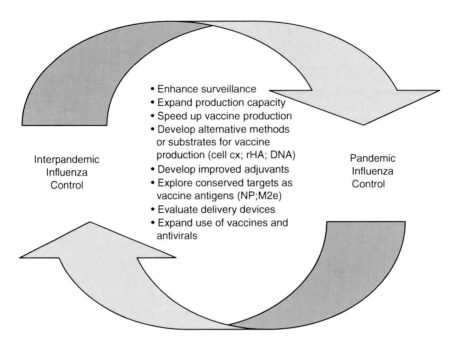

**Fig. 1** Pandemic and interpandemic influenza control: efforts that are underway to improve control of interpandemic influenza and those addressing control of future pandemic influenza are interrelated and mutually beneficial

also contribute to or impede successful control of interpandemic influenza, as outlined in Fig. 1.

# 3   Clinical Trials of Influenza A/H5N1 Vaccines

## 3.1   Approaches to H5 Vaccine Development

A variety of traditional and novel approaches have been and are being explored in the development of H5 vaccines, and several vaccines have been registered and entered into national stockpiles. Clinical trials of H5 subunit and whole-virus vaccines with or without adjuvant and live attenuated vaccines are summarized in Table 2. Note that both egg-based and cell culture-based vaccines have been evaluated, as well as a purified baculovirus-expressed recombinant HA (rHA) vaccine grown in insect cells. Novel approaches, including evaluation of conserved influenza virus proteins as vaccine antigens, adenovirus vectors for antigen delivery, and DNA-based vaccines are also in the early phases of evaluation for control of both pandemic and interpandemic influenza (Ernst et al. 2006; Hoelscher et al. 2006; Kodihalli et al. 1999; Laddy et al. 2007; Watanabe et al. 2008).

**Table 2** Summary of Recent Clinical Trials of Influenza A/H5 Vaccine Candidates

| Type of Vaccine | Reference | Population Studied | Vaccine Description* | Regimen | Dosage(s) (μg HA) | % with HAI Antibody Responses % (dosage) | % with Neut Antibody Responses | Comments |
|---|---|---|---|---|---|---|---|---|
| | | | | | H5 Vaccines | | | |
| Subunit; No Adjuvant | Treanor 2001 | Healthy adults; N=147 | rHA; A/HK/97 | 2 IM doses 21, 28 or 42 days apart | Two doses of 0, 25, 45 or 90; or 90 followed by 10 | N.D. | 17%(25/25) 28% (45/45) 33% (90/10) 52% (90/90) | Response defined as development of a Neut titer ≥80 |
| | Treanor 2006 | Healthy adults; N=451 | SV/E A/ VN/1203 | 2 IM doses 28 days apart | 0, 7.5, 15, 45 or 90 | 0% (0) 13% (7.5) 24% (15) 41% (45) 57% (90) | 0% (0) 7% (7.5) 20% (15) 41% (45) 53% (90) | % with HAI or Neut titer ≥40 assessed |
| | Treanor 2007 | Healthy adults; ≥65 years; N=259 | SV/E A/ VN/1203 | 2 IM doses 28 days apart | 0, 45 or 90 | 35% (45) 46% (90) | N.R. | Response rates similar but slightly lower than among younger adults |
| | Campbell 2007 | Healthy children; 3-9 years; N=125 | SV/E A/ VN/1203 | 2 IM doses 28 days apart | 0 or 45 | 38% (45) | N.R. | Response rates were similar to those seen in young adults |
| | Patel 2006 | Healthy adults; 18-40 years; N=100 | SV/E A/ VN/1203 | 2 IM or ID doses 28 days apart | 3 or 9 ID; 15 or 45 IM | 4% (3) 5% (9) 12% (15) 56% (45) | 12% (3) 9% (9) 20% (15) 32% (45) | Pilot study' serum response rates after two doses |
| | Patel 2006 | Healthy adults who received 2 prior doses; N=77 | SV/E A/ VN/1203 | A single IM booster dose 7 months after dose 2 | 3 or 9 ID; 15 or 45 IM | 15% (3) 25% (9) 6% (15) 62% (45) | 10% (3) 40% (9) 31% (15) 71% (45) | Booster dose study; modestly enhanced responses after the third dose |
| | Goji 2008 | rHA-primed subjects N=37 primed and 103 unprimed | SV/E A/ VN/1203 | A single IM booster dose | 90 IM once | 68% with HAI response vs. 23% of unprimed given 1 dose | 76% with Neut response vs. 10% of unprimed given 1 dose | Responses after priming with rHA vs. a single dose in unprimed subjects |

(continued)

**Table 2** (continued)

| Type of Vaccine | Reference | Population Studied | Vaccine Description* | Regimen | Dosage(s) (μg HA) | % with HAI Antibody Responses % (dosage) | % with Neut Antibody Responses | Comments |
|---|---|---|---|---|---|---|---|---|
| | Zangwill 2008 | Healthy adults; previously given 2 doses N=337 | SV/E A/VN/1203 | A single booster dose IM; 6 months after last dose | 7.5, 15, 45 or 90 | 21% (7.5) 27% (15) 44% (45) 64% (90) | 31% (7.5) 43% (15) 67% (45) 78% (90) | Booster dose study; data are % with HAI ≥40 1 month after 3rd dose |
| | Patel 2008 | Healthy adults; 18-49 years; N=306 | SV/E A/VN/1203 | Two IM or ID doses 28 days apart | 30 | 35% (IM) 42% (ID) | 51% (IM) 57% (ID) | Percent with antibody response after two doses |
| Subunit +/- Aluminum Adjuvant | Bresson 2006 | Healthy adults; 18-40 years; N=300 | SV/E A/VN/1194 | Two IM doses 21 days apart | 7.5, 15 or 30 +/- 600μg AlOH/dose | -/+ AlOH: 43/28% (7.5) 44/44% (15) 53/66% (30) | -/+ AlOH: 20/16% (7.5) 22/18% (15) 27/41% (30) | Variable dose-dependent effect of alum noted |
| | Keitel 2008 | Healthy adults; 18-49 years; N=600 | SV/E \|A/VN/1203 | Two IM doses 28 days apart | 3.75, 7.5, 15, or 45 +/- 600μg AlOH/dose | -/+ AlOH: 2/2% (3.75) 0/14% (7.5) 13/14% (15) 25/33% (45) | N.R. | No meaningful adjuvant effect noted |
| | Brady 2007 | Healthy adults; ≥65 years N=600 | SV/E; \|A/VN/1203 | Two IM doses 28 days apart | 0, 3.75, 7.5, 15 and 45 +/- 600μg AlOH/dose | -/+ AlOH: 6/8% (3.75) 15/12% (7.5) 21/14% (15) 34/33% (45) | N.R. | The 3.75μg dosage level had 300μg of AlOH |
| | Nolan 2008a | Healthy adults; 18-45 years N=400 | SV/E A/VN/1194 | Two IM doses 21 days apart; booster dose 6 months later | 7.5 or 15 +/- AlPO4 | -/+ AlPO4: After 2 Doses: 37/40% (7.5) 37/42% (15) After 3 Doses: 2 28/31% | -/+ AlPO4: After 2 Doses: 37/51% (7.5) 51/54% (15) After 3 Doses: 45/70% (7.5) 65/71% (15) | Neut titer ≥20 after 2 doses |

| | | | | | | | | |
|---|---|---|---|---|---|---|---|---|
| Nolan 2008a | Healthy adults; 18-64 years N=400 | SV/E A/VN/1194 | Two IM doses 21 days apart | 30 or 45 + AlPO4 | 59% (30) 58% (45) | 54% (30) 55% (45) | Neut titer ≥40 after 2 doses |
| Nolan 2008b | Healthy adults; ≥65 years N=201 | SV/E A/VN/1194 | Two IM doses 21 days apart | 30 or 45 + AlPO4 | N.R. | 49% (30) 46% (45) | Neut titer ≥40 after 2 doses |
| Nolan 2008c | Healthy children; 6mo-9yrs N=148 | SV/E A/VN/1194 | Two IM doses 21 days apart | 30 or 45 + AlPO4 | 95% (30) 100% (45) | Combined response rates: 98% | Neut titer ≥40 after 2 doses |
| Krasnilnikov 2008 | Healthy adults; 18-60 years; N=352 | SV/E A/VN/1194 | Two IM doses | 0, 15 or 30 + AlOH | 3% (0) 48% (15) 47% (30) | 3% (0) 78% (15) 78% (30) | HAI response after 2 doses |
| Subunit +/- Oil-in-Water Adjuvant | | | | | | | |
| Nicholson 2001 | Healthy adults; 18-40years; N=65 | PSA/E A/Sgp/97 (H5N3) | Two IM doses 21 days apart | 7.5, 15 and 30 +/- MF59 | -/+ MF59: 0/60% (7.5) 0/45% (15) 9/20% (30) | -/+ MF59: 10/80% (7.5) 18/100% (15) 36/100% (30) | Response rates; Significant adjuvant effect noted |
| Leroux-Roels 2007 | Healthy adults; 18-60 years; N=400 | SV/E A/VN/1194 | Two IM doses 21 days apart | 3.8, 7.5, 15 or 30 +/- AS | -/+ AS: 4/84% (3.8) 16/90% (7.5) 35/96% (15) 43/85% (30) | -/+ AS: 22/86% (3.8) 37/86% (7.5) 53/86% (15) 65/98% (30) | Response Rates; Significant adjuvant effect noted |
| Levie 2008 | Healthy adults; N=265 | SV/E A/VN/1203 | Two IM doses 21 days apart | 1.9, 3.8, 7.5 and 15 with Adj; 7.5 without Adj. | 72% (1.9+) 81% (3.8+) 34% (7.5-) 89% (7.5+) 86% (15+) | 92% (1.9+) 88% (3.8+) 42% (7.5-) 92% (7.5+) 96% (15+) | Response rates; Significant adjuvant effect noted |
| Stephenson 2003 | Healthy adults given 2 doses of vaccine 16 months earlier; N=26 | PSA/E A/Sgp/97 (H5N3) | A single IM booster dose given | | -/+ MF59: 0/67% (7.5) 0/0% (15) 0/69% (30) | -/+ MF59: 0/100% (7.5) 50/100% (15) 100/100% (30) | Small numbers; HAI ≥40 or Neut ≥20 after booster immunization |

(continued)

**Table 2** (continued)

| Type of Vaccine | Reference | Population Studied | Vaccine Description* | Regimen | Dosage(s) (μg HA) | % with HAI Antibody Responses % (dosage) | % with Neut Antibody Responses | Comments |
|---|---|---|---|---|---|---|---|---|
| | Stephenson 2005 | Healthy adults who previously received 3 doses of vaccine; N=26 | PSA/E A/Sgp/97 (H5N3) | NA-blood samples from prior trials were tested | 7.5, 15, or 30 with MF59 (N=150 or without MF59 (N=11) | N.R. | Response rates vs. variants (-/+ MF59): A/Sgp=18/100 A/HK/03= 27/100 A.VN/04=0/43 A/Thai/04=0/71 | Adjuvanted vaccines elicited higher levels of cross-reactive antibodies |
| Subunit +/- Aluminum hydroxide or oil-in-water emulsion | Bernstein 2008 | Healthy adults; N=394 | SV/E A/VN/1203 +/- Aluminum hydroxide or MF59 | Two IM doses 28 days apart | 15, 30 or 45 without adjuvant; 7.5 or 15 with MF59; or 7.5, 15, or 30 with AlOH | 23% (7.5+MF59) 3% (7.5+AlOH) 24% (15-) 63% (15+MF59) 7% (15+AlOH) 18% (30-) 14% (30+AlOH) 29% (45-) | N.R. | HAI antibody response rates reported |
| Whole Virus +/- Aluminum Adjuvant | Tashiro 2007 | Healthy adult males; 20-39 years; N=120 | WV/E A/VN/1194 | Two IM or SC doses 21 days apart | 1.7, 5 or 15 with AlOH | N.R. | IM/SC Response 43/50% (1.7) 71/72% (5) 96/84% (15) | Neut antibody responses |
| | Lin 2006 | Healthy adults; 18-60 years; N=120 | WV/E A/VN/1194 | Two IM doses 28 days apart | 0; or 1.25, 2.5, 5, or 10 + AlOH | 0% (0) 9% (1.25) 17% (2.5) 33% (5) 64% (10) | 0% (0) 9% (1.25) 21% (2.5) 42% (5) 50% (10) | HAI response rates & Neut titer ≥40 after 2 doses |
| | Keitel 2008 | Healthy adults; 18-40 years; N=306 | WV/Vero A/VN/1203 | Two IM doses 28 days apart | 0; 7.5 or 15 +/- AlOH; or 45 | 0 (0) 19 (7.5-) 8 (7.5+) 18 (15-) 8 (15+) 40 (45-) | 0 (0) 17 (7.5-) 6 (7.5+) 27 (15-) 17 (15+) 44 (45-) | Antibody response frequencies after 2 doses |

| | | Population | Vaccine | Regimen | HA dose/adjuvant | Response 1 | Response 2 | Notes |
|---|---|---|---|---|---|---|---|---|
| | Ehrlich 2008 | Healthy adults; 18-45 years; N=270 | WV/Vero A/VN/1203 | Two IM doses 21 days apart | 3.75 or 30 + AlOH; or 7.5 or 15 +/-AlOH | 38% (3.75+) 48% (7.5-) 36% (7.5+) 27% (15-) 15% (15+) 36% (30+) | 69% (3.75+) 76% (7.5-) 64% (7.5+) 71% (15-) 61% (15+) 66% (30+) | Neut titer ≥20 after two doses |
| | Vajo 2007 | Healthy adults; >18 years; N=146 | WV/E A/VN/1194 | One IM dose | 6 + AlPO4 | 67% | N.R. | HAI titer 90 days after a single dose |
| Live Attenuated | Karron 2007 | Healthy young adults; N=21 | LAIV/E A/VN/1203 | Two IN doses 28 days apart | $10^{6.7}TCID_{50}$ | 11% with HAI antibody response after 2 doses | N.R. | Replication "highly restricted" |
| | Rudenko 2007 | Healthy young adults; N=20 | LAIV/E A/duck/Potsdam (H5N2) | Two IN doses | | 47% and 29% with HAI titer rise vs. H5N2 and H5N1, respectively, after 2 doses | 50% with Neut titer rise vs. H5N2 after 2 doses | 55% and 70% shed virus after dose 1 and 2, respectively |

Abbreviations: *HA*, hemagglutinin; *HaI*, hemagglution inhbibition *Neut*, neutralizing; *rHA*, recombinant HA; *HK*, Hong Kong; *IM*, intradermal; *AlOH*, aluminum hydroxide; *AlPO₄*, aluminum E, egg-grown; *VN*, Vitnam; *ID*, intradermal; *AlOH*, aluminum hydroxide; *AlPO₄*, aluminum phosphate; *PSA*, Purified surface antigen; *Sgp*, Singapore; *AS*, adjuvant system; *WV*, whole Virus; *Vero* Vero cell culture-grown; *LAIV*, live attenuated influenza virus

## 3.2 Immunogenicity Endpoints of H5 Vaccine Trials: General Considerations

While the precise determinants of protection against H5 infections are unknown, it is reasonable to extrapolate from the wealth of interpandemic influenza vaccine experience that serum antibodies will confer protection against infection, disease and/or death. Notably, passive protection against disease has been demonstrated in animal models of H5 influenza (Lu et al. 2006a; Hanson et al. 2006). Passive immunization of humans with convalescent plasma also likely conferred protection against death from influenza during the 1918 pandemic (Luke et al. 2006), and this approach was recently used to treat a patient infected with H5N1 virus in China (Zhou et al. 2007). Although a serum HAI antibody level of 40 or greater is generally considered protective, no specific level of antibody invariably confers protection. The levels of neutralizing (Neut) antibody required for protection are even less certain. Finally, in animal models of H5N1 vaccines, protection has been demonstrated even in the absence of a detectable immune response (Hampson 2006; Lipatov et al. 2005, 2006; Govorkova et al. 2006; Ruat et al. 2008).

A generally accepted principle regarding the immunogenicity of HA-based vaccines is that the higher the antibody level, the greater the degree of protection. Nevertheless, specific benchmarks for the assessment of the immunogenicity of candidate pandemic vaccines have been established by European and US regulatory bodies, including the fold change in geometric mean titer (GMT) of antibody after immunization; proportion of subjects with a putative protective HAI titer of $\geq 40$ after immunization; and proportion of subjects responding after immunization (Table 3) (Food and Drug Administration 2006; Committee for Human Medicinal Products 2007). These criteria are based on benchmarks used to assess interpandemic

**Table 3** Recommended serum HAI antibody response profiles for pandemic influenza vaccines

*European guidelines: CPMP/BWP/214/96*
- GMT after immunization / GMT before immunization > 2.5/ > 2[a]
- Titer $\geq$ 40 in >70% / 60% after immunization
- Seroconversion rate > 40% / 30% ($\geq$4-fold rise or increase from <10 before to $\geq$ 40 after immunization)

*Neut Ab also should be measured.*

*US FDA guidelines*
- Titer $\geq$40 in >70% / 60% after immunization[b]
- Seroconversion rate >40% / 30%[b]
  or
- GMT of licensed vaccine / GMT of the new vaccine $\leq$1.5[c]
- Seroconversion rate for licensed vaccine – seroconversion rate of new vaccine $\leq$10%[c]

[a]Target response rates for persons <65 years old in the numerator and for persons $\geq$65 years old in the denominator; GMT = geometric mean titer
[b]Lower bound of two-sided 95% CI
[c]Upper bound of two-sided 95% CI

influenza vaccines; therefore, they will require validation for use in predicting protection against pandemic influenza.

Despite the use of these specific immune response criteria, it remains difficult to compare results between trials because assay procedures lack standardization. Stephenson et al. (2007) have confirmed that there is considerable variability between laboratories when HAI antibody levels are measured on the same samples, underscoring the difficulty of comparing results between trials and laboratories. Variability in measurements of Neut antibody levels is even greater. Use of an antibody standard improved interlaboratory agreement. Efforts are underway to develop international antibody standards in order to reduce the variability observed between laboratories.

## 3.3  Clinical Trials of Influenza A/H5N1 Vaccines: Summary of Results

### 3.3.1  Overview of Vaccine Safety and Reactogenicity

In general, all vaccine constructs evaluated to date have been safe and well tolerated (data not shown). Adjuvanted formulations typically elicit a higher frequency of injection site and occasionally systemic reactions when compared with nonadjuvanted formulations, but the reactogenicity profiles of the vaccines included in Table 2 have been acceptable. The safety of most vaccines has been assessed only in a limited number of subjects, and expanded studies in diverse populations (such as children, pregnant women, the elderly, and persons with underlying health conditions) are needed.

### 3.3.2  Summary of Immunogenicity

With the aforementioned caveats regarding the determination and interpretation of serum antibody responses in mind, results of H5 vaccine trials from around the world are presented in Table 2. In most cases, the data presented are the serum HAI antibody responses or the percent of subjects achieving a putative protective titer (HAI titer $\geq$40) after two doses of vaccine given by the intramuscular (IM) route. For clinical trials of inactivated vaccines conducted in the USA, the clade 1 vaccine strain has been A/Vietnam/1203/04, whereas for trials conducted in Europe, the typical vaccine virus has been the closely related A/Vietnam/1194/04. Exceptions are noted in the table.

Development of the first human H5N1 vaccines relied on novel approaches because traditional egg-based production was thwarted by the lethality of the H5 viruses for chick embryos. Treanor and colleagues assessed the safety and immunogenicity of ascending doses of baculovirus-expressed, clade 0 recombinant

H5 HA among healthy adults (Treanor et al. 2001). Dose-related increases in serum neutralizing antibody responses were noted, but only about half of the subjects developed a serum neutralizing response after the receipt of two 90 μg doses. The findings of this trial presaged the results of subsequent trials of nonadjuvanted subunit vaccines; that is, high dosages of HA are necessary to stimulate antibody responses in most subjects. Subsequent development of inactivated subvirion vaccines was facilitated by the use of reverse genetics techniques, which permitted genetic modification of the HA and plasmid rescue of viruses bearing the HA and NA of the H5N1 virus and the internal genes of A/PR/8/34 (Neumann et al. 1999). These modifications allowed the propagation of these viruses in eggs. In addition, it was observed that traditional HAI antibody assays needed to be modified to use horse red blood cells (RBCs) due to the lack of sensitivity of assays using turkey or chick RBCs (Stephenson et al. 2003c, 2004). Several recent trials conducted among healthy young adult, elderly, and pediatric subjects with nonadjuvanted inactivated subvirion (SV) vaccines have confirmed the relatively poor immunogenicity of the H5 HA (Treanor et al. 2006, 2007; Campbell et al. 2007): about half of the adult subjects given two 90 μg doses and 38% of children aged 2–9 years given two 45 μg doses had serum HAI antibody responses.

In view of the need for high dosages of H5 vaccine antigen, dosage-sparing approaches are being explored. In the United States, only adjuvants containing mineral salts, such as aluminum hydroxide (AlOH), are licensed for human use. Several clinical trials of AlOH-adjuvanted vaccines have been reported. Bresson et al. (2006) described the dose-related immunogenicity of an inactivated SV H5 vaccine adjuvanted with AlOH. HAI antibody response rates were 53% and 66%, respectively, among subjects given two doses of nonadjuvanted or adjuvanted vaccine containing 30 μg of H5 HA. Failure of AlOH to provide clinically meaningful enhancement of immune responses was confirmed in several subsequent trials among healthy young adults and the elderly (Keitel et al. 2008; Brady et al. 2007). In several trials, conclusions regarding the potential for aluminum-containing adjuvants to enhance responses cannot be drawn because nonadjuvanted vaccines were not tested: 54% and 55% of young adult subjects developed a Neut titer $\geq 40$ after two doses of aluminum phosphate (AlPO$_4$)-adjuvanted vaccine containing 30 or 45 μg of HA, respectively, in one study; corresponding response rates among elderly subjects were 49% and 46% (Nolan et al. 2008a, 2008b). Among children between the ages of six months and nine years immunized with the same vaccine, 98% had a Neut antibody titer $\geq 40$ after two doses (Nolan et al. 2008c). In another study, about half of subjects developed an HAI response after two doses of AlOH-adjuvanted vaccine containing 15 or 30 μg of HA in another study (Krasilnikov 2008).

In contrast to the results obtained using mineral salt-containing adjuvants, oil-in-water emulsions appear very promising. Nicholson and colleagues (Nicholson et al. 2001) assessed the immunogenicity of an egg-grown purified surface antigen (PSA) vaccine derived from an apathogenic duck H5N3 virus. Nonadjuvanted formulations containing 7.5, 15, or 30 μg of H5 HA were poorly immunogenic among healthy adults, whereas MF59-adjuvanted formulations containing 7.5 μg of HA stimulated

HAI and Neut antibody responses in six, eight or ten subjects, respectively. In another study, 24%, 7%, and 63% of subjects given nonadjuvanted, AlOH, or MF59-adjuvanted inactivated SV vaccine containing 15 µg of H5N1 HA developed serum HAI response after two doses (Bernstein et al. 2008). A novel adjuvant system (AS) based on an oil-in-water emulsion (AS03) was recently reported to confer significant dosage-sparing effects on an inactivated egg-grown vaccine: adjuvanted SV vaccine containing as little as 3.8 µg of H5N1 HA elicited HAI antibody responses in 84% of subjects (Leroux-Roels et al. 2007). The results of a subsequent phase 3 safety trial of a candidate H5N1 vaccine containing 15 µg of HA formulated with AS03 in over 5,000 healthy adults were recently reported (Rumke et al. 2008). The investigators concluded that the safety and reactogenicity profile of the adjuvanted vaccine are acceptable for immunization against pandemic influenza. Frequencies of injection site and systemic reactions were higher among subjects given the adjuvanted vaccine than those given a licensed influenza vaccine, but no vaccine-related serious adverse events were reported. Levie et al. (2008) have also reported the results of a clinical trial of an inactivated SV vaccine adjuvanted with another novel oil-in-water emulsion: 34% and 89% of subjects given two doses of nonadjuvanted and adjuvanted SV vaccine containing 7.5 µg of H5 HA developed HAI responses, and 72% of subjects given two 1.9 µg doses of adjuvanted vaccine responded. While these results are quite promising, expanded safety databases will be needed before the potential value of these adjuvants can be confirmed.

Another potential dosage-sparing approach involves the use of whole-virus (WV) vaccines. Tashiro and colleagues (2007) assessed the safety and immunogenicity of an AlOH-adjuvanted egg-grown WV vaccine among healthy adults: 96% of subjects given two 15 µg doses by the IM route developed a Neut antibody response. Similarly, 78% of subjects given two 10 µg doses of AlOH-adjuvanted WV H5 vaccine developed HAI antibody responses in another study (Lin et al. 2006). Keitel and colleagues (2008) recently compared the immunogenicity of AlOH-adjuvanted and nonadjuvanted Vero cell culture-grown WV vaccine formulations. In contrast to other studies of inactivated vaccines, the wild-type virus A/Vietnam/1203/04 was used to produce this vaccine. In this study, 18% of subjects given two 15 µg doses of nonadjuvanted vaccine developed an HAI antibody response compared with 8% of those given the adjuvanted vaccine. Using the same vaccine, Kistner noted that 76% of subjects given the 7.5 µg dosage level developed a serum Neut antibody response (Ehrlich 2008). As seen in the Keitel study, inclusion of AlOH reduced the frequencies of response. Note that differences in assays and definitions of responses preclude direct comparisons between the two studies. Finally, 68% of subjects given a single dose of $AlPO_4$-adjuvanted, egg-grown WV vaccine developed an HAI response in a recent study (Vajo et al. 2007). Because WV vaccines have been associated with febrile reactions among children in the past, cautious evaluation of these constructs in the pediatric age groups will be necessary.

Intradermal (ID) immunization has been proposed as a potential antigen-sparing strategy in the past; however, few data support this claim. Previous studies have demonstrated that ID administration of reduced dosages of licensed influenza virus vaccines can elicit responses that are noninferior when compared with higher dosages given IM (Belshe et al. 2004; Kenney et al. 2004). More recently, similar immunogenicity was observed when similar dosages of licensed vaccine were given by the IM or ID route (Belshe et al. 2007). The safety and immunogenicity of ID immunization with an inactivated SV H5 vaccine were assessed in a pilot evaluation in which vaccine was given using the Mantoux technique (Patel et al. 2006). ID administration of two doses of vaccine containing 3 or 9 μg of H5 HA was safe, and immune responses were low and similar to those observed among subjects given 15 μg of HA IM. In a subsequent hypothesis-testing trial, subjects received two 30 μg doses by the ID or the IM route (Patel et al. 2008). Serum HAI antibody response rates were similar in the two groups: 42% and 35% of subjects developed a significant rise in titer after two doses of vaccine in the ID and IM groups, respectively. The potential for alternative methods of ID vaccine delivery remains to be explored.

In addition to exploring immune responses following primary immunization with inactivated vaccines (two-dose regimen), there is a tremendous amount of interest in assessing the ability of the vaccines to prime for immune responses following booster immunization with future variants, as well as for their ability to elicit cross-reactive antibodies. In two studies, the geometric mean titers (GMT) of H5 antibody were significantly higher following a third dose of H5 vaccine than after the second dose of vaccine (Stephenson et al. 2003a,b,c; Zangwill et al. 2008). Goji and colleagues (2007) revaccinated subjects who previously received a rHA vaccine (A/Hong Kong/97-HA) with a SV inactivated A/VN/1203/04 vaccine. Compared with subjects receiving a single 90 μg dose of A/VN/04, those previously immunized with A/HK/97 developed more frequent antibody responses and higher titers of antibody following the heterologous booster immunization. Inclusion of adjuvant has been shown to elicit higher titers of antibody to homologous and heterologous H5 variants following primary and booster immunization (Leroux-Roels et al. 2008; Stephenson et al. 2005).

Live attenuated influenza viruses (LAIV) are currently licensed for preventing interpandemic influenza among healthy persons between the ages of 2 and 49 years. These vaccines are based on reassortment of attenuated influenza A or B viruses with wild-type (wt) viruses that express the HA and neuraminidase (NA) against which antibody is desired (Maassab and DeBorde 1985). Resulting vaccine viruses that bear the six internal genes of the attenuated donor and the HA and NA of the wt virus are selected. Several candidate H5 LAIVs have been developed (Desheva et al. 2006; Sugitan et al. 2006). Karron et al. (2007) evaluated a reassortant of A/Ann Arbor/6/60 (H2N2) and A/Vietnam/1203 (H5N1)—the latter expressing a genetically modified HA—among healthy young adults. Vaccine was well tolerated; however, only 11% of subjects developed an antibody response, and replication of the virus in nasal wash specimens was "highly restricted." Rudenko and her colleagues (Rudenko and Katlinsky 2007) evaluated a reassortant of an A/Leningrad (H2N2) donor and A/duck/Potsdam/86/92 (H5N2) in healthy adults. Forty-seven percent of subjects developed a significant HAI response after two doses, and 55% and 70% of subjects shed virus after the first and second dose, respectively.

# 4 Clinical Trials of Other Potential Pandemic Vaccine Candidates

It is not known which virus will cause the next influenza pandemic, and avian reservoirs harbor 16 HAs, three of which caused pandemics during the twentieth century. Recent human infections caused by H7N7 and H9N2 viruses resulted in the evaluation of candidate vaccines. A phase I clinical trial of an inactivated sub-virion H7N7 vaccine is underway (Cate, personal communication). The safety, infectivity, and immunogenicity of a candidate LAIV (H7N3) among healthy adults was recently reported (Karron et al. 2008). Fifty-five percent of subjects developed a Neut antibody response after any dose of vaccine, and replication of the virus in the respiratory tract was less restricted than with H5N1 or H9N2 candidates (see below). The safety and immunogenicity of WV and subunit inactivated H2N2 and H9N2 vaccines have recently been evaluated (Hehme et al. 2002, 2004). Among healthy adults 18–30 years old who were given 1.9, 3.8, or 7.5 µg doses of a WV H2N2 vaccine formulated with aluminum hydroxide, 80%, 98%, and 87% developed significant antibody responses after two doses, compared with 98% of subjects given 15 µg of a subvirion, nonadjuvanted H2N2 vaccine. Among 18–60 year old subjects given 1.9, 3.8, or 7.5 µg doses of a WV H9N2 vaccine formulated with AlOH, 58%, 54%, and 72% responded after two doses, compared with 65% of subjects who received 15 µg of nonadjuvanted vaccine. Stephenson et al. (2003a,b,c) evaluated subunit and WV H9N2 vaccines among 60 healthy adults. Two IM doses containing 7.5, 15, or 30 µg of H9 HA were given three weeks apart. A number of subjects over the age of 32 years (i.e., born before 1968) were found to possess antibody to the H9 HA before immunization, and these subjects responded to a single dose of vaccine, suggesting that they were primed. Atmar and colleagues (2006) assessed a PSA H9N2 vaccine with or without MF59 adjuvant among 96 healthy young adults. Nonadjuvanted vaccine elicited serum HAI responses in 67%, 50%, 50%, and 67% of subjects given two doses containing 3.75, 7.5, 15, or 30 µg doses, respectively. The corresponding figures for adjuvanted preparations were 92%, 92%, 100%, and 100%. A separate study exploring the age-related differences in immune responses is underway (Atmar, unpublished data).

Finally, a candidate H9N2 LAIV was well tolerated and elicited HAI responses in 92% of healthy adults; however, replication was "highly restricted" (Karron et al. 2009). Additional studies are in progress to explore the variability in infectivity and immunogenicity of the LAIV candidates.

# 5 Summary and Conclusions

Several themes have emerged during the decade of H5N1 vaccine development. Higher than expected dosages of H5-HA are needed to stimulate detectable antibody responses in a majority of subjects; therefore, dosage-sparing approaches will be needed to overcome the poor immunogenicity of the H5 HA. Aluminum-containing adjuvants have conferred little meaningful benefit; however, dosage-dependence and formulation differences will require additional study. Subunit vaccines containing oil-in-water adjuvants and some WV vaccines appear promising, but expanded studies are needed to confirm their safety and immunogenicity. The ability of vaccines to elicit cross-reactive antibodies and to prime efficiently for drifted variants is critical in view of the uncertainty regarding which virus will ultimately spread efficiently among humans. Standardized assays and antibody controls are desirable in order that reasonable comparisons can be made between results of clinical trials being conducted around the world. Finally, an improved understanding of the correlates of protection is sorely needed in order to put results of clinical trials in context.

It is important to bear in mind that no one vaccine type or single vaccine manufacturer will be able to meet the needs of all populations. Some potential advantages and disadvantages of various immunization strategies are outlined in Table 4. Subunit vaccines have a long track record for safety in healthy and high-risk populations, but high dosages are required for H5. Adjuvanted subunit vaccines are antigen sparing, but their safety in large and/or vulnerable populations is unknown. Some whole-virus vaccines appear to be immunogenic at lower dosages or after a single dose; however, their increased reactogenicity among children has been documented in past pandemic trials (Wright et al. 1977). Live attenuated vaccines have the potential to reduce the amount of antigen and/or the need for two doses of vaccine, and may confer enhanced nonspecific resistance during the weeks after immunization. However, safety, poor infectivity of H5 constructs, and the fact that they can only be used in the setting of pandemic spread in the general population are current limitations. Continued evaluation of a variety of approaches is warranted. Solutions to the problems outlined above and improved control of interpandemic influenza will better prepare us for the next pandemic.

**Table 4** Potential advantages and disadvantages of various types of vaccines

| Type of vaccine | Potential advantages | Potential disadvantages |
|---|---|---|
| Subunit | Safety record; broad indications (age to 6 months, underlying diseases, pregnancy); licensed processes | Requirement for two high dosages (H5) |
| Subunit with adjuvant | Antigen sparing | Potential safety concerns; increased reactogenicity; cost |
| Whole virus | Potentially antigen sparing; current/past licensed processes | Reactogenicity in young children |
| Live attenuated | Potentially antigen sparing; broader immune responses; one dose requirement; potential for early protection | Pandemic use only; safety issues (age, health, and immune status); low infectivity; cold chain requirements |

# References

Antonovics J, Hood M, Baker C. (2006) Molecular virology: was the 1918 flu avian in origin. Nature 440(7088):E9–E10

Atmar RL, Keitel WA, Patel SM, Katz JM, She DW, El Sahly H, Pompey J, Cate TR, Couch RB. (2006) Safety and immunogenicity of nonadjuvanted and MF59-adjuvanted influenza A/H9N2 vaccine preparations. Clin Infect Dis 43:1135–1142

Bean WJ, Schell M, Katz J et al. (1992) Evolution of the H3 influenza virus hemagglutinin from human and nonhuman hosts. J Virol 66:1129–1138

Belshe RB, Newman FK, Cannon J, Duane C, Treanor J, Van Hoecke C, Howe BJ, Dubin G. (2004) Serum antibody responses after intradermal vaccination against influenza. N Engl J Med 351:2286–2294

Belshe RB, Newman FK, Wilkins K, Graham IL, Babusis E, Ewell M, Frey SE. (2007) Comparative immunogenicity of trivalent influenza vaccine administered by intradermal or intramuscular route in healthy adults. Vaccine 25:6755–6763

Bernstein D, Edwards K, Dekker C, Belshe R, Talbot H, Graham I, Noah D, He F, Hill H. (2008) Effects of adjuvants on the safety and immunogenicity of an avian influenza H5N1 vaccine in adults. J Infect Dis 197:667–675

Brady RC, Treanor JJ, Atmar RL, Chen WH, Winokur P, Belshe R. (2007) A phase I-II, randomized, controlled, dose-ranging study of the safety, reactogenicity, and immunogenicity of intramuscular inactivated influenza A/H5N1 vaccine given alone or with aluminum hydroxide to healthy elderly adults. Abstract P739 of the Options for the Control of Influenza VI Meeting, Toronto, ON, Canada, 17–23 June 2007

Bresson JL, Perronne C, Launay O, Gerdil C, Saville M, Wood J, Höschler K, Zambon MC. (2006) Safety and immunogenicity of an inactivated split-virion influenza A/Vietnam/1194/2004 (H5N1) vaccine: phase I randomised trial. Lancet 367:1657–1664

Campbell J, Graham I, Zangwill K. (2007) Pediatric H5N1 vaccine trial (online slideshow). http://www.who.int/vaccine_research/diseases/influenza/150207_Campbell.pdf

Chen H, Smith GJ, Li KS et al. (2006) Establishment of multiple sublineages of H5N1 influenza virus in Asia: implications for pandemic control. Proc Natl Acad Sci USA 103:2845–2850

Committee for Human Medicinal Products (CHMP) (2007) Influenza vaccines prepared from viruses with the potential to cause a pandemic and intended for use outside of the core dossier context. CHMP/VWP/263499/06, 1–11

Cunha B. (2004) Influenza: historical aspects of epidemics and pandemics. Infect Dis Clin North Am 18(1):141–55

Desheva JA, Lu XH, Rekstin AR, Rudenko LG, Swayne DE, Cox NJ, Katz JM, Klimov AI. (2006) Characterization of an influenza A H5N2 reassortant as a candidate for live-attenuated and inactivated vaccines against highly pathogenic H5N1 viruses with pandemic potential. Vaccine 24:6859–6866

Ehrlich H, Muller M, Oh H, Tambyah P, Joukhadar C, Montomoli E, Fisher D, Berezuk G, Fritsch S, Low-Baselli A, Vartian N, Bobrovsky R, Pavlova B, Pollabauer E, Kistner O, Barrett P (2008) for the Baxter H5N1 pandemic influenza vaccine clinical study team. A clinical trial of a whole-virus H5N1 vaccine derived from cell culture. N Engl J Med 358:24:2573–2584

Ernst WA, Kim HJ, Tumpey TM, Jansen ADA, Tai W, Cramer DV, Adler-Moore JP, Fujii G. (2006) Protection against H1, H5, H6 and H9 influenza A infection with liposomal matrix 2 epitope vaccines. Vaccine 24:5158–5168

Food and Drug Administration. (2006) Draft guidance for industry on clinical data needed to support the licensure of trivalent inactivated influenza vaccines. Fed Reg 71:12367

Gibbs M, Gibbs A. (2006) Molecular virology: was the 1918 pandemic caused by bird flu. Nature 440:27:E8

Glezen PW. (1996) Emerging infections: pandemic influenza. Epidemiol Rev 18:64–76

Goji N, Nolan C, Hill H, Wolff M, Noah D, Williams T, Rowe T, Treanor J. (2008) Immune responses of healthy subjects to a single dose of intramuscular inactivated influenza A/Vietnam/1203/2004 (H5N1) vaccine after priming with an antigenic variant. J Infect Dis 198:9: 635–641

Govorkova EA, Webby R, Humberd J, Seiler JP, Webster RG. (2006) Immunization with reverse-genetics-produced H5N1 influenza vaccine protects ferrets against homologous and heterologous challenge. J Infect Dis 194:159–167

Guan Y, Poon LL, Cheung CY et al. (2004) H5N1 influenza: a protean pandemic threat. Proc Natl Acad Sci USA 101, 8156–8161

Hampson A. (2006) Ferrets and the challenges of H5N1 vaccine formulation. J Infect Dis 194:143–145

Hanson BJ, Boon ACM, Lim APC, Webb A, Ooi EE, Webby RJ. (2006) Passive immunoprophy-laxis and therapy with humanized monoclonal antibody specific for influenza A H5 hemag-glutinin in mice. Respir Res 7:126

Hehme N, Engelmann H, Kunzel W, Neumeier, Sanger R. (2002) Pandemic preparedness: lessons learnt from H2N2 and H9N2 candidate vaccines. Med Microbiol Immunol 191:203–208

Hehme N, Engelmann H, Kuenzel W, Neumeier E, Saenger R. (2004) Immunogenicity of a monovalent, aluminum-adjuvanted influenza whole virus vaccine for pandemic use. Virus Res 103:16–171

Hoelscher MA, Garg S, Bangari DS et al. (2006) Development of adenoviral-vector-based pandemic influenza vaccine against antigenically distinct human H5N1 strains in mice. Lancet 367:475–481

Karron R, Callahan K, Luke C et al. (2007) Phase I evaluation of live attenuated H9N2 and H5N1 *ca* reassortant vaccines in healthy adults (online slideshow). http://www.who.int/vaccine_research/diseases/influenza/160207_Karron.pdf

Karron R et al. (2008) Clinical evaluation of live attentuated pandemic influenza virus vaccines (online slideshow). http://www.who.int/vaccine_research/diseases/influenza/Karron_infleunza_meeting_140208.pdf

Karron RA, Callahan K, Luke C, Thumar B, McAuliffe J, Schappell E, Joseph T, Coelingh K, Jin H, Kemble G, Murphy BR, Subbarao K. (2009) A live attenuated H9N2 influenza vaccine is well tolerated and immunogenic in healthy adults. J Infect Dis 199:711–16

Keitel W, Campbell J, Treanor J, Walter E, Patel S, He F, Noah D, Hill H. (2008) Safety and immunogenicity of an inactivated influenza A/H5N1 vaccine given with or without aluminum hydroxide to healthy adults: Results of a phase-I-II randomized clinical trial. J Infect Dis 198:11:1309–1315

Kenney RT, Frech SA, Muenz LR, Villar CP, Glenn GM. (2004) Dose sparing with intradermal injection of influenza vaccine. N Engl J Med 351:2295–2301

Kistner O. (2007) Phase 1/2 clinical study with Baxter's H5N1 vaccine: clinical update (online slide-show). http://www.who.int/vaccine_research/diseases/influenza/150207_Kistner_Baxter.pdf

Kodihalli S, Goto H, Kobasa DL, Krauss S, Kawaoka Y, Webster RG. (1999) DNA vaccine encod-ing hemagglutinin provides protective immunity against H5N1 influenza virus infection in mice. J Virol 73:2094–2098

Koopmans M, Wilbrink B, Conyn M, Natrop G, van der Nat H, Vennema H, Meijer A, van Steenbergen J, Fouchier R, Osterhaus A, Bosman A. (2004) Transmission of H7N7 avian influenza A virus to human beings during a large outbreak in commercial poultry farms in the Netherlands. Lancet 363:587–593

Krasilnikov I. (2008) MICROGEN's pandemic vaccine clinical development (online slideshow). http://www.who.int/vaccine_research/diseases/influenza/Krasilnokov_infleunza_meeting_140208.pdf

Laddy DJ, Yan J, Corbitt N, Kobasa D, Kobinger GP, Weiner DB. (2007) Immunogenicity of novel consensus-based DNA vaccines against avian influenza. Vaccine 25:2984–2989

Levie K, Leroux-Roels I, Hoppenbrouwers K, Kervyn A-D, Vandermeulen C, Forgus S, Leroux-Roels G, Pichon S, Kusters I. (2008) An adjuvanted, low-dose, pandemic influenza A (H5N1) vaccine candidate is safe, immunogenic, and induces cross-reactive immune responses in healthy adults. J Infect Dis 198:9:642-649

Leroux-Roels, Barkowski A, Vanwolleghem T, Dramé M, Clement F, Hons E, Devaster JM, Leroux-RoelsG. (2007) Antigen sparing and cross-reactive immunity with an adjuvanted rH5N1 prototype pandemic influenza vaccine: a randomized controlled trial. Lancet 370:580–589

Leroux-Roels I, Bernhard R, Gérard P, Dramé M, Hanon E, Leroux-Roels G. (2008) Broad clade 2 cross-reactive immunity induced by an adjuvanted clade 1 rH5N1 pandemic influenza vac-cine. PLoS One 3(2):e1665

Lin YP, Shaw M, Gregory V et al. (2000) Avian-to-human transmission of H9N2 subtype influenza A viruses: relationship between H9N2 and H5N1 human isolates. Proc Natl Acad. Sci USA 97:9654–9658

Lin J, Zhang J, Dong X, Fang H, Chen J, Su N, Gao Q, Zhang Z, Liu Y, Wang Z, Yang M, Sun R, Li C, Lin S, Ji M, Liu Y, Wang X, Wood J, Feng Z, Wang Y, Yin W. (2006) Safety and immunogenicity of an inactivated adjuvanted whole-virion influenza A (H5N1) vaccine: a phase I randomised controlled trial. Lancet 368, 991–997

Lipatov AS, Webby RJ, Govorkova EA, Krauss S, Webster RG. (2005) Efficacy of H5 influenza vaccines produced by reverse genetics in a lethal mouse model. J Infect Dis 191:1216–1220

Lipatov AS, Hoffman E, Salomon R, Yen HL, Webster RG. (2006) Cross-protectiveness and immunogenicity of influenza A/Duck/Singapore/3/97(H5) vaccines against infection with A/Vietnam/1203/04(H5N1) virus in ferrets. J Infect Dis 194:1040–1043

Lu J, Guo Z, Pan X, Wang G, Zhang D, Li Y, Tan B, Ouyang L, Yu X. (2006a) Passive immunotherapy for influenza A/H5N1 virus infection with equine hyperimmune globulin F (ab′)2 in mice. Respir Res 7:43

Lu X, Edwards LE, Desheva JA, Nguyen DC, Rekstin A, Stephenson I, Szretter K, Cox NJ, Rudenko LG, Klimov A, Katz JM. (2006b) Cross-protective immunity in mice induced by live-attenuated or inactivated vaccines against highly pathogenic influenza A (H5N1) viruses. Vaccine 24:6588–6593

Luke TC, Kilbane EM, Jackson JL, Hoffman SL. (2006) Meta-analysis: convalescent blood products for Spanish influenza pneumonia: a future H5N1 treatment. Ann Intern Med 145:599–609

Ma W, Vincent A, Gramer M, Brockwell C, Lager K, Janke B, Gauger P, Patnayak D, Webby R, Richt J. (2007) Identification of H2N3 influenza A viruses from swine in the United States. Proc Natl Acad Sci USA 104:52:20949–20954

Maassab HF, DeBorde DC. (1985) Development and characterization of cold-adapted viruses for use as live virus vaccines. Vaccine 3:355–69

Neumann G, Watanabe T, Ito H, Watanabe S, Goto H, Gao P, Hughes M, Perez DR, Donis R, Hoffmann E, Hobom G, Kawaoka Y. (1999) Generation of influenza A viruses entirely from cloned cDNAs. Proc Natl Acad Sci USA 96:9345–9350

Nicholson KG, Colgate AE, Podda A, Stephenson I, Wood J, Ypma E, Zambon MC. (2001) Safety and antigenicity of non-adjuvanted and MF59-adjuvanted influenza A/Duck/Singapore/97 (H5N3): a randomized trial of two potential vaccines against H5N1 influenza. Lancet 357:1937–1943

Nolan T, Richmond P, Skeljo M, Pearce G, Hartel G, Formica N, Hoschler K, Bennet J, Ryan D, Papanaoum K, Basser R, Zambon M. (2008a) Phase I and II randomised trials of the safety and immunogenicity of a prototype adjuvanted inactivated split-virus influenza A (H5N1) vaccine in healthy adults. Vaccine 26:4160–4167.

Nolan T, Richmond P, Skeljo M et al.. (2008b) Prototype H5N1 pandemic influenza vaccine: further studies on elderly and children (online slideshow). http://www.who.int/vaccine_research/diseases/influenza/Nolan_infleunza_meeting_140208.pdf

Nolan T, Richmond P, Formica N, Hoschler K, Skeljo M, Stoney T, McVernon J, Hartel G, Sawlwin D, Bennet J, Ryan D, Basser R, Zambon M. (2008c) Safety and immunogenicity of a prototype adjuvanted inactivated split-virus influenza A (H5N1) vaccine in infants and children. Vaccine 26:6383–91.

Patel SM, Atmar RL, El Sahly H, Cate TR, Keitel WA. (2006) A randomized, open-label, phase I clinical trial comparing the safety, reactogenicity, and immunogenicity of booster immunization with inactivated influenza A/H5N1 vaccine administered by the intradermal (ID) or the intramuscular (IM) route among healthy adults. In: 44th Annual Meeting of the Infectious Diseases Society of America, Toronto, ON, Canada, 12–15 Oct 2006

Patel SM, Atmar RL, El Sahly H, Cate TR, Keitel WA. (2008) A phase I/II, randomized, double-blinded placebo-controlled trial to assess the safety, reactogenicity, and immunogenicity of immunization with inactivated subvirion influenza A/H5N1 vaccine administered by the intradermal or the intramuscular route among healthy adults. Abstract S1 of the 11th Annual Conf on Vaccine Research, Baltimore, MD, 5–7 May 2008

Ruat C, Caillet C, Bidaut A, Simon J, Osterhaus A. (2008) Vaccination of macaques with adju-
    vanted formalin-inactivated influenza A virus (H5N1) vaccines: protection against H5N1
    challenge without disease enhancement. J Virol 82:2565–69
Rudenko L, Katlinsky A. (2007) Evaluation of Russian live evaluation of Russian live attenuated
    vaccine H5N2 attenuated vaccine H5N2 in clinical trials (online slideshow). http://www.who.
    int/vaccine_research/diseases/influenza/160207_Rudenko.pdf
Rümke HC, Bayas JM, de Juanes JR, Caso C, Richardus JH, Campins M, Rombo L, Duval X,
    Romanenko V, Schwarz TF, Fassakhov R, Abad-Santos F, von Sonnenburg F, Dramé M,
    Sänger R, Ballou WR. (2008) Safety and reactogenicity profile of an adjuvanted H5N1
    pandemic candidate vaccine in adults within a phase III safety trial. Vaccine 26(19):2378–2388.
    doi:10.1016/j.vaccine.2008.02.068
Shinya K, Ebina M, Yamada S, Ono M, Kasai N, Kawaoka Y. (2006) Influenza virus receptors in
    the human airway. Nature 440:435–436
Stephenson I, Nicholson KG, Colegate A, Podda A, Wood J, Ypma E, Zambon M. (2003a)
    Boosting immunity to influenza H5N1 with MF59-adjuvanted H5N3 A/Duck/Singapore/97
    vaccine in a primed human population. Vaccine 21:1687–1693
Stephenson I, Nicholson KG, Glück R, Mischler R, Newman RW, Palache AM, Verlander NQ,
    Warburton F, Wood JM, Zambon MC. (2003b) Safety and antigenicity of whole virus and
    subunit influenza A/Hong Kong/1073/99 (H9N2) vaccine in healthy adults: phase I randomized
    trial. Lancet 362:1959–1966
Stephenson I, Wood JM, Nicholson KG, Zambon MC. (2003c) Sialic acid receptor specificity on eryth-
    rocytes affects detection of antibody to avian influenza haemagglutinin. J Med Virol 70:391–398
Stephenson I, Wood JM, Nicholson KG, Charlett A, Zambon MC. (2004) Detection of anti-H5
    responses in human sera by HI using horse erythrocytes following MF59-adjuvanted influenza
    A/Duck/Singapore/97 vaccine. Virus Res 103:91–95
Stephenson I, Bugarini R, Nicholson KG, Podda A, Wood JM, Zambon MC, Katz JM. (2005)
    Cross-reactivity to highly pathogenic avian influenza H5N1 viruses after vaccination with non-
    adjuvanted and MF59-adjuvanted influenza A/Duck/Singapore/97 (H5N3) vaccine: a potential
    priming strategy. J Infect Dis 191:1210–1215
Stephenson I, Das RG, Wood JM, Katz JM. (2007) Comparison of neutralising antibody assays
    for detection of antibody to influenza A/H3N2 viruses: an international collaborative study.
    Vaccine 25:4056
Suguitan AL Jr, McAuliffe J, Mills KL, Jin H, Duke G, Lu B, Luke CJ, Murphy B, Swayne DE,
    Kemble G, Subbarao K. (2006) Live, attenuated influenza A H5N1 candidate vaccines provide
    broad cross-protection in mice and ferrets. PLoS Med 3:e360
Tashiro M. (2007) Development of pandemic vaccine in Japan: P I and P II/III clinical studies (online
    slideshow). http://www.who.int/vaccine_research/diseases/influenza/150207_Tashiro.pdf
Taubenberger J, Reid A, Lourens R, Wang R, Guozhong J, Fanning T. (2005) Characterization of
    the 1918 influenza virus polymerase genes. Nature 437:889–893
Treanor JJ, Wilkinson BE, Masseoud F et al.. (2001) Safety and immunogenicity of a recombinant
    hemagglutinin vaccine for H5 influenza in humans. Vaccine 19:1732–1737
Treanor JJ, Campbell James D, Zangwill KM, Rowe T, Wolff M. (2006) Safety and immunogenic-
    ity of an inactivated subvirion influenza A (H5N1) vaccine. N Engl J Med 354:1343–1351
Treanor J, Bernstein D, Edwards K, Zangwill K, Noah D. (2007) Evaluation of inactivated monova-
    lent rga/Vietnam/1203/04 X PR8 subvirion vaccine in healthy elderly adults. Abstract P731 of
    the Options for the Control of Influenza VI Meeting, Toronto, ON, Canada, 17–23 June 2007
Vajo Z, Kosa L, IIdiko V, Jankovics M, Jankovics I. (2007) Inactivated whole virus influenza A
    (H5N1) vaccine. Emerg Infect Dis 13(5):807–808
Watanabe T, Watanabe S, Kim J, Hatta M, Kawaoka Y. (2008) Novel approach to the development
    of effective H5N1 influenza a virus vaccines: use of M2 cytoplasmic tail mutants. J Virol
    82:5:2486–2492
Wright PF, Thompson J, Vaughn WK, Folland DS, Sell SHW, Karzon DT. (1977) Trials of influ-
    enza A/New Jersey/76 virus-vaccine in normal children: overview of age-related antigenicity
    and reactogenicity. J Infect Dis 136:S731–S741

Writing Committee of the Second World Health Organization Consultation on Clinical Aspects of Human Infection with Avian Influenza A. (H5N1) Virus (2008) Update on avian influenza A (H5N1) virus infection in humans. N Engl J Med 358:261–273

Zangwill K, Treanor JJ, Campbell JD, Noah DL, Ryea J. (2008) Evaluation of the safety and immunogenicity of a booster (third) dose of inactivated subvirion H5N1 influenza vaccine in humans. J Infect Dis 197:580–583

Zhou B, Zhong N, Guan Y. (2007) Treatment with convalescent plasma for influenza A(H5N1) infection. N Engl J Med 357:1450

# Considerations for Licensure of Influenza Vaccines with Pandemic and Prepandemic Indications

**Norman W. Baylor and Florence Houn**

**Contents**

**Abstract** With over 409 human cases of avian influenza and over 256 deaths worldwide resulting from infection with avian influenza (H5N1), an influenza pandemic is still a real threat, especially with H5N1 continuing to evolve into antigenically distinct clades. The Food and Drug Administration (FDA) along with other national regulatory authorities (NRAs) recognize the important role that safe

N.W. Baylor (✉) and F. Houn

FDA/CBER/OVRR, HFM-405, 1401 Rockville Pike, Rockville, MD 20852, USA
e-mail: norman.baylor@fda.hhs.gov

R.W. Compans and W.A. Orenstein (eds.), *Vaccines for Pandemic Influenza*,
Current Topics in Microbiology and Immunology 333,
DOI 10.1007/978-3-540-92165-3_22, © Springer-Verlag Berlin Heidelberg 2009

and effective vaccines will play in protecting the public health from the threat of an influenza pandemic. The challenges to the FDA and other NRAs are significant as regulatory agencies pursue the development of new scientific and regulatory criteria to evaluate vaccines against pandemic influenza strains for licensure. To this end, the FDA is actively utilizing current regulatory processes such as accelerated approval and priority review as well as developing the regulatory pathways needed to speed the availability of vaccines against pandemic influenza. In May of 2007, the FDA issued two final guidance documents, one describing the clinical data recommended to support the licensure of annual influenza vaccines, and the other describing the clinical data recommended to support the licensure of pandemic influenza vaccines. These guidances contain specific approaches outlined by the FDA to assist manufacturers in developing new vaccines to increase the supply of safe and effective influenza vaccines for both annual and pandemic use. In this article we define the nomenclature "pandemic" and "prepandemic," describe the regulatory pathway for licensing new influenza vaccines for pandemic and prepandemic use, and outline considerations for evaluating pandemic/prepandemic vaccines that have been formulated using new approaches such as cell culture and non-aluminum salt adjuvants.

# 1   Introduction

The Food and Drug Administration (FDA) is the national regulatory authority in the United States responsible for, among other things, determining whether medical products, including vaccines, are safe and effective for marketing in the USA. The FDA's Center for Biologics Evaluation and Research (CBER) administers the vaccine regulatory program. Licensure for marketing a new vaccine is based on the evaluation of data demonstrating the manufacturer's ability to produce a product in a consistent manner, and the evaluation of results from adequate and well-controlled clinical trials demonstrating the safety and effectiveness of the product.

CBER facilitates the development of new vaccines by providing guidance and scientific advice to vaccine manufacturers and sponsors on how to fulfill the regulatory requirements for the licensure of new vaccines and related products. The regulations impose standards for manufacturing and clinical data. The manufacturing process must be thoroughly documented and include methods for the detection and elimination of adventitious agents, and reliable and sensitive test methods to determine safety, purity, and potency that are applicable to many products including vaccines. For clinical studies, FDA requires endpoints that have meaningful clinical benefit, such as disease reduction or prevention. Vaccine-specific challenges to advance development and evaluation include, inter alia, establishing the correlates of protection necessary for evaluating efficacy (which may obviate a clinical disease prevention endpoint), improving current and developing new assays for potency, or finding animal models that can be used for the evaluation of efficacy when human clinical trials are unethical or post-exposure field trials have not been feasible.

## 2 Background

In the USA, vaccine development and commercialization are complex processes; however, a single set of basic regulatory criteria apply to the evaluation of all new vaccines, regardless of the technology used to produce them. The current legal authority for the regulation of vaccines and other biological products is subject to the provisions of Section 351 of the Public Health Service (PHS) Act and the Federal Food, Drug, and Cosmetic Act. The statutes of the PHS Act are implemented through regulations codified in the Code of Federal Regulations (CFR) (CFR 2007). Title 21 of the CFR, parts 600 through 680, contains regulations specifically applicable to vaccines and other biologicals. Vaccine manufacturers and sponsors must also comply with regulations for investigational new drug applications (IND, 21 CFR, part 312) and Current Good Manufacturing Practices (CGMPs, 21 CFR, parts 210 and 211.9), as well as other regulations.

To obtain a biologics license for marketing a new vaccine under Section 351 of the PHS Act (21 CFR 601.2), an applicant must submit a biologics license application (BLA) to the Director of the CBER; however, this is delegated to the Director of the Office of Vaccines Research and Review (1410.202 of the FDA Staff Manual Guide). The BLA must include data derived from nonclinical laboratory and clinical studies that demonstrate that the manufactured product meets prescribed requirements for safety, purity, and potency. The general considerations for clinical studies to license a vaccine include demonstration through adequate and well-controlled studies of safety and effectiveness (immunogenicity studies based on antibody responses may be used to infer efficacy in certain situations), as well as evaluation of the new vaccine when administered simultaneously with other routinely recommended licensed vaccines. The BLA must also include chemistry, manufacturing, and control (CMC) data that support compliance with standards addressing requirements for: (1) organization and personnel; (2) buildings and facilities; (3) equipment; (4) control of components, containers, and closures; (5) production and process controls; (6) packaging and labeling controls; (7) holding and distribution; and (8) laboratory controls. Furthermore, a full description of manufacturing methods, including data establishing the stability of the product through the dating period; sample(s) representative of the product for introduction or delivery into interstate commerce; summaries of test results performed on the lot(s) represented by the submitted sample(s); specimens of the labels, enclosures, and containers; and the address of each location involved in the manufacture of the biological product should be included in the BLA.

Regulatory mechanisms for advancing new vaccines through FDA's review process have been developed for severe or life-threatening illnesses. These mechanisms include fast-track designation for the product, accelerated approval using surrogate endpoints, and priority review of marketing applications.

The Fast Track program of the FDA is designed to facilitate the development and review of new drugs and biologicals that are intended to treat serious or life-threatening conditions, and that demonstrate the potential to address unmet medical needs. Fast tracking adds the possibility of a "rolling submission" for a marketing application to existing programs. A "rolling submission" allows the applicant to submit agreed-upon sections (e.g., the clinical section or the CMC section) of a marketing application, i.e., a BLA to FDA for review and evaluation as each section is completed, instead of waiting for all data to be collected prior to submitting the complete BLA. An important feature of fast track designation is that it emphasizes the critical nature of frequent and early communication between the FDA and sponsors to improve the efficiency of product development.

Accelerated approval (21 CFR 601.40) may be granted for certain biological products that have been studied for their safety and effectiveness in treating serious or life-threatening illnesses and that provide meaningful therapeutic benefit over existing treatments. The accelerated approval regulations give the FDA flexibility with respect to the types of endpoints that can be relied upon to support marketing approval. In lieu of a clinical endpoint on disease prevention, a surrogate endpoint that is reasonably likely, based on epidemiologic, therapeutic, pathophysiologic or other evidence, to predict clinical benefit may be accepted. This may allow earlier marketing of a vaccine, as the surrogate endpoint may be demonstrated sooner than a clinical endpoint. However, approval using this pathway is subject to the requirement that the applicant study the vaccine to verify and describe its clinical benefit. Postmarketing studies are generally expected to be already underway.

Recently, the option to pursue an accelerated approval pathway for trivalent inactivated influenza vaccines became available to sponsors if a shortage of influenza vaccine exists for the US market at the time the new vaccine is approved. In this case, the FDA interprets the accelerated approval regulation as allowing approval of an influenza vaccine using immunogenicity endpoints instead of influenza disease prevention, because having additional influenza vaccines available would provide a meaningful benefit over an existing shortage situation. FDA has recently licensed three seasonal trivalent influenza vaccines (Fluarix from GlaxoSmithKline, Flulaval from GlaxoSmithKline, and Afluria from CSL Biotherapies) using this regulatory mechanism. The accelerated approval regulations do not affect the quantity or quality of evidence needed to demonstrate effectiveness, safety, or product quality.

Products regulated by CBER are eligible for priority review if they provide a significant improvement in the safety or effectiveness of the treatment, diagnosis, or prevention of a serious or life-threatening disease. The FDA has six months to complete the review of a new BLA receiving a priority review designation. The standard review time for a new BLA, not designated as priority, is ten months.

Other regulatory mechanisms that may be used by FDA to expedite the availability of a new but unlicensed product for use during an emergency such as an influenza pandemic include Emergency Use Authorization (EUA). Upon determination and declaration by the Secretary of the Department of Health and Human Services (DHHS) that a public health emergency (or the potential for one) that affects or has the significant potential to affect national security exists, the Secretary of DHHS can authorize the use of an unlicensed product, if it is reasonable to believe that the product may be effective in diagnosing, treating, or preventing the serious life-threatening disease or condition, if there is no adequate, approved, available alternative, and if the known and potential benefits outweigh the known and potential risks.

The amount of data needed for an EUA will depend on the nature of the product and completed studies, the nature of the emergency, and the adequacy and availability of approved alternatives. An EUA for a product is limited to the duration of a declared emergency (and allows patients to finish treatment courses that they started during the emergency). Once the effective period of the declaration has expired, regulations for INDs apply. The final determination that criteria are met for the issuance of an EUA can only be made after an emergency is declared. Under the EUA, specific conditions of authorization are applied, which may include the requirement to inform healthcare workers or recipients (if feasible) of the EUA status of the product, to identify and communicate significant known and potential risks and benefits from the product, and to provide the option to accept or refuse the product.

# 3 Regulatory Evaluation of Influenza Vaccines

## 3.1 Operational Definitions

Before discussing the various procedures involved in the regulatory evaluation of influenza vaccines, some of the more pertinent operational definitions will be discussed.

### 3.1.1 Seasonal (Annual) Influenza Vaccines

The vaccines currently in distribution for seasonal influenza contain three inactivated influenza viruses (H3N2, H1N1, and a B influenza virus strain). These vaccines are also referred to as trivalent inactivated influenza vaccines (TIV). There is one seasonal trivalent live attenuated vaccine (LAIV) licensed in the USA that contains the same influenza virus strains.

### 3.1.2 Candidate Influenza Vaccines

This is a prospective influenza A virus vaccine which is at the research and clinical development stages and has not been granted marketing licensure by a regulatory agency.

### 3.1.3 Vaccines Against Novel Human Influenza Viruses

This refers to a monovalent vaccine containing a human influenza A virus strain that is not in general circulation among human populations, but where the virus is considered to be a potential threat to humans, and is potentially capable of causing a pandemic. The term "novel" refers to human influenza A viruses, not influenza B viruses. The influenza H5N1 subtypes are considered novel human influenza viruses; other examples include influenza A virus subtypes H7 and H9.

### 3.1.4   Pandemic Influenza Vaccine Indication

FDA may grant a pandemic indication to an influenza vaccine for the active immunization of persons at high risk of exposure to an influenza virus that has the potential to cause an influenza pandemic, and for the active immunization of persons during a pandemic caused by the influenza virus subtype contained in the vaccine provided that substantial evidence is presented to support the intended effect of the vaccine. This vaccine would be used under an emergency declared by the US Secretary of HHS under Section 319(a) of the PHS Act, or in other situations that place persons at high risk of exposure to an influenza virus strain that has the potential to cause a pandemic, such as deployment to an area that has animal-to-human transmission of the avian influenza virus contained in the vaccine.

### 3.1.5   Prepandemic Influenza Vaccine Indication

A prepandemic indication may be granted to an influenza vaccine for active immunization of persons against an influenza virus subtype(s) that has the potential to cause a pandemic as a strategy for pandemic influenza preparedness. This vaccine would be indicated for use in the interpandemic period, well before a pandemic is declared, and will most likely be used for population priming. The immune response elicited would eventually be boosted by an influenza vaccine with a pandemic indication when a pandemic is declared. In the USA, a vaccine with a prepandemic indication is not limited to influenza H5N1.

### 3.1.6   Prime–Boost

Prime–boost is a two-part immunization process. A prime–boost strategy may be employed for influenza vaccines with a pandemic or prepandemic indication. It involves priming the immune system with the first injection of an influenza vaccine containing an influenza strain of pandemic potential. Several weeks, months or years later, another injection of a homologous or heterologous influenza vaccine (boost) is administered. A one- or two-dose boost alone may produce a quicker but weaker immune response as compared to the prime–boost regimen.

## 3.2   Overview of FDA's Licensure of Influenza Vaccines

Currently, all influenza vaccines licensed in the USA are derived from viruses grown in embryonated chicken eggs. Each TIV contains 15 μg of hemagglutinin antigen (HA) from each of the three strains selected for that year's vaccine: two influenza A strains and one influenza B strain. For seasonal influenza vaccines each year, any of the previous three influenza strains in the trivalent vaccine may be replaced with a new strain. Strain changes are based on an evaluation of circulating wild-type strains. The FDA requires submission of a prior approval manufacturing supplement to an existing BLA for strain changes. FDA does not require clinical data for approval of these annual supplements to a BLA for US licensed manufacturers of inactivated influenza vaccine.

Public health experts from national influenza surveillance centers including the Centers for Disease Control and Prevention (CDC), the FDA and the World Health Organization (WHO) annually evaluate worldwide epidemiological data to determine the strains of the viruses that manufacturers will use to make the influenza virus vaccine for annual administration beginning each fall. FDA's Vaccines and Related Biological Products Advisory Committee recommends the strains for the annual TIV for the US. The process of selecting the strains up to manufacturing the final vaccine involves numerous steps in-between, and is a lengthy process that extends approximately from six to eight months.

Annual influenza vaccines are the primary tool for preventing and controlling influenza. FDA policies and strategies for influenza with a pandemic indication have been developed based on the regulatory framework used for seasonal influenza vaccines. On 31 May 2007, the FDA issued two final guidance documents, one describing the clinical data recommended to support the licensure of annual influenza vaccines, and the other describing the clinical data recommended to support the licensure of pandemic influenza vaccines (FDA Guidance for Industry 2007a,b). These guidances contain specific approaches outlined by the FDA to assist manufacturers in developing new vaccines to increase the supply of safe and effective influenza vaccines for both annual and pandemic use.

The FDA recommends that manufacturers submit a BLA for all submissions for initial licensure of pandemic influenza vaccines, including vaccines against novel human influenza viruses. A BLA for a pandemic influenza vaccine should contain CMC data as well as clinical data. Submission of a BLA enables manufacturers to have separate trade names and labeling to distinguish between seasonal and pandemic vaccines, and it also facilitates postmarketing adverse event reporting and collection by differentiating the information collected for each type of vaccine. The amount of data required by the FDA from manufacturers to submit with their pandemic influenza vaccine BLA is dependent upon whether the manufacturer already has a licensed influenza vaccine and intends to use the same manufacturing process for its pandemic vaccine.

## 4 Clinical Data to Support the Licensure of Influenza Vaccines with a Pandemic Indication

The FDA does not require data from clinical trials demonstrating prevention of influenza illness for pandemic influenza vaccine candidates in situations where a manufacturer holds a US license to market and distribute a seasonal inactivated influenza vaccine under either the provisions in 21 CFR 601.2 or the accelerated approval provisions with the vaccine's clinical benefit having been confirmed in a postmarketing study, and where the seasonal influenza vaccine manufacturing process for the production of the pandemic vaccine is used. However, clinical immunogenicity trials are required to determine the appropriate dose and regimen of a candidate pandemic influenza vaccine. These trials should also include an assessment of safety. In addition, FDA will seek agreement from sponsors to conduct postmarketing studies to obtain additional information about the vaccine's safety and effectiveness when the vaccine is used.

Currently, for manufacturers who are not licensed in the USA to produce a seasonal influenza vaccine, the approval of a BLA for an inactivated influenza vaccine indicated for pandemic use will require results from one or more adequate and well-controlled studies designed to meet immunogenicity endpoints utilizing the accelerated approval pathway. Further clinical studies are required after approval to verify the clinical benefit of the vaccine. In addition, all sponsors who seek licensure of an influenza vaccine for a pandemic indication through accelerated approval should provide plans to the FDA to collect additional effectiveness and safety data, such as through epidemiological studies, when the vaccine is used. Safety data must also be collected from subjects enrolled in prelicensure clinical trials intended to support the accelerated approval of a pandemic influenza vaccine.

## 4.1  Assessment of Immunogenicity

A specific hemagglutination inhibition (HI) antibody titer associated with protection against culture-confirmed influenza illness has not been identified to date; however, some studies of influenza infection, including human challenge studies following vaccination, have suggested that HI antibody titers ranging from 1:15 to 1:65 may be associated with protection from illness in 50% of subjects, and higher titers have been shown to be associated with an increase in protection from illness (Hobson et al. 1972; deJong et al. 2003). Evaluations of seroconversion and geometric mean titers (GMT) have been used as measures of vaccine activity (CPMP 1997; Treanor et al. 2002). Antibody response, determined by measuring HI titers, is used as a serological marker of the immunological response to influenza vaccines, and may be appropriate for the evaluation of influenza vaccines with a pandemic indication. The antibody response to HA may be an acceptable surrogate marker of activity that is reasonably likely to predict the clinical benefit of inactivated pandemic influenza vaccines.

FDA recommends that appropriate endpoints include (1) the percentage of subjects achieving an HI antibody titer $\geq$ 1:40, and (2) rates of seroconversion, defined as the percentage of subjects with either a prevaccination HI titer < 1:10 and a postvaccination HI titer > 1:40 or a prevaccination HI titer $\geq$ 1:10 and a minimum fourfold rise in postvaccination HI antibody titer. For adults < 65 years of age and for the pediatric population, the lower bound of the two-sided 95% CI for the percent of subjects achieving an HI antibody titer $\geq$ 1:40 should meet or exceed 70%. For adults 65 years and older, the lower bound of the two-sided 95% CI for the percent of subjects achieving an HI antibody titer $\geq$ 1:40 should meet or exceed 60%. The lower bound of the two-sided 95% CI for the percent of subjects achieving seroconversion for HI antibody should meet or exceed 40% for adults less than 65 years of age and for the pediatric population. For adults 65 years of age and older, the lower bound of the two-sided 95% CI for the percent of subjects achieving seroconversion for HI antibody should meet or exceed 30%.

In the interpandemic setting, it is likely that most subjects will not have been exposed to the influenza virus strain with pandemic potential. Therefore, it is possible that vaccinated subjects may reach both suggested endpoints. Thus, for studies enrolling individuals who are immunologically naïve to the influenza viral strain with pandemic potential, one HI antibody assay endpoint, such as the percentage of participants achieving an HI antibody titer $\geq$ 1:40, may be considered.

Considerable variability can be introduced into the laboratory assay used to measure HI antibodies as a result of a number of factors, including differences in viral strains and red blood cell types, and the presence of nonspecific inhibitors in the assay medium. Thus, suitable controls and assay validation are important for interpreting HI antibody results. Other endpoints and the corresponding immunologic assays, such as virus neutralization, may also be used to support the approval of a BLA for an influenza vaccine with a pandemic influenza indication (Rowe et al. 1999). However, at the present time, such assays have not been standardized and validated, and the surrogate endpoints that may correlate with clinical benefit have not yet been defined. For more detailed descriptions of possible approaches for establishing effectiveness based on immune responses under an accelerated approval, see the guidance documents referenced above (FDA Guidance for Industry 2007a,b).

## 4.2 Live Attenuated Influenza Vaccines for Pandemic Use

Clinical trials for live attenuated influenza vaccines indicated for pandemic use will be required to determine the appropriate dose and regimen as well as the assessment of immunogenicity and safety. Data to support the selected dose and regimen should be based on the evaluation of immune responses elicited by the vaccine. Currently, immune response data following receipt of LAIV are limited; however, use of the accelerated approval pathway for licensure of a LAIV for pandemic use will depend on the identification of an immune surrogate that is reasonably likely to predict clinical benefit. Although LAIV is thought to elicit a variety of immune responses (Belshe et al. 2000), as with TIV, the antibody response to HA may be appropriate for the evaluation of new LAIV for pandemic use. The observed immunogenicity may bridge to the antibody response observed with seasonal LAIV for which clinical efficacy has been demonstrated. The advantages of using HI antibody response as a clinical endpoint are that HI assays are simple and high throughput, they are validated, and the HI antibody titer of 1:40 has been correlated with a reduction in influenza-like illness (ILI). However, the question remains as to whether a 1:40 HI titer is the appropriate protective level for all strains of influenza, particularly those that have pandemic potential. Presently, for the HI antibody assay, FDA recommends the same endpoints for LAIV vaccines for pandemic use as described in the guidance document for inactivated influenza vaccines for pandemic use (FDA Guidance for Industry 2007a,b).

LAIV may also induce protection against disease through immunological mechanisms other than, or in addition to, antibodies to HA. Mucosal antibody may be an option as a clinical endpoint for LAIV for pandemic use since mucosal immunity is thought to be important for protection, and some studies have correlated mucosal IgA antibody in nasal washes with protection (Piedra et al. 2005). However, there are currently no specific IgA antibody titers that have been correlated with reduction in ILI. Other clinical endpoints for LAIV for pandemic use may include measuring neutralizing antibody titers or cell-mediated immunity (CMI), both of which are thought to be important for protection (Greenberg et al. 1975; Couch et al. 1969). Notwithstanding, no specific neutralizing antibody titer nor specific measure of CMI have been correlated with a reduction in ILI. Thus, sponsors may propose alternative endpoints for FDA's consideration. Depending on the endpoints identified as being most likely to correlate with protection, new or improved assays will need to be developed and standardized. Further, identification of the most appropriate selection of clinical endpoints for efficacy analysis will likely include additional studies of seasonal LAIV and possibly additional preclinical studies in animal models.

Clinical studies with live attenuated influenza vaccines for pandemic use performed in advance of an influenza pandemic also present special biosafety considerations. Therefore, sponsors are encouraged to initiate early discussion with the FDA to reach agreement on the size of the safety database needed to support product licensure. The FDA currently recommends that subjects be isolated during the study period to minimize the potential for transmission of the influenza vaccine viral strain. The amount and duration of vaccine strain shedding should be well characterized among all subjects. Contact precautions should be in place for study subjects and study personnel for the duration of shedding. Study personnel should be monitored for possible influenza illness and transmission of the influenza vaccine strain. Study subjects and study personnel with symptoms suggestive of influenza illness should be treated with antiviral agents pending culture or other microbiological results.

# 5 Chemistry, Manufacturing, and Control Requirements for Licensure of Pandemic Influenza Vaccines

Overall, the CMC requirements for influenza vaccines with prepandemic and pandemic indications are similar to the general CMC requirements for other vaccines. However, there are a few CMC issues that are different from those encountered for seasonal influenza vaccines, including the origin of the vaccine strain used for manufacture, the virus inactivation process (for inactivated vaccines), potency determination, and vaccine stability. For pandemic influenza vaccines that are not manufactured in embryonated chicken eggs, there are additional cell substrate issues that must be addressed, such as the detection and removal of adventitious agents and the evaluation of certain cell substrates for tumorigenicity. Other issues that must be addressed include the documentation and characterization of reference viruses derived from reverse genetics that are used for virus seed production. Issues related to inactivation of the virus strain and safety testing are also important to evaluate, especially for influenza viruses with pandemic potential. For licensed manufacturing processes, CBER accepts the inactivation kinetics on the influenza strain itself, avian leukosis virus (ALV) and mycoplasma. For new manufacturers, validating inactivation of the strain, ALV and mycoplasma alone are not sufficient, and inactivation of model viruses may also be required.

## 5.1 Potency

Currently, inactivated seasonal influenza vaccines are assessed for potency using a validated single radial immunodiffusion (SRID) assay method based on the parallel line method using antigen-specific reagents provided by CBER. Moreover, CBER requires that the SRID method provided by CBER's testing labs should be validated by the manufacturer and that all of the technical details should be specified.

## 5.2 Stability

The shelf life of influenza vaccines with prepandemic or pandemic indications for use during the prepandemic period will be dependent on the stability data that are available when the BLA is submitted. For the product licensed for use during the pandemic, there will be a need for a much longer shelf life, as these products may be stockpiled, and due to the limitations of stability data available at the time of approval, there may be a subsequent need to extend the shelf life of the stockpiled product. FDA's experience with influenza vaccines for pandemic use is currently limited to H5N1 vaccines that are present in the US national stockpile mainly as a monovalent bulk concentrate, which will only be formulated in final filled containers in a situation where a pandemic is declared. With the application of alternative technologies for manufacturing influenza vaccines such as the use of cell cultures, the use of expression vectors to make purified HA vaccines, and the use of non-aluminum salt adjuvants, the appropriate stability plan must take into account the differences in these manufacturing methods, and during the development process appropriate storage conditions and an appropriate shelf life will need to be established.

## 6   Enabling New Approaches and Technologies

Manufacturers and researchers are actively studying new manufacturing methods for influenza vaccines for seasonal and pandemic use. These include cell culture-based and recombinant technologies, as well as the use of adjuvants in influenza vaccine formulations.

## 6.1   Adjuvants

Aluminum salt adjuvants, generically referred to as alums, are the only adjuvants currently used in licensed vaccines for human use in the USA, and have been used extensively in the formulation of a number of vaccines against bacterial and viral pathogens. Numerous investigational vaccine adjuvants such as mineral salts, emulsions, synthetic derivatives, and particulate formulations have been studied in humans. Many vaccines have been formulated with adjuvants with the goal of increasing the immune response to the vaccine and decreasing the amount of antigen required per dose. The need to develop more effective influenza vaccines for pandemic use has increased interest in adjuvants. There are currently no licensed adjuvanted influenza vaccines in the USA; however, MF59, an oil/water emulsion, has been used in influenza vaccines licensed in Europe since 1997 (Wadman 2005).

Small studies with inactivated nonadjuvanted H5N1 influenza vaccines have demonstrated that more antigen per dose and more than one dose are necessary to elicit immune responses comparable to those elicited following a single dose of an annual seasonal inactivated influenza vaccine (Fauci 2005). The first H5N1 vaccine licensed was nonadjuvanted and demonstrated poor immunogenicity, requiring two doses of 90 µg of HA antigen to produce neutralizing titers of at least 1:40 in 43% of recipients (Treanor et al. 2006). Recent data suggest that some adjuvants may provide the advantage of antigen sparing; in other words, less antigen can be used in the vaccine formulation. For example, Stephenson et al. (2005) described the development of broadly cross-reactive neutralizing antibodies, with seroconversion rates to A/HongKong/156/97 of 100%, to A/HongKong/213/03 of 100%, to A/Thailand/16/04 of 71%, and to A/Vietnam/1203/04 of 43% in 14 subjects who received the MF59-adjuvanted vaccine, compared with 27%, 27%, 0%, and 0% respectively among 11 subjects who received nonadjuvanted vaccine. Another study using an adjuvanted H5N1 influenza vaccine showed that two doses of A/H5N1 vaccine at antigen concentrations ranging from 7.5 to 45 µg per dose were as high or higher than those reported previously in studies that used 45–90 µg of A/H5 antigen (Bernstein et al. 2008). Other adjuvants are also being clinically tested with influenza antigens.

## 6.2 Regulatory Considerations for Adjuvanted Influenza Vaccines

The immunogenicity induced by the antigen/adjuvant formulation is likely product specific and the result of multiple factors, and currently no data are available that would allow an extrapolation from one antigen/adjuvant formulation to another. The FDA does not currently license adjuvants separately from the vaccine antigen. Moreover, a specific adjuvant concentration in combination with a specific amount of antigen is evaluated by the FDA for safety and efficacy to support the licensure of the antigen/adjuvant vaccine formulation. FDA recommends using a rational approach to combining adjuvants with influenza antigens based on the anticipated clinical benefit of the adjuvanted vaccine formulation over the unadjuvanted formulation. FDA encourages dose-ranging studies to find the optimal combination of both components for enhanced immune response in relationship to adverse events.

Addition of a new adjuvant to a vaccine formulation is considered a new product by the FDA and requires the submission of a new BLA. Because vaccines are administered to healthy individuals including infants and children, and there are potential safety concerns with adjuvants, extensive preclinical safety studies of adjuvants and adjuvanted vaccines, including local reactogenicity and systemic toxicity testing, are required. For vaccines with a prepandemic indication and an adjuvant that has not been licensed for marketing in the USA, there are particular concerns regarding adverse events that, while there may be advantages for a prolonged dosing schedule and determination of safety for a population preparedness strategy, may require a large premarket safety database. Prepandemic use carries less direct benefit to the vaccinated and therefore risks must be thoroughly assessed. The FDA requires data from initial studies that support the medical rationale for adding the adjuvant, such as evidence of enhanced immune response or other advantages, as well as data supporting selection of the dose of the adjuvant itself. Data should be collected to demonstrate the clinical contribution above the unadjuvanted vaccine.

The accelerated approval pathway may be one regulatory pathway to licensing adjuvanted pandemic influenza vaccines; however, designing clinical studies to confirm the clinical benefit of the adjuvanted pandemic vaccine, a requirement of accelerated approval, will be challenging. Since the options for confirming clinical benefit of a pandemic vaccine prior to the outbreak of an actual influenza pandemic are limited, sponsors may consider confirming the efficacy of their adjuvanted pandemic influenza vaccine by studying the efficacy of an adjuvanted seasonal influenza vaccine manufactured by the same process; however, it is recommended that sponsors discuss their plans with the FDA before pursuing these confirmatory studies. It is commonly recognized that the immunogenicity elicited by various HA antigens, especially across subtypes and between clades in a subtype of influenza A, may vary greatly. Therefore, clinical efficacy against influenza disease may also vary greatly for vaccines directed against different influenza A subtypes and clades.

From a public health point of view, the preference would be to have a pandemic influenza vaccine that only requires a single dose that may provide long-term immunogenicity, thus alleviating the need for boosters during waves of the pandemic, and that cross-protects against varying antigenic changes. Adjuvanted pandemic influenza vaccines may be able to meet some of these needs. FDA recommends that clinical studies of adjuvanted pandemic vaccines include exploration of dose optimization for a single administration, duration of immunity, ability to cross-neutralize various antigens, as well as other characteristics. It would be advantageous to do these studies now, in the interpandemic period, to gather information to help characterize these factors.

## 6.3   Cell Culture Technology

Cell culture and recombinant influenza vaccines are also under development, and may eventually replace the use of chicken eggs to produce influenza vaccines. Although egg-based manufacturing has been successful and cost effective, non-egg-based manufacturing has potential advantages in terms of flexibility, and may allow a greater yield of product. The FDA considers the use of a new cell substrate as a new product, and so influenza vaccines manufactured in cell cultures require the submission of a new BLA. Cell culture technology is used in the manufacture of other vaccines, and product characterization is important in using this technology for influenza vaccine manufacture. FDA recently published draft guidance on the characterization and qualification of cell substrates and other biological starting materials used in the production of viral vaccines for the prevention and treatment of infectious diseases (FDA Draft Guidance for Industry 2006). This guidance document outlines FDA's current recommendations on characterizing cell substrates for the production of viral vaccines.

FDA is evaluating other technologies under investigation for potential use in the development of seasonal and pandemic influenza vaccines, such as alternative delivery systems and routes of administration, and the use of immune stimulators. These technologies are at different stages of development, and they all present unique challenges to regulatory authorities.

## 7   Evaluation of Other Vaccination Strategies

Several alternative approaches to vaccinating during the pandemic are under consideration by public health organizations, including the World Health Organization. Waiting to use a vaccine against the actual pandemic strain during a pandemic has the disadvantage that such a vaccine takes time to manufacture and will only be available several months after the pandemic has begun. One strategy under consideration is to use vaccines prepared using influenza viral strains that have pandemic potential in anticipation of the pandemic—influenza vaccines with prepandemic indications—in order to prime the population. This strategy is based on the assumption that these vaccines may elicit memory immune responses with some cross-reactivity with future human pandemic strains, and that such responses could be boosted by a single dose of a pandemic influenza vaccine around the time of the pandemic. The advantage of this may be that it avoids the need for a two-dose immunization regimen during a pandemic, which presents a logistical problem.

Another strategy may be to use an influenza vaccine with a pandemic strain to control outbreaks by terminating the transmission cycle through the combination use of antiviral medication and vaccination, which offers longer-term protection. For such use, the vaccine's ability to generate very high antibody titers as rapidly as possible would be important.

## 7.1  Pandemic vs. Prepandemic Vaccine Indication

Vaccines with a pandemic indication are to be used in situations with high risk of exposure to pandemic influenza strains. These situations include personnel deployed to areas with cases of human-to-human transmission or animal-to-human transmission of the influenza virus subtype, first responders to various outbreaks of the influenza virus subtype, and segments of the general population who may require immunization during a pandemic emergency caused by the influenza virus subtype contained in the vaccine. US health authorities have begun to stockpile vaccines with a pandemic indication, and these authorities, along with input from various healthcare organizations and others, will help to define which situations are "high risk" and appropriate for pandemic vaccine use. There is currently one licensed vaccine with a pandemic indication. This vaccine is a monovalent, inactivated vaccine for the active immunization of persons 18 through 64 years of age at increased risk of exposure to the H5N1 influenza virus subtype A/Vietnam/1203/2004 (H5N1, clade 1), manufactured by Sanofi Pasteur. Until there is a better understanding of H5, and more experience is gained with these influenza virus subtypes, if an actual pandemic is declared or a high-risk situation occurs necessitating the use of this vaccine, but the strain is different, the manufacturer would need to submit to the FDA a supplement to their BLA containing clinical data that supports an optimal dose for the vaccine made with this new strain.

In contrast, vaccines with a prepandemic indication are those intended for the active immunization of persons against an influenza virus subtype(s) that has the potential to cause a pandemic, as a strategy to enhance pandemic influenza preparedness for the population. This approach would involve the licensure of an influenza vaccine that contains one or more potential influenza virus subtypes that could be administered during the current interpandemic period. Such vaccines would need to have a durable memory response that can be boosted. The benefits of disease protection would be theoretical, as the particular immune response generated by the vaccine may not provide direct benefit. Further, in terms of risks, the vaccine must have a low adverse event profile for it to be used in a population preparedness strategy, especially if the pandemic never materializes in the vaccine's lifetime. Also, to be an effective strategy, the memory must last long enough that the population does not need to be frequently boosted. There are no US-licensed influenza vaccines with a prepandemic indication; however, manufacturers have expressed interest in this indication. Currently, more information is needed about what doses are able to elicit a long-term memory response that could be boosted based on heterologous or homologous vaccine strains and given years apart.

The criteria for evaluating the clinical development of both pandemic and prepandemic influenza vaccines should be developed now, in this interpandemic period, so that more information can be obtained though nonclinical and clinical studies of these vaccines. This will mean that, once a pandemic has been declared by the Secretary of HHS under Section 319(a) of the PHS Act, information about how to manufacture and administer effective doses of the appropriate strain of vaccine will be known.

The regulatory framework for evaluating prepandemic influenza vaccines is under active discussion within the FDA, as well as other global regulatory authorities. The clinical development of inactivated influenza vaccines with either a pandemic or prepandemic indication will require different approaches because, as stated above, the risks and benefits, along with the conditions of use, differ for the two indications. Clinical development of the two types of vaccines may vary on the dosing schedule, duration of immunity follow up, and size of the safety database.

Immunogenicity trials for pandemic and prepandemic influenza vaccines should be prospective, randomized, double-blinded, and controlled. Initial clinical studies may start in adults (18–64 years of age), and the fewer exclusion criteria there are, the more likely that the data obtained will be relevant to real-world use. Pediatric and geriatric studies to determine dosing will most likely be needed for the pandemic and prepandemic indications. For prepandemic vaccines, determining the appropriate immunogenicity endpoints to achieve population preparedness for priming and boosting, plus the duration of the study needed to demonstrate boosting, are all considerations. Most likely the boost, administered at least one year later in order to be a reasonable population preparedness strategy, should elicit levels of immune response similar to those stated in FDA's Guidance for Industry (2007b), but this may not be needed for the prime response measured early on.

# 8   Conclusion

There are scientific, technological, and regulatory challenges that remain to be addressed to facilitate the development of effective influenza vaccines for pandemic and prepandemic use. The FDA is actively utilizing current regulatory processes such as fast track, priority review, and accelerated approval to facilitate the availability of vaccines against pandemic influenza, and has made significant strides, as evidenced by the issuance of two guidance documents which outline the regulatory pathway for licensure of influenza vaccines for seasonal and pandemic use. The FDA continues to work closely with its partners in industry, government, academia, and other NRAs to develop additional science-based regulatory strategies to assure the availability of new vaccines for the emerging threat of an influenza pandemic.

**Acknowledgments** Jerry Weir, Marion Gruber, and Theresa Finn for reviewing the manuscript and providing insightful comments, and the entire influenza vaccine review staff in the Office of Vaccines Research and Review, who have worked diligently developing scientific and regulatory pathways for facilitating the availability of new influenza vaccines for seasonal and pandemic use.

# References

Belshe RB, Gruber WC, Mendelman PM et al (2000) Efficacy of vaccination with live attenuated, cold-adapted, trivalent, intranasal influenza virus vaccine against a variant (A/Sydney) not contained in the vaccine. J Pediatr 136:168–175

Bernstein DI, Edwards KM, Decker CL et al (2008) Effects of adjuvants on the safety and immunogenicity of an avian influenza H5N1 vaccine in adults. J Infect Dis 197(5):667–675

Code of Federal Regulations (CFR) (2007) Title 21 Food and Drugs. Office of the Federal Register, National Archives & Records Administration, Washington, DC

Committee for Proprietary Medicinal Products (CPMP) (1997) Note for guidance on harmonisation of requirements for influenza vaccines. CPMP/BWP/214/96. The European Agency for the Evaluation of Medicinal Products (EMEA), London

Couch RB, Douglas RG, Rossen R, Kasel J (1969) Role of secretory antibody in influenza. In: The secretory immune system. US Dept Public Health, Washington, DC, pp 93–112

de Jong JC, Palache AM, Beyer WEP et al. (2003) Haemagglutination-inhibiting antibody to influenza virus. Dev Biol (Basel) 115:63–73

Fauci AS (2005) Testimony before the Committee on Foreign Relations, United States Senate: "Pandemic influenza: the road to preparedness." http://www.hhs.gov/asl/testify/t051109.html

FDA (2006) Draft Guidance for Industry: Characterization and qualification of cell substrates and other biological starting materials used in the production of viral vaccines for the prevention and treatment of infectious diseases. US Food and Drug Administration, Rockville, MD

FDA (2007a) Guidance for Industry: Clinical data needed to support the licensure of seasonal inactivated influenza vaccines. US Food and Drug Administration, Rockville, MD

FDA (2007b) Guidance for Industry: Clinical data needed to support the licensure of pandemic influenza vaccines. US Food and Drug Administration, Rockville, MD

Greenberg SB, Criswell BS, Couch RB (1975) Lymphocyte-mediated cytotoxicity against influenza virus-infected cells. An in vitro method. J Immunol 115:601–603

Hobson D, Curry RL, Beare AS, Ward-Gardner A (1972) The role of serum haemagglutination-inhibiting antibody in protection against challenge infection with influenza A2 and B viruses. J Hyg (Camb) 70:767–777

Piedra PA, Gaglani, MJ, Riggs M et al. (2005) Live attenuated influenza vaccine, trivalent, is safe in healthy children 18 months to 4 years, 5 to 9 years, and 10 to 18 years of age in a community-based, nonrandomized, open-label trial. Pediatrics 16:e397–e407

Rowe T, Abernathy RA, Hu-Primmer J et al. (1999) Detection of antibody to avian influenza A (H5N1) virus in human serum by using a combination of serologic assays. J Clin Microbiol 37:937–943

Stephenson I, Bugarini R, Nicholson KG et al (2005) Cross-reactivity to highly pathogenic avian influenza viruses after vaccination with MF59-adjuvanted influenza A/Duck/Singapore/97 (H5N3) vaccine: a potential priming strategy. J Infect Dis 191:1210–1215

Treanor J, Keitel W, Belshe R et al. (2002) Evaluation of a single dose of half strength inactivated influenza vaccine in healthy adults. Vaccine 20:1099–1105

Treanor JJ, Campbell JD, Zangwill KM, Rowe T, Wolff M (2006) Safety and immunogenicity of
    an inactivated subvirion influenza A (H5N1) vaccine. N Engl J Med 354:1343–1351
Wadman M (2005) Race is on for flu vaccine. Nature 438:23

# Strategies for Broad Global Access to Pandemic Influenza Vaccines

**Kathryn M. Edwards, Adam Sabow, Andrew Pasternak, and John W. Boslego**

## Contents

K.M. Edwards
Sarah H. Sell Professor of Pediatrics, Vanderbilt Vaccine Research Program,
Vanderbilt University School of Medicine, 1161 21st Avenue South,
CCC-5323 Medical Center North, Nashville, TN, 37232, USA
e-mail: kathryn.edwards@vanderbilt.edu

A. Sabow
Associate Partner, Oliver Wyman, 10 South Wacker Drive, 13th Floor,
Chicago, IL, 60606, USA
e-mail: adam.sabow@oliverwyman.com

A. Pasternak
Partner, Oliver Wyman, 10 South Wacker Drive, 13th Floor, Chicago, IL, 60606, USA
e-mail: Andrew_pasternak@yahoo.com

J.W. Boslego (✉)
Director, Vaccine Development Global Program, PATH, 1800 K Street, NW, Suite 800,
Washington, DC, 20006, USA
e-mail: jboslego@path.org

R.W. Compans and W.A. Orenstein (eds.), *Vaccines for Pandemic Influenza*,
Current Topics in Microbiology and Immunology 333,
DOI 10.1007/978-3-540-92165-3_23, © Springer-Verlag Berlin Heidelberg 2009

**Abstract** The global need for a pandemic influenza vaccine is large. High-income countries have stated their intent to provide universal access for pandemic influenza vaccine to their populations. Assuming that a two-dose schedule would be needed, providing universal coverage globally would represent approximately 6.5 billion two-dose courses or 13 billion doses. In the best case scenario, should an outbreak of pandemic influenza occur in the near term, using H5N1 as a proxy for the pandemic virus, the total available doses for the global population within six months of an out break would be only 1.2 billion courses or 2.4 billion doses. In addition, current stockpiles of pandemic influenza vaccine are limited. However, promising developments are occurring with respect to global capacity, technological innovation, and global conviction that offer potential solutions to the problem of pandemic influenza vaccine supply for the world's population.

# 1   Introduction and Overview

From January to June 2007, we conducted a study: (1) to develop strategies for increasing access to pandemic influenza vaccines among developing world populations, and (2) to identify and quantify potential investment opportunities that would increase the global supply of pandemic vaccines. The study included close collaboration with the World Health Organization (WHO) and input from many key constituents from the influenza vaccine community.

The assessment was completed in two phases, as illustrated in Fig. 1. The first phase was the diagnostic phase, which involved three sets of research and analysis: supply/demand mapping, technology economics assessment, and access strategy hypothesis development. These findings informed a second phase that included strategy development as well as strategy evaluation and recommendations. Both of these phases will be discussed in this chapter.

It should be noted that the pandemic influenza vaccine space is rapidly evolving. Thus, the inputs used in our analysis and assessment regarding issues such as product specifications, production methods and yields, manufacturer physical capacity plans, seasonal vaccine demand and pandemic stockpiling efforts have changed

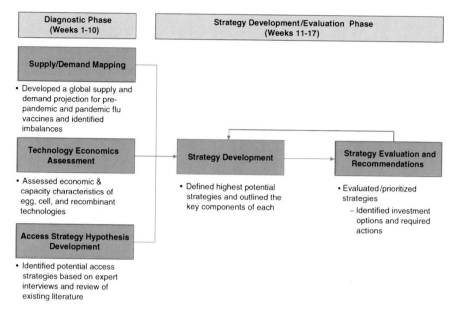

**Fig. 1** Influenza vaccine strategies project phases

since the time this study was conducted, and we expect will continue to do so in the future. Thus, implications and conclusions drawn from this study need to be periodically revisited as the environment continues to evolve. For example, a new study of influenza vaccine capacity was recently completed by Oliver Wyman, in collaboration with the WHO and the IFPMA. The study finds capacity for real-time production of pandemic vaccine has increased considerably in the last two years, but that significant shortages still remain. As such, the strategies discussed in the following commentary are still valid.

# 2 The Diagnostic Phase

## 2.1 Supply and Demand Mapping

To determine the global vaccine supply, the production capacities of all 32 licensed influenza vaccine bulk manufacturing facilities in the world were reviewed as of July 2007. This provided the current capacity available to produce seasonal trivalent vaccine or monovalent pandemic vaccine, since the underlying infrastructure could be used for either product. In addition, manufacturers' plans to either expand existing bulk facilities or build new ones were also determined. Since the time frame for constructing and validating a vaccine bulk production facility is typically 5–7 years, this assessment provided the projected global capacity through 2013.

Data for the supply assessment were obtained by (1) secondary research, including manufacturer websites, press releases, annual reports and analyst reports; (2) interviews with numerous experts in the areas of manufacturing processes, facility design and construction, and specific production technologies; and (3) direct discussions with current and prospective manufacturers of influenza vaccine spanning different technologies and representing the majority of both current and future vaccine capacity.

Global seasonal and pandemic influenza vaccine demand was assessed under different scenarios and strategies utilizing a multistep process. The first step was to build a global demographic dataset based on the world's 187 countries, utilizing multiple databases to size the global population by country and subpopulation (i.e., age, gender, health status, occupation). These data formed the basis of all demand projections, which were based primarily on two key variables—assumed target subpopulations for vaccination and vaccine coverage levels by country—for both seasonal and pandemic vaccines.

The next step involved assessing historic seasonal vaccine coverage rates, based on research conducted by the Macroepidemiology of Influenza Vaccination (MIV) Study Group (2005) and additional expert interviews. These data were combined with the demographic data set to create estimates of the annual number of vaccine doses distributed, by country, from 1997 to 2003. The coverage rate trends observed in these historic data for 55 countries were then used to project seasonal influenza vaccine demand, by country, through 2016.

Finally, we analyzed potential pandemic influenza vaccine demand for different population/country groups, utilizing a range of scenarios. The purpose was to determine the capacity required to serve the developing world under different coverage strategies. The intent was not to determine which specific countries or subpopulations should be immunized upon the outbreak of a pandemic, which is a policy question beyond the scope of our analysis.

## 2.2   *Technology Economics Assessment*

The economics and capacity characteristics of live attenuated and inactivated influenza vaccine produced with egg, cell culture, and recombinant protein technologies were evaluated, since these technologies are currently well advanced in product development. Other technologies with earlier stage candidates, such as universal proteins, viral vectors, and DNA vaccines, were not evaluated. The data to support these technology and economic assessments were gathered through (1) primary research, including direct discussions with the manufacturers/developers as well as discussions with knowledgeable technical experts, and (2) secondary research, including manufacturer disclosures and reports of vaccine technology and clinical trials in the medical literature. These assessments allowed the creation of a dynamic model with potential adjustments for yields, production location, scale, and other manufacturing variables.

## 2.3 Access Strategy Hypothesis Development

Potential pandemic influenza vaccine access strategies were identified. Sources for ideas on pandemic influenza vaccine access strategies included (1) discussions with policymakers from high-income countries who have developed access strategies, such as the United States and the United Kingdom; (2) review of published pandemic preparedness reports from more than 60 countries; (3) consultation with developing country representatives and multilateral organizations; and (4) discussions with influenza vaccine manufacturers and developers.

## 3 Strategy Development and Evaluation Phase

The objectives of the second phase were to identify the most promising strategies for global pandemic vaccine access, to evaluate and prioritize strategies, and then to identify the investment requirements and actions associated with each strategy. This phase involved a series of additional expert interviews; consultations with individuals at the WHO and other key constituents; and discussions related to the specifics of certain technologies and fill/finish capacities.

## 3.1 Available Vaccine Technologies

Based on available data on existing products and those in advanced phases of clinical trials, three main vaccine technologies were evaluated; egg-based, cell culture-based, and recombinant technologies. In addition, both inactivated and live vaccines were considered, as well as several novel new adjuvants shown to markedly enhance immune responses to pandemic vaccines (Bernstein et al. 2008; Leroux-Roels Barkowski et al. 2007). The status of these technologies for both seasonal and pandemic vaccines is illustrated in Fig. 2.

### 3.1.1 Egg-Based Inactivated Influenza Vaccines

In this production process, virus is grown in the allantoic fluid of hen eggs. The harvested fluid is centrifuged and filtered to capture the desired antigen, remove any unwanted material, and concentrate the solution. Depending on the nature of the product, chemical agents may be introduced to disrupt the cell membrane, followed by size exclusion chromatography to further purify the hemagglutinin content. All product variations then undergo an inactivation step where formaldehyde (or a similar agent) is added to kill the virus, followed by a final sterile filtration step to remove any remaining extraneous material and bacteria. The bulk product is then formulated, filled, and packaged to be administered intramuscularly by syringe.

| | Seasonal | | Pandemic | |
|---|---|---|---|---|
| | **Licensed** | **In-development** | **Licensed** | **In-development** |
| **Egg Inactivated** | • CSL, GSK, Novartis, Sanofi, Solvay, and other smaller manufacturers (e.g., Berna Biotech) | • Most existing manufacturers are currently developing improved products<br>• Several emerging suppliers are attempting to enter the market | • GSK (with Alum), Novartis (with MF59), and Sanofi | • CSL (phase 2), GSK (phase 2 with AS03), Solvay (phase 1), and other smaller manufacturers (e.g., Berna Biotech) |
| **Cell Inactivated** | • Solvay (MDCK – free suspension, Netherlands only) and Novartis (MDCK – free suspension) | • Nobilon (phase 1 in 2006 in MDCK), Sanofi (phase 1 in 2006 in Per.C6), GSK (pre-clinical in MDCK), and other small manufacturers (e.g., Vivalis / HepaLife (pre-clinical in embryonic chicken cells)) | • None | • Baxter (start phase 3 in 2007, wild-type in Vero), Nobilon (phase 1 in 2006 in MDCK), Sanofi (phase 1 in 2006 in Per.C6), Solvay (started phase 1 in 2007 in MDCK), GSK (pre-clinical, MDCK), Novartis (pre-clinical MDCK), and other small manufacturers (e.g., Vivalis / HepaLife (pre-clinical in embryonic chicken cells)). |
| **Recombinant** | • None | • Protein Sciences (to start phase 3 in 2007), Novavax (to start phase 1 in 2007), and others (e.g., Lentigen) | • None | • Protein Sciences (pre-clinical) and Novavax (clinical), and others |
| **Live Attenuated** | • MedImmune (egg-based) and Products Immunologicals (egg-based) | • MedImmune (pre-clinical cell-based) and Nobilon (pre-clinical cell-based) | • None | • MedImmune (currently in phase 1 for both egg and cell), Nobilon (pre-clinical in cell), and Products Immunologicals (phase unknown egg-based) |

**Fig. 2** Manufacturers across technologies

### 3.1.2  Cell-Based Inactivated Influenza Vaccines

Host mammalian cells (e.g., Vero, MDCK, Per.C6) are placed into synthetic medium and once the desired cell density is reached, the virus is introduced. Infected cells are harvested by centrifugation, similar to the process described for the egg-based product. A few additional steps are needed to remove host cell DNA. The bulk product is then formulated, filled, and packaged to be administered intramuscularly by syringe. Currently, no cell-based inactivated influenza vaccines are licensed for use in the United States.

### 3.1.3  Recombinant Protein Influenza Vaccines

Recombinant proteins or virus-like particles (VLPs) are also produced in cell-based systems, but require development of an expression plasmid containing the target antigens and a method to produce and secrete the desired antigens. The recombinant product is purified, filtered, and packaged to be administered intramuscularly by syringe. Currently, no recombinant protein influenza vaccines are licensed for use anywhere in the world.

### 3.1.4  Live Attenuated Egg-Based and Cell Culture-Based Vaccines

Production steps for egg-based and cell culture-based live attenuated vaccines are similar to those of their respective inactivated products. However, the master reference strains are attenuated. In addition, egg-based live attenuated vaccines are often produced in specific pathogen-free (SPF) eggs rather than the standard clean eggs used for inactivated vaccine. Although the live attenuated vaccines do not undergo splitting or inactivation, they are filtered, centrifuged and then formulated, filled, and packaged to be administered by intranasal spray or drops.

### 3.1.5  Adjuvants

Several new adjuvants have been developed to improve immunogenicity and facilitate antigen-sparing strategies. Thus far, the most commonly used adjuvant, alum, has yielded conflicting results regarding its ability to enhance pandemic vaccine responses (Brady et al. 2007; Ninomiya et al. 2007). Both Novartis and GlaxoSmithKline (GSK) have developed novel adjuvants that have demonstrated promising results (Bernstein et al. 2008; Leroux-Roels Barkowski et al. 2007).

### 3.1.6  Fill/Finish Capacity

The fill/finish capacity was also evaluated. However, unlike bulk production infrastructure, it is common for manufacturers to fill/finish multiple vaccines within the same facility. Also, depending on whether the vaccine is administered by injection or nasal drops or spray, the fill/finish methods will differ.

## 3.2  The Economics of Influenza Vaccine Production

### 3.2.1  Seasonal Vaccine

Using each of the different technologies, the costs involved in seasonal vaccine production were projected and used to predict costs associated with pandemic vaccine production. All costs were indexed relative to the cost of the conventional egg-grown inactivated vaccine produced in a high-income country during an eight-month production cycle. In addition, for the purposes of these projections, all facilities were assumed to be fully utilized with the exception of required downtime for maintenance. The cost of manufacturing adjuvants was assumed to be negligible relative to the bulk cost.

Figure 3 highlights the comparative costs involved in seasonal vaccine production using the different technologies. The cost per liter projection for the egg-based inactivated vaccine to be used with adjuvant remains the same, but since published data suggest that novel adjuvants can reduce the dosage requirement eightfold, this reduction was included in the projections (Leroux-Roels et al. 2007). For live egg-based vaccines the cost per liter is nearly 2.5 times greater than the egg-based inactivated costs assuming the use of SPF eggs, since these eggs are approximately quadruple the cost of traditional clean eggs used for inactivated vaccine. However, the doses per liter of live vaccine produced are nearly 60 times greater than the egg-based inactivated vaccine.

| Technology | Cost per liter | Doses per liter | Bulk cost per dose |
|---|---|---|---|
| Egg Inactivated | 1,000 | 1,000 | 1,000 |
| Egg Inactivated with Adjuvant | 1,000 | 7,600 | 130 |
| Egg Live | 2,300 | 58,000 | 30 |
| Cell Inactivated | 740 | 450 | 1,750 |
| Cell Inactivated with Adjuvant | 740 | 3,300 | 210 |
| Cell Live | 740 | 25,000 | 30 |
| Recombinant | 710 | 450 - 3,600 | 1,510 - 200 |
| Recombinant with Adjuvant | 710 | 10,000-87,000 | 70 - 10 |

*i.e., bulk cost per dose for cell inactivated with an adjuvant is only 21% of the bulk cost per dose for Egg Inactivated*

All technologies are indexed within each column to the egg inactivated vaccine value, which is set at 1,000. The Egg Inactivated values of 1,000 within each column serve as reference numbers only for each column, and thus do not represent absolute values of cost per liter, doses per liter or bulk cost per dose). Eight-fold enhancement by adjuvant shown by Leroux-Roels Barkowski A et al.

**Fig. 3** Seasonal influenza vaccine cost driver summary

Using a cell culture system, the costs for inactivated vaccine remain at least 50% more than the egg-based inactivated vaccine due to the decreased yields associated with this method. However, the impact of using novel adjuvants for the cell-based technology is assumed to be similar to the egg-based technology, with the doses per liter projected to increase eightfold, as outlined above.

For recombinant technologies, the cost per liter should be relatively comparable to cell-based systems, since the cell-based system is used to generate recombinant vaccines. However, depending on yields achieved in the process, the doses per liter have a large range, from levels comparable to current cell-based systems to a nearly tenfold increase. The most studied recombinant hemagglutinin (Treanor et al. 2007) uses dosage levels of 135 μg, compared to 45 μg for traditional egg-based and cell-based inactivated vaccines. If one were to achieve the higher yields and require lower dosages, then the costs could be similar to levels achieved with live vaccine. Finally, assuming that adjuvants can have a similar impact on recombinant technologies, the cost per liter would remain the same, but doses per liter would increase considerably. Yet, no such vaccines are currently in development.

It was assumed that the vaccine would be formulated, filled, and packaged in a facility of average size and configuration globally with one high-speed, 30K vial per hour line operating for three shifts. In addition, it was assumed that this facility would be located in a high-income country and that the product would be filled in ten-dose vials. Given these assumptions, the projected fill/finish costs are less than those currently experienced by the manufacturers of seasonal influenza vaccines using single-dose, individually packaged vaccine presentations. It is also assumed that the finishing cost per dose will be the same for seasonal and pandemic products.

### 3.2.2 Pandemic Vaccine

To translate seasonal vaccine costs into potential pandemic costs, modifications were needed, as outlined in Fig. 4. The costs per liter would remain the same for cell-based and recombinant technologies since the facilities and materials would be similar for seasonal and pandemic vaccines. However, the cost per liter for egg-based inactivated products increases substantially given the need to biosecure the flocks that produce the eggs. The cost of biosecured eggs is projected to be approximately triple that of clean eggs, similar to the cost of SPF eggs, resulting in an overall doubling of cost per liter.

Bulk courses per liter are also likely to be quite different for the pandemic vaccine based on the manufacturers' experiences with H5N1 as a proxy for a potential pandemic vaccine, even though the pandemic may actually be associated with another strain. For example, for egg-based inactivated vaccines, the H5N1 pandemic doses per liter are considerably smaller than those associated with seasonal production. This is because the antigen yield per liter is between 10% and 80% of the yield obtained with seasonal vaccine. For the purposes of our projections, yields one-third of seasonal levels are assumed. In addition, the dosage required to induce adequate antibody responses to H5N1 vaccine without adjuvants is nearly six times that of seasonal vaccine (Treanor et al. 2006). However, seasonal vaccines are trivalent, while pandemic vaccines are likely to be monovalent, increasing the pandemic yield three times. Finally, two doses of H5N1 vaccine are required to achieve an adequate immune response, given lack of prior exposure to a pandemic strain (Treanor et al. 2006).

Figure 5 summarizes the total cost per course for each technology for pandemic vaccine. The high-cost technologies of egg-based and cell-based inactivated vaccines become even more expensive for pandemic than seasonal strains. In contrast, when using novel adjuvants, both egg-based and cell-based methods become moderately expensive for pandemic, while for recombinant vaccines the costs increase slightly. All of the low-cost technologies, both egg-based and cell-based live and recombinant with the use of novel adjuvants (for which assumptions were made that dosage would be comparable to current adjuvanted inactivated pandemic vaccines, given that no such product is currently being developed), cost even less and are comparable between seasonal and pandemic vaccines (Bernstein et al. 2008; Leroux-Roels Barkowski et al. 2007).

## 4   Pandemic Supply–Demand Scenarios

Two scenarios for pandemic influenza vaccine distribution were evaluated (1) prepandemic production for stockpiling or immunization during the inter-pardemic period, using excess capacity from seasonal influenza vaccine programs, and (2) real-time access using all available capacity, with cessation of seasonal vaccine production.

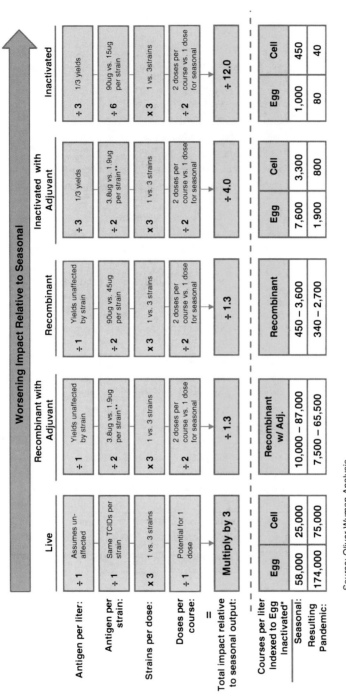

Source: Oliver Wyman Analysis.
Assumptions: Pandemic strain is H5N1 (impacts output of each technology)

*Indexed to Seasonal Egg Inactivated at 1000.
**We have assumed that seasonal inactivated or recombinant vaccines formulated with a novel adjuvant would be characterized by significantly lower
dosage levels than current seasonal products, consistent with the experience for pandemic vaccines
Enhancement from Adjuvant used data fromLeroux-Roels Barkowski A et al.
Dose of avian inactivated obtained from Treanor et al. NEJM.

**Fig. 4** Pandemic influenza vaccine courses per liter translation

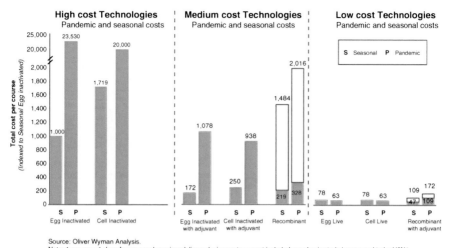

Source: Oliver Wyman Analysis.
Note: 1 course = 1 dose for seasonal vaccine; delivery device costs are not included; pandemic strain is assumed to be H5N1

*The total costs are indexed to seasonal egg inactivated vaccine at 1000. Seasonal vaccine requires one dose and Pandemic vaccine requires two.
Delivery device costs are not included. Recombinant yields and resulting cost per course have been ranged, with the blue bars representing the lowest
yields in the range and the white bars representing the highest yields in the range. Adjuvant enhancement assumed is calculated based on Leroux-Roels
Barkowski A et al.

**Fig. 5** Total pandemic influenza vaccine cost summary

## 4.1 Current Seasonal Supply

In 2007 there were 26 manufacturers with influenza vaccine capacity, but three
manufacturers produced most of the vaccine in use: Novartis, GSK, and Sanofi
Pasteur. Bulk manufacturing facilities for these top three producer are located in the
United States, Canada, the United Kingdom, Germany, France, and Italy. Other
producers include MedImmune, etc. whose facilities are located in United States,
Europe, Australia, Japan, and Russia. Finally, there are a number of manufacturers
in China that produce vaccine for the local Chinese markets.

## 4.2 Seasonal Vaccine Production

For the 2006–2007 influenza season, there were approximately 413 million doses
of seasonal influenza vaccine produced, 407 million of which were inactivated and
6 million of which were live attenuated. Approximately 377 million of these doses
were used in Northern Hemisphere countries, with the remaining 36 million doses
used in Southern Hemisphere countries.

## 4.3 Seasonal Vaccine Capacity

In 2007 global capacity for influenza vaccine production was approximately 826
million seasonal influenza vaccine doses (inactivated and live); this is double the

current production of 413 million doses. For inactivated influenza vaccines alone, the global capacity of approximately 657 million doses was nearly 60% greater than the current production of 407 million doses. The primary factor contributing to this inactivated capacity excess was that most manufacturers produce bulk anti-gen for only 8–9 months of the year, relative to a possible 11 months with one month of required maintenance each year. This largely stems from the fact that most manufacturers only serve Northern Hemisphere markets, for which all bulk production occurs during an 8-9 month time frame, starting with WHO strain iden-tification in January and ending in August/September when packaged vaccines are available for shipment in time for the influenza season. Note that this unused excess could be made available for prepandemic stockpiling or real-time pandemic vaccine production.

## 4.4   Projected Capacity

Based on manufacturers' disclosed expansion plans, influenza vaccine capacity is expected to more than double by 2013, reaching two billion doses globally, of which about 1.5 billion doses will be inactivated vaccine. This growth will come from the expansion of current egg-based production facilities and the construction of new cell-based manufacturing facilities. Five new cell-based manufacturing facilities in the USA alone are expected in response to the government's pandemic contracts. This growth in capacity is shown in Fig. 6.

## 4.5   Seasonal Influenza Vaccine Demand

Seasonal influenza vaccine coverage rates are highest among high and upper mid-dle income countries. As of 2006, only 20 countries among our in-scope set of middle and low-income countries had known vaccination programs for seasonal influenza and, among these countries, the average estimated coverage was low, with only about 25 doses administered per 1,000 population. However, global seasonal influenza vaccine distribution has increased from 160 million doses in 1997 to 310 million doses in 2003, representing a 12% compound annual growth rate.

## 4.6   Projected Seasonal Influenza Vaccine Demand

Projecting similar increases in demand and expansion to other countries, the seasonal influenza vaccine demand is expected to grow somewhat, but to level off in the years ahead. For example, the compound annual growth rate in seasonal vaccine use was 14% in 1997–2001, 8% in 2001–2006 and is expected to be 5%

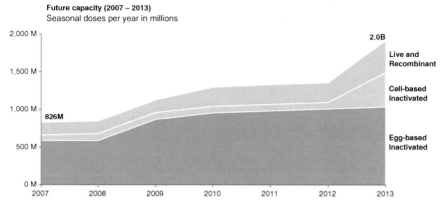

**Future capacity (2007 – 2013)**
Seasonal doses per year in millions

Source: Oliver Wyman Analysis, expert interviews; company statements; UBS Report: "Flu Vaccine Capacity Outstripping Demand" – Nov. 2006
Note: Doses generally equivalent to courses (1 dose / person = 1 course, except for young children, for whom 2 doses = 1 course); assumes
trivalent vaccine (15ug per valent per course).

**Fig. 6** Expected capacity growth by technology

from 2006–2016. The reason for this leveling off is that high and upper middle income countries have largely reached targeted populations. In addition, most developing countries are unlikely to launch new programs since the seasonal influenza disease burden is unknown, the infrastructure needed to deliver vaccines to adults does not exist, and the financial and programmatic issues surrounding annual vaccination are great, with numerous other demands competing for the healthcare resources.

Two more aggressive demand projections were also assessed. The first assumed that all developed countries would reach their target vaccination goals with a near doubling of the current 28% overall population coverage in the USA. The second, even more aggressive, scenario proposed that the influenza vaccination growth rates of nearly 10% would continue over the next decade. As discussed earlier, during the 2006–2007 season, the inactivated vaccine capacity exceeded demand by 250 million doses and is expected to grow as more manufacturing facilities are brought on-line. However, even under the two aggressive demand assumptions, expected inactivated capacity will exceed demand by approximately 710 million doses in 2013. Even more dramatically, if demand grows according to the "base case" projections, manufacturers' excess capacity will exceed 950 million doses in 2013. These estimates assume that newer technologies such as live and recombinant do not capture any of the incremental demand (which is conservative). More likely, capacity for these technologies, particularly live attenuated, will absorb some of this demand, meaning that excess inactivated vaccine capacity will be even larger, with as much as 500 million additional annual doses.

As cited previously, some level of excess capacity is inevitable due to the imbalance between Northern and Southern hemisphere demand. Nonetheless, it is reasonable to ask why market forces would allow for such an imbalance of supply and demand to continue over time. There are several explanations for this special situation. First, the US government is greatly influencing the overall supply situation by funding both expansion of current seasonal influenza capacity in the US and by creating surge capacity for use in a pandemic. This policy has taken the form of direct subsidies to manufacturers to create excess capacity. Second, individual manufacturers have their own objectives and strategies to maximize individual market share. Third, vaccine manufacturing is highly inflexible—capacity takes many years to put in place, regulatory requirements and oversight are significant, and units of capacity exist in large discrete pieces. Thus, while capacity over the long term should be balanced with demand, in the short term and medium term, significant positive and negative imbalances can exist.

Notwithstanding these explanations, it is reasonable to assume that manufacturers with significantly underutilized assets will reduce capacity over time to better match their individual demand levels. Of particular significance will be the start-up of new cell-based facilities, which could trigger a shutdown of older, egg-based facilities if sufficient demand does not exist. Therefore, the ongoing availability of this capacity will likely require some alternative use, perhaps serving prepandemic influenza vaccine demand or demand for other vaccines or biopharmaceuticals that can be produced within this infrastructure.

## 4.7   Prepandemic Influenza Vaccine Supply and Demand

The available seasonal influenza vaccine capacity was translated into the production of pandemic vaccines. Since live attenuated and recombinant pandemic vaccines are not yet adequately developed, the focus remained on inactivated vaccines. Several assumptions were also made: (1) all inactivated egg-based technologies currently available and all inactivated cell-based facilities, when licensed products become available, will be used to produce pandemic vaccines; (2) production yields will be one-third of the levels associated with current seasonal vaccine; (3) manufacturers with novel adjuvants will use them to increase the number of effective doses; (4) inactivated-products without access to novel adjuvants will use alum to potentially reduce their dosage requirements; and (5) stockpiles will be regenerated every two years, but will only be used during an outbreak. These assumptions generated a "base case" scenario.

In contrast, a more aggressive case was also proposed under the following assumptions: (1) all existing egg-based and cell-based facilities are available for pandemic production; (2) production yields will be 80% of current seasonal vaccine levels; (3) widespread access to all the proprietary adjuvants will be available for all vaccines; and (4) stockpiles will not need to be regenerated, or prior to expiration of the stockpiles, individuals will be vaccinated in the prepandemic period.

In addition, estimates were made for high and upper middle income country demand for prepandemic influenza vaccines, in order to determine the remaining capacity that might be available for other countries. To date, a number of governments have announced their intentions to stockpile doses of prepandemic vaccine

**Cumulative H5N1 Courses**

| Year | Base Case | Aggressive Case |
|------|-----------|-----------------|
| 2008 | 111 M | 1.2 B |
| 2009 | 389 M | 3.6 B |
| 2010 | 623 M | 6.3 B |
| 2011 | 688 M | 9.0 B |
| 2012 | 684 M | 11.8 B |
| 2013 | 967 M | 16.4 B |

Source: Oliver Wyman Analysis

**Fig. 7** Cumulative base case and aggressive scenarios for 2007–2013

for their populations, with coverage ranging from 1% to 100% of their populations. However, we are aware of active discussions with various countries for additional stockpiling. As prepandemic influenza vaccines continue to be successfully developed and dosage requirements reduced through the use of adjuvants, it is expected that coverage targets will increase.

As shown in Fig. 7, the opportunity for prepandemic influenza vaccines to be a means for broad global protection is mixed. Under our base case assumptions for prepandemic capacity, a stockpile could be generated that totals ~110 million courses in 2008 and rises to nearly 970 million courses by 2013. However, this is small relative to potential demand in high and upper middle income countries of 1.8 billion and 1.9 billion courses in those years, respectively. In this scenario, the capacity available for other countries is likely to be limited. However, under the more aggressive projections, prepandemic influenza vaccine coverage would be 1.2 billion courses in 2008, rising to 9.0 billion courses cumulatively produced by 2011. This would exceed global need across all countries.

## 4.8   Real-Time Access Pandemic Supply and Demand

As with prepandemic interventions, it is important to express capacity in pandemic influenza vaccine terms. Most of the assumptions for the base case and aggressive cases remain the same for real-time access. However, assumptions regarding stockpile regeneration are not applicable, as real-time access would involve production and administration at the onset of an outbreak. In addition, there are several parameters that need to be incorporated into a real-time access assessment that pertain to the lead time for producing a vaccine based on an emergent strain. These factors are as follows: (1) All manufacturers would have access to reverse genetics to develop reference strains by cloning the desired HA and NA proteins and combining them into plasmids with six additional genes from a backbone strain. This process would be 1–2 weeks faster than classical reassortment. (2) Regulatory authorities would require pathogenicity testing in the base case, which increases the time to produce the first batch of bulk vaccine by six weeks. However, in the aggressive case scenario, pathogenicity testing would be waived. (3) In the base case, cell-based manufacturers would not continuously regenerate biomass and would require an additional six weeks to scale-up biomass. In the aggressive case, biomass would be continuously regenerated, and therefore infection of the biomass with the vaccine strain could begin immediately.

The final key difference between prepandemic and real-time measures relates to the targeted protection time frame (i.e., the time from outbreak to vaccination of the full population). Countries that have signed contracts with manufacturers for real-time access have different time frame targets, ranging from two to six months. Simulation models predict that all countries are likely to experience a first peak of infection within six months of the outbreak, but potentially sooner. For the remainder of this analysis, we will use a six-month protection time frame but will show the sensitivity to shorter time-frame targets. As shown in Fig. 8, even in the aggressive scenario by 2013, only 2.8 billion courses could be produced in a six-month time frame.

It is assumed that high and upper middle income countries would have initial real-time access to the pandemic influenza vaccine capacity in the event of an outbreak. This is based on both the financial resources of these countries and the fact that the vast majority of this vaccine would be produced in high and upper middle income countries.

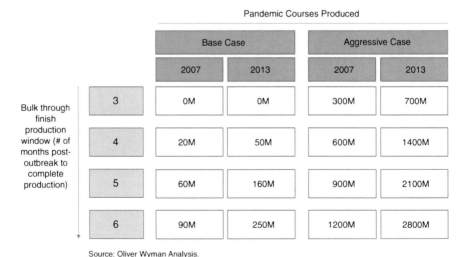

Pandemic Courses Produced

|  | | Base Case | | Aggressive Case | |
|---|---|---|---|---|---|
|  | | 2007 | 2013 | 2007 | 2013 |
| Bulk through finish production window (# of months post-outbreak to complete production) | 3 | 0M | 0M | 300M | 700M |
| | 4 | 20M | 50M | 600M | 1400M |
| | 5 | 60M | 160M | 900M | 2100M |
| | 6 | 90M | 250M | 1200M | 2800M |

Source: Oliver Wyman Analysis.

**Fig. 8** Real-time access pandemic influenza vaccine supply: number of courses

Demand is estimated in high and upper middle income countries based on published pandemic preparedness plans. A review of reports from 63 countries indicates that coverage of broadly prioritized subpopulations such as health workers, military personnel, essential services workers and the elderly would translate into nearly 450 million courses demanded at the onset of an outbreak. However, it is unlikely that these plans provide an accurate portrayal of pandemic demand for several reasons. First, the plans reflect a seasonal vaccination strategy, which may not be appropriate in a pandemic situation. Second, other indicators of governments' intent suggest that high-income countries would secure vaccine for their entire populations in a pandemic. Specifically, of 15 countries known to have entered into contracts with manufacturers for access to pandemic vaccine production capacity to date, ten have contracted for courses to cover their entire populations. If all high and upper middle income countries, including Brazil and Russia, sought vaccines for their entire populations in a pandemic, nearly two billion courses would be required. If China pursued this goal as well, more than three billion courses would be required to serve their populations.

Comparing expected developed-world pandemic vaccine demand with the "base case" and "aggressive case" supply scenarios creates a picture of the expected magnitude of excess/shortages of capacity under different circumstances. As illustrated in Fig. 9, under "base case" supply assumptions, if the pandemic outbreak were to occur in 2008 and high-resource countries sought vaccines for their entire populations, developing countries' demand would not begin to be met for nearly four years after the onset of an outbreak. The populations of these countries would not be fully served for more than ten years. Under our aggressive supply assumptions, if the pandemic outbreak were to occur in 2008, developing-world demand would not begin to be addressed until approximately eight months after the outbreak and

| Year | Base Case | | Aggressive Case | |
| | Time to 1st Dose | Time until Demand Served | Time to 1st Dose | Time until Demand Served |
|---|---|---|---|---|
| 2007 | 5 years | >10 years | 8 months | 25 months |
| 2008 | 4 years | >10 years | 8 months | 24 months |
| 2009 | 3 years | 9 Years | 6 months | 17 months |
| 2010 | 3 years | 9 Years | 6 months | 16 months |
| 2011 | 3 years | 9 Years | 6 months | 16 months |
| 2012 | 3 years | 9 Years | 6 months | 15 months |
| 2013 | 2 years | 6 years | 5 months | 12 months |

Source: Oliver Wyman Analysis.

**Fig. 9** Time frame to provide pandemic influenza vaccine to the developing world

would not be fully served for approximately two years. Even under our most aggressive supply assumptions and assuming the pandemic would not occur until 2013, developing-world demand would not begin to be addressed until five months after the outbreak, and would not be fully served for approximately one year.

## 5  Strategy Recommendations for Pandemic Vaccine: Strategy Definitions

Access strategies for pandemic vaccine have been defined across four dimensions; the populations to be vaccinated, the types of vaccines to be used (including inter-pandemic or real-time interventions), existing or new capacities, and technology. Short-term will be within the next five years and long-term will be greater than five years.

### 5.1  Short-Term Developing World Populations to Be Vaccinated

The developing world population has been divided into three groupings from a pandemic protection perspective, based on interviews with key policy makers and a review of pandemic preparedness reports from different countries. These groupings are as follows.

"The Essentials" (140 million people). This segment of the population includes frontline health workers and essential public service employees. In covering this group, the objectives are to minimize economic and social disruption in addition to preserving their individual well-being. Protecting "The Essentials" either prior to out-break or immediately thereafter would seem to be a requirement of any strategy.

"The Many" (3–5 billion people). This segment includes young children (6–59 months), children and adolescents (5–19 years), and young (20–34 years) and middle-aged (35–59 years) adults. These subpopulations represent the majority of individuals in the developing world, and protecting them would represent an attempt to achieve high impact in terms of mortality and morbidity reduction.

"The Vulnerable" (600 million people). This segment includes older adults (60+) and chronically ill or immunocompromised persons. This group could be viewed as those most in need of protection given their current health condition. Conversely, in a world of limited resources, immunizing vulnerable populations and leaving the healthy unimmunized might not be desirable.

### 5.2  Short-Term Types of Vaccines to Be Used

The analysis in the pandemic supply–demand section demonstrated that real-time access is not a viable intervention in the short term, given current and planned infrastructure and existing technologies. Therefore, the only viable intervention in the short term is a prepandemic approach such as stockpiling or immunization during the inter pandemic period with an H5N1-based adjuvanted vaccine. However, the viability of this approach depends on several factors: providing broad access of novel adjuvants to all manufacturers; achieving high production yields through sharing of operating practices, techniques, and potentially reassorted strains; and immunization during the inter pandemic period or extending stockpile "shelf lives." If these issues are successfully addressed, the entire global population could be protected through prepandemic vaccine measures within a four-year period, subject to the constraints of feasibility and availability of funding.

## 5.3   Short-Term Technology and Capacity

Given the technology and supply constraints that characterize the short-term time frame, options for the technology and capacity elements of an access strategy in the short term are defined a priori. Adjuvanted products are in the process of being licensed for egg-based technologies that, in addition to requiring lower dosage, have demonstrated high levels of direct seroprotection and seroconversion and provide good cross-protection (i.e., protection against other clades). Similarly, questions of capacity are moot in the short-term; capacity used for short-term strategies will have to be the existing or already planned infrastructure, which for bulk production is predominantly egg-based inactivated, located primarily in high-income countries such as the USA, UK, and Germany.

## 5.4   Longer-Term Strategies

In the longer term, live attenuated and recombinant/VLP-based vaccines are among the most attractive options for consideration. With these technologies added to the mix, there are different roles for each of the existing and newer technologies to play given their economic, clinical, and other distinguishing characteristics, as follows.

*Inactivated (egg or cell).* The proven track of inactivated technologies for H5N vaccine production, combined with issues around using live attenuated vaccines in the inter-pandemic period, suggests that inactivated vaccines are likely mc attractive for inter-pandemic interventions. Even with the use of novel adjuvants, inactivated vaccines still are likely to be the most costly vaccines to produce. For egg-based systems, any prepandemic or real-time access will need to consider the merit of biosecuring flocks given the avian nature of H5N1, the most threatening strain today. In addition, the complexity of cell-based manufacturing systems may make cell-based technologies less suitable for new bulk capacity created in the developing world by emerging suppliers with less experience in these systems.

*Live attenuated.* Live attenuated vaccines have low bulk production costs and have the capability to generate a sufficient number of courses in a six-month window to serve large portions of the global population following the onset of an outbreak. This distinction reflects both the high output per production run and the potential to generate a sufficiently immunogenic response with a single dose. Therefore, this technology is suitable for a real-time access solution that could serve "The Many." It is less suitable, however, for a longer-term solution involving prepandemic measures, since immunization with a live attenuated vaccine prior to outbreak could itself trigger a pandemic through reassortment with circulating strains. For real-time access, the potential for multidose, dropper-based administration offers advantages in reduced reliance on trained personnel in-country and lower cost/ stockpiling requirements for the delivery device. However, live attenuated technology requires further development for pandemic purposes, as H5N1-based live attenuated vaccines have not yet proven successful clinically. Furthermore, live-attenuated vaccines may not be as safe within "The Vulnerable" segment as inactivated vaccines.

*Recombinant protein.* Recombinant proteins have the potential to be relatively low-cost and more affordable for providing access to large populations. However, like live-attenuated vaccines, further development of recombinant protein vaccines is needed to achieve high production yields and effectiveness against H5N1. Like inactivated technologies, recombinant proteins would still require syringes and a two-dose course.

## 5.5   Long-Term Interventions and Populations

Given the previously described considerations, a combination of interventions with specific technologies is recommended for each of the three population groups; "The Essentials," "The Many," and "The Vulnerable."

Prepandemic interventions with either stockpiling or immunization are recommended for the "The Essentials" using inactivated technologies, and potentially recombinant proteins over time. Even with broad access to real-time capacity, the first vaccine doses are not expected to be available for three-to-four months after an outbreak given technical limitations, so protection of "The Essentials" with interpandemic interventions will still be critical. Individuals in this segment could be immunized multiple times based on the most likely pandemic strain in circulation at the time (e.g., H2, H5, H7, or H9), or with a combination vaccine if one were developed.

Real-time access for "The Many" and "The Vulnerable" is realized using live attenuated and potentially recombinants. Providing prepandemic protection on an ongoing basis to large segments of the global population is less necessary if a viable real-time access solution exists. As noted above, live attenuated is most suited for mass-scale, real-time access, given its cost advantages. However, exploring recombinant protein technologies is also attractive as a risk-spreading strategy for two reasons. First, it is a low-cost alternative in case the development of a live attenuated pandemic vaccine is not successful. Second, recombinant protein vaccines offer an alternative solution for "The Vulnerable" in case live attenuated vaccines are shown to be unsafe in this population.

## 5.6   Long-Term Capacity (New vs. Existing, Location, and Number)

Like the short-term approach, existing capacity could be leveraged to provide prepandemic interventions to "The Essentials." However, providing real-time access to the vast majority of the global populations will require new bulk production capacity. Given that the majority of the demand for pandemic vaccine will be located in the developing world, consideration should be given to locating bulk production capacity in these countries. In terms of the number of facilities, there will be a trade-off between investment efficiency and diversification. More facilities for a given amount of global demand mean smaller facilities, resulting in higher investment cost per dose. As an example, 16 live attenuated production facilities to serve three billion courses of pandemic demand would require an investment of more than triple what would be spent on four facilities to serve the same level of demand. One could argue, however, that fewer facilities could lead to nationalization by countries with production facilities. A possible solution that balances these considerations is 4–8 facilities in total, with one located each in India and China (given their large size) and the other 2–6 located in smaller countries in different regions to provide more regionally balanced access. The successful execution of this strategy will rely on identifying ways of economically maintaining the operations of these facilities in the inter-pandemic period if demand does not exist.

## 5.7   Further Considerations Within and Across Short-Term and Longer-Term Strategies

While different, both short-term and longer-term strategies are mutually reinforcing, and pursuing one without the other is not ideal. Specifically, protecting global populations from H5N1 in the short term using existing capacity may not allow for protection against other strains that may ultimately emerge as the source of a pandemic. In addition, H5N1 strains may drift over time such that protection afforded by current H5N1-based vaccines (even adjuvanted) may not be sufficiently effective. On the other hand, enabling real-time access for large portions of the global population to vaccines based on the strain that has become the source of a pandemic is not an option in the next five years, given the requirements for further development of the appropriate technologies and the time frames for new capacity build-out. Pursuing both paths in parallel provides the greatest opportunity to minimize the impact of an influenza pandemic.

# 6 Investment and Implementation Considerations

## *6.1 Investment*

In the short term, the total cost is highly sensitive to product and operations parameters. In the best case, sufficient courses of H5N1 vaccine to serve the five billion people in the developing world can be produced for $1–5 billion, since no additional capital investment is required because existing excess capacity is used.

Costs to implement the longer-term strategy are more complicated to estimate. These costs include upfront investment to build new bulk facilities, the annual cost of producing vaccines for prepandemic use for "The Essentials" group, the annual cost to operate bulk facilities during the prepandemic period, and the cost to produce doses and provide delivery devices for broad developing world coverage. The total estimated costs for this long-term strategy over a 25-year period would be $1–5 billion, with the range driven by assumptions for use of new live attenuated and recombinant protein capacity during the prepandemic period.

## *6.2 Implementation Considerations*

Implementation of both the short-term and longer-term access strategies will require a concerted and carefully orchestrated effort. First and foremost, excellent communication is needed to build broad consensus among the key constituents, such as manufacturers, developing world governments, donors, and agencies with critical responsibilities. Media reports and published deliberations among various stakeholders indicate that broad consensus does not exist among all constituencies. Implementing a carefully orchestrated communication plan to achieve broad-based buy-in, followed by a thoughtfully designed implementation plan that addresses the wide range of required activities across the areas of supply, demand, and finance are required for implementation of the outlined strategies. Although the challenges are many, the stakes are too high to ignore pursuit of both the short- and the long-term pandemic vaccination goals.

**Acknowledgments** This study was conducted by PATH, an international health nongovernmental organization (NGO), and Oliver Wyman, a management consulting firm.

# References

Bernstein DI, Edwards KM, Dekker CL, Belshe R, Talbot HK, Graham IL, Noah DL, He F, Hill H. (2008) Effects of adjuvants on the safety and immunogenicity of an avian influenza H5N1 vaccine in adults. J Infect Dis 197(5):667–675

Brady RC, Treanor JJ, Atmar RL, Chen WH, Winokur P, Belshe R. (2007) A Phase I–II, randomized, controlled, dose-ranging study of the safety, reactogenicity, and immunogenicity of intramuscular inactivated influenza A/H5N1 vaccine given alone or with aluminum hydroxide to healthy adults. In: Abstracts of the Options for the Control of Influenza VI Meeting, Toronto, ON, Canada, 17–23 June 2007, p 204

Leroux-Roels Barkowski A, Vanwolleghem T et al. (2007) Antigen sparing and cross-reactive immunity with an adjuvanted rH5N1 prototype pandemic influenza vaccine: a randomized controlled trial. Lancet 370:580–589

Macroepidemiology of Influenza Vaccination (MIV) Study Group. (2005) The macro-epidemiology of influenza vaccination in 56 countries, 1997–2003. Vaccine 23(44):5133–5143

Ninomiya A, Imai M, Tashiro M, Odagiri T. (2007) Inactivated influenza H5N1 whole-virus vaccine with aluminum adjuvant induces homologous and heterologous protective in a mouse model. Vaccine 25:3554–3560

Treanor JJ, Campbell JD, Zangwill KM, Rowe T, Wolff M. (2006) Safety and immunogenicity of an inactivated subvirion influenza A (H5N1) vaccine. New Engl J Med 354:1343–1351

Treanor JJ, Schiff GM, Hayden FG, Brady RC, Hay CM, Meyer AL, Holden-Wiltse J, Liang H, Gilbert A, Cox M. (2007) Safety and immunogenicity of a baculovirus-expressed hemagglutinin influenza vaccine: a randomized controlled trial. JAMA 297(14):1577–1582

# Prioritization of Pandemic Influenza Vaccine: Rationale and Strategy for Decision Making

Benjamin Schwartz and Walter A. Orenstein

## Contents

## 1  Introduction

Few catastrophes can compare with the global impact of a severe influenza pandemic. The 1918–1919 pandemic was associated with more than 500,000 deaths in the USA and an estimated 20–40 million deaths worldwide, though some place the global total much higher. In an era when infectious disease mortality had been steadily decreasing, the 1918–1919 pandemic caused a large spike in overall population mortality, temporarily reversing decades of progress. The US Department of Health and Human Services, extrapolating from the 1918–1919 pandemic to the current US population size and demographics, has estimated that a comparable pandemic today would result in almost two million deaths.

Vaccination is an important component of a pandemic response. Public health measures such as reduction of close contacts with others, improved hygiene, and

B. Schwartz (✉)
National Vaccine Program Office, US Department of Health and Human Services, Centers for Disease Control and Prevention, 1600 Clifton Rd NE, Mailstop E-05, Atlanta, GA 30333, USA
*New affiliation*: Care USA, Atlanta, Georgia
e-mail: bxs1@cdc.gov

W.A. Orenstein
Department of Medicine and Emory Vaccine Center, Emory University School of Medicine, 1462 Clifton Rd, Atlanta, GA 30322
*New affiliation*: Gates Foundation, Seattle, Washington

R.W. Compans and W.A. Orenstein (eds.), *Vaccines for Pandemic Influenza*,                    495
Current Topics in Microbiology and Immunology 333,
DOI 10.1007/978-3-540-92165-3_24, © Springer-Verlag Berlin Heidelberg 2009

respiratory protection with facemasks or respirators can reduce the risk of exposure and illness (Germann et al. 2006; Ferguson et al. 2006), but would not reduce susceptibility among the population. Prophylaxis with antiviral medications also may prevent illness but depends on the availability of large antiviral drug stockpiles and also does not provide long-term immunity. By contrast, immunization with a well-matched pandemic vaccine would provide active immunity and represent the most durable pandemic response. However, given current timelines for the development of a pandemic influenza vaccine and its production capacity, vaccine is likely not to be available in sufficient quantities to protect the entire population before pandemic outbreaks occur, and thus potentially limited stocks may need to be prioritized. This chapter reviews information on influenza vaccine production capacity, describes approaches used in the USA to set priorities for vaccination in the setting of limited supply, and presents a proposed strategy for prioritization.

An influenza pandemic occurs with the introduction and spread of a new influenza A virus subtype among people. Although some cross-protection against antigenically different influenza viruses within a subtype occurs following prior infection or vaccination, the entire population is likely to be susceptible to an influenza A virus subtype that has not circulated (or has not circulated recently) among people. Consequently, in an influenza pandemic, rates of illness are higher, severity is greater, and the distribution of mortality is more widespread compared with seasonal influenza (Simonsen et al. 1998). Given the susceptibility of the entire population, the goal of the United States' pandemic vaccination program is to offer vaccination to everyone living in the USA.

There are several potential approaches to implementing pandemic influenza vaccination when vaccine supplies are inadequate to rapidly vaccinate the entire population: vaccine could be administered on a "first come, first served" basis or could be targeted first to individuals and groups based on specified criteria. Criteria for targeting in other mass vaccination campaigns have included geographic area (e.g., group A meningococcus in the African meningitis belt), exposure or proximity to a case (e.g., smallpox), age (e.g., polio), risk of infection (e.g., *H. influenzae* type b), risk of complications from infection (e.g., seasonal influenza), risk for transmitting infection (e.g., rubella), or (most often) a combination of these factors. Targeting has been justified as providing earliest protection to those who are most vulnerable to infection, most at risk of severe or fatal disease, or whose protection may prevent or reduce further transmission (Heymann and Aylward 2006). When vaccine supply, the capacity to administer it, or funding is limited, so that the optimal strategy— rapid universal vaccination—is impossible to implement, targeting mass vaccination becomes more important to achieve the best possible outcomes.

## 2  Efforts to Avoid the Need for Prioritization

In the 2005 *National Strategy for Pandemic Influenza*, the President defined a goal of establishing domestic manufacturing capacity that produces sufficient vaccine to vaccinate the entire US population within six months of the emergence of a virus with pandemic potential (The White House 2005). To achieve this, over $1 billion has been allocated (1) to expand domestic egg-based influenza vaccine production, (2) to support advanced development of new vaccine production technologies, such as growth of influenza virus in cultured cells or development of recombinant vaccines, and (3) to support the advanced development of "antigen-sparing" approaches, such as new adjuvants, that can stimulate a more robust immune response, allowing manufacturers to reduce the amount of antigen in each dose and formulating the antigen produced into more vaccine doses.

Until the promise of these approaches is realized, however, pandemic influenza vaccine supply is likely to be far less than pandemic response needs. For the 2007–2008 influenza season, most of the influenza vaccine administered in the USA was produced in other countries, and these sources of supply may not be available during a pandemic. Moreover, the amount of antigen needed to achieve a protective immune response could be substantially greater for a pandemic virus compared with seasonal influenza viruses. A clinical trial of an unadjuvanted candidate H5N1 vaccine showed that two doses containing 90 µg of hemagglutinin antigen were needed to achieve an immune response that may correlate with protection in more than half of healthy adult recipients (Treanor et al. 2006). This per dose concentration is sixfold higher than the quantities of hemagglutinin antigen included for each strain in the seasonal trivalent inactivated vaccines (TIV), and twofold higher than the total hemagglutinin in a standard dose of TIV. Since two doses of the H5N1 vaccine were needed to achieve adequate immunogenicity, the quantity of antigen needed to immunize an adult would be 12-fold higher than the amount of antigen to vaccinate against a seasonal strain. Initial trials with other candidate H5N1 vaccines that contain alum, novel lipid-based adjuvants, or that use the inactivated whole virus documented immunogenicity with two doses of 30 µg, 3.8, and 10µg, respectively (Bresson et al. 2006; Leroux-Roels et al. 2007; Lin et al. 2006). While additional studies are needed, these results suggest a potentially wide range of antigen quantities needed in different vaccine formulations, which will directly impact how quickly the population can be effectively vaccinated in the event of a pandemic. Vaccine supply, therefore, would depend on the production capacity for different vaccine formulations at the time a pandemic occurs. Under some scenarios, vaccine supply would be very limited, whereas under others, assuming success in evaluating and licensing new formulations and producing them in the USA, supply may be robust.

The time required to develop, license, and manufacture pandemic influenza vaccine is also an important variable. Using current technologies, at least 20 weeks would be required from the time the pandemic virus was identified until the first vaccine doses become available. Depending on a combination of factors, including where the pandemic begins, how quickly it is detected, the effectiveness of containment measures and the season, the first US pandemic wave may occur before any pandemic vaccine becomes available or after sufficient lead time such that vaccination is already widespread (Ferguson et al. 2005; Longini et al. 2005). In the 1957 pandemic, the first US cases occurred in June but no community outbreak occurred until August and the first pandemic wave did not peak until the end of October; by this time almost half of the approximately 60 million vaccine doses eventually produced had been delivered. By contrast, in 1968, the pandemic was not recognized until later in the year, and at the time initial US outbreaks began few persons had been vaccinated (Schwartz and Wortley 2006). Because influenza vaccine production capacity, vaccine formulation, and the time from pandemic recognition to onset of US outbreaks all are uncertain for the next pandemic, we are unable to predict how many people will be vaccinated before pandemic disease is widespread. Thus, prioritizing who is vaccinated earlier and who later will best target available supply to achieve national pandemic response goals.

## 3   Pandemic Response Goals and Principles for Setting Vaccination Priorities

US pandemic response goals include slowing the spread of pandemic disease and reducing the health, societal, and economic impacts of the pandemic (The White House 2005). The approach to using a limited supply of pandemic vaccine may differ depending on which goals are considered most important. Results of mathematical models suggest that vaccinating school-aged children can best reduce transmission of influenza, slowing disease spread and reducing overall community attack rates (Germann et al. 2006). While studies of vaccination for seasonal influenza support a strategy of vaccinating children to protect others in the community through herd immunity (Monto et al. 1969; Piedra et al. 2005; Reichert et al. 2001), uncertainty in the amount of vaccine that will be available or its timeliness make reliance on trying to induce indirect protection a risky strategy. Hospitalizations and deaths from pandemic illness can be reduced by directly vaccinating those at highest risk for these severe outcomes. Based on age-specific mortality rates in the 1957 and 1968 pandemics, vaccinating persons 65 years old would have prevented substantially more deaths compared with vaccinating other age groups, despite the lower vaccine efficacy among the elderly (in 1918 this would not have been the case because of the high mortality rate among young adults).

Another approach to reduce the health impacts of a pandemic would be to vaccinate healthcare workers so that they can continue to provide care to others. In an unmitigated pandemic, the demand for healthcare services will be overwhelming at a time when healthcare workers may be out of work due to illness, the need to care for sick family members, or because they are afraid of becoming infected at the workplace. A survey of county health department workers in Maryland found that 46% of respondents indicated they would not report to work in a pandemic. In a multivariable analysis, confidence in one's personal safety was significantly associated with a willingness to work (Balicer et al. 2006). Whether response to a survey is predictive of actual behavior is unclear; anecdotally, virtually all healthcare workers in Toronto reported to work during the SARS outbreak, despite the fear associated with a new disease and the spread that occurred within hospitals. Whether vaccinating healthcare providers to maintain effective care or vaccinating those at highest risk of illness would better reduce the health impacts of a pandemic is unknown.

The potential societal and economic impacts of a pandemic are associated with pandemic severity, although even in a severe pandemic these impacts cannot accurately be predicted. Historical experience does not provide a guide, as a severe pandemic has not occurred for almost a century. A report by the US Department of Homeland Security's National Infrastructure Advisory Committee (NIAC) analyzed the components of 14 critical infrastructure sectors that would be essential to society in a pandemic and the workforce needed to maintain those products and services (National Infrastructure Advisory Council 2007). The report identifies significant interdependency between sectors, expresses concern about the maintenance of supply chains, many of which stretch overseas, and emphasizes the importance and challenges of implementing a targeted vaccination program. Of the approximately 85 million workers in these sectors, 16.9 million were defined by NIAC as essential in a pandemic. About nine million of these workers are in the healthcare and emergency services (emergency medical services, law enforcement, and fire protection) sectors. In other sectors, the proportion of the workforce defined as critical ranges from almost 50% in the nuclear sector to less than 5% of the food and agriculture sector. Because the availability of pandemic vaccine before disease outbreaks is not assured, business planning includes other measures such as "social distancing," improved hygiene, use of facemasks or respirators, and possibly antiviral drug prophylaxis to protect workers in essential operations.

# 4    US Efforts to Define Pandemic Influenza Vaccination Priorities

Because of the uncertainties about the severity and epidemiology of the next pandemic, vaccine supply, and the best approach to using vaccine to reduce health, societal and economic impacts, there is no scientific method to define the optimal use of pandemic influenza vaccine. In 2005, a working group from two US advisory committees, the Advisory Committee on Immunization Practices (ACIP) and the National Vaccine Advisory Committee (NVAC) met to develop a pandemic vaccine prioritization strategy. The working group considered the epidemiology and impacts of pandemics, the groups at highest risk for complications and death from influenza, vaccine efficacy, critical societal functions, and ethical issues. The prioritization strategy proposed by the committees included vaccinating groups defined in tiers and subtiers, depending on vaccine supply. Groups that were prioritized for earliest vaccination included healthcare workers, manufacturers of pandemic vaccine and antiviral drugs, and persons at high risk of severe illness and death. Personnel in critical infrastructure sectors other than healthcare were prioritized after these groups, which include over 100 million persons. This strategy was published in the Department of Health and Human Services' pandemic plan to provide guidance to state planners and stimulate further discussions (US Department of Health and Human Services 2005).

Shortly after publication of the plan, a federal working group was created to reassess and potentially revise pandemic vaccine prioritization guidance. Factors contributing to the decision to reassess the recommendations included a shift in national pandemic planning assumptions to a more severe pandemic scenario extrapolated from the 1918 pandemic (Table 1); recognition that the HHS guidance did not include groups that could be considered for prioritization such as border protection personnel or the military; a broader understanding of the risk to essential services stimulated by the NIAC report; and a series of public engagement meetings convened by the CDC, where participants identified protecting essential community services as the most important goal for pandemic vaccination rather than protecting those who are at highest risk (Public Engagement Pilot Project on Pandemic Influenza 2005). The federal working group process included consideration of the scientific issues reviewed in the earlier prioritization process, assessment of mathematical modeling results, and discussion with public health officials, critical infrastructure providers and homeland and national security experts. Recognizing that science alone cannot define the best approach to pandemic vaccine prioritization, key elements of the process were consideration of ethical issues, input from the public and stakeholders, and a formal decision analysis.

Ethical input into the working group process was achieved through the participation of public and private sector ethicists and an analysis conducted by the Ethics Subcommittee of CDC's Advisory Committee to the Director (Ethics Subcommittee of the Advisory Committee to the Director, CDC 2007). A strategy of targeting pandemic influenza vaccination to reduce health, societal and economic impacts was considered ethically appropriate. Although a strict utilitarian principle could not be applied because of uncertainty about what strategy would provide the most benefit, targeting protection of society in a broad sense was given higher priority than protecting individuals at high risk of complications from influenza. Fairness and equity are important principles where everyone is recognized to have equal value, and all

**Table 1** National pandemic planning assumptions. Note that planning for some responses such as nonpharmaceutical community mitigation strategies is done across a range of pandemic severities, as defined by the pandemic severity index (CDC, Community Mitigation Guidance)

- Universal susceptibility to the pandemic influenza virus
- Clinical and healthcare impacts absent effective mitigation strategies
  - Clinical illness attack rate of 30% (rates highest among school-aged children, about 40%, and declining with age); US national estimate: 90,000,000 cases
  - Care seeking by about half of those who are clinically ill
  - Hospitalization of 11% of clinical cases; US national estimate: 9,900,000
  - Case fatality rate of 2.1%; US national estimate: 1,900,000
- Risk groups for severe illness and death will depend on the pandemic virus and are likely to include infants, pregnant women, persons with chronic and immunosuppressive medical conditions, and the elderly
- Outbreaks will last 6–8 weeks in affected communities; effective use of nonpharmaceutical community mitigation strategies (e.g., social distancing) will prolong community outbreaks but reduce their overall magnitude
- Multiple waves of illness will occur, with each wave lasting 2–3 months

persons within a targeted group should have similar access to vaccination. Reciprocity, which posits that protection should be afforded to those who assume increased risk in an occupation that benefits society, also was considered important, and a reasonable corollary to healthcare providers' "duty of care" where one is committed to provide care even in settings that increase personal risk. Procedural ethical principles of inclusiveness and transparency were met through a process of engaging with the public and stakeholders in meetings, and through a request for comments posted in the *Federal Register* and on the government's pandemic influenza website.

The goal of the public and stakeholder meetings was to identify the objectives of a pandemic vaccination program that participants felt were most important to pursue. Public meetings were held in two demographically different communities with participants recruited by community groups. Stakeholder representatives from government, healthcare, business, and community organizations participated in a third meeting. Each meeting included initial presentations to educate partici-pants on influenza and influenza vaccine, pandemics, and the rationale for vaccine prioritization. Participants discussed potential objectives of pandemic vaccination in small groups and then met in a plenary session where the objectives were dis-cussed further. Finally, participants rated the importance of each of ten proposed objectives using a seven-point Likert scale ranging from "extremely important" (a score of 7) to "not important" (a score of 1). Despite the differences between groups in terms of geographic location, demographic characteristics, and occupa-tional background, the values expressed at each meeting were similar (Table 2).

**Table 2** Importance of pandemic vaccination program objectives based on scores assigned by participants at public engagement meetings in Las Cruces, New Mexico, and Nassau County, New York, and a stakeholders meeting in Washington, DC. Scores were assigned from a seven-point Likert scale ranging from 7 = extremely important to 1 = not at all important

| Vaccination goal: To protect… | Public meetings | | Stakeholders meeting: Washington D.C. | Average score |
| | Las Cruces | Nassau County | | |
| --- | --- | --- | --- | --- |
| People working to fight pandemic and provide care | 6.7 | 6.0 | 6.8 | 6.5 |
| People providing essential community services | 5.9 | 5.7 | 6.5 | 6.0 |
| People most vulnerable due to jobs | 5.8 | 5.6 | 5.9 | 5.8 |
| Children | 5.9 | 5.7 | 4.9 | 5.5 |
| People most likely to spread virus to unprotected | 5.3 | 5.3 | 4.6 | 5.1 |
| People protecting homeland security | 4.6 | 5.2 | 4.7 | 4.8 |
| People most likely to get sick or die | 4.5 | 4.8 | 4.8 | 4.7 |
| People most likely to be protected by the vaccine | 4.5 | 5.1 | 4.0 | 4.5 |
| People keeping pandemic out of the USA | 4.3 | 5.3 | 3.3 | 4.3 |
| People providing essential economic services | 3.0 | 4.2 | 4.5 | 3.9 |

Key outcomes of this process included the importance of achieving multiple objectives with the pandemic vaccination program, the value given to protecting critical services and exposed workers, and the preference for vaccinating children before those who are most likely to become sick or to die—older adults and those who have underlying medical conditions.

Results from the public and stakeholder engagement process provide insight into the values and preferences of the population but do not translate directly into a prioritization strategy for pandemic vaccine. We therefore conducted a formal decision analysis to assess the priority of different population groups. We identified 53 potential target groups for pandemic vaccination defined by their occupation or by their age and health status. The degree to which each group met each of the ten vaccination program objectives was then assessed and scored: how well each group met objectives related to occupational role or exposure was scored by representatives on the federal working group; for objectives where clinical trial or epidemiological data can be used to assess how well a group met an objective, scoring was done by influenza experts from CDC and academic medical centers. The score assigned to each group for each objective was then weighted by the average rating of the objective's importance from the public engagement and stakeholders meetings (Table 2). A total score was calculated for each group as the sum of the objective scores multiplied by their weights for the ten vaccination program objectives, as described by $S_x = O_1 w_1 + O_2 w_2 + \ldots + O_{10} w_{10}$, where $S_x$ is the total score for group x; $O_{1-10}$ are the scores the group received for each of the ten objectives; and $w_{1-10}$ are the weights for each of the objectives.[1]

As an example, medical care practitioners received high scores from the working group for objectives of fighting the pandemic and providing care, providing an essential community service, being vulnerable due to their jobs, and being at risk of spreading infection to those who are unprotected (their patient population). Because most healthcare workers are healthy adults who would respond well to vaccination, they also received high scores for the objective of being most likely to be protected by the vaccine. Medical care practitioners score lower for providing essential economic services, protecting homeland and national security, and being most likely to get sick or die (as some may have underlying medical conditions or be 65 years old or older). This group would receive no points for keeping the pandemic out of the USA or being children.

Based on this analysis, groups scoring highest for vaccination were front-line public health workers involved in the pandemic response (for example, providing vaccinations), medical care practitioners, emergency medical service personnel, law enforcement personnel, and emergency relief workers. Occupational groups invariably scored higher than general population groups defined by their age and health status because more of the ten program objectives were relevant (i.e., they would receive some score for objectives related to one's occupational role and exposure risk as well as one's age- and health-related risk of influenza, ability to be protected by vaccination, and potential role in disease spread). By contrast, general population groups received no score for the occupationally-related objectives. To control for this difference, we stratified potential vaccination target groups into four categories: those that provide healthcare and community support services; those that provide critical infrastructure services; those that protect homeland and national security; and the general population. Within these categories, target groups were clustered based on their scores, with breakpoints between clusters defined by difference between scores. Groups scoring highest among each of these categories are shown in Table 3.

# 5  US Pandemic Vaccine Prioritization

The US pandemic vaccine prioritization guidance incorporates both the tier structure from the guidance included in the 2005 HHS pandemic plan and the target group categorization used in the decision analysis. Reflecting the similar value placed by the public on protecting persons who provide pandemic healthcare, who maintain essential community services or are at high occupational risk, and protecting children, each of the highest vaccination tiers for a severe pandemic includes groups from each category (Table 4). Generally, the specific groups included in each tier track closely with the results of the decision analysis. Some groups, such as deployed military forces and those who provide support for their mission, are placed in a higher tier in recognition that they may be affected in a pandemic earlier than persons in the USA due to their geographical locations, their increased risk because of crowded living conditions, and the impact of illness on their ability to function effectively. In some critical infrastructure sectors, target groups are prioritized in a lower tier because their expected occupational burden would likely decrease in a pandemic (e.g., passenger transportation), they can largely be protected by changes in work practices such as teleworking, and/or the workforce or work is "fungible;"

**Table 3** Summary of groups with the highest prioritization scores from the decision analysis on proposed pandemic vaccination target groups for a severe pandemic. Results are stratified into four strata that correspond to categories included in the proposed guidance

| Category | Groups with highest prioritization scores |
|---|---|
| Health care and community support services | • Front-line public health emergency responders<br>• Medical care practitioners (inpatient and outpatient facilities)<br>• Emergency relief workers |
| Critical infrastructure | • Emergency response services (law enforcement, fire, emergency medical services)<br>• Pandemic vaccine and antiviral drug manufacturers |
| National and homeland security | • Military (active duty)<br>• National guard<br>• Border protection personnel |
| General population | • Children (all ages)<br>• Household contacts of vulnerable persons<br>• Persons with underlying medical conditions that increase their risk of severe or fatal influenza (18–64 years old) |

**Table 4** US strategy for pandemic influenza vaccine prioritization. Vaccination tiers are color coded (*red* = Tier 1; *orange* = Tier 2; *yellow* = Tier 3; *green* = Tier 4; *blue* = Tier 5). An *unshaded box* for an occupationally defined group indicates that the group is not specifically targeted at that level of pandemic severity, and persons from those groups would be vaccinated as part of the general population

| Category | Target group | Estimated number[a] | Severe | Moderate | Less severe |
|---|---|---|---|---|---|
| Homeland and national security | Deployed and mission critical pers. | 700,000 | | | |
| | Essential support and sustainment pers. | 650,000 | | | |
| | Intelligence services | 150,000 | | | |
| | Border protection personnel | 100,000 | | | |
| | National Guard personnel | 500,000 | | | |
| | Other domestic national security pers. | 50,000 | | | |
| | Other active duty and essential suppt. | 1,500,000 | | | |
| Health care and community support services | Public health personnel | 300,000 | | | |
| | Inpatient health care providers | 3,200,000 | | | |
| | Outpatient and home health providers | 2,500,000 | | | |
| | Health care providers in LTCFs | 1,600,000 | | | |
| | Community suppt. and emergency mgt. | 600,000 | | | |
| | Pharmacists | 150,000 | | | |
| | Mortuary services personnel | 50,000 | | | |
| | Other important health care personnel | 300,000 | | | |
| Critical infra-structure | Emergency services sector personnel (EMS, law enforcement, and fire services) | 2,000,000 | | | |
| | Mfrs of pandemic vaccine and antivirals | 50,000 | | | |
| | Communications/IT, electricity, nuclear, oil and gas, and water sector personnel Financial clearing and settlement pers. Critical operational and regulatory government personnel | 2,150,000 | | | |
| | Banking and finance, chemical, food and agriculture, pharmaceutical, postal and shipping, and transportation sector personnel Other critical government personnel | 3,400,000 | | | |
| General population | Pregnant women | 3,100,000 | | | |
| | Infants and toddlers 6–35 months old | 10,300,000 | | | |
| | Household contacts of infants <6 months | 4,300,000 | | | |
| | Children 3–18 years with high risk cond. | 6,500,000 | | | |
| | Children 3–18 years without high risk | 58,500,000 | | | |
| | Persons 19–64 with high risk cond. | 36,000,000 | | | |
| | Persons 65 years old | 38,000,000 | | | |
| | Healthy adults 19–64 years old | 123,350,000 | | | |

[a]Estimates are rounded to the closest 50,000. Occupational target group population sizes may change as plans are developed further for implementation of the pandemic vaccination program

that is, the impact of absenteeism or reduced function can be mitigated by the redundancy within the sector (e.g., trucking, food processing).

Workers in infrastructure sectors are targeted for early pandemic vaccination to maintain the essential services they provide in recognition of the interdependencies between sectors. Healthcare, for example, relies on the sectors that provide electricity, clean water, communications, information technology, transportation, pharmaceuticals, food, and chemicals. In a less severe pandemic, however, historical experience suggests that these services are unlikely to be substantially affected. In both the 1957 and 1968 pandemics, essential services were maintained without targeting pandemic vaccination. Therefore, the US strategy differs for severe, moderate, and less severe pandemics, with some of the occupational groups not targeted in moderate and less severe pandemics, and those workers being vaccinated with their age and health status group in the general population category. Pandemic severity is classified using the Pandemic Severity Index, which defines five categories based on the case fatality rate of pandemic illness (CDC 2007). A Category 1 pandemic, defined by a case fatality rate of <0.1%, would result in a mortality only slightly greater than a severe seasonal influenza epidemic, and the proposed US vaccine prioritization guidance for less severe pandemics (Categories 1 and 2) is formulated to be more similar to recommendations for annual influenza vaccination.

# 6  Pandemic Vaccine Prioritization in Other Industrialized Countries

Pandemic vaccine prioritization strategies developed in other industrialized countries are generally based on similar ethical principles and target similar groups to those in the US plan. While healthcare providers and those critical to a pandemic response are the groups targeted first in many plans, workers in other infrastructure sectors may not be targeted. This may reflect national planning assumptions for a less severe pandemic, lower predicted rates of worker absenteeism, and a belief that infrastructures can be protected by planning to protect workers using nonpharmaceutical interventions and antiviral medications to treat or prevent illness. Some countries, such as Canada or Australia, which have substantial domestic influenza vaccine manufacturing capacity and small populations, may choose not to prioritize vaccination because of the ability to vaccinate everyone over several months. To our knowledge, only the US strategy explicitly presents different vaccine targeting based on pandemic severity, although every country is likely to reassess and potentially modify their national plan based on the epidemiology of the pandemic.

# 7  Future Needs

Prioritizing pandemic vaccination addresses only a single component of planning an effective pandemic influenza vaccination program. Plans are also needed on how the vaccine supply will be allocated among the states or other jurisdictions, how it will be distributed, and how the program will be implemented. Key implementation issues include the method of identifying persons who are in target groups, validation at the vaccination site, vaccine administration and tracking, and monitoring for the occurrence of adverse events. A major problem could be having to turn away persons who are panicked about the severity of a pandemic yet do not meet the criteria for vaccination at that time under the prioritization strategy. Currently, no comparable program exists and each step will need to be planned and tested in preparedness exercises. Effective communications also will be important. While substantial public involvement in the development of the vaccine prioritization strategy increases the chance that the approach will be acceptable to the public, communications goals will be to assure the public that the entire population will have the opportunity to be vaccinated, to communicate the rationale for prioritization and the prioritization strategy, and to inform people when it is their turn to be vaccinated.

Rationing of healthcare is not an issue that most Americans have had to face in the past. Outside of military settings, healthcare services generally have not been limited by availability as much as by economic or geographic factors. Prioritizing pandemic influenza vaccine introduces a new paradigm. The approach taken by US planners considering science, ethics, and public values and preferences creates a model for how such rationing can take place. Nevertheless, the optimal solution is to pursue preparedness activities that will obviate the need to prioritize. Ongoing programs to increase influenza vaccine production capacity, to stretch vaccine supply through the use of new adjuvants, and to develop influenza vaccines targeted at antigens that are conserved across the different influenza A subtypes may all lead to a time when pandemic influenza vaccine prioritization will be unnecessary.

# References

Balicer RD, Omer SB, Barnerr DJ, Everly GS Jr (2006) Local public health workers' perceptions toward responding to an influenza pandemic. BMC Pub Health 6:99. doi:10.1 1186/1471-2458-6-99

Bresson J-L, Perronne C, Launay O, Gerdil C, Saville M, Wood J, Hoschler K, Zambon MC (2006) Safety and immunogenicity of an inactivated split-virion influenza A/Vietnam/1194/2004 (H5N1) vaccine: phase 1 randomised trial. Lancet 367:1657–1664

CDC (2007) Community strategy for pandemic influenza mitigation. http://www.pandemicflu.gov/plan/community/commitigation.html. Accessed 26 Feb 2008

Ethics Subcommittee of the Advisory Committee to the Director, CDC (2007) Ethical guidelines in pandemic influenza. http://www.cdc.gov/od/science/phec/panFlu_Ethic_Guidelines.pdf. Accessed 26 Feb 2008

Ferguson NM, Cummings DA, Cauchemez S, Fraser C, Riley S, Meeyai A, Iamsirithaworn S, Burke DS (2005) Strategies for containing an emerging influenza pandemic in Southeast Asia. Nature 437:209–214

Ferguson NM, Cummings DA, Fraser C, Cajka JC, Cooley PC, Burke DS (2006) Strategies for mitigating an influenza pandemic. Nature 442:448–452

Germann TC, Kadau K, Longini IM Jr, Macken CA (2006) Mitigation strategies for pandemic influenza in the United States. Proc Natl Acad Sci USA 103:5935–5640

Heymann DL, Aylward RB (2006) Mass vaccination: when and why. Curr Top Microbiol Immunol 304:1–16

Leroux-Roels I, Borkowski A, Vanwolleghem T, Drame M, Clement F, Hons, E, Devaster J-M, Leroux-Roels G (2007) Antigen sparing and cross-reactive immunity with an adjuvanted rH5N1 prototype pandemic influenza vaccine: a randomised controlled trial. Lancet 370:580–589

Lin J, Zhang J, Dong X, Fang H, Chen J, Su N, Gao Q, Zhang Z, Liu Y, Wang Z, Yang M, Sun R, Li C, Lin S, Ji M, Liu Y, Wang X, Wood J, Feng Z, Wang Y, Yin W (2006) Safety and immunogenicity of an inactivated adjuvanted whole-virion influenza A (H5N1) vaccine: a phase 1 randomised controlled trial. Lancet 368:991–997

Longini IM Jr, Nizam A, Xu S, Ungchusak K, Hanshaoworakul W, Cummings DA, Halloran ME (2005) Containing pandemic influenza at the source. Science 309:1083–1087

Monto AS, Davenport FM, Napier JA, Francis T Jr (1969) Effect of vaccination of a school-aged population upon the course of an A2/Hong Kong influenza epidemic. Bull World Health Org 41:537–542

National Infrastructure Advisory Council (2007) The prioritization of critical infrastructure in a pandemic influenza outbreak in the United States: final report and recommendations by the council. http://www.dhs.gov/xlibrary/assets/niac/niac-pandemic-wg_v8-011707.pdf. Accessed 26 Feb 2008

Piedra PA, Gagliani MJ, Kozinetz CA, Riggs M, Griffith M, Fewlass C, Watts M, Hessel C, Cordova J, Glezen WP (2005) Herd immunity in adults against influenza-related illnesses with use of the trivalent-live attenuated influenza vaccine (CAIV-T) in children. Vaccine 23:1540–1548

Public Engagement Pilot Project on Pandemic Influenza (2005) Citizen voices on pandemic flu choices. http://www.hhs.gov/nvpo/PEPPPI/PEPPPICompleteFinalReport.pdf. Accessed 26 Feb 2008

Reichert TA, Sugaya N, Fedson DS, Glezen WP, Simonsen L, Tashiro M (2001) The Japanese experience with vaccinating schoolchildren against influenza. N Engl J Med 344:889–896

Schwartz B, Wortley P (2006) Mass vaccination for annual and pandemic influenza. Curr Top Microbiol Immunol 304:131–152

Simonsen L, Clarke MJ, Schoenberger LB, Arden NH, Cox NJ, Fukuda K (1998) Pandemic versus epidemic influenza mortality: a pattern of changing age distribution. J Infect Dis 178:53–60

The White House (2005) National strategy for pandemic influenza. http://www.whitehouse.gov/homeland/pandemic-influenza.html. Accessed 26 Feb 2008

Treanor JJ, Campbell JD, Zangwill KM, Rowe T, Wolff M (2006) Safety and immunogenicity of an inactivated subvirion influenza A. (H5N1) vaccine. N Engl J Med 354:1343–1351

US Department of Health and Human Services (2005) Pandemic influenza plan: Appendix D. http://www.hhs.gov/pandemicflu/plan/appendixd.html. Accessed 26 Feb 2008

# Index

Printed by Books on Demand, Germany